Isotope Hydrology: A Practical Approach

Isotope Hydrology: A Practical Approach

Editor: Teddy Reynolds

www.callistoreference.com

Callisto Reference,
118-35 Queens Blvd., Suite 400,
Forest Hills, NY 11375, USA

Visit us on the World Wide Web at:
www.callistoreference.com

ISBN: 978-1-64116-578-5 (Hardback)

Cataloging-in-Publication Data

Isotope hydrology : a practical approach / edited by Teddy Reynolds.
      p. cm.
Includes bibliographical references and index.
ISBN 978-1-64116-578-5
1. Radioisotopes in hydrology. 2. Hydrology. 3. Stable isotopes. I. Reynolds, Teddy.
GB656.2.R34 I76 2022
551.480 28--dc23

Table of Contents

Preface

Isotope hydrology is a field of hydrology which is concerned with the utilization of isotope dating to estimate the age and origins of water. It also studies the movement of water within the hydrologic cycle. Water consists of water molecules carrying unique fingerprints located in part on different proportions of oxygen and hydrogen isotopes. Isotope hydrology helps in creating an elaborated picture of earth's water resources. It is primarily involved in using stable isotopes to determine the age of ice or snow. Its techniques are also used for conserving water supplies, controlling pollution and mapping aquifers. This book outlines the processes and applications of isotope hydrology in detail. It will also provide interesting topics for research which interested readers can take up. Those in search of information to further their knowledge will be greatly assisted by this book.

This book is a comprehensive compilation of works of different researchers from varied parts of the world. It includes valuable experiences of the researchers with the sole objective of providing the readers (learners) with a proper knowledge of the concerned field. This book will be beneficial in evoking inspiration and enhancing the knowledge of the interested readers.

In the end, I would like to extend my heartiest thanks to the authors who worked with great determination on their chapters. I also appreciate the publisher's support in the course of the book. I would also like to deeply acknowledge my family who stood by me as a source of inspiration during the project.

Editor

Characterisation of stable isotopes to identify residence times and runoff components in two meso-scale catchments in the Abay/Upper Blue Nile basin, Ethiopia

S. Tekleab[1,2,3,4], J. Wenninger[1,4], and S. Uhlenbrook[1,4]

[1]UNESCO-IHE Institute for Water Education, Department of Water Science and Engineering, P.O. Box 3015, 2601 DA Delft, the Netherlands
[2]Addis Ababa University, Institute for Environment, Water and Development, P.O. Box 1176, Addis Ababa, Ethiopia
[3]Hawassa University, Institute of Technology, Department of Irrigation and Water Resources Engineering, P.O. Box 5, Hawassa, Ethiopia
[4]Delft University of Technology, Faculty of Civil Engineering and Applied Geosciences, Water Resources Section, P.O. Box 5048, 2600 GA Delft, the Netherlands

Correspondence to: S. Tekleab (siraktekleab@yahoo.com)

Abstract. Measurements of the stable isotopes oxygen-18 (^{18}O) and deuterium (^{2}H) were carried out in two meso-scale catchments, Chemoga (358 km^2) and Jedeb (296 km^2) south of Lake Tana, Abay/Upper Blue Nile basin, Ethiopia. The region is of paramount importance for the water resources in the Nile basin, as more than 70 % of total Nile water flow originates from the Ethiopian highlands. Stable isotope compositions in precipitation, spring water and streamflow were analysed (i) to characterise the spatial and temporal variations of water fluxes; (ii) to estimate the mean residence time of water using a sine wave regression approach; and (iii) to identify runoff components using classical two-component hydrograph separations on a seasonal timescale.

The results show that the isotopic composition of precipitation exhibits marked seasonal variations, which suggests different sources of moisture generation for the rainfall in the study area. The Atlantic–Indian Ocean, Congo basin, Upper White Nile and the Sudd swamps are the potential moisture source areas during the main rainy (summer) season, while the Indian–Arabian and Mediterranean Sea moisture source areas during little rain (spring) and dry (winter) seasons. The spatial variation in the isotopic composition is influenced by the amount effect as depicted by moderate coefficients of determination on a monthly timescale (R^2 varies from 0.38 to 0.68) and weak regression coefficients (R^2 varies from 0.18 to 0.58) for the altitude and temperature effects. A mean altitude effect accounting for -0.12 ‰/100 m for ^{18}O and -0.58 ‰/100 m for ^{2}H was discernible in precipitation isotope composition.

Results from the hydrograph separation on a seasonal timescale indicate the dominance of event water, with an average of 71 and 64 % of the total runoff during the wet season in the Chemoga and Jedeb catchments, respectively.

Moreover, the stable isotope compositions of streamflow samples were damped compared to the input function of precipitation for both catchments. This damping was used to estimate mean residence times of stream water of 4.1 and 6.0 months at the Chemoga and Jedeb catchment outlets, respectively. Short mean residence times and high fractions of event water components recommend catchment management measures aiming at reduction of overland flow/soil erosion and increasing of soil water retention and recharge to enable sustainable development in these agriculturally dominated catchments.

1 Introduction

Environmental isotopes as tracers are commonly applied for examination of runoff generation mechanisms on different spatial and temporal scales (e.g. Uhlenbrook et al., 2002; Laudon et al., 2007; Didszun and Uhlenbrook, 2008). Isotope tracer studies are used for hydrograph separations (Sklash and Farvolden, 1979; Buttle, 1994), provide additional information for identifying source areas and flow pathways under different flow conditions, and estimate the mean residence time of a catchment (Soulsby et al., 2000; Uhlenbrook et al., 2002; McGuire et al., 2005; McGuire and McDonnell, 2006; Soulsby and Tetzlaff, 2008). The application of isotopes in catchment hydrology studies has been carried out in small experimental catchments to meso-scale catchments (e.g. Mc-Donnell et al., 1991; Uhlenbrook et al., 2002; Tetzlaff et al., 2007a) and large-scale catchments (Taylor et al., 1989; Liu et al., 2008).

Only few studies have been undertaken to characterise water cycle components using stable isotopes in Ethiopia (e.g. Rozanski et al., 1996; Kebede, 2004; Levin et al., 2009; Kebede and Travi, 2012). The results from these studies indicate that stable isotope composition of precipitation is only little affected by the typical dominant controls: amount, altitude and continental effects. Despite low mean annual temperature and high altitude, the Ethiopian meteoric water (e.g. the Addis Ababa station of the International Atomic Energy Agency (IAEA)) exhibits less negative isotopic composition as compared to the East African meteoric water (e.g. Nairobi, and Dar es Salaam) (Levin et al., 2009; Kebede and Travi, 2012). The literature suggests different reasons for the less negative isotopic composition of the Addis Ababa station in Ethiopia. For instance, Joseph et al. (1992) indicated that the moisture from the Indian Ocean that results in an initial stage of condensation vapour, which did not undergo a major rain-out fractionation effect, is likely the main reason for higher isotopic composition. Levin et al. (2009) hypothesised that the less negative isotopic composition in Addis Ababa is due to advection of recycled moisture from the Congo basin and the Sudd wetland. Rozanski et al. (1996) and Darling and Gizaw (2002) show that the increased sea surface temperature at the moisture source, and evaporation condition at the sources, are attributed to the less negative isotopic composition for the Ethiopian meteoric water as compared to the East African meteoric water.

Despite the paramount importance of water resources in the Abay/Upper Blue Nile area for the whole Nile basin, the usefulness of stable isotope data for catchment hydrological studies is largely unexplored. Very little is known about the use of stable isotopes for hydrological studies in the region. However, stable isotopes have the potential to provide enormous benefit with respect to hydrological process understanding and sustainable planning of water resource management strategies and policies in such data-scarcs areas (Hrachowitz et al., 2011a). This potential has been demonstrated

in numerous other regions worldwide (Kendall and Caldwell, 1998; Kendall and Coplen, 2001; Gibson et al., 2005; Barthold et al., 2010; Kirchner et al., 2010).

The use of isotope tracer techniques to understand mean residence times (MRTs) and residence time distributions (RTDs) has received a lot of attention (Rodgers et al., 2005a; McGuire and McDonnell, 2006). They are used to gain a better understanding of flow path heterogeneities (Dunn et al., 2007), to get insights into the internal processes of hydrological systems, and are used as a tool for hydrological model construction and evaluation (Uhlenbrook and Leibundgut, 2002; Wissmeier and Uhlenbrook, 2007; Hrachowitz et al., 2011b). Furthermore, they can be used as fundamental catchment descriptors, providing information about the storage, flow pathways and sources of water (McGuire and McDonnell, 2006), and are used for conceptualising the differences in hydrological processes by comparing different catchments (McGuire et al., 2005; Soulsby et al., 2006; Tetzlaff et al., 2009). The steady state assumption of fluxes for estimating the mean residence time has been commonly used, although time-invariant mean residence times do not exist naturally in real-world catchments. To circumvent this issue, a few recent studies developed a method for estimating time-variable mean residence times (e.g. van der Velde et al., 2010; Botter et al., 2011; Heidbüchel et al., 2012; Hrachowitz et al., 2013).

In the present study, the investigation of meteoric water in the source of the Abay/Upper Blue Nile basin is undertaken as the basis for characterisation and better understanding of the dominant runoff components. This is used as a baseline study for future hydrological studies using environmental isotopes in the region. The main objectives of this study are (a) to characterise the spatio–temporal variations in the isotopic composition in precipitation, spring and stream water; (b) to estimate the mean residence time of stream water; and (c) to separate the hydrograph on a seasonal timescale in the two adjacent meso-scale catchments, Chemoga and Jedeb in the Abay/Upper Blue Nile basin.

2 Study area

The Chemoga and Jedeb rivers are tributaries of the Abay/Upper Blue Nile, located south of Lake Tana, and extend between approximately $10°10'$ and $10°40'$ N latitude and $37°30'$ and $37°54'$ E longitude. Both rivers originate from the Choke Mountains at an elevation of 4000 m a.s.l. (see Fig. 1). The climate in these catchments has a distinct seasonality with three seasons: (i) summer as the main rainy season from June to September, (ii) winter as the dry season from October to February, and (iii) spring as the short rainy season from March to May (NMSA, 1996).

The long-term average annual temperature over the period of 1973–2008 at Debre Markos weather station is about $16.3 °C$. The mean precipitation ranges between 1342 and

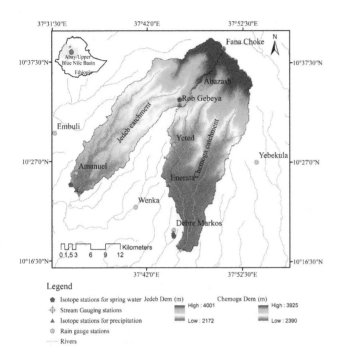

Figure 1. Location of the study area indicating the network of rain gauges, streamflow gauges and sampling points for stable isotopes of precipitation, surface water and spring water. The red dot within the Ethiopian map (inset, top left) indicates the location of the Chemoga and Jedeb catchments.

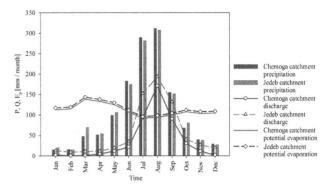

Figure 2. Intra-annual variability of hydro-climate data for the period 2008–2010 showing similar climate and distinct streamflow response in the Chemoga and Jedeb catchments. P, Q and E_p on the y axis stand for precipitation, discharge and potential evaporation, respectively.

$1434\,\mathrm{mm\,a^{-1}}$ (1973–2010) in the lower and upper parts of the catchments.

In these two catchments subsistence farming is commonly practiced. The farmers rely on rain-fed agriculture for their livelihoods. Barley, oats and potato are the main crops grown in the upland area, whereas wheat, tef and maize are grown in the middle and lower parts of the catchments. According to the studies by Bewket and Sterk (2005) and Teferi et al. (2010, 2013), the land use in the Chemoga catchment had been subjected to changes before the 1950s. The major change was the increase in cultivated area at the expense of open grazing area and a slight increase in plantation forest cover due to eucalyptus plantations. Natural vegetation cover can hardly be found in both catchments. A recent study by Teferi et al. (2013) showed that 46 % of the Jedeb catchment experienced transitions from one land cover to another over the last 52 years. Nowadays, about 70 % of the land is used for agriculture and 3 % is forest plantations, and the remaining percentages are utilised for other land uses (pasture land, bare land with shrubs and bushes).

3 Methodology

3.1 Hydro-meteorological data collection

Streamflow data sets are based on manual water level mea-

surements (daily at 6 a.m. and 6 p.m.) at Chemoga and Jedeb gauging stations from 1 July 2009 to 31 October 2011. Based on the stage discharge relationships rating curves were developed using regression models. A network of ten manual rain gauges was established in July 2009. Consequently, daily precipitation data were collected from these stations over the same period as the stream flows. Similarly, the mean daily minimum and maximum temperature data at Debre Markos station were obtained from the Ethiopian National Meteorological Agency. Regionalisation of the temperature data at Debre Markos station was used to estimate the temperature at Enerata, Rob Gebeya, Fana Choke, and Yewla stations based on a decrease of 0.6 °C in temperature per 100 m increase in altitude. The catchment average precipitation amount, catchment mean annual temperature, potential evaporation, and isotopic composition of precipitation were computed using the Thiessen polygon method. Due to the limited climatic data availability, the potential evaporation was computed using the Hargreaves method (Hargreaves and Samani, 1982). The method was selected due to the fact that other meteorological data (e.g. humidity, solar radiation, wind speed, etc.) were missing and only temperature data were available for the study catchments. The intra-annual variability of hydro-climatic data within the catchments is shown in Fig. 2. Furthermore, detailed descriptions of the hydro-meteorological data and isotope sampling sites are presented in Tables 1 and 2, respectively.

3.2 Field measurements and sampling

To characterise the spatial and temporal variability of stable isotope composition in precipitation, spring discharge and

Table 1. Descriptions of hydro-meteorological characteristics of investigated catchments (2008–2010). P, Q, E_p, and E stand for catchment average precipitation, runoff, potential evaporation and actual evaporation, respectively.

Catchment	Area (km^2)	Mean annual values (mm a^{-1})				Mean annual temperature (°C)
		P	Q	E_p	E	T
Chemoga	358	1303	588	1338	715	13.9
Jedeb	296	1306	692	1384	614	15

Table 2. Description of isotope sample locations and total number of samples taken during the investigation period (August 2008–August 2011).

Sample type	Location name	Abbreviation	Elevation (m a.s.l.)	Number of samples	Investigation period
Precipitation	Debre Markos	P-DM	2515	58	Aug 2008–Aug 2011
Precipitation	Enerata	P-EN	2517	24	Jul 2009–Aug 2011
Precipitation	Rob Gebeya	P-RG	2962	53	Oct 2008–Aug 2011
Precipitation	Fana Choke	P-FC	3993	41	Jul 2009–Aug 2011
Precipitation	Yewla	P-YW	2219	46	Jul 2009–Aug 2011
Spring	Debre Markos	S-DM	2339	64	Aug 2008–Aug 2011
Spring	Rob Gebeya	S-RG	2820	53	Oct 2008–Aug 2011
Spring	Yewla	S-YW	2255	59	Jul 2009–Aug 2011
Stream	Chemoga	Q-CH	2402	83	Jul 2009–Jul 2011
Stream	Jedeb	Q-JD	2190	98	Jul 2009–Jul 2011

streamflow, field investigations were undertaken from August 2008 until August 2011. At five different locations, plastic funnels were used to collect the precipitation samples for the analysis of isotopic signature in precipitation. The samples were collected on a bi-weekly basis.

The rainfall sample collectors have a capacity of 10 litres fitted with a vertical funnel with a mesh on top to avoid dirt and a long plastic tube to minimise evaporation out of the collection device according to the IAEA (2009) technical procedure for sampling. Spring water was sampled at three locations at different altitudes on a weekly basis, and the two weekly samples were mixed and taken for the analysis. Streamflow was sampled at the outlet of the Chemoga and Jedeb rivers on a weekly basis. During sampling, the water was filled into 2 mL glass bottles and closed immediately to avoid fractionation due to evaporation. Details about isotope sample locations and investigation periods are given in Table 2.

3.3 Laboratory analysis

All water samples were analysed at UNESCO-IHE (Delft, the Netherlands) using an LGR liquid-water isotope analyser. The stable isotopic composition of oxygen-18 and deuterium are reported using the δ notation, defined according to the Vienna Standard Mean Ocean Water (VSMOW) with δ^{18}O and δ^2H. The accuracy of the LGR liquid-water isotope analyser

measurements was 0.2‰ for δ^{18}O and 0.6‰ for δ^2H, respectively.

3.4 Hydrograph separation on a seasonal timescale

The classical steady state mass balance equations of water and tracer fluxes in a catchment were used in this study to separate the hydrograph into different components. The assumptions used for the hydrograph separations and the basic concepts are described in detail by e.g. Sklash and Farvolden (1979), Wels et al. (1991) and Buttle (1994).

The mass balance equation used for a time-based two-component separation using (^{18}O) as a tracer can be described as

$$Q_T = Q_E + Q_{Pe} \tag{1}$$

$$C_T Q_T = C_E Q_E + C_{Pe} Q_{Pe}, \tag{2}$$

where Q_T is the total runoff [m^{3}s^{-1}], and Q_E [m^3 s^{-1}] and Q_{Pe} [m^3 s^{-1}] are the runoff event and pre-event components, respectively. C_T is the total concentration of tracer observed in total runoff [‰VSMOW], and C_E [‰VSMOW] and C_{Pe} [‰VSMOW] are the tracer concentrations in event and pre-event water, respectively. Combining Eqs. (1) and (2), the contribution of event water and pre-event water to the total runoff can be estimated as

$$Q_E = Q_T \left(\frac{C_T - C_{Pe}}{C_E - C_{Pe}} \right) \tag{3}$$

$$Q_{Pe} = Q_T \left(\frac{C_T - C_E}{C_{Pe} - C_E} \right). \tag{4}$$

The precipitation isotopic composition was weighted based on the cumulative incremental weighting approach as outlined by McDonnell et al. (1990):

$$\delta^{18}O = \frac{\sum\limits_{i=1}^{n} p_i \delta_i}{\sum\limits_{i=1}^{n} p_i}, \tag{5}$$

where p_i and δ_i denote the rainfall amount and δ value, respectively.

Similarly, the monthly discharge isotopic composition in the rivers was weighted using Eq. (6).

$$\delta^{18}O = \frac{\sum\limits_{i=1}^{n} Q_i \delta_i}{\sum\limits_{i=1}^{n} Q_i}, \tag{6}$$

where Q_i [m^3 s^{-1}] is the daily volumetric flow rate and δ_i [‰] is the isotopic composition of the streamflow.

Due to the distinct seasonality, the precipitation during the dry (winter) and little rain (spring) seasons does not contribute significantly to the total streamflow, neither as surface nor as subsurface flow. This is due to the fact that the precipitation in these seasons mostly evaporates without producing direct runoff or recharging the groundwater (Kebede and Travi, 2012). To account for the effects of seasonality on the results of hydrograph separation, the end member signature is not taken as a constant value throughout the whole seasons. Consequently, the pre-event water isotopic composition was taken as the monthly isotopic values at each month during the dry (winter) and spring seasons. At the same time, to see the effects of different pre-event end member concentrations on the results of the wet season hydrograph separation, three different end members were estimated: first, the average values for the whole dry season isotope concentration were taken as all end members, second, the average value of the isotopic concentration in the month of February, which represents the baseflow in the rivers, was considered as an end member, and third, the average isotopic concentration of combined dry and spring season concentrations.

The event water $\delta^{18}O$ end member was taken as the weighted mean isotopic composition of precipitation, in each month for the investigated period. The differences in isotopic composition for event water vary from -6.37 to -4.24‰, and pre-event water from -0.25 to 0.62‰ is adequate for the hydrograph separation in these catchments based on the

assumptions of classical hydrograph separation described in Buttle (1994).

Hydrograph separation using isotope technique is prone to error due to the uncertainty in the estimation of end member concentrations (e.g. Genereux, 1994; Uhlenbrook and Hoeg, 2003). In this study the uncertainty in the two-component separations during the wet season June to September is evaluated based on the Gaussian error propagation technique according to Eq. (7) (e.g. Genereux, 1994).

$$W_y = \sqrt{(\frac{\partial y}{\partial x1} W_{x1})^2 + (\frac{\partial y}{\partial x2} W_{x2})^2 + \ldots + (\frac{\partial y}{\partial xn} W_{xn})^2}, \tag{7}$$

where W represents the uncertainty in the variables indicated in the subscript, assuming that y is a function of the variables $x_1, x_2, \ldots, x_n$ and the uncertainty in each variable is independent of the uncertainty in the others (Genereux, 1994). The uncertainty in y is related to the uncertainty in each of the subscript variables by using Eq. (7). Application of Eq. (7) into Eq. (4) gives the propagated total uncertainty related to the different component computed using Eq. (8).

$$W = \left\{ \left[\frac{(C_E - C_T)}{(C_E - C_{Pe})^2} \times W_{C_{Pe}} \right]^2 + \left[\frac{(C_T - C_{Pe})}{(C_E - C_{Pe})^2} \times W_{CE} \right]^2 \right.$$
$$\left. + \left[\frac{-1}{(C_E - C_{Pe})} \times W_{CT} \right]^2 \right\}^{\frac{1}{2}}, \tag{8}$$

where W is the total uncertainty or error fraction related to each component, and W_{CPe}, W_{CE}, and W_{CT} are the uncertainty in the pre-event, event and total stream water, respectively. The uncertainties related to each component are computed by multiplying the standard deviations by t values from the Student's t distribution at the confidence level of 70 % (Genereux, 1994).

3.5　Estimation of mean residence time

The mean residence time of stream water in a catchment is commonly computed using lumped parameter black box models described in Maloszewski and Zuber (1982). However, the application of this method to short data records and coarse spatial and temporal sampling leads to inaccurate estimates of parameters and tracer mass imbalance if the timescale of residence time distribution is larger than the input data (McGuire and McDonnell, 2006).

Hence, due to the short record length and coarse frequency of spatial and temporal tracer sampling, in this study the mean residence time is estimated based on the sine wave approach fitting the seasonal $\delta^{18}O$ variation in precipitation and streamflow (e.g. McGuire et al., 2002; Rodgers et al., 2005a; Tetzlaff et al., 2007b). The method gives indicative first approximation estimates of mean residence times (Soulsby et al., 2000; Rodgers et al., 2005a). The predicted $\delta^{18}O$ can be defined as

$$\delta = C_o + A \left[\cos \left(ct - \varphi \right) \right], \tag{9}$$

where δ is the predicted $\delta^{18}O$ [‰] composition, C_o is the weighted mean annual measured $\delta^{18}O$ [‰], A is the annual amplitude of predicted $\delta^{18}O$ [‰], c is the angular frequency constant $(0.017214\,\mathrm{rad\,d^{-1}})$, t is the time in days after the start of the sampling period and φ is the phase lag of predicted $\delta^{18}O$ in radians. Furthermore, Eq. (9) can be evaluated using sine and cosine terms in a periodic regression analysis (Bliss, 1970) as

$$\delta = C_o + \beta_{\cos}\cos(ct) + \beta_{\sin}\sin(ct). \qquad (10)$$

The estimated regression coefficients $\beta_{\cos}$ and $\beta_{\sin}$ are used to compute the amplitude in input and output signals $\left(A = \sqrt{\beta^2\cos + \beta^2\sin}\right)$, and consequently the phase lag $\tan\varphi = \left|\frac{\beta_{\sin}}{\beta_{\cos}}\right|$.

The mean residence time from the fitted sine wave in input and output signals was estimated as

$$T = c^{-1}\left[\left(\frac{A_2}{A_1}\right)^{-2} - 1\right]^{0.5}, \qquad (11)$$

where T is the mean residence time [d], A_1 is the amplitude of precipitation $\delta^{18}O$ [‰], A_2 is the amplitude of streamflow $\delta^{18}O$ [‰], and c is defined in Eq. (9).

4 Results and discussion

4.1 Meteoric water lines

The plot representing the relationship between $\delta^{18}O$ and δ^2H isotopic composition for precipitation is shown in Fig. 3. The spring and river water isotopic compositions are also plotted in the same figure for comparison. The spatial distribution of $\delta^{18}O$ and δ^2H compositions of precipitation varies considerably along the elevation gradient. Fana Choke station located at the highest elevation has more negative isotope compositions than the Yewla lower altitude station. The air masses lifted at the higher altitude with lower air temperature and higher relative humidity, due to orographic effects, could be possible reasons for the elevation-dependent variations of isotope composition. The difference in the isotopic composition at these two stations was evaluated using the Wilcoxon signed rank statistical test. The test results show that the difference in the isotopic values at the two locations is statistically significant ($p = 0.020$) evaluated at the 95 % confidence level. The scatter of the isotopic composition from the global meteoric water line might be related to the effect of the evaporation of falling rain drops, condensation in the cloud and different moisture sources over different seasons (Dansgard, 1964; Gat, 1996; Levin et al., 2009; Kebede and Travi, 2012). However, from the plot of the relationship between $\delta^{18}O$ and δ^2H isotopic composition for precipitation, the effect of evaporation is insignificant and plots along the local

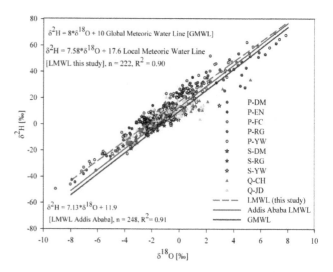

Figure 3. Relationship between $\delta^{18}O$ and δ^2H for precipitation, stream and spring water in the study area. The abbreviations in the legend are described in Table 2.

evaporation line. This means that the isotope values do not deviate from the local and global water lines.

The representation of the isotope values with the local meteoric water line was compared with that of the Addis Ababa Meteoric Water Line (MWL) produced using Global Network of Isotopes in Precipitation (GNIP) samples (Fig. 3). The relationships of $\delta^{18}O$ and δ^2H composition of the present study all exhibit an almost similar slope to that of the Addis Ababa station. However, it has a higher intercept/deuterium excess than the Addis Ababa GNIP station. The higher d excess values > 10 ‰ in both Addis Ababa and the study area are attributed to land surface–atmosphere interaction through transpired moisture contribution (Gat et al., 1994).

Furthermore, it is shown that the precipitation waters with more positive isotopic values are derived from the winter and spring season precipitation, whereas those plotting at the more negative end of the LMWL are derived from summer precipitation. The river water isotopic values in the Chemoga and Jedeb catchments exhibit little variation along the LMWL. These variations indicate that the waters for both catchments derived mostly from the summer precipitation. The spring waters at Debre Markos and Rob Gebeya exhibit negative isotopic composition as compared to Yewla, which is located towards the positive end of the LMWL.

4.2 Spatio–temporal variation of isotope composition in precipitation, spring water and streamflow

4.2.1 Isotope composition of precipitation

The results of the measured isotopic composition of precipitation samples exhibit marked spatial and seasonal variations (Fig. 4). The precipitation at Yewla station (lowest altitude) shows less negative $\delta^{18}O$ and δ^2H values in contrast

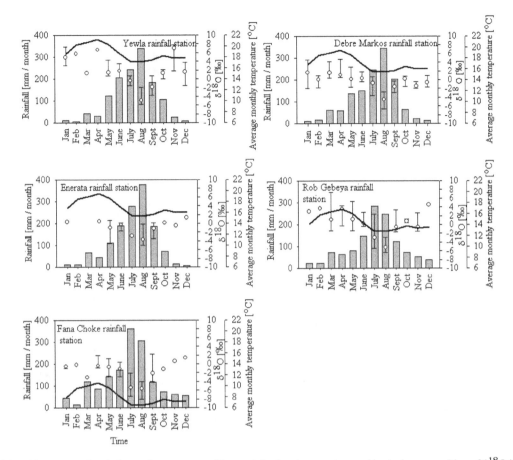

Figure 4. Spatial and intra-annual variations of average monthly precipitation, temperature and isotopic composition of $\delta^{18}O$ in precipitation. The error bar of the isotopic measurements stand for the standard deviation. Missing error bars for some months are due to limited isotope samples. The grey bar, black solid line and open circle with error bar are the precipitation, temperature and isotopic composition, respectively. For the isotopic composition, the open circle and the lower and upper error bars indicate the median and the 25th and 75th percentiles for the raw (non-weighted) precipitation isotope sample data, respectively.

to more negative isotopic composition at Fana Choke station (highest altitude). This shows the anticipated isotopic composition influenced by the altitude effect (Dansgaard, 1964; Rozanski et al., 1993). Nevertheless, the altitude effect varies temporally over different seasons depending on the moisture source, amount and trajectories of air mass bringing precipitation and local meteorological settings (Aravena et al., 1999). For instance, the seasonal isotopic composition relationship with elevation along the gradient during different seasons shows more negative values of isotopic composition at higher altitudes (Fig. 5). The altitude effect accounts for -0.12 and $-0.58\,‰$ per 100 m increase in altitude for $\delta^{18}O$ and $\delta^{2}H$, respectively. This is consistent with an earlier finding by Kebede and Travi (2012), who found more negative values of $\delta^{18}O$ by $-0.1\,‰$ per 100 m in the higher elevations of the Upper Blue Nile plateau.

The $\delta^{18}O$ and $\delta^{2}H$ composition is also affected by the precipitation amount effect (Dansgaard, 1964; Rozanski et al., 1993). Figures 6 and 7 illustrate the $\delta^{18}O$ and $\delta^{2}H$ composition of precipitation at sampling stations, which show moder-

ate regression coefficients ranging from ($R^2 = 0.36–0.68$, p value varies from 0.001 to 0.007) for precipitation and ($R^2 = 0.26–0.39$, p value varies from 0.001 to 0.007) for temperature. This suggests that the amount effect at each sampling location is important for the variation in isotopic composition in the area in addition to other factors. However, our results are in contrast to the earlier studies by Kebede (2004) and Kebede and Travi (2012), who reported weak relationships between rainfall amounts and isotopic compositions in the northwestern Ethiopian plateau.

Moreover, multiple linear regression models are used to show the effect of monthly precipitation and mean monthly temperature on $\delta^{18}O$ and $\delta^{2}H$ isotopic composition of precipitation in the Chemoga and Jedeb catchments, respectively. The multiple regression models for $\delta^{2}H$ composition in the Chemoga and Jedeb catchments are described as

$$-\ \delta^{2}H = -0.096P + 2.093T + 0.736\ (R^2 = 0.74,\ n = 28,\ p = 0.001\ \text{for precipitation and}\ p = 0.121\ \text{for temperature})\ \text{in the Chemoga catchment, and}$$

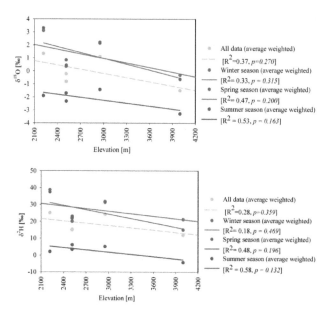

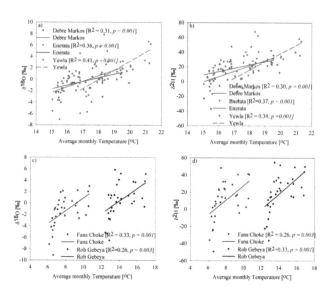

Figure 6. Relationships between amount-weighted isotopic composition of precipitation samples at different stations with monthly average air temperature over the investigation period.

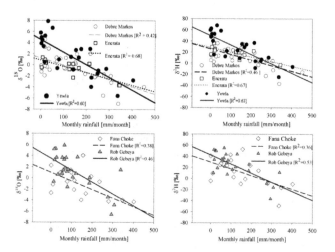

Figure 5. Relationships between average and seasonal amount weighted isotopic composition of precipitation with elevation at five precipitation sampling stations.

$$-\ \delta^2 H = -0.116P + 2.414T - 1.374\ (R^2 = 0.76, n = 28,$$

$p = 0.001$ for precipitation and $p = 0.175$ for temperature evaluated at the 5 % significance level) in the Jedeb catchment,

where P in the regression equation is the monthly precipitation (mm month^{-1}) and T is the mean monthly average temperature (°C). These results from the multiple linear regression models also show that the amount effect has a more dominant role in the variations in the isotopic composition in the study area than the temperature effect.

The seasonal variations in the isotopic composition of precipitation are observed among the stations. For instance, the winter seasonal mean weighted δ^{18}O composition of precipitation has, with a value of $-0.41\,\text{‰}$, a negative isotope value at Fana Choke at the higher altitude and has positive isotope values of $3.08\,\text{‰}$ at the lowest altitude (Yewla station). During the spring season the mean weighted δ^{18}O composition is $-0.62\,\text{‰}$ at Fana Choke and $3.3\,\text{‰}$ at Yewla. Similarly, during summer a more negative isotopic composition of $-3.28\,\text{‰}$ is observed at Fana Choke and a relatively less negative composition $-1.9\,\text{‰}$ is observed at Yewla.

In the Chemoga catchment the mean weighted seasonal isotopic compositions of δ^{18}O and δ^2H in precipitation during winter, spring and summer are 0.72 and 24.85‰, 0.86 and 23.71‰, and -2.09 and 2.36‰, respectively. Obviously the summer seasonal isotopic compositions in both δ^{18}O and δ^2H have more negative values than in the winter and spring seasons owing to the different moisture sources and the local meteorological settings like precipitation, air temperature, and humidity. In comparison to the Chemoga catchment, the mean weighted seasonal isotopic composition in the Jedeb

Figure 7. Relationship between amount-weighted isotopic composition of precipitation samples at different stations with monthly precipitation amounts at respective stations during the investigation period.

catchment shows consistently less negative δ^{18}O and δ^2H isotopic values of 1.48 and 29.23‰, 1.65 and 28.30‰, and -1.93 and 2.74‰ in the winter, spring and summer seasons, respectively. This implies that the less negative isotopic values of precipitation are likely related to different temperatures and altitudes in the Jedeb catchment.

4.2.2 Isotope composition of spring water

The isotopic composition of spring water at the three locations shows distinct variability ranging from -8.5 to 13.5‰ and -4.1 to 2.9‰ for δ^2H and δ^{18}O, respectively (see Table 3). The spring waters of Debre Markos (at an elevation of 2339 m a.s.l.) and Rob Gebeya (at 2820 m a.s.l.) exhibit more negative isotopic compositions as compared to the isotopic composition of spring water of Yewla (at an elevation of 2255 m a.s.l.), which showed less negative isotopic compositions. The mean raw isotopic composition of spring water indicates a wide range at the three locations. The observed mean isotopic variation at the three locations ranged from -0.6 to $5.5\,‰$ for δ^2H and -2.1 to $-0.7\,‰$ for $\delta^{18}O$. The mean values for $\delta^{18}O$ at Debre Markos and Rob Gebeya exhibit a similar isotopic composition of $-2.1\,‰$.

It is interesting that the isotopic compositions for the springs at Debre Markos and Rob Gebeya follow similar patterns and exhibit no major distinction in their isotope composition (Fig. 8). This indicates that the spring water isotopic composition for both springs derived from the same altitude range of the recharge area or different areas with the same mean elevation. The mean seasonal isotopic variations at the three spring locations during the winter season ranged between -0.2 and $4.7\,‰$, and between -2.1 and $-0.7\,‰$ for δ^2H and $\delta^{18}O$, respectively. During the spring season the mean seasonal isotopic values ranged between -2.6 and $6.3\,‰$, and between -2.4 and $-0.5\,‰$ for δ^2H and $\delta^{18}O$, respectively. During the summer season the mean seasonal isotopic variations ranged between 0.2 and $5.9\,‰$, and between -2.0 and $-0.9\,‰$ for δ^2H and $\delta^{18}O$, respectively. This implies that the isotopic values are more negative during the spring season and positive during the winter and summer seasons.

The mean winter and spring seasonal $\delta^{18}O$ isotopic compositions of precipitation exhibit values greater than $0\,‰$ for all five stations (see Sect. 4.2.1), except the more negative values at the highest altitude (Fana Choke station). In contrast to the precipitation signature during these seasons, the spring waters exhibit a more negative isotopic composition. This suggests that the spring water during the winter and spring seasons is merely derived from summer season precipitation. The highlands seem to be the main recharge area of the spring (see Fig. 10). This shows that the winter and spring season precipitation does not contribute significantly to recharging the groundwater. This finding is in agreement with the previous studies in the region (e.g. Kebede et al., 2003; Kebede and Travi, 2012). They pointed out that during the dry and spring seasons most of the water is evaporated without contributing to the groundwater recharge. Furthermore, the fact that the spring water isotope signal is more damped compared to the river water isotope signal gives a hint that the spring water could be a mixture of old water components having longer residence times than the river water.

4.2.3 Isotopic composition of river water

The mean volume weighted $\delta^{18}O$ isotope value for the Chemoga catchment was $-1.4\,‰$, and the δ^2H composition was $2.7\,‰$. During the winter or dry seasons the mean iso-

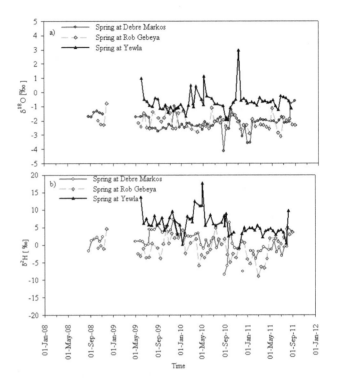

Figure 8. Temporal isotopic variability of three different non-weighted spring water samples: **(a)** $\delta^{18}O$ and **(b)** δ^2H composition over different investigation periods.

tope compositions were 0.1 and $6.2\,‰$ for $\delta^{18}O$ and δ^2H, respectively. The mean isotope compositions during the spring or little rainy seasons were -0.7 and $11.7\,‰$ for $\delta^{18}O$ and δ^2H, respectively. The summer or main rainy season mean isotope compositions were -2.3 and $-3.3\,‰$, respectively. This implies that the summer isotope composition always exhibited more negative values than the winter and spring seasons.

In the Jedeb catchment the mean volume weighted $\delta^{18}O$ and δ^2H compositions in river water are -0.6 and $4.9\,‰$, respectively. For the dry season (winter) the mean $\delta^{18}O$ and δ^2H values were 0.12 and $6.3\,‰$, respectively. For the little rainy season (spring) the mean $\delta^{18}O$ and δ^2H values were -0.3 and $8.5\,‰$, and for the summer long rainy season -1.3 and $1.7\,‰$, respectively. These results show that in all seasons except for the δ^2H composition in the spring season, the Jedeb river water exhibits more positive isotope composition as compared to the Chemoga river. Moreover, the damped response of the isotope signature during the summer season in the Jedeb river as compared to the Chemoga river might suggest that the differences in catchment storage have a relatively higher mean residence time.

Field visits during three years also supportedthe hypothesis that during the dry season in the Jedeb catchment the flow was sustained, while in the Chemoga catchment the dry season flow in the river was occasionally not sustained. Thus, the two catchments have different storage capacities. This

Table 3. Mean, range and standard deviation of δ^2H and δ^{18}O [‰, VSMOW] amount-weighted concentration for precipitation and volume weighted for discharge and non-weighted for spring water during different investigation periods.

Description	Mean [‰, VSMOW]		Minimum [‰, VSMOW]		Maximum [‰, VSMOW]		Standard deviation [‰, VSMOW]	
	δ^2H	δ^{18}O	δ^2H	δ^{18}O	δ^2H	δ^{18}O	δ^2H	δ^{18}O
Precipitation at Yewla	22.5	1.0	−25.5	−5.7	67.8	7.8	24.7	3.5
Precipitation at Debre Markos	12.3	−0.6	−37.1	−7.0	74.6	4.6	18.4	2.4
Precipitation at Enerata	15.4	−0.8	−7.1	−3.7	29.2	1.3	11.9	1.7
Precipitation at Rob Gebeya	23.0	1.0	−36.5	−6.5	54.8	5.9	21.8	3.1
Precipitation at Fana Choke	8.5	−1.8	−49.3	−9.1	48.3	2.2	25.5	3.0
Chemoga catchment precipitation	15.5	−0.4	−29.8	−6.4	40.2	3.4	16.7	2.2
Jedeb catchment precipitation	18.3	0.13	−28.4	−6.1	45.9	4.4	18.6	2.5
Chemoga discharge	2.7	−1.4	−15.5	−3.9	19.8	0.8	9.2	1.5
Jedeb discharge	4.9	−0.6	−3.3	−3.5	13.1	0.8	4.6	1.1
Yewla spring water	5.7	−0.7	−8.0	−3.9	13.5	2.9	4.0	0.9
Debre Markos spring water	0.1	−2.1	−8.5	−4.1	7.2	−0.6	5.8	0.6
Rob Gebeya spring water	−0.6	−2.1	−9.0	−3.1	5.1	−0.8	3.3	0.5

in turn seems to be related to the differences in hydrologic behaviour (see Fig. 9). A water balance study by Tekleab et al. (2011) in these catchments has also shown their hydrological differences in terms of partitioning the available water on an annual timescale. The plot of the annual evaporation ratio (the ratio of mean annual evaporation to mean annual precipitation) versus the aridity index (the ratio of mean annual potential evaporation to mean annual precipitation) in a Budyko curve demonstrated a higher evaporation ratio in the Chemoga catchment than in the Jedeb catchment.

Figure 9 presents the temporal variations in δ^{18}O and δ^2H for the Chemoga and Jedeb catchments. In the figure the isotope composition streamflow during the main rainy season reflects damped characteristics (decreases in the amplitude of the streamflow isotope signals) as compared to the fluctuations in precipitation as were observed by e.g. McDonnell et al. (1990), Buttle (1994), and Soulsby et al. (2000). These investigations indicate that the damping behaviour of the isotope signal in streamflow is due to the fact that the pre-event or old water component of the groundwater is a mixture of many past precipitation events and resulted in an isotopic concentration which is higher than the precipitation composition during storm events. The same holds true during summer months, when rainfall generates the highest flows; in the hydrological year 2010 at both catchments, the isotope composition of river water exhibits a damped response as compared to the precipitation responses (Fig. 9 inset figures).

4.3 Potential moisture source areas for the study area

Table 3 presents the mean, minimum, maximum and standard deviation of the amount-weighted precipitation and volume-weighted discharge data and the non-weighted composition for spring water. It demonstrates that more negative isotopic

compositions values are observed during the main rainy season from June to September and that the less negative values are observed during the winter and spring seasons. This is obviously related to the multiple moisture sources (e.g. the Atlantic–Congo vegetation Sudd swamp, and the Indian Ocean) and to the local meteorological processes (e.g. localised precipitation, air temperature, and humidity) (Kebede and Travi, 2012).

Seasonal variations in the isotopic composition of different water samples are shown in Fig. 10. The isotope values in different seasons might suggest different potential moisture sources bringing precipitation into the study area. These different moisture sources are investigated by mapping the potential source areas of precipitation for different seasons. Figure 11 presents these source areas of precipitation for different seasons, whereby the starting points of the trajectories in the study area were computed using the HYSPLIT (Hybrid Single Particle Lagrangian Integrated Trajectory) model developed by NOAA (National Oceanic and Atmospheric Administration) at the Air Resources Laboratory (www.arl.noaa.gov/HYSPLIT_info.php). The model computes the trajectories by tracing back an air package for 14 days in different seasons.

It is shown that during the three seasons the source areas for starting points of moisture trajectories into the study area are different. During the main rainy season, i.e. the summer, the Atlantic Ocean, Indian Ocean, the White Nile and the Congo basin are the potential source areas of precipitation in the study area. However, in the spring and winter seasons, the potential source areas of moisture origin that are responsible for generating the little precipitation in study area are the Arabian Sea–Mediterranean Sea and to some extent the Indian Ocean. These results are in agreement with the

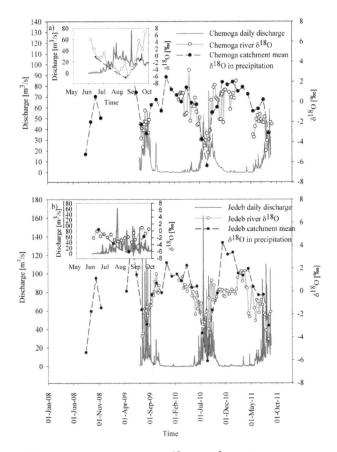

Figure 9. Temporal variations in $\delta^{18}O$ and $\delta^2 H$ composition in precipitation and river discharge along with the daily flow rate for **(a)** the Chemoga catchment and **(b)** the Jedeb catchment. The inset figures are the details for summer season discharge and isotope composition for precipitation and streamflow for the hydrological year 2010.

earlier studies with regard to the source areas from the Atlantic Ocean, Indian Ocean, and the Congo basin (e.g. Levin et al., 2009; Kebede and Travi, 2012).

To date many research findings have not reached a consensus on the origin of common moisture source areas to the northern Ethiopian highlands. Mohamed et al. (2005) indicated that the moisture flux for the northern Ethiopian plateau has mainly Atlantic origin. However, a recent moisture transport study by Viste and Sorteberg (2013) in the Ethiopian highlands reported that the moisture flow from the Gulf of Guinea, the Indian Ocean and from the Mediterranean region across the Red Sea and the Arabian Peninsula are identified as the main sources of moisture transport in the region. According to their study, the largest contribution to the moisture transport into the northern Ethiopian highland was attributed to the air travelling from the Indian Ocean and from the Mediterranean region across the Red Sea and the Arabian Peninsula.

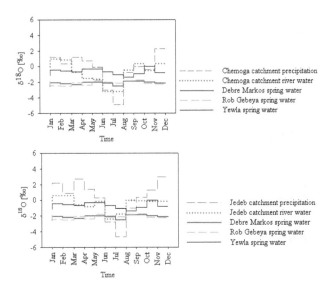

Figure 10. Monthly $\delta^{18}O$ [‰] isotopic variation in amount-weighted precipitation, volume-weighted discharge and non-weighted spring water; **(a)** Chemoga and **(b)** Jedeb catchment for the period July 2009–August 2011.

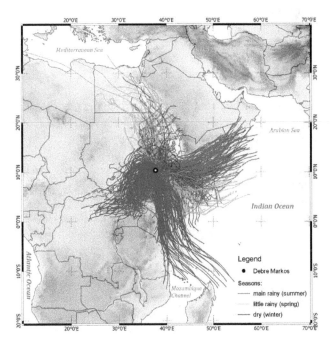

Figure 11. Potential source areas of precipitation to the study area in different seasons. Lines indicate the starting points of 14-day backward calculated trajectories. The black dot indicates the location of the study area.

4.4 Hydrograph separation on a seasonal timescale

The results of the two-component seasonal hydrograph separations reveal that the event water fraction is more dominant than the pre-event component in both catchments, in particular during the rainy season (Fig. 12). The proportion of the

Table 4. Proportion of runoff components in the Chemoga and Jedeb catchments during the wet season.

Month	Chemoga catchment		Jedeb catchment	
	Event water (%)	Pre-event water (%)	Event water (%)	Pre-event water (%)
Jul 2009	94.59	5.41	96.46	3.54
Aug 2009	79.36	20.64	68.44	31.56
Sep 2009	32.03	67.97	69.34	30.66
Jun 2010	57.45	42.55	89.30	10.70
Jul 2010	99.72	0.28	58.28	41.72
Aug 2010	64.62	35.38	34.91	65.09
Sep 2010	38.69	61.31	31.23	68.77
Jun 2011	67.67	32.33	67.56	32.44
Jul 2011	88.75	11.25	62.06	37.94
Aug 2011	89.88	10.12	64.86	35.14

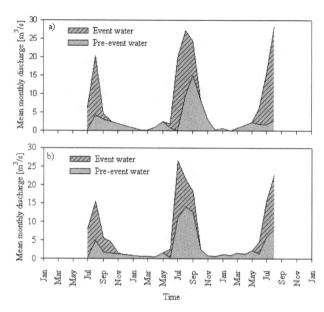

Figure 12. Two-component hydrograph separations using $\delta^{18}O$ as a tracer in the (**a**) Chemoga and (**b**) Jedeb meso-scale catchments on a seasonal timescale over the period July 2009–August 2011.

summer (rainy season) monthly variation in the event water component varies from 32 to 99 % with an average of 71 % in the Chemoga catchment and from 31 to 96 % with an average of 64 % in the Jedeb catchment over three different wet seasons of the investigation period (Table 4). Obviously, pre-event water is almost the sole contribution during the other seasons.

The average proportions of the different runoff components due to different end members (i.e. the whole dry season average concentration and the average of dry and little rainy season concentrations) exhibit a higher proportion of event water but vary from 62 to 67 and from 33 to 38 % for the pre-event water in the Chemoga catchment. In the Jedeb catchment the event water component varies from 52 to 55 % and the pre-event water varies from 45 to 48 %, respectively, due to different end members. These relatively small ranges of flow component contribution give a hint of the robustness of the method.

The proportion of the new water during the three wet seasons in both catchments is more towards the rising limb of the hydrograph. This implies that the new water component is generated via surface hydrological flow pathways in both of the catchments, and has a greater proportion than the pre-event component. Surface runoff generation starts immediately after the rainfall event in agricultural fields, in grazing lands and on bare lands. The existing gully formation as a result of severe erosion from different land use also corroborates the results of the isotope study, which shows the dominance of the event water proportion in these catchments. This is also supported by the observed flashy behaviour of small catchments (Temesgen et al., 2012). However, it is noted that the pre-event water dominates after the main event water peak.

Furthermore, it can be assumed that the high percentage of event water in both catchments is due to the low infiltration rate and the compaction of the top soil in the agri-

cultural lands. Research in the vicinity of these catchments also suggests that the effect of a plough pan due to long years of ploughing activities reduces the infiltration capacity of the soil (Temesgen et al., 2012). The effect of topography, soil physical parameters and land degradation could also be another factor in the large proportion of the event water component (Teferi et al., 2013). Nonetheless, during winter (dry season) and spring (small rainy season) the river water at both catchments is solely derived from the groundwater recharged during the rainy (summer) season. Similar studies in the US from small agriculturally dominated catchments showed that event water has a large proportion of runoff components due to low infiltration rates of agriculturally compacted soils (Shanley et al., 2002).

Past studies in different regions showed that the pre-event water is the dominant runoff component during the event (e.g. Sklash and Farvolden, 1979; Pearce et al., 1986; McDonnell, 1990; Mul et al., 2008; Hrachowitz et al., 2011a; Munyaneza et al., 2012). However, the results from the present study showed that event water is the dominant runoff component. The possible reasons could be an agriculturally dominated catchment with a high soil erosion affected area, steep slopes (e.g. varying between 2 % in the lower part of the catchment and more than 45 % in the upper part) and high seasonality of climate; the event water proportion is the dominant runoff component during the wet season on the seasonal timescale. Thus, due to these factors, the event water proportion is the dominant runoff component during the wet season on the seasonal timescale.

Uncertainty analysis of the hydrograph separations

Table 5 presents the results of the uncertainty analysis for the seasonal hydrograph separation at the 70 % (approximately one standard deviation) confidence interval. The isotope concentrations in the month of February (low flow) pre-event end member were selected for the uncertainty estimation over the whole wet season. The uncertainty results from this end member concentration are relatively low compared to the sensitivity analysis made for the different pre-event end member concentrations.

The average uncertainty terms arising from the pre-event, event and river water for the three wet season periods accounted for 61, 7, and 32 % for the Chemoga catchment and 51, 4, and 45 % for the Jedeb catchment. This suggests that most of the uncertainty stems from the event water component. Genereux (1994) pointed out that the greater uncertainty can mainly be attributed to the proportions that contribute the higher runoff components.

The error in hydrograph separation originates from different sources (see Uhlenbrook and Hoeg, 2003 for details). Based on the uncertainty results, the proportion of the different components using different pre-event end member concentrations gives only a range of values, not the exact number. Thus, due to spatial and temporal variation in the end member concentrations, the classical hydrograph separations methods give only a qualitative description of the runoff components and their variable contributions in time (Uhlenbrook and Hoeg, 2003).

4.5 Estimation of mean residence times

Preliminary estimation of mean residence time was obtained using the model described in Eqs. (9)–(11) and results are provided in Table 6. Based on the seasonal variation in $\delta^{18}O$ both in precipitation and streamflow, the mean residence times are estimated as 4.1 and 6.0 months in the Chemoga and Jedeb catchments, respectively (see Fig. 13). The goodness of fit of the observed streamflow output isotope signal is moderate, as is shown by the coefficients of determination (R^2 varies from 0.47 to 0.66). The method is appropriate for the short record length and coarse frequency of spatial and temporal tracer sampling. Indeed, the method gives indicative first approximation estimates of mean residence times and the level of fit is in line with previous studies (Soulsby et al., 2000; Rodgers et al., 2005a). The results of short mean residence times in both catchments are in line with the hydrograph separations, which indicate more surface runoff generation than base flow contribution during the storm events in these steep headwater catchments.

The results of preliminary estimation of mean residence time are plausible and anticipated from steep agriculturally dominated catchments with little adoption of soil and water conservation measures, which enhance more surface runoff

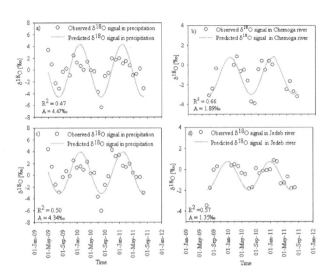

Figure 13. Fitted sine wave regression models to $\delta^{18}O$ values for precipitation and river water (**a**) and (**b**) in the Chemoga, and (**c**) and (**d**) in the Jedeb river. Inside the figures, R^2 is the coefficient of determination, and "A" is the amplitude for the input and output isotope signals.

generation (Temesgen et al., 2012). Consequently, the surface condition or the responsiveness of the soil due to the plough pan effect influences the ability of the soil to infiltrate the given rainfall amount to recharge the groundwater system (Tekleab et al., 2013). Nonetheless, this result cannot be generalised to other regions with agriculturally dominated catchments. Other factors like the soil infiltration and retention conditions, slope, drainage network and other parameters and processes could also alter/influence the runoff generation mechanism. However, in these case study catchments, the overland flow is the dominant runoff component.

Furthermore, study in the northwestern Ethiopian plateau reported that the groundwater in the area is characterised by shallow and rapid circulation leading to a young age of the groundwater system, which may considered to be the cause of the drying out of groundwater wells after prolonged droughts (Kebede, 2004). However, the groundwater age distribution is not yet fully understood in the area, and needs further research.

The mean residence time in a catchment varies depending on topography, soil types, land cover, and geologic properties (McGlynn et al., 2003; Tetzlaff, et al., 2007a, 2009). The estimated mean residence times in this study are in agreement with similar meso-scale catchments around the world. For instance, Rodgers et al. (2005a) found a residence time of 6.8 months in the Scotland Feugh meso-scale catchment, which is dominated by responsive soils. In a similar isotope study in Scotland in a nested meso-scale catchment (the catchment area varies from 10 to 231 km^2), the estimated mean residence time varies from 1 month to 14 months (Rodgers et al., 2005b). Uhlenbrook et al. (2002) found the residence times

Table 5. Percentage of total uncertainty in event, pre-event and stream water concentrations during the wet season in the Chemoga and Jedeb catchments.

Month	Chemoga catchment			Jedeb catchment		
	Event water (%)	Pre-event water (%)	River water (%)	Event water (%)	Pre-event water (%)	River water (%)
Jul 2009	96.90	0.02	3.08	96.3	0.00	3.69
Aug 2009	54.21	2.67	43.12	25.6	3.34	71.05
Sep 2009	41.37	58.63	0.00	58.7	1.75	39.51
Jun 2010	14.41	7.35	78.24	76.2	0.13	23.64
Jul 2010	72.80	0.00	27.20	55.9	3.35	40.72
Aug 2010	94.67	2.96	2.37	44.5	9.25	46.23
Sep 2010	11.00	10.21	78.79	29.2	12.93	57.92
Jun 2011	82.86	14.03	3.11	36.6	3.00	60.43
Jul 2011	67.72	0.29	31.99	39.8	3.83	56.39
Aug 2011	72.22	0.01	27.77	42.7	3.16	54.15
Average	61.00	9.00	30.00	51.00	4.00	45.00

Table 6. Amount-weighted mean precipitation and flow $\delta^{18}O$ composition, estimated amplitude, phase lag φ and mean residence time in the Chemoga and Jedeb catchments over the period July 2009–August 2011.

Description	Mean annual measured $\delta^{18}O$ [‰, VSMOW]	Amplitude [‰, VSMOW]	Phase lag φ [radian]	Mean residence time [months]
Chemoga catchment precipitation	−0.58	4.47	1.11	
Chemoga discharge	−1.34	1.89	0.01	4.1
Jedeb catchment precipitation	0.40	4.34	1.12	
Jedeb discharge	−0.70	1.35	0.01	6.0

of 24 to 36 months for the shallow groundwater, and 6 to 9 years for the deep ground water in the Brugga (40 km²) black forest meso-scale catchment in Germany. Although the climate, topography, land use, soil and geology of the catchments in the present study are different from those of other investigated catchments, the estimated mean residence times are comparable.

It is apparent from the above discussion that the mean residence time is not directly dependent on the catchment size. However, the mean residence times might be longer for small headwater catchments (McGlynn et al., 2003). It is also noted that the mean residence time of a catchment is influenced by the heterogeneity in climatic setting, topography and geology (Hrachowitz et al., 2009), landscape controls, particularly soil cover (Soulsby et al., 2006), percent coverage responsive soil using catchment soil maps (Soulsby and Tetzlaff, 2008), and topography and soil drainage conditions in different geomorphic provinces (Tetzlaff et al., 2009). A recent study by Heidbüchel et al. (2012) showed that the mean residence time is better estimated using time-variable functions. Thus, the residence time depends not only on the catchment characteristics but also on time varying climate inputs. Nevertheless, investigating all these different controlling processes on

mean residence time estimation was beyond the scope of the present study. Therefore, there is a need for further research that takes the influence of catchment heterogeneity and time varying input on the estimation of mean residence time into account in catchments of the Nile basin.

5 Conclusions

Characterisations of stable isotope composition of precipitation, spring and river water along different altitude gradients were undertaken with the aim of preliminarily estimating the mean residence time and runoff component contributions on a seasonal timescale. The results show that precipitation, stream and spring waters exhibit noticeable spatial and temporal variations in stable $\delta^{18}O$ and δ^2H composition in the study area.

The results further demonstrate that the meteoric water in the study area is influenced by the amount and to a lesser extent by the altitude and temperature effects. The climatic seasonality, which is dominated by different moisture sources, along with the local meteorological conditions play a significant role in the isotopic composition of rainfall in the area.

The analyses of isotope results reveal the dominance of event water and short mean residence times in both of the catchments. From the point of view of managing the water resources and the importance of the available soil water for consumptive use of the crops, catchment management aiming at reducing overland flow/soil erosion and increasing soil moisture storage and recharge has paramount importance for the farmers residing in these catchments.

It should be noted that in the light of the data availability the estimated mean residence times and seasonal hydrograph separation represent first approximations. Consequently, for more reliable estimates of the mean residence times and runoff contributions there is a need for further research with finer resolution sampling during storm events and long-term isotope tracer data collection at different spatial and finer temporal scales (e.g. daily and hourly) that will improve our understanding of how these catchments function. It is noteworthy that the applied methods were used for the first time in the region that has critical regional importance regarding the water resources in the Nile. Thus, the results can be used as a baseline for further hydrological studies for a better understanding of the dominant runoff components in the future.

Acknowledgements. The study was carried out as a project within the integrated research programme "In search of sustainable catchments and basin-wide solidarities in the Blue Nile River Basin Hydrosolidarity", which was funded by the Foundation for the Advancement of Tropical Research (WOTRO) of the Netherlands Organisation for Scientific Research (NWO), UNESCO-IHE and Addis Ababa University. We also thank the Ethiopian Ministry of Water Resources and Energy for providing the hydrological data and the National Meteorological Agency for providing the weather data. We would like to thank Melesse Temesgen, who facilitated the data collection initially and made the precipitation sample collectors. We would also like to thank Solomon Takele, Lakachew Alemu, Limenew Mihrete, Mengistu Abate, Chekolech Mengistie, and Derseh Gebeyehu for isotope sample collection of precipitation, spring and stream waters. Our thanks extend to Jorge Valdemar Zafra Cordova, who analysed part of the isotope samples at the UNESCO-IHE laboratory. The research would not have been possible without their kind co-operation in collecting the samples and laboratory analysis.

Edited by: F. Fenicia

References

Aravena, R., Suzuki, O., Pena, H., Grilli, A., Pollastri, A., and Fuenzalida, H.: Isotopic composition and origin of the precipitation in Northern Chile, Appl. Geochem., 14, 411–422, 1999.

Barthold, F. K., Wu, J., Vache, K. B., Schneider, K., Frede, H. G., and Breuer, L.: Identification of geographic runoff sources in a data sparse region: hydrological processes and the limitations of tracer-based approaches, Hydrol. Process, 24, 2313–2327, doi:10.1002/hyp.7678, 2010.

Bewket, W. and Sterk, G.: Dynamics land cover and its effect on the streamflow on the Chemoga watershed in the Blue Nile basin, Ethiopia, Hydrol. Processes., 19, 445–458, 2005.

Bliss, C. I.: Periodic Regressions Statistics in Biology, McGraw-Hill Book Co., New York, USA, 219–287, 1970.

Botter, G., Bertuzzo, E., and Rinaldo, A..: Catchment residence and travel time distributions: The master equation, Geophys. Res. Lett., 38, L11403, doi:10.1029/2011GL047666, 2011.

Buttle, J. M.: Isotope hydrograph separations and rapid delivery of pre-event water from drainage basins, Prog. Phys. Geogr., 18, 16–41, 1994.

Dansgaard, W.: Stable isotopes in precipitation, Tellus, 16, 436–468, 1964.

Darling, W. G. and Gizaw, B.: Rainfall–groundwater isotopic relationships in eastern Africa: the Addis Ababa anomaly. Study of environmental change using isotopic techniques, C&S Papers Series, IAEA, 489–490, 2002.

Didszun, J. and Uhlenbrook, S.: Scaling of dominant runoff generation processes: Nested catchments approach using multiple tracers, Water Resour. Res., 44, W02410, doi:10.1029/2006WR005242, 2008.

Dunn, S. M., McDonnell, J. J., and Vache, K. B.: Factors influencing the residence time of catchment waters: A virtual experiment approach, Water Resour. Res., 43, W06408, doi:10.1029/2006WR005393, 2007.

Gat, J. R., Bowser, C. J., and Kendall, C.: The contribution of evaporation from the Great Lakes to the continental atmosphere: Estimate based on stable isotope data, Geophys. Res. Lett., 21, 557–560, 1994.

Gat, J. R.: Oxygen and Hydrogen isotopes in the hydrologic cycle, Annu. Rev. Earth Planet. Sci., 24, 225–262, 1996.

Genereux, D. P.: Quantifying uncertainty in tracer based hydrograph separation, Water Resour. Res., 34, 915–919, 1998.

Gibson, J. J., Edwards, T. W. D., Birks, S. J., Amour, N. A. St., Buhay, W. M., McEachern, P., Wolfe, B. B., and Peters, D. L.: Progress in isotope tracer hydrology in Canada, Hydrol. Process., 19, 303–327, doi:10.1002/hyp.5766, 2005.

Hargreaves, G. H. and Samani, Z. A.: Estimating potential evaporation, J. Irrig. Drain. E.-ASCE, 108, 225–230, 1982.

Heidbüchel, I., Troch, P. A., Lyon, S. W., and Weiler, M.: The master transit time distribution of variable flow systems, Water Resour. Res., 48, W06520, doi:10.1029/2011WR011293, 2012.

Hrachowitz, M., Soulsby C., Tetzlaff, D., Dawson, J. J. C., Dunn, S. M., and Malcolm, I. A.: Using longer-term data sets to understand transit times in contrasting headwater catchments, J. Hydrol., 367, 237–248, doi:10.1016/j.jhydrol.2009.01.001, 2009.

Hrachowitz, M., Bohte, R., Mul, M. L., Bogaard, T. A., Savenije, H. H. G., and Uhlenbrook, S.: On the value of combined event runoff and tracer analysis to improve understanding of catchment functioning in a data-scarce semi-arid area, Hydrol. Earth Syst. Sci., 15, 2007–2024, doi:10.5194/hess-15-2007, 2011a.

Hrachowitz, M., Soulsby, C., Tetzlaff, I., and Malccolm, A.: Sensitivity of mean transit time estimates to model conditioning and data availability, Hydrol. Processes, 25, 980–990, 2011b.

Hrachowitz, M., Savenije, H., Bogaard, T. A., Tetzlaff, D., and Soulsby, C.: What can flux tracking teach us about water age distribution patterns and their temporal dynamics?, Hydrol. Earth Syst. Sci., 17, 533–564, doi:10.5194/hess-17-533-2013, 2013.

Hargreaves, G. H. and Samani Z. A.: Estimating potential evaporation, J. Irrig. Drain. Eng., 108, 225–230, 1982.

International Atomic Energy Agency.: IAEA-WMO Programme on Isotopic Composition of Precipitation: Global Network of Isotopes In Precipitation (GNIP) Technical procedure for sampling, 2009.

Joseph, A., Frangi, P., and Aranyossy, J. F.: Isotopic composition of Meteoric water and groundwater in the Sahelo-Sudanese Zone, J. Geophys. Res., 97, 7543–7551, 1992.

Kebede, S.: Approaches isotopique et geochimique pour l'etudedes eaux souterraines et des lacs: Examples du haut bassin du Nil Bleu et du rift Ethiopien [Environmental isotopes and geochemistry in groundwater and lake hydrology: cases from the Blue Nile basin, main Ethiopian rift and Afar, Ethiopia], PhD thesis, University of Avignon, France, 2004, p. 162, 2004.

Kebede, S. and Travi, Y.: Origin of the $\delta^{18}O$ and δ^2H composition of meteoric waters in Ethiopia, Quatern. Int., 257, 4–12, doi:10.1016/j.quaint.2011.09.032, 2012.

Kebede, S., Travi, Y., Alemayehu, T., Ayenew, T., and Aggarwal, P.: Tracing sources of recharge to ground waters in the Ethiopian Rift and bordering plateau: isotopic evidence, Fourth International Conference on Isotope for Groundwater Management, IAEA-CN-104/36, 19–22 May, Vienna, IAEA, 19–20, 2003.

Kendall, C. and Caldwell, E. A.: Fundamentals of isotope geochemistry, in: Isotope Tracers in Catchment Hydrology, edited by: Kendall, C. and McDonnel, J. J., Elsevier Science, Amsterdam, 51–86, 1998.

Kendall, C. and Coplen, T. B.: Distribution of oxygen −18 and deuterium in river waters across the United States, Hydrol. Process., 15, 1363–1393, doi:10.1002/hyp.217, 2001.

Kirchner, J. W., Tetzlaff, D., and Soulsby, C.: Comparing chloride and water isotopes as hydrological tracers in two Scottish catchments, Hydrol. Process., 24, 1631–1645, doi:10.1002/hyp.7676, 2010.

Laudon, H., Sjöblom, V., Buffam, I., Seibert, J., and Mörth, M.: The role of catchment scale and landscape characteristics for runoff generation of boreal stream, J. Hydrol., 344, 198–209, doi:10.1016/j.jhydrol, 2007.

Levin, N. E., Zipser, E. J., and Cerling, T. E.: Isotopic composition of waters from Ethiopia and Kenya: Insight into moisture sources for eastern Africa, J. Geophys. Res., 114, D23306, doi:10.1029/2009JD012166, 2009.

Liu, Y., Fan, N., An, S., Bai, X., Liu, F., Xu, Z., Wang, Z., and Liu, S.: Characteristics of water isotopes and hydrograph separation during the wet season in the Heishui River, China, J. Hydrol., 353, 314–321, 2008.

Maloszewski, P. and Zuber, A.: Determining the turnover time of groundwater systems with the aid of environmental tracers, Models and their applicability, J. Hydrol., 57, 207–231, 1982.

McDonnell, J. J.: A rational for old water discharge through macro pores in a steep humid catchment, Water Resour. Res., 26, 2821–2832, 1990.

McDonnell, J. J., Bonell, M., Stewart, M. K., and Pearce, A. J.: Deuterium variations in storm rainfall-Implications for stream hydrograph separation, Water Resour. Res., 26, 455–458, 1990.

McDonnell, J. J., Stewart, M. K., and Owens, I. F.: Effect of catchment scale subsurface mixing on stream isotopic response, Water Resour. Res., 27, 3065–3073, 1991.

McGlynn, B., McDonnell, J., Stewart, M., and Seibert, J.: On the relationship between catchment scale and stream water mean residence time, Hydrol. Process., 17, 175–181, 2003.

McGuire, K. J., DeWalle, D. R., and Gburek, W. J.: Evaluation of mean residence time in subsurface waters using oxygen-18 fluctuations during drought conditions in the semi-Appalachains, J. Hydrol., 261, 132–149, 2002.

McGuire, K. J., McDonnell, J. J., Weiler, M., Kendall, C., McGlynn, C. L., Welker, J. L., and Siebert, J.: The role of topography on catchment scale water residence time, Water Resour. Res., 41, W05002, doi:10.1029/2004WR003657, 2005.

McGuire, K. J. and McDonnell, J. J.: A review and evaluation of catchment transit time modeling, J. Hydrol., 330, 543–563, 2006.

Mohamed, Y. A., van den Hurk, B. J. J. M., Savenije, H. H. G., and Bastiaanssen, W. G. M.: Hydroclimatology of the Nile: results from a regional climate model, Hydrol. Earth Syst. Sci., 9, 263–278, doi:10.5194/hess-9-263-2005, 2005.

Mul, M. L., Mutiibwa, K. R., Uhlenbrook, S., and Savenije, H. H. G.: Hydrograph separation using hydrochemical tracers in the Makanya catchment, Tanzania, Phys. Chem. Earth., 33, 151–156, 2008.

Munyaneza, O., Wenninger, J., and Uhlenbrook, S.: Identification of runoff generation processes using hydrometric and tracer methods in a meso-scale catchment in Rwanda, Hydrol. Earth Syst. Sci., 16, 1991–2004, doi:10.5194/hess-16-1991, 2012.

NMSA (National Meteorological service Agency): Climatic and agroclimatic resources of Ethiopia, NMSA Meteorological Research Report Series, Vol. 1, No. 1, Addis Ababa, p. 137, 1996.

Pearce, A. J., Stewart, M. K., and Sklash, M. G.: Storm runoff generation in humid headwater catchments. Where does the water come from?, Water Resour. Res., 22, 1263–1272, 1986.

Rodgers, P., Soulsby, C., Waldron, S., and Tetzlaff, D.: Using stable isotope tracers to assess hydrological flow paths, residence times and landscape influences in a nested mesoscale catchment, Hydrol. Earth Syst. Sci., 9, 139–155, doi:10.5194/hess-9-139-2005, 2005a.

Rodgers, P., Soulsby, C., and Waldron, S.: Stable isotope tracers as diagnostic tools in up scaling flow path understanding and residence time estimates in a mountainous mesoscale catchment, Hydrol. Process. 19, 2291–2307, 2005b.

Rozanski, K., Araguás-Araguás, L., and Gonfiantini, R.: Isotopic patterns in modern global precipitation, in climate change in Continetal Isotopic Records, Geophysical Monograph 78, American Geophysical Union, 1–36, 1993.

Rozanski, K., Araguás-Araguás, L., and Gonfiantini, R.: Isotope patterns of precipitation in the east African region, in: The Liminology, Climatology and Paleoclimatology of the East African Lakes, edited by: Johnson, T. C. and Odada, E., Gordon and Breach, Toronto, 79–93, 1996.

Shanley, J. B., Kendall, C., Smith, T. E., Wolock, D. M., and McDonnell, J. J.: Controls on old and new water contributions to stream flow at some nested catchments in Vermont, USA, Hydrol. Processes., 16, 589–609, 2002.

Sklash, M. G. and Farvolden, R. N.: The role of groundwater in storm runoff, J. Hydrol., 43, 45–65, doi:10.1016/0022-1694(79)90164-1, 1979.

Soulsby, S. and Tetzlaff, D.: Towards simple approaches for mean residence time estimation in ungauged basins using tracers and soil distributions, J., Hydrol., 363, 60–74, 2008.

Soulsby, C., Malcolm, R., Helliwell, R., Ferrier, R. C., and Jenkins, A.: Isotope hydrology of the Allt a' Mharcaidh catchment, Cairngorms, Scotland: implications for hydrological pathways and residence times, Hydrol. Processes., 14, 747–762, 2000.

Soulsby, C., Tetzlaff, D., Dunn, S. M., and Waldron, S.: Scaling up and out in runoff process understanding: insight from nested experimental catchment studies, Hydrol. Processes., 20, 2461–2465, 2006.

Taylor, C. B., Wilson, D. D., Borwn, L. J., Stewart, M. K., Burdon, R. J., and Brailsford, G. W.: Sources and flow of North Canterbury plains ground water, New Zealand, J. Hydrol., 106, 311–340, 1989.

Teferi, E., Uhlenbrook, S., Bewket, W., Wenninger, J., and Simane, B.: The use of remote sensing to quantify wetland loss in the Choke Mountain range, Upper Blue Nile basin, Ethiopia, Hydrol. Earth Syst. Sci., 14, 2415–2428, doi:10.5194/hess-14-2415-2010, 2010.

Teferi, E., Bewket, W., Uhlenbrook, S., and Wenninger, J.: Understanding recent land use and land cover dynamics in the source region of the Upper Blue Nile, Ethiopia: Spatially explicit statistical modeling of systematic transitions, Agr. Ecosyst. Environ., 165, 98–117, 2013.

Tekleab, S., Uhlenbrook, S., Mohamed, Y., Savenije, H. H. G., Temesgen, M., and Wenninger, J.: Water balance modeling of Upper Blue Nile catchments using a top-down approach, Hydrol. Earth Syst. Sci., 15, 2179–2193, doi:10.5194/hess-15-2179-2011, 2011.

Tekleab, S., Mohamed, Y., Uhlenbrook, S., and Wenninger, J.: Hydrologic responses to land cover change, the case of Jedeb meso-scale catchment, Abay/Upper Blue Nile basin, Ethiopia, Hydrol. Process., doi:10.1002/hyp.9998, in press, 2013.

Temesgen, M., Uhlenbrook, S., Simane, B., van der Zaag, P., Mohamed, Y., Wenninger, J., and Savenije, H. H. G.: Impacts of conservation tillage on the hydrological and agronomic performance of *Fanya juus* in the upper Blue Nile (Abbay) river basin, Hydrol. Earth Syst. Sci., 16, 4725–4735, doi:10.5194/hess-16-4725-2012, 2012.

Tetzlaff, D., Waldron, S., Brewer, M. J., and Soulsby, C.: Assessing nested hydrological and hydochemical behaviour of a mesoscale catchment using continuous tracer data, J. Hydrol., 336, 430–443, 2007a.

Tetzlaff, D., Soulsby, C., Waldron, S., Malcolm, I. A., Bacon, P. J., Dunn, S. M., and Lilly, A.: Conceptualization of runoff processes using GIS and tracers in a nested mesoscale catchment, Hydrol. Processes., 21, 1289–1307, 2007b.

Tetzlaff, D., Seibert, J., and Soulsby, C.: Inter-catchment comparison to assess the influence of topography and soils on catchment transit times in a geomorphic province; in a Cairngorm Mountains Scotland, Hydrol. Processes., 23, 1874–1886, 2009.

Uhlenbrook, S. and Hoeg, S.: Quantifying uncertainties in tracer based hydrograph separations: a case study for two, three and five component hydrograph separations in a mountainous catchment, Hydrol. Processes., 17, 431–453, 2003.

Uhlenbrook, S. and Leibundgut, C.: Process-oriented catchment modeling and multiple-response validation, Hydrol. Processes., 16, 423–440, doi:10.1002/hyp.330, 2002.

Uhlenbrook, S., Frey, M., Leibundgut, C., and Maloszewski, P.: Hydrograph separations in a mesoscale mountainous basin at event and seasonal time scales, Water Resour. Res., 38, 1–13, 2002.

Van der Velde Y., de Rooij, G. H., Rozemeijer, J. C., Van Geer, F. C., and Broers, H. P.: Nitrate response of a lowland catchment: On the relation between stream concentration and travel time distribution dynamics,. Water Resour. Res., 46, W11534, doi:10.1029/2010WR009105, 2010.

Viste, E. and Sorteberg, A.: Moisture transport into the Ethiopia highlands, Int. J. Climatol., 33, 249–263, 2013.

Wels, C., Cornett, R. J., and Lazerte, B. D.: Hydrograph separation: a comparison of geochemical and isotopic tracers, J. Hydrol., 122, 253–274, 1991.

Wissmeier, L. and Uhlenbrook, S.: Distributed, high-resolution modeling of ^{18}O Signals in a meso-scale catchment, J. Hydrol., 332, 497–510, doi:10.1016/j.jhydrol.2006.08.003, 2007.

2

Identification of dominant hydrogeochemical processes for groundwaters in the Algerian Sahara supported by inverse modeling of chemical and isotopic data

Rabia Slimani[1], **Abdelhamid Guendouz**[2], **Fabienne Trolard**[3], **Adnane Souffi Moulla**[4], **Belhadj Hamdi-Aïssa**[1], **and Guilhem Bourrié**[3]

[1]Ouargla University, Fac. des Sciences de la Nature et de la Vie, Lab. Biochimie des Milieux Désertiques, 30000 Ouargla, Algeria
[2]Blida University, Science and Engineering Faculty, P.O. Box 270, Soumaâ, Blida, Algeria
[3]INRA – UMR1114 EMMAH, Avignon, France
[4]Algiers Nuclear Research Centre, P.O. Box 399, Alger-RP, 16000 Algiers, Algeria

Correspondence to: Rabia Slimani (slm_rabia@yahoo.fr)

Abstract. Unpublished chemical and isotopic data taken in November 1992 from the three major Saharan aquifers, namely the Continental Intercalaire (CI), the Complexe Terminal (CT) and the phreatic aquifer (Phr), were integrated with original samples in order to chemically and isotopically characterize the largest Saharan aquifer system and investigate the processes through which groundwaters acquire their mineralization. Instead of classical Debye–Hückel extended law, a specific interaction theory (SIT) model, recently incorporated in PHREEQC 3.0, was used. Inverse modeling of hydrochemical data constrained by isotopic data was used here to quantitatively assess the influence of geochemical processes: at depth, the dissolution of salts from the geological formations during upward leakage without evaporation explains the transitions from CI to CT and to a first end member, a cluster of Phr (cluster I); near the surface, the dissolution of salts from sabkhas by rainwater explains another cluster of Phr (cluster II). In every case, secondary precipitation of calcite occurs during dissolution. All Phr waters result from the mixing of these two clusters together with calcite precipitation and ion exchange processes. These processes are quantitatively assessed by the PHREEQC model. Globally, gypsum dissolution and calcite precipitation were found to act as a carbon sink.

1 Introduction

A scientific study published in 2008 (OECD, 2008) showed that 85 % of the world population lives in the driest half of the earth. More than 1 billion people residing in arid and semi-arid areas of the world have access to only few or nonrenewable water resources. The North-Western Sahara Aquifer System (NWSAS) is one of the largest confined reservoirs in the world and its huge water reserves are essentially composed of an old component. It is represented by two main deep aquifers, the Continental Intercalaire and the Complexe Terminal. This system covers a surface of more than 1 million km^2 (700 000 km^2 in Algeria, 80 000 km^2 in Tunisia and 250 000 km^2 in Libya). Due to the climatic conditions of Sahara, these formations are poorly renewed: about 1 billion m^3 yr^{-1} essentially infiltrated in the piedmont of the Saharan Atlas in Algeria, as well as in the Jebel Dahar and Jebel Nafusa in Tunisia and Libya, respectively. However, the very large extension of the system as well as the great thickness of the aquifer layers has favored the accumulation of huge water reserves. Ouargla basin is located in the middle of the NWSAS and thus benefits from groundwater resources (Fig. 1) which are contained in the following three main reservoirs (UNESCO, 1972; Eckstein and Eckstein, 2003; OSS, 2003, 2008):

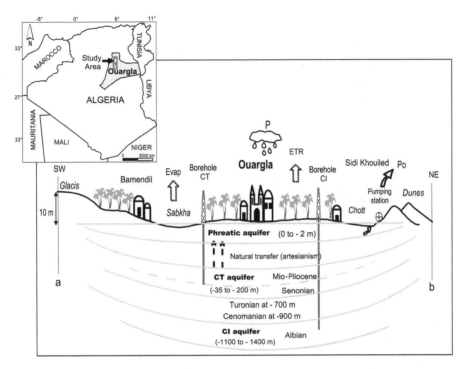

Figure 1. Location and schematic relations of aquifers in Ouargla. Blue lines represent limits between aquifers, and the names of aquifers are given in bold letters; as the limit between Senonian and Mio-Pliocene aquifers is not well defined, a dashed blue line is used. Names of villages and cities are given in roman (Bamendil, Ouargla, Sidi Khouiled), while geological/geomorphological features are in italic (glacis, sabkha, chott, dunes). Depths are relative to the ground surface. Letters "a" and "b" refer to the cross section (Fig. 2) and to the localization map (Fig. 3).

- At the top, the phreatic aquifer (Phr), in the Quaternary sandy gypsum permeable formations of Quaternary, is almost unexploited, due to its extreme salinity $(50 \, \text{g} \, \text{L}^{-1})$.

- In the middle, the Complexe Terminal (CT) (Cornet and Gouscov, 1952; UNESCO, 1972) is the most exploited and includes several aquifers in different geological formations. Groundwater circulates in one or the two lithostratigraphic formations of the Eocene and Senonian carbonates or in the Mio-Pliocene sands.

- At the bottom, the Continental Intercalaire (CI), hosted in the lower Cretaceous continental formations (Barremian and Albian), mainly composed of sandstones, sands and clays. It is only partially exploited because of its significant depth.

The integrated management of these groundwaters is presently a serious issue for local water resources managers due to the large extension of the aquifers and the complexity of the relations between them. Several studies (Guendouz, 1985; Fontes et al., 1986; Guendouz and Moulla, 1996; Edmunds et al., 2003; Guendouz et al., 2003; Hamdi-Aïssa et al., 2004; Foster et al., 2006; OSS, 2008) started from chemical and isotopic information (^{2}H, ^{18}O, ^{234}U, ^{238}U, ^{36}Cl) to characterize the relationships between aquifers. In particular, such studies focused on the recharge of the deep CI aquifer system. These investigations especially dealt with water chemical facies, mapped iso-contents of various parameters and reported typical geochemical ratios ($[SO_4^{2-}]/[Cl^-]$, $[Mg^{2+}]/[Ca^{2+}]$) as well as other correlations. Minerals–solutions equilibria were checked by computing saturation indices with respect to calcite, gypsum, anhydrite and halite, but processes were only qualitatively assessed. The present study aims at applying for the first time ever in Algeria, inverse modeling to an extreme environment featuring a lack of data on a scarce natural resource (groundwater). New data were hence collected in order to characterize the hydrochemical and the isotopic composition of the major aquifers in the Saharan region of Ouargla. New possibilities offered by progress in geochemical modeling were used. The objective was also to identify the origin of the mineralization and the water-rock interactions that occur along the flow path. More specifically, inverse modeling of chemical reactions allows one to select the best conceptual model for the interpretation of the geochemical evolution of Ouargla aquifer system. The stepwise inversion strategy involves designing a list of scenarios (hypotheses) that take into consideration the most plausible combinations of geochemical processes that may occur within the studied medium. After resolving the scenarios in a stepwise manner, the one that provides the best conceptual geochemical model is then se-

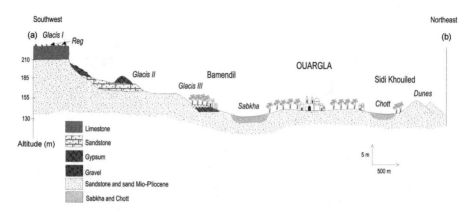

Figure 2. Geologic cross section in the region of Ouargla. The blue pattern used for chott and sabkha correspond to the limit of the saturated zone.

lected, which allowed Dai et al. (2006) to optimize simultaneously transmissivities and geochemical transformations in a confined aquifer. Inverse modeling with PHREEQC 3.0 was used here in a different way (only on geochemical data but for several aquifers) to account for the modifications of the composition of water along the flow path. At least two chemical analyses of groundwater at different points of the flow path, and a set of phases (minerals and/or gases) which potentially react while water circulates, are needed to operate the program (Charlton et al., 1996).

A number of assumptions are inherent to the application of inverse geochemical modeling: (i) the two groundwater analyses from the initial and the final boreholes should represent groundwater that flows along the same flow path; (ii) dispersion and diffusion do not significantly affect groundwater chemistry; (iii) a chemical steady state prevails in the groundwater system during the time considered; and (iv) the mineral phases used in the inverse calculation are or were present in the aquifer (Zhu and Anderson, 2002). The soundness or the validity of the results depends on a valid conceptualization of the groundwater system, on the validity of the basic hydrogeochemical concepts and principles, on the accuracy of model input data and on the level of understanding of the geochemical processes occurring in the area (Güler and Thyne, 2004; Sharif et al., 2008). These requirements are fulfilled in the region of Ouargla, which can be considered as a "window" to the largest Saharan aquifer, and thus one of the largest aquifers in the world in a semi-arid to hyper-arid region subject to both global changes: urban sprawl and climate change. The methodology developed here and the data collected can easily be integrated in the PRECOS framework proposed for the management of environmental resources (Trolard et al., 2016).

2 Methodology

2.1 Presentation of the study area

The study area is located in the northeastern desert of Algeria (lower Sahara) (Le Houérou, 2009) near the city of Ouargla (Fig. 1), 31°54' to 32°1' N and 5°15' to 5°27' E, with a mean elevation of 134 (m a.s.l.). It is located in the quaternary valley of Oued Mya basin. The present climate belongs to the arid Mediterranean type (Dubief, 1963; Le Houérou, 2009; ONM, 1975/2013), as it is characterized by a mean annual temperature of 22.5 °C, a yearly rainfall of 43.6 mm yr^{-1} and a very high evaporation rate of 2138 mm yr^{-1}.

Ouargla's region and the entire lower Sahara has experienced during its long geological history alternating marine and continental sedimentation phases. During Secondary era, vertical movements affected the Precambrian basement, causing, in particular, collapse of its central part, along an axis passing approximately through the Oued Righ Valley and the upper portion of the Oued Mya Valley. According to Furon (1960), a epicontinental sea spread to the Lower Eocene of northern Sahara. After the Oligocene, the sea gradually withdrew. It is estimated at present that this sea did not reach Ouargla and transgression stopped at the edge of the bowl (Furon, 1960; Lelièvre, 1969). The basin is carved into Mio-Pliocene (MP) deposits, which alternate with red sands, clays and sometimes marls; gypsum is not abundant and dated from the Pontian era (during the MP) (Cornet and Gouscov, 1952; Dubief, 1953; Ould Baba Sy and Besbes, 2006). The continental Pliocene consists of a local limestone crust with puddingstone or lacustrine limestone (Fig. 2), shaped by eolian erosion into flat areas (regs). The Quaternary formations are lithologically composed of alternating layers of permeable sand and relatively impermeable marl (Aumassip et al., 1972; Chellat et al., 2014).

The exploitation of Mio-Pliocene aquifer is ancient and at the origin of the creation of the oasis (Lelièvre, 1969; Moulias, 1927). The piezometric level was higher (145 m a.s.l.)

but overexploitation at the end of the 19th century led to a catastrophic decrease of the resource, with presently more than 900 boreholes (ANRH, 2011).

The exploitation of the Senonian aquifer dates back to 1953 at a depth between 140 and 200 m, with a small initial rate of approximately $9\,L\,s^{-1}$; two boreholes have been exploited since 1965 and 1969, with a total flow rate of approximately $42\,L\,s^{-1}$, for drinking water and irrigation.

The exploitation of the Albian aquifer dates back to 1956; presently, two boreholes are exploited:

- El Hedeb I, 1335 m deep, with a flow rate of $141\,L\,s^{-1}$;

- El Hedeb II, 1400 m deep, with a flow rate of $68\,L\,s^{-1}$.

2.2 Sampling and analytical methods

The sampling programme consisted of collecting samples along transects corresponding to directions of flow for both the Phr and CT aquifers, while it was possible to collect only eight samples from the CI. A total of 107 samples were collected during a field campaign in 2013 along the main flow path of Oued Mya. Of these, 67 were from piezometers tapping the phreatic aquifer, 32 from CT wells and the last 8 from boreholes tapping the CI aquifer (Fig. 3). Analyses of Na^+, K^+, Ca^{2+}, Mg^{2+}, Cl^-, SO_4^{2-} and HCO_3^- were performed by ion chromatography at Algiers Nuclear Research Center (CRNA). Previous and yet unpublished data (Guendouz and Moulla, 1996) sampled in 1992 are used here too: 59 samples for the Phr aquifer, 15 samples for the CT aquifer and 3 samples for the CI aquifer for chemical analyses and data of ^{18}O and ^3H (Guendouz and Moulla, 1996).

2.3 Geochemical method

PHREEQC was used to check minerals–solutions equilibria using the specific interaction theory (SIT), i.e., the extension of Debye–Hückel law by Scatchard and Guggenheim incorporated recently in PHREEQC 3.0 (Parkhurst and Appelo, 2013). Inverse modeling was used to calculate the number of minerals and gases' moles that must, respectively, dissolve or precipitate/degas to account for the difference in composition between initial and final water end members (Plummer and Back, 1980; Kenoyer and Bowser, 1992; Deutsch, 1997; Plummer and Sprinckle, 2001; Güler and Thyne, 2004; Parkhurst and Appelo, 2013). This mass balance technique has been used to quantify reactions controlling water chemistry along flow paths (Thomas et al., 1989). It is also used to quantify the mixing proportions of end-member components in a flow system (Kuells et al., 2000; Belkhiri et al., 2010, 2012).

Inverse modeling involves designing a list of scenarios (modeling setups) that take into account the most plausible combinations of geochemical processes that are likely to occur in our system. For example, the way to identify whether calcite dissolution/precipitation is relevant or not consists of

solving the inverse problem under two alternate scenarios: (1) considering a geochemical system in which calcite is present and (2) considering a geochemical system without calcite. After simulating the two scenarios, it is usually possible to select the setup that gives the best results as the solution to the inverse modeling according to the fit between the modeled and observed values. Then one can conclude whether calcite dissolution/precipitation is relevant or not. This stepwise strategy allows us to identify the relevance of a given chemical process by inversely solving the problem through alternate scenarios in which the process is either participating or not (Dai et al., 2006).

In the geochemical modeling inverse, soundness of results is dependent upon valid conceptualization of the system, validity of basic concepts and principles, accuracy of input data and level of understanding of the geochemical processes. We use the information from the lithology, general hydrochemical evolution patterns, saturation indices and mineral stability diagrams to constrain the inverse models.

3 Results and discussion

Tables 1 to 4 illustrate the results of the chemical and the isotopic analyses. Samples are ordered according to an increasing electric conductivity (EC), and this is assumed to provide an order for increasing salt content. In both the phreatic and CT aquifers, temperature is close to 25 °C, while for the CI aquifer, temperature is close to 50 °C. The values presented in Tables 1 to 5 are raw analytical data that were corrected for defects of charge balance before computing activities with PHREEQC. As analytical errors could not be ascribed to a specific analyte, the correction was made proportionally. The corrections do not affect the anion-to-anion mole ratios, such as for $[HCO_3^-]/([Cl^-]+2[SO_4^{2-}])$ or $[SO_4^{2-}]/[Cl^-]$, whereas they affect the cation-to-anion ratio, such as for $[Na^+]/[Cl^-]$.

3.1 Characterization of chemical facies of the groundwater

Piper diagrams drawn for the studied groundwaters (Fig. 4) broadly show a scatter plot dominated by a sodium chloride facies. However, when going into small details, the widespread chemical facies of the Phr aquifer are closer to the NaCl cluster than those of the CI and CT aquifers. Respectively, $CaSO_4$, Na_2SO_4, $MgSO_4$ and NaCl are the most dominant chemical species (minerals) that are present in the phreatic waters. This sequential order of solutes is comparable to that of other groundwater occurring in north Africa and especially in the neighboring area of the chotts (depressions where salts concentrate by evaporation) Merouane and Melrhir (Vallès et al., 1997; Hamdi-Aïssa et al., 2004).

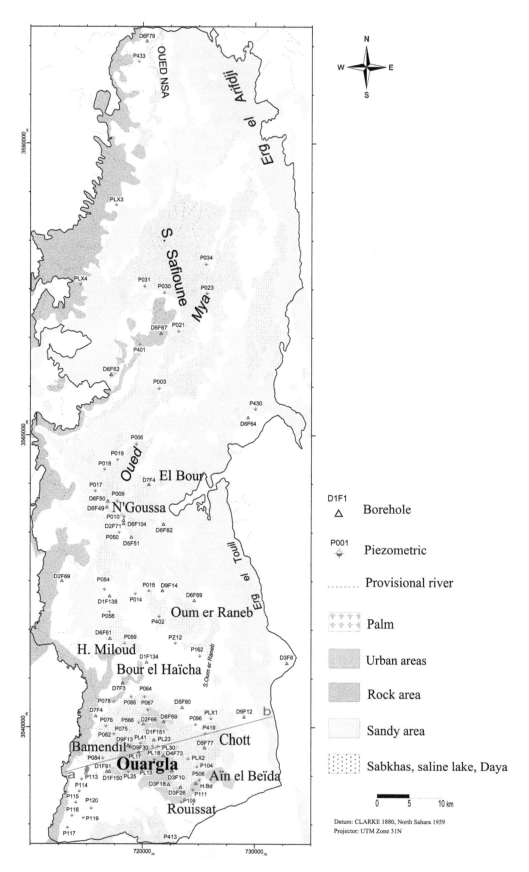

Figure 3. Location map of sampling points.

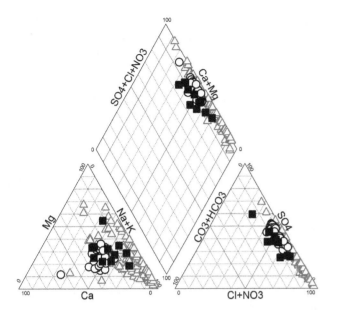

Figure 4. Piper diagram for the Continental Intercalaire (filled squares), Complexe Terminal (open circles) and phreatic (open triangles) aquifers.

3.2 Spatial distribution of the mineralization

The salinity of the phreatic aquifer varies considerably depending on the location (namely, the distance from wells or drains) and time (due to the influence of irrigation) (Fig. 5a).

Its salinity is low around irrigated and fairly well-drained areas, such as the palm groves of Hassi Miloud, just north of Ouargla (Fig. 3), that benefit from freshwater and are drained to the sabkha Oum El Raneb. However, the three lowest salinity values are observed in the wells of the Ouargla palm grove itself, where the Phr aquifer water table is deeper than 2 m.

Conversely, the highest salinity waters are found in wells drilled in the chotts and sabkhas (a sabkha is the central part of a chott where salinity is the largest) (Safioune and Oum er Raneb) where the aquifer is often shallower than 50 cm.

The salinity of the CT (Mio-Pliocene) aquifer (Fig. 5b) is much lower than that of the Phr aquifer and ranges from 1 to $2\,g\,L^{-1}$; however, its hardness is larger and it contains more sulfate, chloride and sodium than the waters of the Senonian formations and those of the CI aquifer. The salinity of the Senonian aquifer ranges from 1.1 to $1.7\,g\,L^{-1}$, while the average salinity of the CI aquifer is $0.7\,g\,L^{-1}$ (Fig. 5c).

A likely contamination of the Mio-Pliocene aquifer by phreatic groundwaters through casing leakage in an area where water is heavily loaded with salt, and therefore particularly aggressive, cannot be excluded.

3.3 Saturation indices

The calculated saturation indices (SIs) reveal that waters from CI at 50 °C are close to equilibrium with respect to calcite, except for three samples that are slightly oversaturated.

They are, however, all undersaturated with respect to gypsum (Fig. 6).

Moreover, they are oversaturated with respect to dolomite and undersaturated with respect to anhydrite and halite (Fig. 7).

Waters from the CT and phreatic aquifers show the same pattern, but some of them are more largely oversaturated with respect to calcite, at 25 °C.

However, several phreatic waters (P031, P566, PLX4, PL18, P002, P023, P116, P066, P162 and P036) that are located in the sabkhas of Safioune, Oum er Raneb, Bamendil and Aïn el Beïda's chott are saturated with gypsum and anhydrite. This is in accordance with highly evaporative environments found elsewhere (UNESCO, 1972; Hamdi-Aïssa et al., 2004; Slimani, 2006).

No significant trend of SI from south to north upstream and downstream of Oued Mya (Fig. 7) is observed. This suggests that the acquisition of mineralization is due to geochemical processes that have already reached equilibrium or steady state in the upstream areas of Ouargla.

3.4 Change of facies from the carbonated cluster to the evaporites' cluster

The facies shifts progressively from the carbonated cluster (CI and CT aquifers) to the evaporites' cluster (Phr aquifer) with an increase in sulfates and chlorides at the expense of carbonates (SI of gypsum, anhydrite and halite). This is illustrated by a decrease of the $[HCO_3^-]/([Cl^-]+2[SO_4^{2-}])$ ratio (Fig. 8) from 0.2 to 0 and of the $[SO_4^{2-}]/[Cl^-]$ ratio from 0.8 to values smaller than 0.3 (Fig. 9) while salinity increases. Carbonate concentrations tend towards very small values, while it is not the case for sulfates. This is due to both gypsum dissolution and calcite precipitation. Chlorides in groundwater may come from three different sources: (i) ancient sea water entrapped in sediments, (ii) dissolution of halite and related minerals that are present in evaporite deposits and (iii) dissolution of dry fallout from the atmosphere, particularly in these arid regions (Matiatos et al., 2014; Hadj-Ammar et al., 2014).

The $[Na^+]/[Cl^-]$ ratio ranges from 0.85 to 1.26 for the CI aquifer, from 0.40 to 1.02 for the CT aquifer and from 0.13 to 2.15 for the Phr aquifer. The measured points from the three considered aquifers are linearly scattered with good approximation around the unity slope straight line that stands for halite dissolution (Fig. 10). The latter appears as the most dominant reaction occurring in the medium. However, at very high salinity, Na^+ seems to swerve from the straight line towards smaller values.

A further scrutiny of Fig. 10 shows that CI waters are very close to the 1 : 1 line. CT waters are enriched in both Na^+ and Cl^- but slightly lower than the 1 : 1 line while phreatic waters are largely enriched and much more scattered. CT waters are closer to the seawater mole ratio (0.858), but some lower values imply a contribution from a source of

Table 1. Field and analytical data for the Continental Intercalaire aquifer.

Locality	Lat.	Long.	Elev.	Date	EC	T	pH	Alk.	Cl⁻	SO₄²⁻	Na⁺	K⁺	Mg²⁺	Ca²⁺	Br⁻
		(m)			(mS cm^{-1})	(°C)					(mmol L^{-1})				
Hedeb I	3534750	723986	134	9 Nov 2012	2.0	46.5	7.6	3.5	5.8	6.8	10.7	0.6	2.5	3.3	0.034
Hedeb I	3534750	723986	134	1992	1.9	49.3	7.3	0.4	5.8	1.1	5.7	0.2	0.8	0.5	
Hedeb II	3534310	724290	146	1992	2.0	47.4	7.6	0.6	6.2	1.2	5.1	0.2	1.3	0.8	
Aouinet Moussa	3548896	721076	132	1992	2.2	48.9	7.5	1.3	6.5	1.9	5.6	0.2	1.1	1.2	
Aouinet Moussa	3548896	721076	132	22 Feb 2013	2.2	48.9	7.5	3.2	9.8	3.9	6.3	0.7	5.7	1.3	
Hedeb I	3534750	723986	134	11 Dec 2010	2.2	49.3	7.3	1.9	12.4	4.6	10.7	0.7	3.8	2.3	
Hedeb II	3534310	724290	146	11 Dec 2010	2.2	47.4	7.6	2.1	13.1	5.5	13.9	0.5	4.5	1.4	
Hassi Khfif	3591659	721636	110	24 Feb 2013	2.4	50.5	6.8	2.9	14.3	5.2	10.8	0.8	3.4	4.6	0.033
Hedeb I	3534750	723986	134	27 Feb 2013	2.0	46.5	7.6	3.5	15.1	7.7	11.8	0.5	5.6	5.2	
Hassi Khfif	3591659	721636	110	9 Nov 2012	2.2	50.1	7.6	3.3	15.3	7.8	12.2	0.6	5.8	4.9	
El Bour	3560264	720366	160	22 Feb 2013	2.9	54.5	7.3	2.6	18.6	6.2	20.6	0.7	4.8	1.4	

chloride other than halite or from entrapped seawater. Conversely, a $[Na^+]/[Cl^-]$ ratio larger than 1 is observed for phreatic waters, which implies the contribution of another source of sodium, most likely sodium sulfate, that is present as mirabilite or thenardite in the chotts and the sabkha areas.

The $[Br^-]/[Cl^-]$ ratio ranges from 2×10^{-3} to 3×10^{-3}. The value of this molar ratio for halite is around 2.5×10^{-3}, which matches the aforementioned range and confirms that halite dissolution is the most dominant reaction taking place in the studied medium.

In the CI, CT and Phr aquifers, calcium originates both from carbonate and sulfate (Figs. 11 and 12). Three samples from the CI aquifer are close to the $[Ca^{2+}]/[HCO_3^-]$ $1:2$ line, while calcium sulfate dissolution explains the excess of calcium. However, nine samples from the Phr aquifer are depleted in calcium, and plotted under the $[Ca^{2+}]/[HCO_3^-]$ $1:2$ line. This cannot be explained by precipitation of calcite, as some are undersaturated with respect to that mineral, while others are oversaturated.

In this case, a cation exchange process seems to occur and lead to a preferential adsorption of divalent cations, with a release of Na^+. This is confirmed by the inverse modeling that is developed below and which implies Mg^{2+} fixation and Na^+ and K^+ releases.

Larger sulfate values observed in the phreatic aquifer (Fig. 12) with $[Ca^{2+}]/[SO_4^{2-}] < 1$ can be attributed to a Na-Mg sulfate dissolution from a mineral bearing such elements. This is, for instance, the case of bloedite.

3.5 Isotope geochemistry

The CT and CI aquifers exhibit depleted and homogeneous ^{18}O contents, ranging from -8.32 to -7.85%. This was already previously reported by many authors (Edmunds et al., 2003; Guendouz et al., 2003; Moulla et al., 2012). On the other hand, ^{18}O values for the phreatic aquifer are widely dispersed and vary between -8.84 and 3.42% (Table 4). Waters located north of the virtual line connecting, approximately, Hassi Miloud to Sebkhet Safioune, are found more enriched in heavy isotopes and are thus more evaporated. In that area, the water table is close to the surface and mixing of both CI and CT groundwaters with phreatic ones through irrigation is nonexistent. Conversely, waters located south of Hassi Miloud up to Ouargla city show depleted values. This is the clear fingerprint of a contribution to the Phr waters from the underlying CI and CT aquifers (Gonfiantini et al., 1975; Guendouz, 1985; Fontes et al., 1986; Guendouz and Moulla, 1996).

Phreatic waters result from a mixing of two end members. Evidence for this is given by considering the $[Cl^-]$ and ^{18}O relationship (Fig. 13). The two clusters are (i) a first cluster of ^{18}O-depleted groundwater (Fig. 14), and (ii) another cluster of ^{18}O-enriched groundwater with positive values and a high salinity. The latter is composed of phreatic waters occurring in the northern part of the study region.

Table 2. Field and analytical data for the Complexe Terminal aquifer.

Locality	Site	Aquifer	Lat.	Long.	Elev.	Date	EC	T	pH	Alk.	Cl⁻	SO₄²⁻	Na⁺	K⁺	Mg²⁺	Ca²⁺	Br⁻
			(m)	(m)			(mS cm⁻¹)	(°C)		(mmol L⁻¹)							
Bamendil	D7F4	M	3 560 759	720 586	296	20 Jan 2013	2.0	20.1	7.9	1.6	10.1	5.8	9.9	0.7	3.9	2.5	
Bamendil	D7F4	M	3 560 759	720 586	296	1992	2.0	21.1	8.2	0.9	10.6	3.5	10.6	0.1	2.3	1.8	
Ifri	D1F151	S	3 538 891	721 060	204	1992	2.7	23.5	7.0	1.3	10.7	2.7	8.0	0.7	2.3	2.1	
Said Otba	D2F66	S	3 540 257	720 085	216	1992	2.3	24.0	8.0	1.4	11.0	4.7	11.5	0.2	2.1	3.3	
Oglat Larbaâ	D6F64	M	3 566 501	729 369	177	1992	2.3	18.0	7.9	1.4	11.4	6.8	11.6	2.3	2.0	4.6	
El Bour	D4F94	M	3 536 245	722 641	100	27 Jan 2013	3.1	26.2	7.4	1.6	12.8	6.8	5.2	1.9	1.6	9.1	
Said Otba I	D2F71	S	3 557 412	718 272	211	1992	2.3	24.2	8.2	1.5	13.5	5.7	15.0	0.3	3.3	2.6	
Debiche	D6F61	M	3 547 557	717 067	173	26 Jan 2013	2.2	23.9	7.7	1.8	14.2	8.4	12.6	0.7	5.4	4.4	
Rouissat III	D3F10	S	3 535 068	722 352	248	1992	3.1	26.1	7.3	2.4	14.3	6.9	13.1	0.4	3.4	5.4	0.034
Said Otba I	D2F71	S	3 557 412	718 272	212	26 Jan 2013	5.6	25.1	7.3	2.4	14.3	6.9	13.1	0.4	3.4	5.4	
Rouissat III	D3F10	S	3 535 068	722 352	248	20 Jan 2013	2.3	18.9	8.0	1.6	15.2	8.6	12.6	1.6	5.8	4.3	
Ifri	D1F151	S	3 538 891	721 060	204	27 Jan 2013	2.4	22.9	7.8	1.7	15.4	8.3	13.7	0.2	5.2	4.8	
Said Otba	D2F66	S	3 540 257	720 085	216	31 Jan 2013	2.4	24.9	7.9	2.2	16.1	8.6	16.5	0.7	4.9	4.3	0.033
Oglat Larbaâ	D6F64	M	3 566 501	729 369	177	31 Jan 2013	2.4	23.7	7.6	2.3	16.3	8.6	13.6	0.7	5.9	5.0	
SAR Mekhadma	D1F91	S	3 536 757	717 822	221	3 Feb 2013	2.5	25.8	7.7	3.4	16.5	8.5	16.1	0.7	5.3	4.9	
Sidi Kouiled	D9F12	S	3 540 855	729 055	329	24 Jan 2013	2.6	21.3	8.1	4.6	16.8	8.8	16.1	0.8	6.2	5.0	0.033
Aïn N'Sara	D6F50	S	3 559 323	716 868	255	25 Jan 2013	3.4	25.7	7.4	2.0	16.9	9.7	15.9	0.3	3.4	7.9	
A. Louise	D4F73	S	3 537 523	721 904	310	26 Jan 2013	2.6	24.0	7.5	2.0	17.4	9.1	13.9	2.0	5.8	5.1	
Ghâzalet A. H	D6F79	M	3 598 750	720 356	119	2 Feb 2013	2.8	22.5	7.5	3.5	17.4	9.3	16.6	0.6	6.2	4.9	
Aïn Moussa II	D9F30	S	3 537 814	719 665	220	2 Feb 2013	7.5	23.9	7.5	2.4	17.5	8.2	17.3	0.4	3.1	6.5	0.033
Aïn N'Sara	D6F50	S	3 559 323	716 868	255	2 Feb 2013	2.6	23.8	7.6	2.1	17.7	9.2	15.5	1.1	6.1	4.7	
H. Miloud	D1F135	M	3 547 557	717 067	173	3 Feb 2013	2.8	21.6	7.5	3.3	17.9	9.2	16.5	1.0	6.2	4.9	
El Bour	D6F97	S	3 540 936	715 816	169	25 Jan 2013	2.6	19.9	8.0	2.1	17.9	9.3	15.8	1.6	5.8	4.7	
H. Miloud	D1F135	M	3 547 557	717 067	173	1992	2.1	22.7	8.1	2.8	18.1	5.7	16.6	0.5	3.6	4.3	
N'Goussa El Hou	D6F51	S	3 556 256	718 979	198	31 Jan 2013	2.9	22.9	7.5	2.0	18.4	9.6	17.1	0.5	6.2	5.0	
El Koum	D6F67	S	3 573 694	721 639	143	21 Jan 2013	3.1	22.9	8.1	3.5	18.4	9.7	17.9	0.3	6.5	5.1	
El Koum	D6F67	S	3 573 694	721 639	143	1992	2.5	25.0	7.6	1.5	18.8	7.2	10.2	3.4	5.0	5.8	
Itas	D1F150	M	3 536 186	717 046	93	21 Jan 2013	3.7	23.9	7.5	1.5	18.8	7.1	10.1	3.4	5.0	5.8	
Aïn Moussa V	D9F13	M	3 538 409	718 680	210	8 Feb 2013	2.4	25.3	7.2	2.3	19.4	9.4	18.8	0.4	3.3	7.6	0.034
El Bour	D4F94	M	3 536 245	722 641	100	1992	2.3	21.2	7.9	1.6	20.1	7.2	12.1	2.6	5.8	5.2	
Rouissat I	D3F18	M	3 535 564	722 498	80	26 Jan 2013	3.1	23.0	8.1	3.2	21.2	11.1	19.6	0.9	7.1	6.0	
Rouissat I	D3F18	M	3 535 564	722 498	80	1992	2.0	20.0	7.8	1.7	21.7	8.5	17.7	1.2	5.1	6.0	
Station de Pompage chott	D5F80	S	3 541 656	723 521	224	4 Feb 2013	3.3	24.5	8.2	3.9	22.1	11.9	19.9	2.1	7.6	6.3	
Chott Palmeraie	D5F77	S	3 538 219	725 541	243	5 Feb 2013	3.4	24.6	7.5	3.3	22.3	12.1	20.9	1.2	8.3	5.8	
Bour el Aïcha	D1F134	M	3 545 533	720 391	86	5 Feb 2013	3.4	22.2	7.3	4.1	23.2	12.2	21.2	1.5	8.6	6.0	
Abazat	D2F69	M	3 552 504	712 786	137	3 Feb 2013	3.5	24.6	7.6	2.2	24.7	12.7	21.1	1.7	8.5	6.5	
Garet Chemia	D1F113	S	3 536 174	716 808	213	28 Jan 2013	4.1	28.0	7.3	2.2	25.9	9.5	25.4	0.6	3.6	7.2	0.037
Frane	D6F62	M	3 570 175	717 133	167	27 Jan 2013	3.8	24.2	7.9	2.3	25.9	13.5	22.6	0.6	8.9	7.2	

Table 2. Continued.

Locality	Site	Aquifer	Lat.	Long.	Elev.	Date	EC	T	pH	Alk.	Cl⁻	SO₄²⁻	Na⁺	K⁺	Mg²⁺	Ca²⁺	Br⁻
			(m)				(mS cm⁻¹)	(°C)					(mmol L⁻¹)				
Oum er Raneb	D6 F69	M	3 540 451	721 919	216	25 Jan 2013	4.2	24.1	7.0	2.6	27.9	8.7	22.9	0.6	4.4	8.0	0.035
N'Goussa El Hou	D6F51	S	3 556 256	718 979	198	1992	3.1	23.2	8.0	2.6	28.4	8.6	23.1	0.6	4.5	8.0	
H. Miloud Benyaza	D1F138	M	3 551 192	717 042	89	28 Jan 2013	3.8	25.2	7.6	2.4	28.4	14.2	23.9	1.7	10.0	7.1	0.037
Aïn Larbaâ	D6F49	M	3 558 822	716 799	156	28 Jan 2013	3.9	23.7	7.3	2.2	28.9	9.0	23.9	0.5	5.0	7.7	
H. Miloud Benyaza	D1F138	M	3 551 192	717 042	89	1992	2.9	22.8	7.5	2.2	28.9	9.1	23.9	0.5	5.0	7.7	
Rouissat	D3F8	M	3 545 470	732 837	332	3 Feb 2013	4.4	25.4	7.5	1.7	29.8	8.3	22.8	1.2	6.2	6.1	
Rouissat	D3F8	M	3 545 470	732 837	332	1992	6.2	25.3	7.2	1.7	29.8	8.3	22.9	1.2	6.2	6.1	
Aïn El Arch	D3F26	M	3 534 843	723 381	93	1992	5.1	25.1	7.4	1.6	34.7	8.9	24.0	0.9	8.4	6.5	
Station de Pompage chott	D5F80	S	3 541 656	723 521	224	1992	3.7	25.4	7.7	2.3	42.2	13.5	36.8	1.1	7.4	9.7	

M = Mio-Pliocene aquifer; S = Senonian aquifer.

Cluster I represents the waters from CI and CT whose isotopic composition is depleted in ^{18}O (average value around $-8.2‰$) (Fig. 13). They correspond to an old water recharge (paleo-recharge), whose age, estimated by means of ^{14}C, exceeds 15 000 yr BP (Guendouz, 1985; Guendouz and Michelot, 2006). Thus, it is not a water body that is recharged by recent precipitation. It consists of CI and CT groundwaters and partly of phreatic waters and can be ascribed to an upward leakage favored by the extension of faults near Amguid El Biod dorsal.

Cluster II, observed in Sebkhet Safioune, can be ascribed to the direct dissolution of surficial evaporitic deposits conveyed by evaporated rainwater.

Evaporation alone cannot explain the distribution of data that is observed (Fig. 13). Evidence for this is given in a semi-logarithmic plot (Fig. 14), as classically obtained according to the simple approximation of the Rayleigh equation (cf. Appendix):

$$\delta^{18}O \approx 1000 \times (1-\alpha)\log\left[Cl^-\right] + k,$$
$$\approx -\epsilon \log\left[Cl^-\right] + k, \qquad (1)$$

where α is the fractionation factor during evaporation, $\epsilon \equiv -1000 \times (1-\alpha)$ is the enrichment factor and k is a constant (Ma et al., 2010; Chkir et al., 2009). CI and CT waters are better separated in the semi-logarithmic plot because they are differentiated by their chloride content. According to Eq. (1), simple evaporation gives a straight line (solid line in Fig. 14). The value of ϵ used is the value at 25 °C, which is equal to -73.5.

P115 is the only sample that appears on the straight evaporation line (Fig. 14). It should be considered as an outlier since the rest of the samples are all well aligned on the logarithmic fit derived from the mixing line of Fig. 13.

The phreatic waters that are close to cluster I (Fig. 13) correspond to groundwaters occurring in the edges of the basin (Hassi Miloud, piezometer P433) (Fig. 14). They are low mineralized and acquire their salinity via two processes, namely dissolution of evaporites along their underground transit up to Sebkhet Safioune and dilution through upward leakage by the less-mineralized waters of the CI and CT aquifers (for example, Hedeb I for CI and D7F4 for CT) (Fig. 14) (Guendouz, 1985; Guendouz and Moulla, 1996).

The rates of the mixing that are due to upward leakage from CI to CT towards the phreatic aquifer can be calculated by means of a mass balance equation. It only requires knowing the δ values of each fraction that is involved in the mixing process.

The δ value of the mixture is given by

$$\delta_{mix} = f \times \delta_1 + (1-f) \times \delta_2, \qquad (2)$$

where f is the fraction of the CI aquifer, $1-f$ the fraction of the CT and δ_1, δ_2 are the respective isotope contents.

Table 3. Field and analytical data for the phreatic aquifer.

Locality	Site	Lat.	Long.	Elev.	Date	EC	T	pH	Alk.	Cl⁻	SO₄²⁻	Na⁺	K⁺	Mg²⁺	Ca²⁺	Br⁻
		(m)				(mS cm⁻¹)	(°C)					(mmol L⁻¹)				
Bou Khezana	P433	3 597 046	719 626	118	20 Jan 2013	2.1	22.7	9.2	1.6	12.0	7.3	13.0	1.0	4.3	2.8	
Bou Khezana	P433	3 597 046	719 626	118	1992	2.0	22.1	8.9	1.5	12.0	6.9	11.6	0.9	4.4	2.9	
Hassi Miloud	P059	3 547 216	718 358	124	27/01/2013	2.1	23.9	8.2	1.9	13.0	7.3	12.6	1.3	4.4	3.4	0.024
Aïn Kheir	PL06				1992	4.0	23.8	7.5	1.9	14.2	17.9	15.9	0.6	10.6	7.5	
Hassi Naga	PLX3	3 584 761	717 604	125	20 Jan 2013	2.9	23.0	8.1	2.0	17.7	9.4	16.6	0.9	5.8	5.0	0.031
	LTP 30				1992	4.1	23.7	7.1	5.3	18.2	10.0	24.3	0.4	1.4	8.1	
Maison de culture	PL31	3 537 988	720 114	124	1992	2.5	23.8	8.1	1.5	18.9	7.8	26.1	0.6	2.1	3.0	
El Bour	P006	3 564 272	719 421	161	1992	3.0	23.4	7.9	1.3	19.0	7.7	12.4	2.7	5.3	5.3	
Hassi Miloud	P059	3 547 216	718 358	124	1992	2.8	23.5	7.8	2.3	20.8	9.4	34.2	4.3	1.4	0.9	
Oglat Larbaâ	P430	3 567 287	730 058	139	24 Jan 2013	4.5	27.5	8.3	3.3	22.1	12.4	21.8	2.6	8.6	5.5	
Maison de culture	PL31	3 537 988	720 114	124	28 Jan 2013	3.7	22.2	8.2	4.2	22.6	8.6	28.4	2.2	4.0	3.2	
Frane El Koum	P401	3 572 820	719 721	112	20 Jan 2013	3.4	27.5	7.5	2.2	23.3	13.4	21.8	1.9	8.3	6.3	0.032
Gherbouz	PL15	3 537 962	718 744	134	1992	2.5	23.5	7.7	3.0	23.5	14.0	50.6	2.8	1.0	0.3	
Bour El Haïcha	P408	3 544 999	719 930	110	1992	2.4	23.5	7.8	2.4	24.2	13.2	41.9	6.1	2.3	0.8	
Station d'épuration	PL30	3 538 398	721 404	130	1992	5.5	23.8	7.4	3.0	24.3	21.2	24.3	0.9	20.2	2.2	
Frane Ank Djemel	P422	3 575 339	718 875	109	20 Jan 2013	4.1	24.2	8.4	4.4	25.3	9.5	23.7	1.8	4.2	7.9	0.025
Route Aïn Beïda	PLX2	3 537 323	724 063	127	1992	4.7	23.6	7.2	2.0	25.7	10.4	14.8	0.2	9.3	7.4	
H. Chegga	PLX4	3 577 944	714 428	111	20 Jan 2013	4.1	25.2	7.6	3.0	26.2	9.8	24.0	2.3	5.0	7.5	0.033
Hassi Miloud	P058	3 547 329	716 520	129	27 Jan 2013	3.7	24.6	8.1	3.0	27.7	10.6	19.0	2.3	9.1	6.6	0.033
Route Aïn Moussa	P057	3 548 943	717 353	133	1992	5.3	23.4	7.7	1.3	28.2	11.5	17.6	2.0	11.5	5.8	
Route El Goléa	P115	3 533 586	714 060	141	1992	2.6	23.7	7.6	2.8	28.8	14.5	58.7	0.1	0.8	0.7	
Mekhadma	PL05	3 537 109	718 419	137			23.9	7.8	1.7	30.9	16.7	24.9	1.0	15.7	4.5	
Polyclinique Bel Abbès	PL18	3 537 270	721 119	119	31/01/2013	4.7	22.2	7.9	1.8	31.2	15.4	21.3	3.9	11.2	8.4	
H. Chegga	PLX4	3 577 944	714 428	111	1992	4.5	23.7	7.6	1.5	31.5	10.1	20.1	5.9	7.5	6.5	
Route El Goléa	P116	3 532 463	713 715	117	1992	5.6	23.7	7.6	1.4	31.9	12.8	22.2	0.8	10.6	8.0	
Gherbouz	PL15	3 537 962	718 744	134	21 Jan 2013	4.7	23.3	8.2	1.8	32.4	14.6	27.8	0.8	6.8	10.8	
Route El Goléa	P117	3 531 435	713 298	111	1992	4.7	23.7	7.7	1.5	32.8	12.8	30.2	1.0	9.2	5.7	
Route Aïn Moussa	P057	3 548 943	717 353	133	26 Jan 2013	5.7	26.2	7.6	2.5	33.5	11.9	27.7	5.9	6.0	7.6	
Ecole Paramédicale	P032	3 538 478	720 170	131	21 Jan 2013	5.7	22.9	8.2	2.0	33.6	12.1	29.2	3.3	6.4	8.2	
Direction des Services Agricoles	PL10	3 537 055	719 746	114	1992	6.1	23.7	7.7	1.3	35.0	13.5	8.6	1.9	19.4	7.2	
Route El Goléa	P117	3 531 435	713 298	111	3 Feb 2013	5.5	25.0	7.7	3.3	35.4	13.8	37.1	3.0	8.4	5.7	
Route El Goléa	P116	3 532 463	713 715	117	3 Feb 2013	5.8	22.5	8.1	1.7	36.3	11.6	28.5	3.2	6.8	8.4	
Station d'épuration	PL30	3 538 398	721 404	130	31 Jan 2013	5.3	25.1	7.8	4.1	38.4	14.6	28.5	4.5	11.6	8.1	
Hassi Debiche	P416	3 581 097	730 922	106	24 Jan 2013	5.5	23.7	8.8	0.3	38.6	18.0	22.3	0.9	4.8	21.3	
Direction des Services Agricoles	PL10	3 537 055	719 746	114	28 Jan 2013	5.5	24.6	8.4	2.4	38.8	16.9	36.9	1.9	9.1	9.2	
Hôpital	LTPSN2	3 538 292	720 442	132	27 Jan 2013	6.1	25.4	7.8	1.6	39.7	11.7	36.0	8.4	5.1	6.0	
Parc SONACOM	PL28	3 536 077	719 558	134	21 Jan 2013	6.1	24.5	8.1	1.8	39.8	11.8	30.6	5.2	7.1	8.5	
Bour El Haïcha	P408	3 544 999	719 930	110	27 Jan 2013	6.2	23.1	8.1	1.8	42.0	19.1	27.5	13.2	13.4	8.1	
Route Aïn Moussa	P056	3 549 933	717 022	128	1992	7.6	23.6	7.9	0.6	42.1	10.7	18.9	1.9	12.6	9.3	
Route Aïn Moussa	P056	3 549 933	717 022	128	26 Jan 2013	6.0	24.6	7.6	2.2	42.5	17.9	32.1	8.0	12.5	8.1	

Table 3. Continued.

Locality	Site	Lat.	Long.	Elev.	Date	EC	T	pH	Alk.	Cl⁻	SO₄²⁻	Na⁺	K⁺	Mg²⁺	Ca²⁺	Br⁻
		(m)		(m)		(mS cm⁻¹)	(°C)					(mmol L⁻¹)				
Ecole Okba B. Nafaa	PL41	3 538 660	719 831	127	31 Jan 2013	6.3	24.1	7.7	2.1	44.9	13.2	36.2	11.8	6.3	6.7	
Parc Hydraulique	P419	3 539 494	725 605	132	31 Jan 2013	7.0	26.4	7.8	2.1	45.1	14.4	41.4	10.8	6.0	6.9	
Parc Hydraulique	PL13	3 536 550	720 200	123	21 Jan 2013	7.2	24.5	7.5	3.2	47.8	14.5	44.4	10.6	6.4	6.6	
Mekhadma	PL25	3 536 230	718 708	129	21 Jan 2013	7.6	27.1	7.9	1.8	48.0	14.5	42.9	6.6	7.4	7.6	
Saïd Otba	P506	3 535 528	725 075	126	4 Feb 2013	8.3	24.3	8.1	1.7	52.6	14.6	42.8	11.0	7.5	7.8	
Saïd Otba	P506	3 535 528	725 075	126	1992	6.7	23.3	7.5	1.8	54.4	17.6	33.3	4.1	22.2	5.2	
Mekhadma	P566	3 540 433	719 661	115	27 Jan 2013	9.0	24.6	7.6	1.7	62.5	15.2	71.6	3.0	4.6	6.1	
Mekhadma	PL17	3 536 908	718 511	130	21 Jan 2013	9.4	24.5	8.1	3.4	63.2	15.6	77.2	2.5	4.1	5.1	
Palm. Gara Krima	P413	3 530 116	722 775	130	4 Feb 2013	10.1	30.2	7.9	1.6	63.6	21.5	88.3	4.1	4.2	4.7	
Mekhadma	PL25	3 536 230	718 708	129	1992	9.5	23.7	8.0	0.6	75.6	10.6	10.2	2.6	32.9	9.5	
Saïd Otba (Bab Sbaa)	P066	3 542 636	718 957	126	1992	7.8	23.5	7.6	1.5	80.2	12.5	45.9	2.5	23.6	5.9	
CEM Malek Bennabi	PL03	3 540 010	725 738	130	1992	7.3	23.9	7.6	3.1	84.1	30.6	108.6	2.2	10.2	9.0	
Entreprise nationale de télévision	PL21	3 536 074	721 268	128	1992	9.7	23.8	7.3	4.5	84.3	23.7	61.6	3.8	33.5	1.9	
Hôtel Transat	PL23	3 538 419	720 950	126	28 Jan 2013	15.0	24.2	8.2	4.5	86.6	16.7	79.9	3.2	14.5	6.9	
Entreprise nationale de télévision	PL21	3 536 074	721 268	128	28 Jan 2013	16.4	25.7	7.5	2.0	99.9	17.4	85.5	5.7	15.7	7.6	
Mekhadma	PL05	3 537 109	718 419	137	21 Jan 2013	16.8	24.8	7.6	2.0	101.3	17.7	85.9	5.9	16.7	7.6	
Beni Thour	PL44	3 536 039	721 673	134	1992	4.7	23.9	7.2	2.7	109.8	67.2	134.7	5.7	42.0	8.8	
Tazegrart	PLSN1	3 537 675	719 416	125	22 Jan 2013	17.1	24.9	8.0	3.4	114.2	18.1	92.9	12.8	16.9	7.8	
CEM Malek Bennabi	PL03	3 540 010	725 738	130	27 Jan 2013	10.8	23.1	7.5	3.3	117.3	14.7	116.4	2.1	9.0	7.2	
El Bour	P006	3 564 272	719 421	161	3 Feb 2013	18.3	23.6	7.8	6.3	131.9	18.1	96.3	8.6	27.1	8.0	
Aïn Moussa	P015	3 551 711	720 591	103	1992	12.4	23.6	7.7	2.4	134.7	28.2	73.0	3.1	52.4	6.3	
Station de Pompage	PL04	3 541 410	723 501	138	27 Jan 2013	19.0	26.4	7.9	4.0	138.0	16.7	108.8	13.1	19.5	8.7	
Drain Chott Ouargla	D. Ch						23.9	7.7	2.7	142.2	24.5	96.31	3.2	44.2	3.0	
Beni Thour	PL44	3 536 039	721 673	134	28 Jan 2013	20.2	25.8	7.8	5.0	153.0	17.7	125.9	6.3	22.8	8.1	
CNMC (Société nationale des matériaux de construction)	PL27	3 535 474	718 407	126	21 Jan 2013	21.2	24.8	8.1	1.7	169.4	18.4	130.3	4.9	27.8	8.6	
Bamendil	P076	3 540 137	716 721	118	26 Jan 2013	22.3	27.2	7.6	4.3	171.5	17.1	130.8	6.3	28.0	8.8	
N'Goussa	P041	3 559 563	716 543	135	26 Jan 2013	25.9	24.5	8.2	8.0	208.6	13.4	198.9	3.6	11.8	8.8	
N'Goussa	P009	3 559 388	717 707	123	26 Jan 2013	27.5	28.4	8.4	11.5	208.8	15.8	195.1	2.7	18.7	9.0	
	LTP16				1992	11.5	23.8	7.5	3.8	213.4	48.6	147.9	7.5	75.3	4.3	
	P100				1992	17.2	23.6	7.6	3.4	235.0	46.4	264.8	4.7	25.6	5.6	
Chott Adjadja Aven	PLX1	3 540 758	726 115	132	28 Jan 2013	32.9	23.4	8.0	4.4	245.6	20.9	141.4	26.9	44.6	17.7	
Route Frane	P003	3 569 043	721 496	134	2 Feb 2013	31.0	23.5	8.0	6.9	252.7	17.9	208.2	9.4	30.0	10.0	
El Bour–N'Goussa	P007	3 562 236	718 651	129	26 Jan 2013	30.1	28.4	7.8	5.4	254.5	15.5	209.2	10.4	28.8	7.5	
Route Aïn Beïda	PLX2	3 537 323	724 063	127	21 Jan 2013	43.3	25.7	8.1	5.2	262.2	93.0	270.4	15.5	62.8	21.7	
Aïn Moussa	P015	3 551 711	720 591	103	25 Jan 2013	32.0	22.7	8.0	2.9	263.0	15.4	206.9	6.6	32.1	9.9	
Aïn Moussa	P402	3 549 503	721 514	138	25 Jan 2013	60.0	28.7	8.6	7.7	313.2	93.9	442.8	23.3	12.6	10.2	
Route Frane	P001	3 572 148	722 366	127	1992		23.6	8.4	4.0	323.6	58.1	331.4	5.0	49.8	4.0	
Aïn Moussa	P014	3 551 466	719 339	131	1992		23.4	7.3	4.0	337.0	64.3	328.7	5.5	62.4	5.5	

Table 3. Continued.

Locality	Site	Lat. (m)	Long. (m)	Elev.	Date	EC (mS cm⁻¹)	T (°C)	pH	Alk.	Cl⁻	SO₄²⁻	Na⁺	K⁺	Mg²⁺	Ca²⁺	Br⁻
												(mmol L⁻¹)				
N'Goussa	P019	3 562 960	717 719	113	2 Feb 2013	60.6	27.8	7.7	6.0	356.2	96.0	432.5	29.8	21.0	26.2	
N'Goussa	P018	3 562 122	716 590	110	26 Jan 2013	61.1	26.2	8.4	6.5	372.4	82.3	347.1	22.6	60.7	26.6	
Aïn Moussa	P014	3 551 466	719 339	131	25 Jan 2013	49.0	25.2	7.9	1.8	399.7	21.1	389.3	2.4	19.0	7.4	
Route Sedrata	P113	3 535 586	714 576	105	3 Feb 2013	62.2	24.8	8.2	6.0	414.8	83.8	362.7	33.3	70.2	26.5	
N'Goussa	P009	3 559 388	717 707	123	1992		23.3	7.8	2.4	426.9	57.8	393.8	9.1	59.1	12.0	
Route Frane	P001	3 572 148	722 366	127	2 Feb 2013	66.2	28.3	7.2	6.5	468.7	101.5	350.3	26.0	116.2	35.3	
Sebkhet Safioune	P031	3 577 804	720 172	120	1992		23.8	7.3	6.3	481.8	43.4	326.8	12.6	94.2	23.6	
Sebkhet Safioune	P031	3 577 804	720 172	120	2 Feb 2013	76.0	27.9	8.1	5.9	500.3	110.3	470.5	28.7	79.1	35.5	
Route Frane	P002	3 570 523	722 028	108	1992		23.8	7.8	6.3	522.4	183.0	653.8	10.0	104.7	11.0	
Sebkhet Safioune	P030	3 577 253	721 936	130	1992		23.5	7.7	4.4	527.7	123.5	533.8	11.6	106.2	10.7	
Oum Raneb	P012	3 554 089	718 612	114	25 Jan 2013	64.1	30.3	7.8	7.8	534.3	20.9	529.6	6.4	19.7	4.7	
Oum Raneb	P012	3 554 089	718 612	114	1992		23.4	7.5	2.7	539.4	60.6	413.6	5.6	112.8	9.4	
Ank Djemel	P423	3 540 881	723 178	102	31 Jan 2013	90.8	23.5	7.5	6.2	636.5	101.3	495.5	38.3	125.8	30.3	
Said Otba Chott	P096	3 540 265	724 729	111	1992		23.6	7.7	3.7	645.1	78.5	357.3	6.0	208.4	12.9	
Sebkhet Safioune	P030	3 577 253	721 936	130	03/02/2013	64.7	23.1	7.8	3.7	671.8	90.3	742.9	16.0	41.5	7.7	
N'Goussa	P017	3 560 256	715 781	130	26 Jan 2013	100.1	31.0	7.1	3.8	679.3	114.1	597.8	10.7	125.9	26.3	
Ank Djemel	P021	3 573 943	723 161	105	1992		23.6	7.4	4.2	700.8	154.5	605.7	53.6	163.1	14.2	
Station de Pompage	PL04	3 541 410	723 501	138	1992		23.6	7.4	2.4	716.3	34.8	560.1	7.0	99.6	11.0	
Route Frane	P002	3 570 523	722 028	108	2 Feb 2013	62.8	26.9	7.6	1.7	748.5	62.6	651.5	14.7	77.7	27.3	
Said Otba chott	P096	3 540 265	724 729	111	3 Feb 2013	68.3	25.9	8.7	1.2	771.0	53.1	615.9	23.5	69.6	50.4	
N'Goussa	P019	3 562 960	717 719	113	1992		23.3	7.7	2.4	779.1	77.1	711.5	9.2	95.6	12.1	
Said Otba (Bab Sbaa)	P066	3 542 636	718 957	126	3 Feb 2013	150.6	26.2	7.2	12.3	799.1	283.0	1249.7	19.0	270.9	18.1	
Ank Djemel	P021	3 573 943	723 161	105	24 Jan 2013	82.3	29.6	7.6	2.4	800.4	94.4	824.0	11.0	53.4	25.4	
N'Goussa	P018	3 562 122	716 590	110	1992		23.3	7.5	1.2	818.7	81.0	244.2	49.5	319.4	24.8	
Oum Raneb	P162	3 546 133	725 129	98	25 Jan 2013	160.0	30.7	7.2	2.4	842.8	289.9	1309.9	13.3	33.5	17.7	
Route Sedrata	P113	3 535 586	714 576	105	1992		23.7	7.7	2.8	954.9	124.9	997.5	13.3	86.7	11.7	
Oum Raneb	PZ12	3 547 234	722 931	110	5 Feb 2013	114.9	27.4	7.4	2.9	980.1	15.5	930.8	7.5	23.9	14.2	
Hôtel Transat	PL23	3 538 419	720 950	126	1992		23.5	7.4	3.0	1103.3	94.5	707.8	19.1	270.9	13.3	
Sebkhet Safioune	P023	3 577 198	725 726	99	1992		23.3	7.4	2.3	1177.0	91.1	1058.2	11.7	133.5	12.4	
Sebkhet Safioune	P034	3 579 698	725 633	97	5 Feb 2013	130.0	34.9	8.1	1.8	1189.1	14.7	1055.1	18.3	56.4	17.4	
Sebkhet Safioune	P023	3 577 198	725 726	99	5 Feb 2013	117.9	29.4	8.2	1.9	1209.3	15.6	1129.4	8.4	42.9	10.2	
Chott Adjadja	PLX1	3 540 758	726 115	132	1992		23.6	8.0	3.8	1296.7	134.0	1458.7	5.2	48.0	4.3	
Sebkhet Safioune	P063	3 545 586	725 667	99	1992		23.5	7.5	1.9	1379.4	139.6	1257.4	18.6	182.3	10.0	
	LTP06				1992		23.8	7.6	7.8	1638.7	712.1	2621.6	41.6	190.5	13.3	
Bamendil	P076	3 540 137	716 721	118	1992		23.5	7.7	5.7	1743.6	143.4	1321.9	26.9	331.4	12.3	
El Bour–N'Goussa	P007	3 562 236	718 651	129	1992		23.3	7.7	1.4	1860.5	91.6	1434.7	26.2	278.8	13.3	
Sebkhet Safioune	P063	3 545 586	725 667	99	5 Feb 2013	178.9	26.7	7.7	1.4	1887.9	92.9	1455.8	26.7	282.9	13.4	
	P044				1992		23.4	7.8	4.5	2106.1	18.3	1765.5	27.3	171.2	6.5	
	P093				1992		23.6	7.5	1.5	2198.6	182.1	1957.5	29.5	278.2	10.4	
	P042				1992		23.4	7.6	1.1	2330.9	101.2	1963.7	52.2	248.1	11.2	
	P068				1992		23.5	7.5	3.4	2335.7	222.1	2302.3	26.8	219.9	7.2	

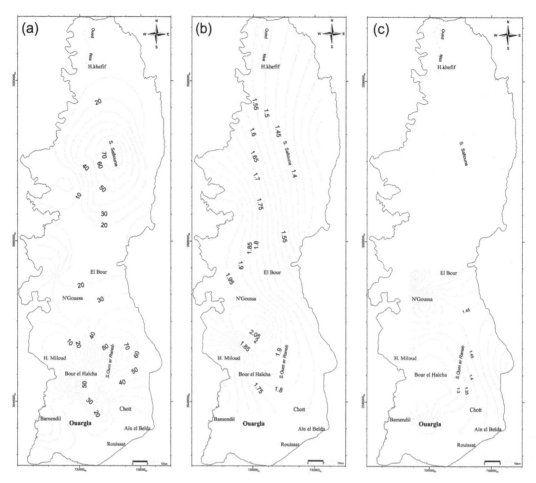

Figure 5. Contour maps of the salinity (expressed as global mineralization) in the aquifer system of the **(a)** phreatic aquifer and **(b, c)** Complexe Terminal (with **b** indicating Mio-Pliocene and **c** indicating Senonian); figures are isovalues of global mineralization (values in $g\,L^{-1}$).

Average values of mixing fractions from each aquifer to the phreatic waters computed by means of Eq. (2) gave the rates of 65 % for the CI aquifer and 35 % for the CT aquifer.

A mixture of a phreatic water component that is close to cluster I (i.e., P433) with another component which is rather close to cluster II (i.e., P039) (Figs. 13 and 14), for an intermediate water with a $\delta^{18}O$ signature ranging from -5 to -2‰, gives mixture fraction values of 52 % for cluster I and 48 % for cluster II. Isotope results will be used to independently cross-check the validity of the mixing fractions derived from an inverse modeling involving chemical data (see Sect. 3.6).

Turonian evaporites are found to lie in between the CI deep aquifer and the Senonian and Miocene formations bearing the CT aquifer. CT waters can thus simply originate from ascending CI waters that dissolve Turonian evaporites, a process which does not involve any change in ^{18}O content. Conversely, phreatic waters result, to a minor degree, from evaporation and mostly from dissolution of sabkha evaporites by ^{18}O-enriched rainwater and mixing with CI/CT waters.

Tritium content of water

Tritium contents of the Phr aquifer are relatively small (Table 4), they vary between 0 and 8 TU. Piezometers PZ12, P036 and P068 show values close to 8 TU; piezometers P018, P019, P416, P034, P042 and P093 exhibit values ranging between 5 and 6 TU; and the rest of the samples' concentrations are lower than 2 TU.

These values are dated back to November 1992, so they are old values and they are considered high comparatively to what is expected to be found nowadays. In fact, at present times, tritium figures have fallen lower than 5 TU in precipitation measured in the northern part of the country.

Tritium content of precipitation was measured as 16 TU in 1992 on a single sample that was collected from the National Agency for Water Resources station in Ouargla. A major part of this rainfall evaporates back into the atmosphere that is unsaturated in moisture. Consequently, enrichment in tritium happens as water evaporates back. The lightest fractions (isotopes) are the ones that escape first, enriching the remaining fraction in tritium. The 16 TU value would thus

Table 3. Continued.

Locality	Site	Lat. (m)	Long. (m)	Elev. (m)	Date	EC (mS cm⁻¹)	T (°C)	pH	Alk.	Cl⁻	SO₄²⁻	Na⁺	K⁺	Mg²⁺	Ca²⁺	Br⁻
										(mmol L⁻¹)						
Oum Raneb	PZ12	3547234	722931	110	1992	23.3	23.3	7.6	2.2	2405.6	109.9	2178.6	25.2	199.4	12.7	
Hassi Debiche	P416	3581097	730922	106	1992	23.3	23.3	7.8	4.3	2433.7	178.9	2361.1	24.3	196.1	9.2	
N'Goussa	P041	3559563	716543	135	1992	23.4	23.4	7.9	2.1	2599.7	324.6	2879.0	44.6	152.8	11.0	
Sebkhet Safioune	P034	3579698	725633	97	1992	23.3	23.3	7.8	1.9	2752.0	134.1	2616.8	24.4	180.1	10.5	
Sebkhet Safioune	P039				1992	23.4	23.4	6.9	1.9	4189.5	201.4	4042.6	17.9	257.8	9.2	
Sebkhet Safioune	P074				1992	23.5	23.5	6.5	4.2	4356.5	180.9	2759.9	57.4	930.1	22.6	
Sebkhet Safioune	P037				1992	23.4	23.4	6.9	1.5	4953.8	184.5	4611.1	2.9	347.6	7.9	
Sebkhet Safioune	P036				1992	23.4	23.4	7.5	1.4	4972.8	108.1	4692.2	36.8	221.1	9.6	

For longitude and latitude, the reference is UTM 31 projection for north Sahara 1959 (CLARKE 1880 ellipsoid).

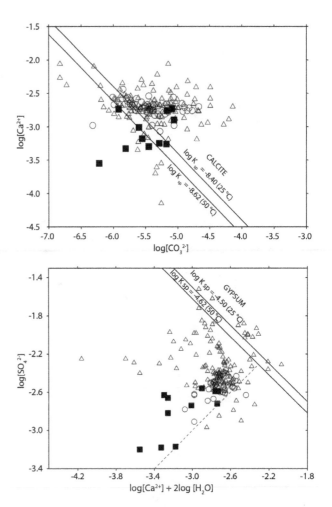

Figure 6. Equilibrium diagrams of calcite (top panel) and gypsum (bottom panel) for the Continental Intercalaire (filled squares), Complexe Terminal (open circles) and phreatic (open triangles) aquifers. Equilibrium lines are defined as $\log\{Ca^{2+}\} + \log\{CO_3^{2-}\} = \log K_{sp}$ for calcite and $\log\{Ca^{2+}\} + 2\log\{H_2O\} + \log\{SO_4^{2-}\} = \log K_{sp}$ for gypsum.

correspond to a rainy event that had happened during the field campaign (5 and 6 November 1992). It is the most representative value for that region and for that time. Unfortunately, all the other stations (Algiers, Ankara and Tenerife) (Martinelli et al., 2014) are subject to a completely different climatic regime and (besides the fact that they have more recent values) can absolutely not be used for our case. Therefore, all the assumptions based on recent tritium rain values do not apply to this study.

Depleted contents in ^{18}O and low tritium concentrations for phreatic waters fit the mixing scheme well and confirm the contribution from the older and deeper CI/CT groundwaters. The affected areas were clearly identified in the field and correspond to locations that are subject to a recycling and a return of irrigation waters whose origin are CI/CT boreholes. Moreover, the mixing that is clearly brought to light by

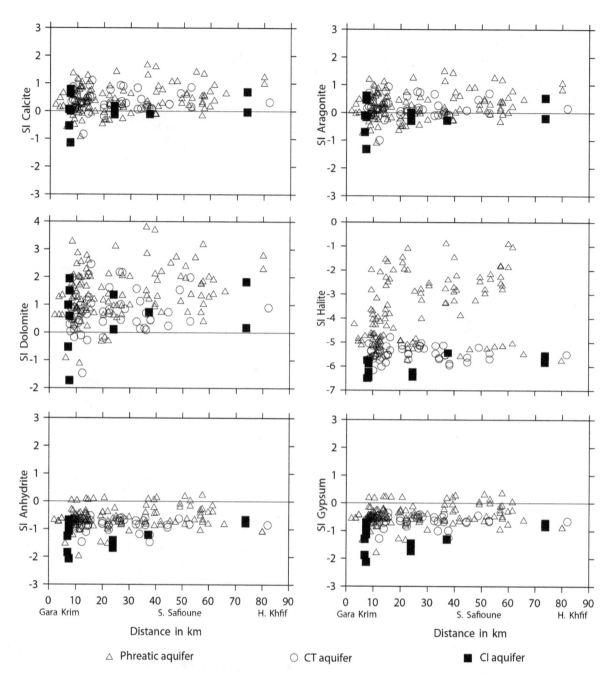

Figure 7. Variation of saturation indices with distance from south to north in the region of Ouargla.

the Cl⁻ versus ¹⁸O diagrams (Figs. 13 and 14) could partly derive from an ascending drainage from the deep and confined CI aquifer (exhibiting depleted homogenous ¹⁸O contents and very low tritium), a vertical leakage that is favored by the Amguid El Biod highly faulted area (Guendouz and Moulla, 1996; Edmunds et al., 2003; Guendouz et al., 2003; Moulla et al., 2012).

3.6 Inverse modeling

We assume that the relationship between ¹⁸O and Cl⁻ data

obtained in 1992 is stable with time, which is a logical assumption as times of transfer from CI to both CT and Phr are very long. Considering both ¹⁸O and Cl⁻ data, CI, CT and Phr data populations can be categorized. The CI and CT do not show appreciable ¹⁸O variations and can be considered as a single population. The Phr samples consist, however, of different populations: cluster I, with δ¹⁸O values close to −8 and small Cl⁻ concentrations, more specifically less

Table 4. Isotopic data ^{18}O and ^{3}H and chloride concentration in the Continental Intercalaire, Complexe Terminal and phreatic aquifers (sampling campaign in 1992).

Piezometer	Cl⁻ (mmol L⁻¹)	$\delta^{18}O$ (‰)	^{3}H (UT)	Piezometer	Cl⁻ (mmol L⁻¹)	$\delta^{18}O$ (‰)	^{3}H (UT)	Piezometer	Cl⁻ (mmol L⁻¹)	$\delta^{18}O$ (‰)	^{3}H (UT)
					Phreatic aquifer						
P007	1860.5	−2.5	0	PL15	23.5	−7.85	0.6(1)	P074	4356.4	3.4	6.8(8)
P009	426.9	−6.6	1.2(3)	P066	80.2	−8.1	0.8(1)	PL06	14.2	−8.1	1.0(2)
P506	54.4	−6.8	1.6(3)	PL23	1103.3	−6.1	0	PL30	24.3	−7.48	2.4(4)
P018	818.7	−2.9	6.2(11)	P063	1379.3	−3.4	8.7(15)	P002	522.4	−5.7	0.6(1)
P019	779.1	−4.7	5.6(9)	P068	2335.6	−3.1	8.8(14)	PL21	84.3	−7.7	1.2(2)
PZ12	2405.5	−2.3	8.1(13)	P030	527.7	−6.6	2.4(4)	PL31	18.9	−7.4	1.6(3)
P023	1176.9	−2.6	0.2(1)	P076	1743.5	−5.6	2.8(5)	P433	12.0	−8.8	0
P416	2433.7	−7.9	5.9(9)	P021	700.7	−5.2	2.6(4)	PL03	84.1	−7.4	1.7(3)
P034	2752.0	−1.8	5.7(9)	PL04	716.3	−2.9		PL44	109.8	−8.8	1.0(2)
P036	4972.7	3.3	2.1(4)	P093	2198.5	−2.6	5.1(8)	PL05	30.9	−7.4	1.9(3)
P037	4953.8	3.1	1.8(3)	P096	645.1	−6.1	4.8(8)	P408	24.2	−7.9	0
P039	4189.5	1.0	2.2(4)	PLX1	1296.6	−5.6	1.1(2)	P116	31.9	−7.2	1.1(2)
P041	2599.7	−0.6	7.3(13)	PLX2	25.7	−7.6	1.3(2)	LTP 16	213.4	−7.5	1.6(3)
P044	2106.1	−4.5	2.7(5)	P015	134.7	−6.8	3.0(5)	P117	32.8	−6.9	0.1
P014	336.9	−6.9	2.8(5)	P001	323.6	−4.7	2.5(4)	PL10	35.0	−7.3	0.2(1)
P012	539.3	−6.4	2.2(4)	P100	235.0	−5.8	0	PL25	75.6	−7.4	0.9(2)
P042	2330.8	2.1	6.0(10)	P056	42.1	−7.0	2.9(5)	LTP30	18.2	−7.5	1.1(2)
P006	19.0	−6.6	0.5(1)	P113	954.9	−4.8	0.8(2)	LTP06	1638.6	−1.9	2.8(5)
P057	28.2	−7.3	1.1(2)	PLX4	31.5	−7.1	0.3(1)	P031	481.8	−6.1	3.0(5)
P059	20.8	−7.8	0	P115	28.8	−2.5	6.8(12)				

Borehole	Cl⁻ (mmol L⁻¹)	$\delta^{18}O$ (‰)	^{3}H (UT)	Borehole	Cl⁻ (mmol L⁻¹)	$\delta^{18}O$ (‰)	^{3}H (UT)	Borehole	Cl⁻ (mmol L⁻¹)	$\delta^{18}O$ (‰)	^{3}H (UT)
					Complexe Terminal aquifer						
D5F80	42.2	−7.9		D1F138	28.9	−8.1	0.7(1)	D2F71	13.5	−8.2	0.6(1)
D3F8	29.8	−8.1	1.4(2)	D3F18	21.7	−8.2	0.2(1)	D7F4	10.6	−8.3	0.1(1)
D3F26	34.7	−8.0	0.8(1)	D3F10	14.3	−7.9	1.5(2)	D2F66	11.0	−8.3	
D4F94	20.1	−8.2	0.6(1)	D6F51	28.4	−7.9	0.7(1)	D1F151	10.8	−8.3	0.4(1)
D6F67	18.8	−8.2	3.7(6)	D1F135	18.1	−8.1	1.1(2)	D6F64	11.4	−8.3	4.3(7)

					Continental Intercalaire aquifer						
Hadeb I	5.8	−8.0	0	Hadeb II	6.2	−7.9	0.1(1)	Aouinet Moussa	6.5	−7.9	1.1(2)

than 35 mmol L⁻¹; cluster II, with $\delta^{18}O$ values larger than 3 and very large Cl⁻ concentrations, more specifically larger than 4000 mmol L⁻¹ (Table 5); intermediate Phr samples resulting from mixing between clusters I and II (mixing line in Fig. 13, mixing curve in Fig. 14) and from evaporation of cluster I (evaporation line in Fig. 14). The mass-balance modeling has shown that relatively few phases are required to derive observed changes in water chemistry and to account for the hydrochemical evolution in Ouargla's region. The mineral phases' selection is based upon geological descriptions and analysis of rocks and sediments from the area (OSS, 2003; Hamdi-Aïssa et al., 2004).

The inverse model was constrained so that mineral phases from evaporites including gypsum, halite, mirabilite, glauberite, sylvite and bloedite were set to dissolve until they reach saturation, and calcite and dolomite were set to precipitate once they reached saturation. Cation exchange reactions of Ca^{2+}, Mg^{2+}, K^+ and Na^+ on exchange sites were included in the model to check which cations are adsorbed or desorbed during the process. Dissolution and desorption contribute as positive terms in the mass balance, as elements are released in solution. On the other hand, precipitation and adsorption contribute as negative terms, while elements are removed from the solution. $CO_{2(g)}$ dissolution is considered by PHREEQC as a dissolution of a mineral, whereas $CO_{2(g)}$ degassing is dealt with as if it were a mineral precipitation.

Inverse modeling leads to a quantitative assessment of the different solutes' acquisition processes and a mass balance for the salts that are dissolved or precipitated from CI, CT and Phr groundwaters (Fig. 14, Table 6), as follows:

– Transition from CI to CT involves gypsum, halite and sylvite dissolution, and some ion exchange, namely calcium and potassium fixation on exchange sites against magnesium release, with a very small and quite negligible amount of $CO_{2(g)}$ degassing. The maximum elemental concentration fractional error equals 1 %. The

Table 5. Statistical parameters for the Continental Intercalaire (CI), Complexe Terminal (CT) and phreatic (Phr) aquifer samples selected on the basis of $\delta^{18}O$ and Cl^- data (see text).

Aquifer	Size	Parameter	EC ($mS\,cm^{-1}$)	T (°C)	pH	Alk.	Cl^-	SO_4^{2-}	Na^+ ($mmol\,L^{-1}$)	K^+	Mg^{2+}	Ca^{2+}
CI	11	Average	2.2	49.0	7.5	2.3	11.0	4.7	10.3	0.5	3.6	2.4
CI	11	SD	0.3	2.0	0.2	1.0	4.6	2.5	4.6	0.2	2.0	1.8
CT	50	Average	3.2	23.0	7.8	2.3	20.0	8.9	17.0	1.0	5.5	5.6
CT	50	SD	1.1	2.4	0.4	0.8	7.0	2.6	6.0	0.8	2.2	1.7
Phr cluster I	30	Average	3.9	24.0	7.9	2.3	24.7	11.8	24.2	2.1	7.2	5.3
Phr cluster I	30	SD	1.3	1.3	0.4	1.0	6.9	3.4	11.0	1.7	5.0	2.7
Phr cluster II	3	Average		23.4	7.0	2.4	4761.0	158.0	4021.0	32.4	500.0	13.0
Phr cluster II	3	SD		0.1	0.5	1.6	350.0	43.0	1093.0	28.0	378.0	8.0

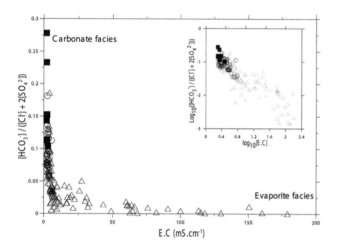

Figure 8. Change from carbonate facies to evaporite from the Continental Intercalaire (filled squares), Complexe Terminal (open circles) and phreatic (open triangles) aquifers.

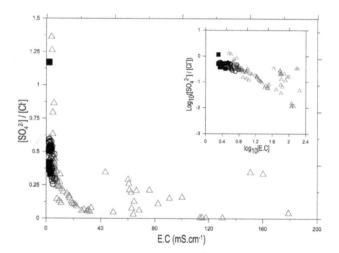

Figure 9. Change from sulfate facies to chloride from the Continental Intercalaire (filled squares), Complexe Terminal (open circles) and phreatic (open triangles) aquifers.

model consists of a minimum number of phases (i.e., six solid phases and $CO_{2(g)}$); another model also implies dolomite precipitation with the same fractional error.

- Transition from CT to an average water component of cluster I involves dissolution of halite, sylvite and bloedite from Turonian evaporites, with a very tiny calcite precipitation. The maximum fractional error in elemental concentration is 4 %. Another model implies $CO_{2(g)}$ escape from the solution, with the same fractional error. Large amounts of Mg^{2+} and SO_4^{2-} are released within the solution (Sharif et al., 2008; Li et al., 2010; Carucci et al., 2012).

- The formation of Phr cluster II can be modeled as being a direct dissolution of salts from the sabkha by rainwater with positive $\delta^{18}O$; the most concentrated water (P036 from Sebkhet Safioune) is taken here for cluster II, and pure water is taken as rainwater. In a de-

scending order of amount, halite, sylvite, gypsum and huntite are the minerals that are the most involved in the dissolution process. A small amount of calcite precipitates while some Mg^{2+} is released versus K^+ fixation on exchange sites. The maximum elemental fractional error in the concentration is equal to 0.004 %. Another model implies dolomite precipitation with some more huntite dissolving, instead of calcite precipitation, but salt dissolution and ion exchange are the same. Huntite, dolomite and calcite stoichiometries are linearly related, so both models can fit field data, but calcite precipitation is preferred compared to dolomite precipitation at low temperature.

- The origin of all phreatic waters can be explained by a mixing in variable proportions of cluster I and cluster II. For instance, waters from cluster I and cluster II can easily be separated by their $\delta^{18}O$, respectively, close to

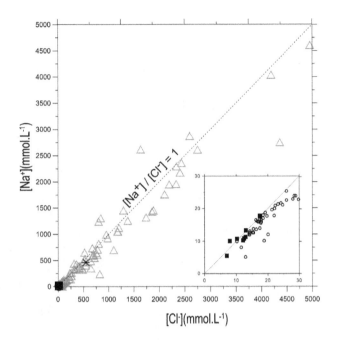

Figure 10. Correlation between Na^+ and Cl^- concentrations in the Continental Intercalaire (filled squares), Complexe Terminal (open circles) and phreatic (open triangles) aquifers. Seawater composition (star) is $[Na^+] = 459.3\,\mathrm{mmol\,L^{-1}}$ and $[Cl^-] = 535.3\,\mathrm{mmol\,L^{-1}}$ (Stumm and Morgan, 1999, p. 899).

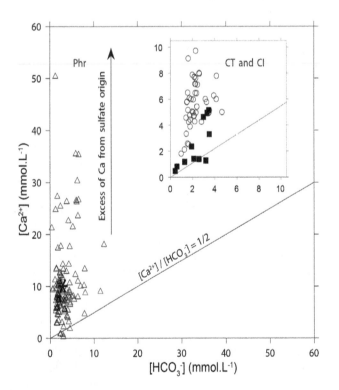

Figure 11. Calcium versus HCO_3^- diagram in the Continental Intercalaire (filled squares), Complexe Terminal (open circles) and phreatic (open triangles) aquifers. Seawater composition (star) is $[Ca^{2+}] = 10.2\,\mathrm{mmol\,L^{-1}}$ and $[HCO_3^-] = 2.38\,\mathrm{mmol\,L^{-1}}$ (Stumm and Morgan, 1999, p. 899).

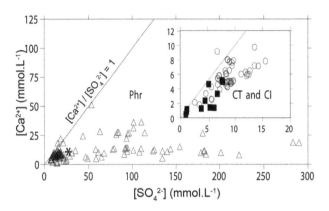

Figure 12. Calcium versus SO_4^{2-} diagram in the Continental Intercalaire (filled squares), Complexe Terminal (open circles) and phreatic (open triangles) aquifers. Seawater composition (star) is $[Ca^{2+}] = 10.2\,\mathrm{mmol\,L^{-1}}$ and $[SO_4^{2-}] = 28.2\,\mathrm{mmol\,L^{-1}}$ (Stumm and Morgan, 1999, p. 899).

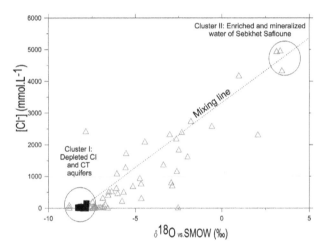

Figure 13. Chloride concentration versus $\delta^{18}O$ in the Continental Intercalaire (filled squares), Complexe Terminal (open circles) and phreatic (open triangles) aquifers from Ouargla.

−8 and +3.5‰ (Figs. 13 and 14). Mixing the two clusters is of course not an inert reaction, but rather results in the dissolution and the precipitation of minerals. Inverse modeling is then used to compute both mixing rates and the extent of matter exchange between soil and solution. For example, a phreatic water (piezometer P068) with intermediate values ($\delta^{18}O = -3$ and $[Cl^-] \simeq 2\,\mathrm{M}$) is explained by the mixing of 58 % water from cluster I and 42 % from cluster II. In addition, calcite precipitates, Mg^{2+} fixes on exchange sites against Na^+ and K^+, and gypsum dissolves, as does a minor amount of huntite (Table 6). The maximum elemental concentration fractional error is 2.5 % and the mixing fractions' weighted $\delta^{18}O$ is −3.17‰, which is very close to the measured value (−3.04‰). All the other models, making use of a minimum number of phases, and not taking into consid-

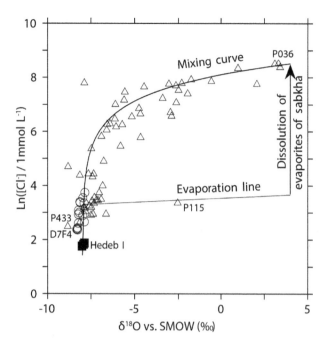

Figure 14. Log [Cl⁻] concentration versus δ^{18}O in Continental Intercalaire (filled squares), Complexe Terminal (open circles) and phreatic (open triangles) aquifers from Ouargla.

eration ion exchange reactions are not found compatible with isotope data. Mixing rates obtained with such models are, for example, 98 % of cluster I and 0.9 % of cluster II, which leads to a δ^{18}O = (−7.80‰) which is quite far from the real measured value (−3.04‰).

The main types of groundwaters occurring in the Ouargla basin are thus explained and could quantitatively be reconstructed. An exception is, however, sample P115, which is located exactly on the evaporation line of Phr cluster I. Despite numerous attempts, it could not be quantitatively rebuilt. Its ^{3}H value (6.8) indicates that it is derived from a more or less recent water component with very small salt content, most possibly affected by rainwater and some preferential flow within the piezometer. As this is the only sample on this evaporation line, there remains doubt regarding its significance.

Globally, the summary of mass transfer reactions occurring in the studied system (Table 6) shows that gypsum dissolution results in calcite precipitation and $CO_{2(g)}$ dissolution, thus acting as an inorganic carbon sink.

4   Conclusions

Two of the aquifers studied in this work, Complexe Terminal and Continental Intercalaire, are the main aquifers of Sahara, by extent (thousands of kilometers from the recharge area to the Gulf of Gabès) and time of transfer (thousands of years). The last one, the phreatic aquifer, is a shallow aquifer. The

chemical facies of these aquifers have long been qualitatively described. Our results quantitatively explain, for the first time, the processes that occur during upward leakage through interaction between solution and the mineral constituents of the aquifers and ultimately by mixing with surface waters. The hydrochemical study of the aquifer system occurring in Ouargla's basin allowed us to identify the origin of its mineralization. Waters exhibit two different facies: sodium chloride and sodium sulfate for the phreatic aquifer (Phr), sodium sulfate for the Complexe Terminal (CT) aquifer and sodium chloride for the Continental Intercalaire (CI) aquifer. Calcium carbonate precipitation and evaporite dissolution explain the facies change from carbonate to sodium chloride or sodium sulfate that is recorded. However, reactions imply many minerals with common ions, deep reactions without evaporation, as well as shallow processes affected by both evaporation and mixing. Those processes are separated by considering both chemical and isotopic data, and quantitatively explained making use of an inverse geochemical modeling. The latter was applied, for the first time ever, in Algeria, to an extreme environment featuring a lack of data on a scarce natural resource such as Saharan groundwater. The populations of the region rely on this resource for their daily drinkable water as well as for agriculture, which mainly consists of date production and some vegetables that grow within the date-palm groves. Results obtained through inverse modeling could help water resources managers, both at the local and the regional scales, to gather the necessary information for an integrated management of that vital resource. Moreover, and regarding the large geographic scale of the aquifers, such a pilot study could be taken as a supporting work to further investigations elsewhere in similar regions. The present study leads to the main result that phreatic waters do not originate simply from infiltration of rainwater and dissolution of salts from the sabkhas. Conversely, Phr waters are largely influenced by the upwardly mobile deep CT and CI groundwaters, with fractions of the latter interacting with evaporites from the Turonian formations. Phreatic water occurrence is explained as a mixture of two end-member components: cluster I, which is very close to CI and CT, and cluster II, which is highly mineralized and results from the dissolution by rainwater of salts from the sabkhas. At depth, CI leaks upwardly and dissolves gypsum, halite and sylvite, with some ion exchange, to give waters of the CT aquifer composition. CT transformation into Phr cluster I waters involves the dissolution of Turonian evaporites (halite, sylvite and bloedite) with minor calcite precipitation. At the surface, direct dissolution by rainwater of salts from sabkhas (halite, sylvite, gypsum and some huntite) with precipitation of calcite and Mg^{2+}/K^+ ion exchange results in cluster II Phr composition. All phreatic groundwaters result from a mixing of cluster I and cluster II water that is accompanied by calcite precipitation, fixation of Mg^{2+} on ion exchange sites against the release of K^+ and Na^+. Moreover, some $CO_{2(g)}$ escapes from the solution at depth, but dissolves much more at the surface.

Table 6. Summary of mass transfer for geochemical inverse modeling. Phases and thermodynamic database are from PHREEQC 3.0 (Parkhurst and Appelo, 2013).

Phases	Stoichiometry	CI/CT	CT/Phr I	Rainwater/P036	PhrI/PhrII 60/40 %
Calcite	$CaCO_3$	–	-6.62×10^{-6}	-1.88×10^{-1}	-2.26×10^{-1}
$CO_{2(g)}$	CO_2	-6.88×10^{-5}	–	8.42×10^{-4}	5.77×10^{-4}
Gypsum	$CaSO_4 \cdot 2H_2O$	4.33×10^{-3}	–	1.55×10^{-1}	1.67×10^{-1}
Halite	$NaCl$	7.05×10^{-3}	3.76×10^{-3}	6.72×10^{0}	1.28×10^{0}
Sylvite	KCl	2.18×10^{-3}	1.08×10^{-3}	4.02×10^{-1}	–
Bloedite	$Na_2Mg(SO_4)_2 \cdot 4H_2O$	–	1.44×10^{-3}	–	–
Huntite	$CaMg_3(CO_3)_4$	–	–	4.74×10^{-2}	5.65×10^{-2}
Ca ion exchange	CaX_2	-1.11×10^{-3}	–	–	–
Mg ion exchange	MgX_2	1.96×10^{-3}	–	1.75×10^{-1}	-2.02×10^{-1}
Na ion exchange	NaX	–	–	–	3.92×10^{-1}
K ion exchange	KX	-1.69×10^{-3}	–	-3.49×10^{-1}	1.20×10^{-2}

Values are in $mol\,kg^{-1}$ H_2O. Positive-phase (mass entering solution) and negative-phase (mass leaving solution) mole transfers indicate dissolution and precipitation, respectively; this indicates no mass transfer.

The most complex phenomena occur during the dissolution of Turonian evaporites while CI leaks upwardly towards CT, and from Phr I to Phr II, while the transition from CT to Phr I implies a very limited number of phases. Globally, gypsum dissolution and calcite precipitation processes both act as an inorganic carbon sink.

Appendix A

According to a simple Rayleigh equation, the evolution of the heavy isotope ratio in the remaining liquid R_l is given by

$$R_l \approx R_{l,0} \times f_l^{\alpha-1}, \tag{A1}$$

where f_l is the fraction of remaining liquid and α the fractionation factor.

The fraction of remaining liquid is derived from chloride concentration, as chloride can be considered conservative during evaporation: all phreatic waters are undersaturated with respect to halite, which precipitates only in the last stage. Hence, the following equation holds:

$$f_l \equiv \frac{n_{w,1}}{n_{w,0}} = \frac{[Cl^-]_0}{[Cl^-]_1}. \tag{A2}$$

By taking natural logarithms, one obtains

$$\ln R_l \approx (1-\alpha) \times \ln[Cl^-] + \text{constant}. \tag{A3}$$

As, by definition,

$$R_l \equiv R_{SD} \times \left(1 + \frac{\delta^{18}O}{1000}\right), \tag{A4}$$

one has

$$\ln R_l \equiv \ln R_{SD} + \ln\left(1 + \frac{\delta^{18}O}{1000}\right),$$
$$\approx \ln R_{SD} + \frac{\delta^{18}O}{1000}. \tag{A5}$$

Hence, with base 10 logarithms,

$$\delta^{18}O \approx 1000(1-\alpha)\log[Cl^-] + \text{constant},$$
$$\approx -\epsilon \log[Cl^-] + k, \tag{A6}$$

where, as classically defined, $\epsilon = 1000(\alpha-1)$ is the enrichment factor.

Competing interests. The authors declare that they have no conflict of interest.

Acknowledgements. The authors wish to thank the staff members of the National Agency for Water Resources in Ouargla (ANRH) and the Laboratory of Algerian Waters (ADE) for the support provided to the Technical Cooperation programme within which this work was carried out. Analyses of ^{18}O were funded by the project CDTN/DDHI (Guendouz and Moulla, 1996). The support of University of Ouargla and of INRA for travel grants of R. Slimani and G. Bourrié are gratefully acknowledged too.

Edited by: M. Giudici

References

ANRH: Inventaire des forages de la Wilaya de Ouargla, Rapport technique, Agence Nationale des Ressources Hydrauliques, Direction régionale Sud, Ouargla, 2011.

Aumassip, G., Dagorne, A., Estorges, P., Leèvre-Witier, P., Mahrour, F., Nesson, C., Rouvillois-Brigol, M., and Trecolle, G.: Aperçus sur l'Évolution du paysage quaternaire et le peuplement de la région de Ouargla, Libyca, Cnrpah, Tome XX, Algérie, 205–257, 1972.

Belkhiri, L., Boudoukha, A., Mouni, L., and Baouz, T.: Application of multivariate statistical methods and inverse geochemical modeling for characterization of groundwater – A case study: Ain Azel plain (Algeria), Geoderma, 159, 390–398, 2010.

Belkhiri, L., Mouni, L., and Boudoukha, A.: Geochemical evolution of groundwater in an alluvial aquifer: Case of El Eulma aquifer, East Algeria, J. Afr. Earth Sci., 66–67, 46–55, 2012.

Carucci, V., Petitta, M., and Aravena, R.: Interaction between shallow and deep aquifers in the Tivoli Plain (Central Italy) enhanced by groundwater extraction: A multi-isotope approach and geochemical modeling, Appl. Geochem., 27, 266–280, doi:10.1016/j.apgeochem.2011.11.007, 2012.

Charlton, S., Macklin, C., and Parkhurst, D.: PhreeqcI – A graphical user interface for the geochemical computer program PHREEQC, Rapport technique, US Geological Survey Water-Resources, USA, 1996.

Chellat, S., Bourefis, A., Hamdi-Aïss, a. B., and Djerrab, A.: Paleoenvironemental reconstitution of Mio-pliocenes sandstones of the lower-Sahara at the base of exoscopic and sequential analysis, Pensee J., 76, 34–51, 2014.

Chkir, N., Guendouz, A., Zouari, K., Hadj Ammar, F., and Moulla, A.: Uranium isotopes in groundwater from the continental intercalaire aquifer in Algerian Tunisian Sahara (Northern Africa), J. Environ. Radioact., 100, 649–656, doi:10.1016/j.jenvrad.2009.05.009, 2009.

Cornet, A. and Gouscov, N.: Les eaux du Crétacé inférieur continental dans le Sahara algérien: nappe dite "Albien", in: Congrès géologique international, vol. tome II, Alger, p. 30, 1952.

Dai, Z., Samper, J., and Ritzi, R.: Identifying Geochemical Processes By Inverse Modeling of Multicomponent Reactive Transport in the Aquia Aquifer, Geosphere, 2, 210–219, 2006.

Deutsch, W.: Groundwater Chemistry-Fundamentals and Applications to Contamination, Lewis Publishers, New York, 1997.

Dubief, J.: Essai sur l'hydrologie superficielle au Sahara, Direction du service de la colonisation et de l'hydraulique, Service des Études scientifiques, Algérie, 1953.

Dubief, J.: Le climat du Sahara, Hors-série, Institut de recherches sahariennes, Algérie, 1963.

Eckstein, G. and Eckstein, Y.: A hydrogeological approach to transboundary ground water resources and international law, Am. Univers. Int. Law Rev., 19, 201–258, 2003.

Edmunds, W., Guendouz, A., Mamou, A., Moulla, A., Shand, P., and Zouari, K.: Groundwater evolution in the continental intercalaire aquifer of southern Algeria and Tunisia: trace element and isotopic indicators, Appl. Geochem., 18, 805–822, 2003.

Fontes, J., Yousfi, M., and Allison, G.: Estimation of long-term, diffuse groundwater discharge in the northern Sahara using stable isotope profiles in soil water, J. Hydrol., 86, 315–327, 1986.

Foster, S., Margat, J., and Droubi, A.: Concept and importance of nonrenewable resources, no. 10 in IHP-VI Series on Groundwater, UNESCO, Paris, 2006.

Furon, R.: Géologie de l'Afrique, 2nd Edn., Payot, Paris, 1960.

Gonfiantini, R., Conrad, G., Fontes, J.-C., Sauzay, G., and Payne, B.: Étude isotopique de la nappe du Continental Intercalaire et de ses relations avec les autres nappes du Sahara septentrional, Isotop. Tech. Groundw. Hydrol., 1, 227–241, 1975.

Guendouz, A.: Contribution á létude hydrochimique et isotopique des nappes profondes du Sahara nord-est septentrional, Algérie, PhD thesis, Université d'Orsay, Orsay, France, 1985.

Guendouz, A. and Michelot, J.: Chlorine-36 dating of deep groundwater from northern Sahara, J. Hydrol., 328, 572–580, 2006.

Guendouz, A. and Moulla, A.: Étude hydrochimique et isotopique des eaux souterraines de la cuvette de Ouargla, Algérie, Rapport technique, CDTN/DDHI, Algérie, 1996.

Guendouz, A., Moulla, A., Edmunds, W., Zouari, K., Shands, P., and Mamou, A.: Hydrogeochemical and isotopic evolution of water in the complex terminal aquifer in Algerian Sahara, Hydrogeol. J., 11, 483–495, 2003.

Güler, C. and Thyne, G.: Hydrologic and geologic factors controlling surface and groundwater chemistry in Indian wells – Owens valley area, southeastern California, USA, J. Hydrol., 285, 177–198, 2004.

Hadj-Ammar, F., Chkir, N., Zouari, K., Hamelin, B., Deschamps, P., and Aigoun, A.: Hydrogeochemical processes in the Complexe Terminal aquifer of southern Tunisia: An integrated investigation based on geochemical and multivariate statistical methods, J. Afr. Earth Sci., 100, 81–95, 2014.

Hamdi-Aïssa, B., Vallès, V., Aventurier, A., and Ribolzi, O.: Soils and brines geochemistry and mineralogy of hyper arid desert playa, Ouargla basin, Algerian Sahara, Arid Land Res. Manage., 18, 103–126, 2004.

Kenoyer, G. and Bowser, C.: Groundwater chemical evolution in a sandy aquifer in northern Wisconsin, Water Resour. Res., 28, 591–600, 1992.

Kuells, C., Adar, E., and Udluft, P.: Resolving patterns of ground water flow by inverse hydrochemical modeling in a semiarid Kalahari basin, Trac. Model. Hydrogeol., 262, 447–451, 2000.

Le Houérou, H.: Bioclimatology and biogeography of Africa, Springer Verlag, Berlin, Heidelberg, 2009.

Lelièvre, R.: Assainissement de la cuvette de Ouargla, rapports géo-hydraulique no. 2, Ministère des Travaux Publique et de la construction, Algérie, 1969.

Li, P., Qian, H., Wu, J., and Ding, J.: Geochemical modeling of groundwater in southern plain area of Pengyang County, Ningxia, China, Water Sci. Eng., 3, 282–291, 2010.

Ma, J., Pan, F., Chen, L., Edmunds, W., Ding, Z., Zhou, K., He, J., Zhoua, K., and Huang, T.: Isotopic and geochemical evidence of recharge sources and water quality in the Quaternary aquifer beneath Jinchang city, NW China, Appl. Geochem., 25, 996–1007, 2010.

Martinelli, G., Chahoud, A., Dadomo, A., and Fava, A.: Isotopic features of Emilia-Romagna region (North Italy) groundwaters: Environmental and climatological implications, J. Hydrol., 519, 1928–1938, doi:10.1016/j.jhydrol.2014.09.077, 2014.

Matiatos, I., Alexopoulos, A., and Godelitsas, A.: Multivariate statistical analysis of the hydrogeochemical and isotopic composition of the groundwater resources in northeastern Peloponnesus (Greece), Sci. Total Environ., 476–477, 577–590, doi:10.1016/j.scitotenv.2014.01.042, 2014.

Moulias, D.: L'eau dans les oasis sahariennes, organisation hydraulique, régime juridique, PhD thesis, Algérie, 1927.

Moulla, A., Guendouz, A., Cherchali, M.-H., Chaid, Z., and Ouarezki, S.: Updated geochemical and isotopic data from the Continental Intercalaire aquifer in the Great Occidental Erg sub-basin (south-western Algeria), Quatern. Int., 257, 64–73, 2012.

OECD: OECD Environmental Outlook to 2030, Tech. Rep. 1, Organisation for Economic Cooperation and Development, Paris, 2008.

ONM: Bulletins mensuels de relevé des paramtres climatologiques en Algérie, Office national météorologique, Ouargla, Algérie, 1975/2013.

OSS: Système aquifère du Sahara septentrional, Tech. rep., Observatoire du Sahara et du Sahel, Tunis, 2003.

OSS: Systme aquifère du Sahara septentrional (Algérie, Tunisie, Libye): gestion concertée d'un bassin transfrontalier, Tech. Rep. 1, Observatoire du Sahara et du Sahel, Tunis, 2008.

Ould Baba Sy, M. and Besbes, M.: Holocene recharge and present recharge of the Saharan aquifers – a study by numerical modeling, in: International symposium – Management of major aquifers, Dijon, France, 2006.

Parkhurst, D. and Appelo, C.: Description of Input and Examples for PHREEQC (Version 3) – A computer program for speciation, batch-reaction, one-dimensional transport, and inverse geochemical calculations, Tech. Rep. 6, US Department of the Interior, US Geological Survey, http://wwwbrr.cr.usgs.gov/ (last access: 15 November 2016), 2013.

Plummer, L. and Back, M.: The mass balance approach: application to interpreting the chemical evolution of hydrological systems., American Journal of Science, 280, 130–142, 1980.

Plummer, L. and Sprinckle, C.: Radiocarbon dating of dissolved inorganic carbon in groundwater from confined parts of the upper Floridan aquifer, Florida, USA, J. Hydrol., 9, 127–150, 2001.

Sharif, M., Davis, R., Steele, K., Kim, B., Kresse, T., and Fazio, J.: Inverse geochemical modeling of groundwater evolution with emphasis on arsenic in the Mississippi River Valley alluvial aquifer, Arkansas (USA), J. Hydrol., 350, 41–55, doi:10.1016/j.jhydrol.2007.11.027, 2008.

Slimani, R.: Contribution à l'évaluation d'indicateurs de pollution environnementaux dans la région de Ouargla: cas des eaux de rejets agricoles et urbaines, MS thesis, Université de Ouargla, Ouargla, 2006.

Stumm, W. and Morgan, J.: Aquatic Chemistry: Chemical Equilibria and Rates in Natural Waters, John Wiley and Sons, New York, 1999.

Thomas, J., Welch, A., and Preissler, A.: Geochemical evolution of ground water in Smith Creek valley – a hydrologically closed basin in central Nevada, USA, Appl. Geochem., 4, 493–510, 1989.

Trolard, F., Bourrié, G., Baillieux, A., Buis, S., Chanzy, A., Clastre, P., Closet, J.-F., Courault, D., Dangeard, M.-L., Di Virgilio, N., Dussouillez, P., Fleury, J., Gasc, J., Géniaux, G., Jouan, R., Keller, C., Lecharpentier, P., Lecroart, J., Napoleone, C., Mohammed, G., Olioso, A., Reynders, S., Rossi, F., Tennant, M., and Lopez, J. D. V.: The PRECOS framework: Measuring the impacts of the global changes on soils, water, agriculture on territories to better anticipate the future, J. Environ. Manage., 181, 590–601, doi:10.1016/j.jenvman.2016.07.002, 2016.

UNESCO: Projet ERESS, Étude des ressources en eau du Sahara septentrional, Tech. Rep. 10, UNESCO, Paris, 1972.

Vallès, V., Rezagui, M., Auque, L., Semadi, A., Roger, L., and Zouggari, H.: Geochemistry of saline soils in two arid zones of the Mediterranean basin. I. Geochemistry of the Chott Melghir-Mehrouane watershed in Algeria, Arid Soil Res. Rehabil., 11, 71–84, 1997.

Zhu, C. and Anderson, G.: Environmental Application of Geochemical Modeling, Cambridge University Press, Cambridge, 139, 596–597, 2002.

3

Quantification of anthropogenic impact on groundwater-dependent terrestrial ecosystem using geochemical and isotope tools combined with 3-D flow and transport modelling

A. J. Zurek[1], S. Witczak[1], M. Dulinski[2], P. Wachniew[2], K. Rozanski[2], J. Kania[1], A. Postawa[1], J. Karczewski[1], and W. J. Moscicki[1]

[1] AGH University of Science and Technology, Faculty of Geology, Geophysics and Environmental Protection, Krakow, Poland
[2] AGH University of Science and Technology, Faculty of Physics and Applied Computer Science, Krakow, Poland

Correspondence to: A. J. Zurek (zurek@agh.edu.pl)

Abstract. Groundwater-dependent ecosystems (GDEs) have important functions in all climatic zones as they contribute to biological and landscape diversity and provide important economic and social services. Steadily growing anthropogenic pressure on groundwater resources creates a conflict situation between nature and man which are competing for clean and safe sources of water. Such conflicts are particularly noticeable in GDEs located in densely populated regions. A dedicated study was launched in 2010 with the main aim to better understand the functioning of a groundwater-dependent terrestrial ecosystem (GDTE) located in southern Poland. The GDTE consists of a valuable forest stand (Niepolomice Forest) and associated wetland (Wielkie Błoto fen). It relies mostly on groundwater from the shallow Quaternary aquifer and possibly from the deeper Neogene (Bogucice Sands) aquifer. In July 2009 a cluster of new pumping wells abstracting water from the Neogene aquifer was set up 1 km to the northern border of the fen. A conceptual model of the Wielkie Błoto fen area for the natural, pre-exploitation state and for the envisaged future status resulting from intense abstraction of groundwater through the new well field was developed. The main aim of the reported study was to probe the validity of the conceptual model and to quantify the expected anthropogenic impact on the studied GDTE. A wide range of research tools was used. The results obtained through combined geologic, geophysical, geochemical, hydrometric and isotope investigations provide strong evidence for the existence of upward seepage of groundwater from the deeper Neogene aquifer to the shallow Quaternary aquifer supporting the studied GDTE. Simulations of the groundwater flow field in the study area with the aid of a 3-D flow and transport model developed for Bogucice Sands (Neogene) aquifer and calibrated using environmental tracer data and observations of hydraulic head in three different locations on the study area, allowed us to quantify the transient response of the aquifer to operation of the newly established Wola Batorska well field. The model runs reveal the presence of upward groundwater seepage to the shallow Quaternary aquifer of the order of $440\,m^3\,d^{-1}$. By the end of the simulation period (2029), with continuous operation of the Wola Batorska well field at maximum permissible capacity (ca. $10\,000\,m^3\,d^{-1}$), the direction of groundwater seepage will change sign (total change of the order of $900\,m^3\,d^{-1}$). The water table drawdown in the study area will reach ca. 30 cm. This may have significant adverse effects on functioning of the studied GDTE.

1 Introduction

There is a growing awareness among policy makers, legislators, water resources managers and researchers of the important environmental and socio-economic functions of groundwater-dependent ecosystems (GDEs) as reflected, for example, in the environmental legislation of the European Union (Kløve et al., 2011b; EC, 2000, 2006). Human needs and GDEs appear as two, sometimes conflicting, groundwater uses (Wachniew et al., 2014) which need to be man-

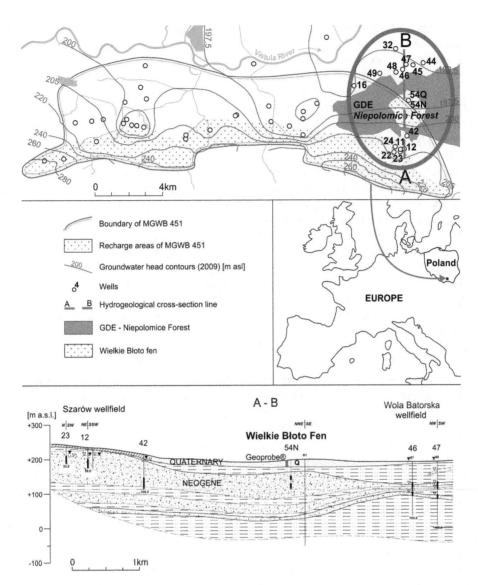

Figure 1. Hydrogeological map and cross-section of the Bogucice Sands (Neogene) aquifer (major groundwater basin – MGWB 451). The study area is marked by a red oval. Open circles mark the position of pumping wells. Cross-section according to Górka et al. (2010).

aged in an integrated, multidisciplinary manner (Kløve et al., 2011b). Groundwater exploitation, climatic and land-use changes, pollution as well as other pressures on groundwater quantity and quality affect functions of the GDEs, yet the relationships between groundwater systems and the performance of dependent ecosystems are not fully understood (Kløve et al., 2011a, b, 2014). The great diversity of GDEs stems primarily from space and time variations of groundwater supply to those ecosystems. Various classifications of GDEs (Hatton and Evans, 1998; Sinclair Knight Merz, 2001; EC, 2003, 2011; Boulton, 2005; Pettit et al., 2007; Dresel et al., 2010; Kløve et al., 2011a; Bertrand et al., 2012) reflect this diversity. The basic division includes the terrestrial (GDTE; e.g. wet forests, riparian zones, wetlands) and

aquatic (GDAE; e.g. springs, lakes, rivers with hyporheic zones, lagoons) GDEs.

Sustainable management of GDEs requires that their vulnerability to anthropogenic impacts is assessed (Wachniew et al., 2014). A conceptualization of GDE vulnerability must include understanding of two factors: (i) the degree of ecosystem reliance on groundwater (Hatton and Evans, 1998), and (ii) groundwater availability to the ecosystem (Sinclair Knight Merz, 2001). Consequently, a substantial component of conceptual models, on which vulnerability assessments are based (EC, 2010), is related to the identification of the origin and pathways and quantification of groundwater fluxes to GDEs.

The presented study was aimed at comprehensive investigation of groundwater dependence of a terrestrial ecosystem

(GDTE) consisting of valuable forest stand and associated wetland, located in the south of Poland (Fig. 1). The central hypothesis of the presented work was that the studied GDTE relies not only on the shallow, unconfined Quaternary aquifer but indirectly also on groundwater originating from deeper confined aquifer, underlying the Quaternary cover and separated from it by an aquitard of variable thickness. Consequently, the presented study was addressing flow paths and water ages of the deeper aquifer and its connectivity with the shallow Quaternary aquifer. An important additional objective was the quantification of the potential risk to the studied GDTE associated with the operation of a nearby cluster of water-supply wells exploiting the deeper aquifer. The deeper aquifer is an important source of drinking water for the local population, intensely exploited for several decades now.

A suite of tools were applied to address the problems outlined above. Two monitoring wells were drilled in the centre of the studied GDTE to obtain direct information on the vertical extent and geologic structure of the Quaternary cover and the deeper aquifer. The drillings were supplemented by geophysical prospecting. DC resistivity sounding was used to obtain additional information about the spatial extent and thickness of the confining layer separating the Quaternary cover and the underlying deeper Neogene aquifer. Ground-penetrating radar surveys supplied information about the thickness of peat layers in the area of GDTE. Hydrometric measurements, carried out over a 2-year period on the Dluga Woda stream draining the area of GDTE and supported by chemical and isotope analyses of stream water, were used to quantify the expected contribution of groundwater seepage from the deeper aquifer to the water balance of the Dluga Woda catchment. The seepage was further characterized by dedicated Geoprobe® sampling of the Quaternary cover enabling vertical stratification of environmental tracers and water chemistry within the Quaternary cover to be assessed. The hydrochemical evolution and age of water in the Neogene (Bogucice Sands) aquifer was characterized using chemical and isotope data (water chemistry, stable isotopes of water (^{2}H and ^{18}O), tritium (^{3}H), isotopes of carbon (^{14}C, ^{13}C)) accompanied by geochemical modelling (PHREEQC and NET-PATH). Finally, the 3-D flow and transport model available for the Bogucice Sands aquifer was used to quantify the expected impact of enhanced exploitation of the aquifer on the status of the studied GDTE.

The presented study focusing on the interaction between the Bogucice Sands (Neogene) aquifer and the associated GDTE is a follow-up of the earlier work concerned mostly with the dynamics and geochemical evolution of groundwater in the deeper aquifer (Zuber et al., 2005; Witczak et al., 2008; Dulinski et al., 2013).

2 The study area

The study area is located in the south of Poland, in the vicin-

ity of Krakow agglomeration (Fig. 1). The studied GDTE consists of valuable forest stand – the Niepolomice Forest, and associated wetland – the Wielkie Błoto fen. The Niepolomice Forest is a relatively large (ca. 110 km²) lowland forest complex. This relict of once vast forests occupying southern Poland is protected as a Natura 2000 Special Protection Area "Puszcza Niepołomicka" (PLB120002) which supports bird populations of European importance. The Niepolomice Forest contains also several nature reserves and the European bison breeding centre and has important recreational value as the largest forest complex in the vicinity of Krakow agglomeration.

The Wielkie Błoto fen located in the western part of Niepolomice Forest (Figs. 1 and 2) comprises a separate Natura 2000 area (Torfowisko Wielkie Błoto, PLH120080), a significant habitat of endangered butterfly species associated with wet meadows. It contains different types of peat deposits with variable thickness (Fig. 2). Due to drainage works carried out mostly after the Second World War, the uppermost peat layers were drained and converted to arable land (Lipka, 1989; Łajczak, 1997; Lipka et al., 2006). In recent decades the agriculture usage of Wielkie Błoto was greatly reduced and the studied GDTE is returning nowadays to its natural state.

The climate of the study area has an intermediate character between oceanic and continental, with mean annual temperature of 8.2 °C. Mean annual precipitation rate amounts to 725 mm whereas the mean actual evapotranspiration in the Niepolomice Forest area reaches 480 mm. The annual mean runoff fluctuates around 245 mm. The regional runoff is related to the drainage system of Vistula River and its tributaries (see Fig. 1). The Wielkie Błoto fen area and the adjacent parts of Niepolomice Forest are drained by the Dluga Woda stream with 8.2 km² of gauged catchment area (Fig. 2).

Depth to the water table in the study area is generally small, with wetlands and marshes occurring in several parts of the Niepolomice Forest. The dependence of Niepolomice Forest stands on groundwater is enhanced by low available water capacity and low capillary rise of the soils supporting the forest (Łajczak, 1997; Chełmicki et al., 2003). Depth to water table was used as a basis for defining an index quantifying dependency of Niepolomice Forest on groundwater. This dependency relies on rooting depth and the depth to local water table. It also influences the typology of forest and type of plant cover associated with GDTE (Schaffers and Sýkora, 2000; Pettit et al., 2007; GENESIS, 2012; Hose et al., 2014). The rooting depth must be smaller than the depth to the water table. Otherwise water-clogging occurs and roots cannot respire due to excess of water in the soil profile. Three classes of GDTE susceptibility to changes of water table depth were proposed (Fig. 2): (i) class A: very strongly dependent (depth of water table ranging from 0.0

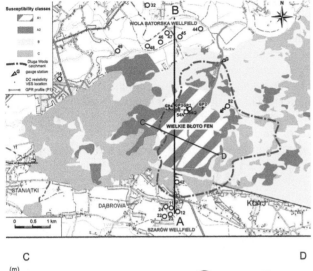

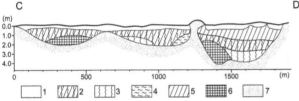

Figure 2. Upper panel: map of the study area showing the western part of Niepolomice Forest and Wielkie Błoto fen. GDTE susceptibility classes based on the depth to water table: A – very strongly dependent (0.0 to 0.5 m); A1 – wetland ecosystem; A2 – forest ecosystem; B – strongly dependent (0.5 to 2.0 m) forest ecosystem; C – weakly dependent (>2.0 m) forest ecosystem. Lower panel: cross-section through Wielkie Błoto fen according to Lipka (1989). 1 – mineralized peat soil; 2 – tall sedge-reed peat; 3 – reed peat; 4 – *Sphagnum* peat; 5 – tall sedge peat; 6 – gyttja; 7 – sand.

to 0.5 m), (ii) class B: strongly dependent (0.5 to 2.0 m), and (iii) class C: weakly dependent (>2.0 m). Forest stands growing on areas where depth to water table exceeds 2 m utilize mostly soil moisture and are weakly dependent on groundwater level fluctuations. Forest stands on areas with shallower water table are more susceptible to changes in groundwater level, regardless of their direction (Forest Management Manual, 2012).

Groundwater is stored in the study area in the upper, phreatic aquifer associated with Quaternary sediments and in the confined, deeper aquifer composed of Neogene marine sediments (Bogucice Sands). The unsaturated zone consists mainly of sands and loess of variable thickness, from a fraction of a metre in wetland areas to approximately 30 m in the recharge area of the deeper aquifer layers.

The Bogucice Sands (Neogene) aquifer covers an area of ca. 200 km^2 and belongs to the category of major groundwater basins (MGWB) in Poland (Kleczkowski et al., 1990). It is located on the border of the Carpathian Foredeep Basin and belongs to the Upper Badenian (Middle Miocene). The aquifer (MGWB no. 451) is composed mainly of unconsolidated sands, locally sandstones with carbonate cement. The

variable percentage of carbonate cement, up to 30 %, affects the hydraulic conductivity of water-bearing horizons. The aquifer is underlined by impermeable clays and claystones of the Chodenice Beds (Porebski and Oszczypko, 1999). To the north, the aquifer is progressively covered by mudstones and claystones with thin sandstone interbeds. Palaeoflow directional indicators suggest proximity to deltaic shoreline. The mean thickness of the aquifer is approximately 100 m, with two water-bearing horizons (see Fig. 1). The hydrogeology of the aquifer can be considered in three areas: (i) the recharge area related to the outcrops of Bogucice Sands in the south, (ii) the central confined area generally with artesian water, and (iii) the northern discharge area in the Vistula River valley. Groundwater movement takes place from the outcrops in the south, in the direction of the Vistula River valley (Fig. 1) where the aquifer is drained by upward seepage through semi-permeable aquitard. The recharge of groundwater, related to outcropping lithology, is of the order of 8 to 28 % of annual precipitation.

The principal economic role of Bogucice Sands aquifer is to provide potable water for public and private users. Estimated safe yield of the aquifer is approximately 40 000 m^{3}d^{-1}, with typical well capacities of 4 to 200 m^{3}h^{-1} (Kleczkowski et al., 1990; Witczak et al., 2008; Górka et al., 2010). Hospitals and food processing plants also exploit some wells. Yield of the aquifer is insufficient to meet all present and emerging needs and, as a consequence, licensing conflicts arise between water supply companies and industry about the amount of water available for safe exploitation.

In the pre-exploitation era, artesian water existed most probably on the entire confined area of the aquifer. Intensive exploitation decreased the hydraulic head in some areas causing downward seepage. In the area of Wielkie Błoto fen the aquifer is exploited by Szarów well field located in the south (wells Nos. 11, 12, 22–24, 42 in Figs. 1 and 2). In July 2009 a cluster of six new water-supply wells (Wola Batorska well field – wells Nos. 44–49 in Figs. 1 and 2) exploiting deeper aquifer layers was set up close to the northern border of Niepolomice Forest. There is a growing concern that intense exploitation of this new well field may lead to lowering of hydraulic head in the western part of the Niepolomice Forest area.

The available geological information (Porebski and Oszczypko, 1999; Górka et al., 2010), supplemented by the results of previous work on the dynamics and geochemical evolution of groundwater in the Bogucice Sand aquifer (Zuber et al., 2005; Witczak et al., 2008), provided the basis for construction of a conceptual model illustrating the interaction between shallow Quaternary aquifer and the deeper Neogene aquifer in the area of studied GDTE and suggesting the possible impact of intense exploitation of groundwater by the Wola Batorska well field (Fig. 3). Figure 3a presents the presumed natural state of this interaction. The Wielkie Błoto fen represents in this model a local discharge area for both the shallow and the deeper aquifer. Artesian conditions in

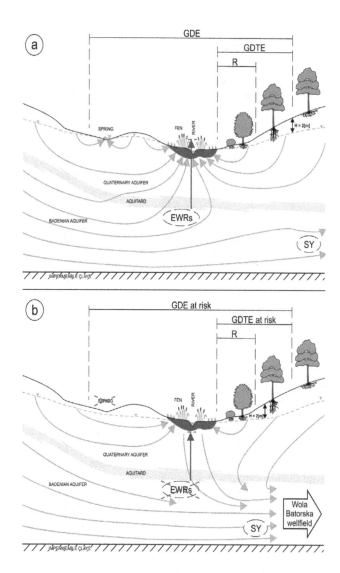

Figure 3. Conceptual model of the Wielkie Błoto fen. (**a**) Natural state; (**b**) envisaged future status as a result of intense exploitation of the Wola Batorska well field. GDE – groundwater-dependent ecosystem; GDTE – groundwater-dependent terrestrial ecosystem; R – riparian forest; EWRs – environmental water requirements; SY – safe yield of the aquifer exploited by the Wola Batorska well field.

the deeper aquifer combined with the relatively thin aquitard layer separating both aquifers may lead to upward seepage of deeper groundwater, contributing to the water balance of the studied GDTE. The envisaged future status of the shallow/deep aquifer interaction as a result of intense exploitation of the Wola Batorska well field is shown in Fig. 3b. It is expected that intensive pumping of the deeper aquifer by the well field localized close to the northern border of Niepołomice Forest (wells Nos. 44–49 in Fig. 2), exceeding its safe yield, may modify the groundwater flow field in the area of Wielkie Błoto fen in such a way that the upward leakage will be stopped or significantly reduced, thus leading to a lowering of the water table and endangering the environmental water requirements of the studied GDTE.

3   Materials and methods

A suite of different methods was applied to address two major questions posed by the conceptual model presented in Fig. 3, i.e. the existence of upward seepage of groundwater from the deeper Neogene aquifer to the shallow Quaternary aquifer and its role in the water balance of Wielkie Błoto fen, and quantification of the expected impact of intense exploitation of the deeper aquifer by the Wola Batorska well field on groundwater flow in the study area, in particular on the postulated upward seepage of groundwater. Four major areas of investigation were pursued: (i) verifying the available information on the vertical extent and geologic structure of the shallow Quaternary aquifer and the deeper Neogene aquifer in the area of the studied GDTE through direct (drillings) and indirect (geophysical prospecting) observations, (ii) assessing, through hydrometric observations, the water balance of the Dluga Woda stream draining the area of Wielkie Błoto fen, (iii) extensive sampling of surface water and groundwater in the study area for chemical and isotope analyses, aimed at quantifying the dynamics of water flow and tracing the postulated upward seepage of groundwater, and (iv) modelling of expected changes in groundwater flow in the study area, in response to intense pumping by the Wola Batorska well field.

The Geoprobe® direct push device (Model 420M) was used to perform vertical profiling of the Quaternary cover in the area of Wielkie Błoto, combined with sampling of water at different depths. Water samples were collected at GP1, GP3 and GP4 sites (Fig. 2). Site GP2 did not yield enough water for sampling. In addition, soil cores were collected at GP1, GP2 and GP3 sites (Fig. 4). The Geoprobe® profiling verified the position of the aquitard separating the shallow and deep aquifer in the study area. PVC screened pipes with a 2.5 inch outside diameter were installed in GP1, GP2 and GP3 for subsequent observations of water table.

In July 2014, two monitoring wells were drilled in the centre of the Wielkie Błoto fen (see Fig. 2). The well no. 54N reached the depth of 97.5 m penetrating the Quaternary cover and reaching the deeper Neogene aquifer (see Figs. 1 and 4). The second well (no. 54Q) was drilled to the depth of 8 m. Both wells were screened (see Table 1) and water samples were collected for chemical and isotope analyses. Also, measurements of hydraulic heads in both Quaternary and Neogene aquifers were made.

Geoprobe® profiling and direct drillings were supplemented by geophysical prospecting. Surface DC resistivity sounding surveys were used as a reconnaissance tool to assess the depth and thickness of clay and claystone layers separating the shallow aquifer from the deeper aquifer in the area of Wielkie Błoto fen. The vertical electrical sounding (VES) surveys with the Schlumberger array (Koefoed, 1979) were applied at 11 locations, linked to the locations of Geoprobe® profiling (Fig. 2). Quantitative interpretation of the apparent resistivity as a function of electrode spac-

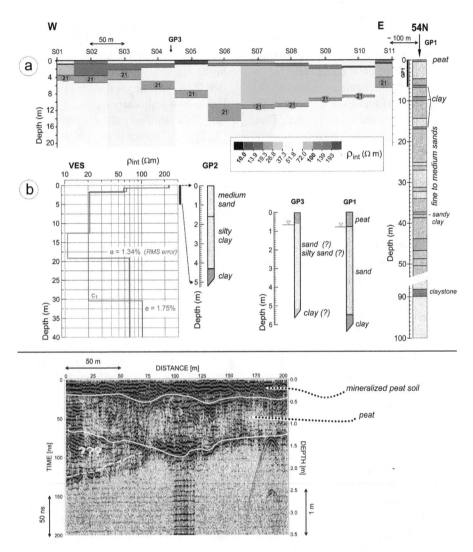

Figure 4. Upper panel: the results of vertical electrical sounding (VES). **(a)** 1-D interpreted resistivity section (S01–S11). Clay layer marked in blue. **(b)** Two variants of interpreted resistivity vertical profile based on VES sounding in the vicinity of GP 2 site. C1 corresponds to fixed resistivity of clay equal to 21 Ohmm. Also shown are the geological logs of Geoprobe® soil cores at GP1, GP2 and GP3 and the borehole drilled in the centre of the Wielkie Błoto fen (54N). Lower panel: GPR echogram along the P1 profile shown in Fig. 2 (see text for details).

ing (VES curve) was performed with the aid of RESIS and IPI2WIN software (Mościcki, 2005; Bobachev, 2010). The ground-penetrating radar (GPR) method (Daniels, 2004) was applied to assess the thickness of peat layers in the area of Wielkie Błoto fen. This method has been successfully used in the past to delineate the location of peat layers in various environments (e.g. Warner et al., 1990; Slater and Reeve, 2002; Plado et al., 2011). A short electromagnetic pulse generated by a transmitting antenna of georadar propagates in the shallow subsurface and is reflected back from the geological strata which differ in relative dielectric permittivity (ε_r), defined as the ratio of the measured dielectric permittivity to the dielectric permittivity of vacuum. Peat layers are characterized by very high ε_r values (60–75) while sandy layers show ε_r values of the order of 10–15, depending on actual water content. The GPR surveys were performed with

a ProEx System (MALA Geoscience), using offset configuration with co-polarized 250 MHz centre frequency shielded antenna. Seven separate GPR profiles with total length of approximately 1400 m were obtained (see Fig. 2).

In order to detect the hydraulic response of the aquifer to the operation of the new pumping wells located in Wola Batorska, systematic observations of the hydraulic head in well no. 32, situated ca. 1 km north of the Wola Batorska well field, were performed (see Fig. 2). Initially, the depth of water table was measured manually using a water level meter. Starting from 4 July 2012, automatic recording of the position of the water table using a pressure transducer was performed. Well no. 32 maintained artesian conditions prior to establishing the cluster of the new water-supply wells.

The network for collecting water samples for chemical and isotope analyses is shown in Fig. 2. Field procedures for hydrochemical sampling were similar to those described by Salminen et al. (2005). Unfiltered water was collected in 500 mL polyethylene bottles for major ion analysis. Filtered water was acidified using HNO_3 to $pH < 2$ and collected to new hardened polyethylene 100 mL bottles for major, minor and trace components. For pH and Eh field determinations, two laboratory-calibrated instruments were used. They were immerged in the pumped water until equilibrium was reached and a minimal difference between both instruments was recorded. Then, the mean value of both readings was taken as the accepted value. Alkalinity was measured in the field by titration method. Inductively coupled plasma mass spectrometry (ICP-MS) and other routine methods were used for determination of the chemical composition of the water samples collected during the study (exploratory boreholes, water-supply wells, Dluga Woda stream, Geoprobe® samples). Samples of water for isotope analysis were collected using established protocols. Isotope and chemical data were also obtained for the "Anna Spring" (no. 52 in Fig. 2). Groundwater appearance at this site is linked to a badly sealed borehole drilled in 1970s for seismic prospecting.

Samples for chemical and isotope analysis of the Dluga Woda stream were collected at gauge station G (Fig. 2) over the 2-year period from August 2011 to August 2013, at roughly monthly intervals (see Table 4). The purpose of this monitoring activity was the identification and quantification of the expected contribution of the upward seepage from the Bogucice Sands (Neogene) aquifer to the total discharge of the Dluga Woda stream draining the Wielkie Błoto area. Initially, both the stages and flow rates of the Dluga Woda were recorded. They provided the basis for constructing the rating curve of the stream. Subsequently, a pressure transducer was installed for continuous stream-level monitoring, starting from June 2012. A discharge hydrograph of the Dluga Woda stream was then generated for the entire observation period.

Tritium (^3H) and radiocarbon (^{14}C) concentrations in the analysed groundwater samples (water and total dissolved inorganic carbon pool, respectively) were measured at the AGH University of Science and Technology in Krakow by electrolytic enrichment followed by liquid scintillation spectrometry for tritium, and benzene synthesis followed by liquid scintillation spectrometry for ^{14}C. Tritium concentrations are reported in tritium units (TU) (1 TU corresponds to the ratio $^3H / ^1H$ equal 10^{-18}). Radiocarbon content is reported in percent of modern carbon (pMC), following recommendations of Stuiver and Polach (1977) and Mook and van der Plicht (1999). The stable isotope composition of water ($\delta^{18}O$, δ^2H) and TDIC pool ($\delta^{13}C$) was determined in the same laboratory by dual-inlet isotope-ratio mass spectrometry and reported on V-SMOW and V-PDB scales (Coplen, 1996). Typical uncertainties of ^3H, ^{14}C, $\delta^{18}O$, δ^2H and $\delta^{13}C$ analyses were of the order of 0.5 TU, 0.7 pMC, 0.1 ‰, 1 ‰ and 0.1 ‰, respectively. Dissolved carbonates in the analysed groundwater samples were precipitated in the field from ca. 60 L

of water following the established procedures (Florkowski et al., 1975; Clark and Fritz, 1997).

Chemical composition of groundwater samples collected in the recharge area was modelled using PHREEQC (Version 2.18) geochemical code (Parkhurst and Appelo, 1999). Piston-flow radiocarbon ages of groundwater in the confined part of the studied system were calculated for using NETH-PATH code (Fontes et al., 1979; Plummer et al., 1994).

The existing 3-D numerical model of the Bogucice Sands aquifer was employed to investigate the impact of Wola Batorska well field on groundwater flow in the area of Wielkie Błoto fen. The MODFLOW-2000 code for simulation of flow (Harbaugh et al., 2000) and MT3DMS code (Zengh and Wang, 1999) for modelling mass transport, both incorporated in the Visual MODFLOW 2011.1 Pro (Schlumberger Water Services, 2011) were used. The finite-difference grid consisted of five layers with 27 225 rectangular cells (250×250 m, 72 rows and 129 columns). The longitudinal dispersivity (α_L) was assumed to be 50 m. Although the selected size of computational cells did not satisfy the criterion $\Delta x < 2\alpha_L$ required for avoiding the numerical dispersion (Kinzelbach, 1986), its influence was reduced by applying the total-variation-diminishing method (TVD – Zheng and Wang, 1999; Hill and Tiedeman, 2007). The MODFLOW River Package was used to simulate the exchange of water between the aquifer and the surface water with head-dependent seepage interaction. The agreement between calculated and observed heads was satisfactory. Hydraulic heads were maintained in subsequent calibrations of the transport model with the aid of tracer data (Zuber et al., 2005; Witczak et al., 2008). In this process the hydraulic conductivity and the aquifer thickness were modified in individual grid cells, without changing adopted transmissivity values. The changes of aquifer thickness were constrained by available geological information. Three water-bearing layers were distinguished in the model: one layer in the shallow Quaternary aquifer and two layers in the deeper Neogene aquifer. Transient flow simulations were performed by the model, with quarterly pumping rates of Wola Batorska well field during the period July 2009–September 2013 and with maximum permitted capacity of 10 080 m^3 d^{-1}, starting from the end of 2014 and continuing to the end of 2029.

4   Results and discussion

4.1   Delineation of vertical structure of the Quaternary cover and the Neogene aquifer in the area of Wielkie Błoto fen

The geological structure of the Neogene (Bogucice Sands) aquifer in the area of Wielkie Błoto fen, emerging from the results of Geoprobe® profiling and exploratory drillings, is shown in Fig. 4. At site GP1 and GP3 unconsolidated sands reach the thickness of 5.5 and 5 m, respectively. Below

Table 1. Environmental tracer data for groundwater samples collected in the study area (n.m.: not measured).

Site/Well no.	Depth[a] (m)	$\delta^2 H$ (‰)	$\delta^{18}O$ (‰)	d-excess (‰)	Tritium (TU)	$\delta^{13}C_{TDIC}$ (‰)	$^{14}C_{TDIC}$ (pMC)
			Szarów				
Well no. 11	49.5–60.1	−70.3	−9.75	7.7	9.0	−14.1	64.6
Well no. 12	44.5–63.6	−70.1	−9.93	9.3	1.1	−12.8	63.6
Well no. 22	48.0–60.0	−69.4	−9.81	9.1	16.1	n.m.	n.m.
Well no. 23	33.0–50.0	−68.5	−9.84	10.2	0.7	n.m.	n.m.
Well no. 24	45.9–58.4	−72.1	−10.03	8.1	15.2	n.m.	n.m.
Well no. 42	70.0–95.0	−69.2	−9.68	8.2	<0.3	−12.2	48.5
			Wola Batorska[b]				
Well no. 44	98.0–144.0	−75.7	−10.19	5.8	<0.3	−10.2	3.2
Well no. 45	75.0–149.0	−78.3	−10.67	7.1	<0.3	n.m.	n.m.
Well no. 46	63.0–131.0	−79.9	−10.86	7.0	<0.3	−10.4	1.3
Well no. 47	69.0–132.0	−79.2	−10.89	7.9	<0.3	n.m.	n.m.
Well no. 48	79.0–131.0	−80.2	−10.83	6.4	<0.3	n.m.	n.m.
Well no. 49	72.0–146.0	−78.2	−10.71	7.5	<0.3	−9.1	3.0
Well no. 16	107.5–143.1	−69.7	−10.03	10.5	<0.3	−13.3	32.1
Well no. 32	90.9–102.0	−76.8	−10.93	10.6	<0.3	−10.6	<0.7
			Wielkie Błoto area				
GP1-A	1.6	−70.8	−10.07	9.8	8.1	n.m.	n.m.
GP1-B	2.8	−68.2	−9.65	9.0	5.4	n.m.	n.m.
GP1-C	4.6	−71.0	−10.10	9.8	0.9	n.m.	n.m.
GP3-A	1.6	−61.9	−8.83	8.7	10.1	n.m	n.m.
GP3-B	3.1	−69.3	−9.86	9.6	1.4	n.m.	n.m.
GP4-A	1.6	−64.4	−9.09	8.3	6.5	n.m.	n.m.
GP4-B	4.0	−69.6	−9.67	7.8	2.1	−14.3	57.2[c]
"Anna Spring"(no. 52)[d]	∼ 30	−67.6	−9.55	8.8	<0.3	−12.9	36.9
Well no. 54Q	2.0–6.0	−70.5	−9.90	8.7	5.8	−16.8	56.0
Well no. 54N	22.5–85.5	−71.4	−10.11	9.5	<0.3	−12.7	30.0

[a] Screen position in the production wells; maximum depth for Geoprobe® sampling; [b] isotope data reported for wells Nos. 44, 46 and 49 are arithmetic averages of the results obtained in three consecutive sampling campaigns carried out in June 2010, July 2012 and October 2013; [c] analysed using AMS technique; [d] badly sealed borehole drilled in the 1970s for seismic prospecting (see Figs. 1 and 2).

this depth, mudstones and claystones start to appear, making deeper penetration of Geoprobe® not possible. At both locations a thin layer of peat at ca. 50–70 cm was identified. The water table was located at the same depth. At GP2 site peat was absent and the mudstone layer began at only 1.8 m depth.

Interpretation of the apparent resistivity profiles from VES surveys was performed for 11 sections located near GP1, GP3 and GP4 sites. The interpreted values of resistivity (ρ_{int}) obtained on the basis of VES curves are presented in Fig. 4 in the form of depth profiles of ρ_{int} along the 11 studied sections (S01–S11, Fig. 2). The selection of VES curves was aided by additional measurements performed in the vicinity of GP2 site which allowed us to select a fixed resistivity value (21 Ω m) representing clay layers in the profile. It is worth noting that the clay layer (blue) seen in the upper panel of Fig. 4 is very thin in some places (less than ca. 1 m), with possible discontinuities facilitating hydraulic contact of deeper aquifer layers with the shallow Quaternary aquifer.

The interpreted resistivity of the strata lying above the clay layer roughly corresponds to sand with high water content. The uppermost layer is characterized by distinctly higher resistivity which can be linked to the presence of peat (see discussion below).

The monitoring wells drilled in July 2014 confirmed the results obtained from VES profiling. The simplified geological profile of well no. 54N shown in Fig. 4 revealed that the thickness of the aquitard separating the shallow Quaternary aquifer and the deeper Neogene aquifer is rather small. Three mudstone layers were identified in the profile. The thickness of the largest layer does not exceed one metre and occurs at the depth of 9 m.

The total length of the GPR profiles obtained in the area of Wielkie Błoto exceeded 1400 m. Here only one echogram representing the distance of approximately 200 m (profile P1 in Fig. 2) is discussed. Based on the data available for the soil core collected at GP1 site, electromagnetic wave velocity in

Table 2. Physico-chemical parameters of groundwater samples collected in the study area (n.m.: not measured).

Site/Well no.	Temp. (°C)	pH	SEC (μS cm^{-1})	Ca (mg L^{-1})	Mg (mg L^{-1})	Na (mg L^{-1})	K (mg L^{-1})	HCO$_3$ (mg L^{-1})	Cl (mg L^{-1})	SO$_4$ (mg L^{-1})
					Szarów					
Well no. 11	11.5	7.1	733	116	16.5	10.1	1.27	384	23.3	41.1
Well no. 12	11.5	7.1	646	107	15.8	6.91	1.19	394	7.38	21.8
Well no. 22	11.0	7.0	607	105	17.1	7.54	1.42	410	8.82	19.2
Well no. 23	11.5	7.5	906	138	17.4	20.5	1.51	340	45.1	87.3
Well no. 24	11.0	7.1	542	96.8	15.6	6.53	1.59	306	20.4	57.3
Well no. 42	11.6	7.1	n.m.	117	18.4	25.7	1.49	350	15.5	77.2
					Wola Batorska					
Well no. 44	12.7	8.7	745	6.28	1.71	170	2.67	395	28.2	<3.0
Well no. 45	12.4	8.1	780	19.2	5.13	160	4.69	448	51.0	<3.0
Well no. 46	13.2	8.3	768	16.5	3.48	153	3.27	388	49.7	<3.0
Well no. 47	14.1	8.1	824	23.6	4.98	174	3.70	436	60.4	<3.0
Well no. 48	16.8	8.1	855	16.9	3.68	178	4.43	468	28.5	8.10
Well no. 49	12.5	8.7	1152	10.3	2.67	250	4.12	429	79.8	21.0
Well no. 16	13.0	7.4	1313	80.4	16.6	160	7.28	413	200.0	19.2
Well no. 32	12.0	8.3	717	5.62	1.68	139	5.84	324	65.3	0.59
					Wielkie Błoto area					
GP1-A	11.2	6.0	352	50.0	3.72	11.7	2.21	76.3	13.9	81.1
GP1-B	11.8	7.1	899	78.4	14.7	65.8	5.44	276	57.5	112.0
GP1-C	12.3	7.6	960	46.9	13.8	123	7.92	278	119.0	68.0
GP3-A	12.3	7.2	549	97.3	9.36	8.90	2.20	221	25.9	66.9
GP3-B	15.3	8.2	1150	48.4	17.9	163	7.88	398	121.0	44.2
GP4-A	11.8	6.6	568	99.8	10.6	9.46	1.85	298	21.1	39.2
GP4-B	11.6	8.8	1054	136	32.7	51.9	8.49	473	43.3	129.0
"Anna Spring" (no. 52)	9.0	7.7	390	40.2	8.72	31.9	6.17	288	5.74	<3.0
Well no. 54Q	12.6	6.5	706	81.6	16.0	48.7	8.09	207	51.2	141
Well no. 54N	12.5	7.4	1576	15.7	16.3	276	9.56	424	244	26.0

peat was set at $v = 3.6$ cm nsec^{-1}. This value was then used to construct the depth scale of the echogram presented in the lower panel of Fig. 4. A peat layer located between ca. 0.4 and 1.2 m can be distinguished. The boundary located at approximately 0.4 m depth can be linked to degraded mineralized peat soil, also visible in the vertical cross-section shown in the lower panel of Fig. 2. Due to high attenuation of the signal, the border between sands and clays seen in the apparent resistivity profile presented in the upper panel of Fig. 4 could not be identified.

4.2 Geochemical evolution and age of groundwater in the Neogene aquifer

Table 1 summarizes environmental tracer data obtained for water samples collected in the production wells of the Bogucice Sands (Neogene) aquifer and during Geoprobe® survey of the Quaternary cover in the area of Wielkie Błoto fen. The corresponding physico-chemical parameters are summarized in Table 2.

Deuterium and oxygen-18 isotope composition of the water in the production wells located in the study area and tapping the Bogucice Sands (Neogene) aquifer is shown in Fig. 5a in δ^2H–δ^{18}O space, against the background of global and local meteoric water lines and the mean isotopic composition of modern recharge. As seen in Fig. 5a, all wells located in the eastern part of the recharge area (Szarów well field, wells Nos. 11, 12, 22, 23, 24) cluster around the mean isotopic composition of modern recharge. All of them contain tritium, testifying to the recent origin of groundwater in this area. Radiocarbon content was measured in two wells (Nos. 11, 12) and shows values around 64 pMC, in the range of radiocarbon concentrations measured in other wells located in the recharge area of the Bogucice Sands aquifer (see Fig. 1). Reduced concentrations of radiocarbon in recharge waters containing tritium result from geochemical evolution of TDIC reservoir in these waters (see e.g. Dulinski et al., 2013). Stable isotope composition of water in well no. 42, located ca. 1 km north of the Szarów well field also belongs to this cluster of points in Fig. 5a. This water is devoid of tritium and reveals reduced radiocarbon concentration (ca. 48 pMC) pointing to its pre-bomb (Holocene) age. The same applies to newly drilled monitoring well tapping the Neogene aquifer (no. 54N). Its radiocarbon content (30 pMC) reflects a gradual aging of the water along the flow lines starting in the recharge area (Szarów well field). Well no. 54Q tapping the Quaternary aquifer shows significant concentration of tritium and reduced radiocarbon content, in agreement with expectations. The stable isotope composition of waters collected during Geoprobe® survey from different levels of the Quaternary cover scatter along the local meteoric water line reflects seasonal variations of δ^2H and δ^{18}O in local precipitation (see Sect. 4.4).

The stable isotope composition of water in wells belonging to the newly established well field in Wola Batorska reveals a systematic shift towards more negative $\delta^2 H$ and $\delta^{18}O$ values, clearly indicating recharge in a colder climate (Rozanski, 1985; Zuber et al., 2004). This groundwater does not contain tritium and shows low radiocarbon content, of the order of few pMC, also suggesting a glacial age of this water (see discussion below). In well no. 32, located ca. 1 km north of the Wola Batorska well field, the radiocarbon content of TDIC reservoir drops below the detection limit (< 0.7 pMC) suggesting a significant increase in age of the groundwater, while maintaining the characteristic stable isotope signature of this water indicating recharge in a cold climate (Fig. 5). In contrast, the stable isotope composition of water in well no. 16, located ca. 2 km southwest of Wola Batorska well field (see Fig. 2) shows higher radiocarbon content (32.1 pMC), lack of tritium and stable isotope composition of water suggesting its Holocene origin. The same applies to "Anna Spring" (no. 52) located in the forest, east of Wielkie Błoto fen.

Waters in the recharge zone of the investigated part of the Bogucice Sands (Neogene) aquifer (wells Nos. 11, 12, 22, 23, 24) reveal almost neutral, uniform pH values between 7.00 and 7.13 and are dominated by HCO_3^-, Ca^{2+} and Mg^{2+} ions (Table 2). They show average TDIC content around 7.2 mmol L^{-1}. Saturation indices with respect to calcite are close to zero indicating full development of carbonate mineralization. The partial pressure of CO_2 controlling the observed carbonate chemistry calculated from the available chemical data varies between 0.018 and 0.032 atm, in agreement with partial pressures of soil CO_2 observed close to the study area (Dulinski et al., 2013). These waters contain elevated concentrations of sulfate ions, most probably originating from industrial pollution of the regional atmosphere during the second half of the 20th century.

Waters exploited by the Wola Batorska well field (wells Nos. 44–49) are dominated by HCO_3^-, Na^+ and Cl^- ions. The TDIC content is reduced by ca. 0.3 mmol L^{-1} when compared to waters from the recharge area. The Ca^{2+} and Mg^{2+} content is also reduced, while significantly higher Na^+ concentrations are recorded (Table 2). These waters reveal elevated pH values (8.07–8.82) and are supersaturated with respect to both calcite and dolomite.

The observed patterns of geochemical evolution of groundwater in the studied part of the Bogucice Sands (Neogene) aquifer reflect its marine origin. Gradual freshening of the aquifer continuing since the Miocene involves ion exchange processes between the solution and the aquifer matrix. Waters dominated by Ca^{2+} and Mg^{2+} ions, while penetrating the aquifer, exchange those ions in favour of Na^+ ions which are released to the solution. Presence of this process is supported by Fig. 6 which shows the relationship between deficit of Ca^{2+} and Mg^{2+} ions with respect to the sum of HCO_3^- and SO_4^{2-} ions and the excess of Na^+ and K^+ ions over the Cl^- ions. The data points in Fig. 6 cluster along the

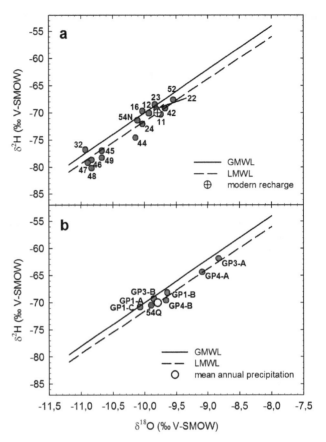

Figure 5. (a) $\delta^2 H$–$\delta^{18}O$ relationship for groundwater samples representing the Neogene aquifer, collected in the study area. Mean isotopic composition of modern recharge of the aquifer is also shown. GMWL – global meteoric water line; LMWL – local meteoric water line (monthly precipitation at Krakow station, ca. 15 km northwest of the study area, collected during the period 1975–2013). (b) $\delta^2 H$–$\delta^{18}O$ relationship for groundwater samples representing shallow Quaternary aquifer underlying the Wielkie Błoto fen (see Figs. 1, 2 and Table 1).

charge equilibrium line confirming that chemical evolution of groundwater in the investigated part of Bogucice Sands aquifer is dominated by cation exchange processes. This conclusion is supported by Piper diagram shown in Fig. 7.

Inverse modelling of radiocarbon ages of groundwater in wells Nos. 16, 32, 42, 44, 46, 49, 52 and 54N was performed using NETPATH code. First, the input solution representing the recharge area in the investigated part of Bogucice Sands (Neogene) aquifer was calculated. The calculations were based on chemical and isotope data available for wells Nos. 11, 12, 22, 23 and 24. Equal contribution of waters from those wells to the final solution was assumed. Carbon isotope parameters of the input solution were calculated as the mean values of respective parameters in individual waters contributing to the final solution, weighted by the size of carbonate reservoirs (TDIC). The calculated input carbon isotope values characterizing TDIC reservoir in the solution

Table 3. Radiocarbon piston-flow ages of groundwater in the confined zone of the investigated part of the Neogene aquifer, calculated using NETPATH code.

Well no.	Measured $\delta^{13}C_{TDIC}$ (‰ V-PDB)	Computed $\delta^{13}C_{TDIC}$ (‰ V-PDB)	Measured ^{14}C content [a] (pMC)	^{14}C age (ka)	Constraints	Phases
16	−13.3	−12.9	32.2	5	C, Ca, Mg,	calcite, dolomite,
32	−10.6	−10.6	<0.7	>36	K, Na, S, Cl	CO_2 gas, halite,
42[b]	−12.2	−12.2	48.5	2		sylvite, gypsum,
44	−10.2	−10.2	2.9	25		exchange, CH_2O,
46	−10.5	−10.7	0.8	34		Mg/Na exchange
49	−9.3	−9.3	2.2	26		
52	−12.8	−12.9	36.9	6		
54N[c]	−12.7	−12.7	30.0	6		

[a] Carbon isotope and chemical analyses of water samples in collected in wells Nos. 44, 46 and 49 in June 2010 were used for NETPATH inverse calculations and determination of radiocarbon ages; [b] isotope exchange between solid carbonates and water solution to reconcile the computed $\delta^{13}C$ of TDIC with observed value (0.2–0.3 mmol L^{-1} of exchanged carbon) was required only for well no. 42; [c] carbon isotope and chemical analyses of water sample collected in July 2014 were used for NETPATH inverse calculations and determination of radiocarbon ages.

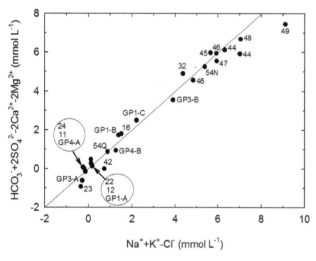

Figure 6. The relationship between deficit of Ca^{2+} and Mg^{2+} ions with respect to the sum of HCO_3^- and SO_4^{2-} ions and the excess of Na^+ and K^+ ions over the Cl^- ions in groundwater samples representing the part of the Neogene aquifer located in the study area and the shallow Quaternary aquifer in the area of Wielkie Błoto fen (see Table 2).

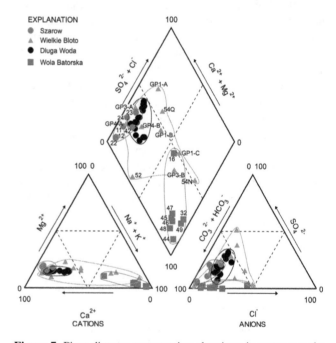

Figure 7. Piper diagram representing chemistry in water samples collected in the study area.

were $\delta^{13}C = -13.4‰$, $^{14}C = 64.1$ pMC. The solution determined in this way was then used as the initial solution in inverse calculations using NETPATH code. Parameters used in the calculations (constraints and phases) and the resulting radiocarbon ages are summarized in Table 3. The calculated radiocarbon ages vary from ca. 2 ka for water in well no. 42 located close to the recharge area of the aquifer, up to the age in excess of ca. 36 ka for well no. 32 located in most distant, northern part of the aquifer. Radiocarbon ages of three wells representing Wola Batorska well field (Nos. 44, 46 and 49) reveal radiocarbon ages between 25 and 34 ka, confirming the glacial origin of water in this well field, already apparent from the stable isotope data presented in Fig. 5. Well

no. 54N, the "Anna Spring" (no. 52) and well no. 16 reveal mid-Holocene groundwater ages.

It is apparent from the above discussion that groundwater which eventually penetrates the confining layer and reaches the shallow Quaternary aquifer in the area of Wielkie Błoto fen should have distinct chemical and isotopic characteristics. In particular, it should be characterized by reduced Ca^{2+} and Mg^{2+} and elevated Na^+ content when compared to young groundwater present in the Quaternary cover. It should also have elevated pH values (around 8). This water does not contain tritium and is of Holocene age. The Holocene age of this water implies that its stable isotope signature will be

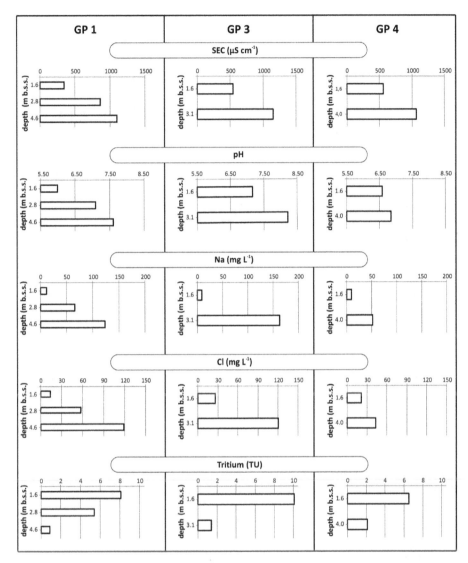

Figure 8. Depth stratification of pH, conductivity, Cl, Na and tritium content in the shallow Quaternary aquifer underlying the Wielkie Błoto fen and adjacent parts of the Niepołomice Forest. Water samples were collected with the aid of Geoprobe® device. Location of sampling sites (GP1, GP3, GP4) is shown in Fig. 2.

non-distinguishable from the mean isotopic composition of present-day precipitation in the area.

4.3　Isotope and chemical stratification of shallow Quaternary aquifer in the area of Wielkie Błoto fen

To investigate the isotopic and chemical stratification of groundwater in the shallow Quaternary aquifer underlying Wielkie Błoto fen, and to detect the eventual contribution coming from the deeper Neogene aquifer, a dedicated sampling campaign using Geoprobe® device was carried out in October 2011. Isotope and chemical data obtained for water samples collected during this campaign are summarized in Tables 1 and 2, respectively.

As seen in Fig. 5b and Table 1, stable isotope composition of Geoprobe® water samples varies significantly with depth

and location. The deepest points cluster around the mean isotopic composition of local precipitation suggesting that the observed variability of δ^2H and $\delta^{18}O$ in the upper portions of the profiles stems from strong seasonality of δ^2H and $\delta^{18}O$ signal in local precipitation, surviving in the upper part of the Quaternary cover and converging towards the mean isotopic signature of the local precipitation at the bottom of this cover.

Vertical profiles of tritium content and selected chemical parameters are summarized in Fig. 8. A distinct reduction of tritium content with depth, accompanied by increase of pH, conductivity and concentration of major ions (Cl and Na) is apparent. The observed increase of pH, conductivity and concentration of major ions (Cl and Na) with depth in the shallow Quaternary aquifer, accompanied by reduction of tritium content, strongly suggest that upward seepage

Table 4. Physico-chemical parameters of Dluga Woda stream monitored on a monthly basis during the period August 2011–August 2013 (n.m.: not measured).

Date	Flow rate ($L\,s^{-1}$)	$\delta^2 H$ (‰)	$\delta^{18}O$ (‰)	Tritium content (TU)	SEC ($\mu S\,cm^{-1}$)	pH	Cl ($mg\,L^{-1}$)	Na ($mg\,L^{-1}$)
16 Jul 2011	38.5	n.m.	n.m.	n.m.	562	7.75	25.1	17.2
29 Aug 2011	10.1	−62.7	−8.70	5.9	559	7.86	33.2	31.9
25 Sep 2011	1.0	n.m.	n.m.	n.m.	635	7.94	44.1	33.1
29 Oct 2011	12.4	−62.5	−8.81	9.1	628	7.87	46.5	41.8
29 Nov 2011	36.7	−65.9	−9.17	7.7	599	7.75	34.8	18.1
30 Dec 2011	46.7	−61.9	−9.21	7.6	548	7.72	25.2	17.1
28 Jan 2012	23.1	−68.4	−9.76	9.0	628	7.23	35.0	17.8
29 Feb 2012	167	−68.7	−9.78	7.8	456	6.84	24.9	11.5
31 Mar 2012	97.5	−65.8	−9.12	8.2	506	7.41	46.3	16.7
30 Apr 2012	37.3	−64.7	−8.57	6.2	560	7.59	31.9	19.4
30 May 2012	5.0	−62.7	−8.17	4.0	681	7.79	50.6	31.7
22 Jun 2012	14.2	−55.5	−7.72	8.1	556	7.76	41.5	35.1
30 Jul 2012	73.1	−46.7	−6.92	9.4	460	7.57	23.8	12.2
31 Aug 2012	1.5	−56.7	−7.48	5.0	768	8.27	54.0	18.1
28 Sep 2012	3.4	−55.5	−7.54	5.4	736	8.12	53.4	44.9
30 Oct 2012	79.3	−65.9	−9.27	6.3	566	7.62	33.0	15.2
29 Nov2012	42.2	−63.8	−9.01	5.4	573	7.75	30.1	20.5
29 Dec 2012	89.4	−76.2	−9.39	7.0	567	7.48	33.6	17.2
06 Mar 2013	128.3	−70.2	−9.97	5.7	501	7.44	31.2	n.m.
20 Apr 2013	71.3	−71.7	−10.01	6.2	505	7.69	24.2	17.0
08 Jun 2013	180.7	−67.6	−9.62	8.6	356	7.02	13.5	15.4
06 Jul 2013	49	−68.0	−9.20	6.1	495	7.62	26.3	18.8
09 Aug 2013	1.2	−63.0	−8.26	5.6	692	8.00	51.7	44.3

of groundwater from deeper, confined Neogene aquifer indeed takes place in the area of Wielkie Błoto (Fig. 8). The chemical data of Geoprobe® water samples are also plotted in Figs. 6 and 7. They are consistent with the geochemical evolution of groundwater in the Neogene (Bogucice Sands) aquifer, discussed in the previous section.

4.4 Water balance of the Dluga Woda catchment

The catchment of Dluga Woda stream comprises Wielkie Błoto fen and adjacent parts of the Niepolomice Forest (see Fig. 2). Physico-chemical parameters of the stream water (flow rate, temperature, pH, major ions, stable isotopes of water and tritium content) were monitored on a monthly basis over the 2-year period (August 2011–August 2013) with the main aim of detecting and quantifying the possible contribution of groundwater seeping from the deeper, confined aquifer to the shallow aquifer, in the total discharge of the Dluga Woda stream. The results are summarized in Table 4. Chemical data for the Dluga Woda stream are shown also on a Piper diagram (Fig. 7).

Figure 9 shows temporal variations of $\delta^{18}O$ and tritium content in the Dluga Woda stream presented against the background of seasonal variability of those parameters in local monthly precipitation. It is apparent from Fig. 9a that

the strong seasonality of $\delta^{18}O$ in precipitation survives during transport through the watershed and is visible in the Dluga Woda $\delta^{18}O$ record. However, the amplitude of seasonal changes of $\delta^{18}O$ is significantly reduced: from approximately 5‰ seen in precipitation to ca. 1.5‰ in the stream water. Maloszewski et al. (1983) have shown that the mean transit time of purely sinusoidal isotope input signal through a hydrological system characterized by an exponential distribution of transit times can be expressed by the following equation:

$$MTT = \frac{1}{2\pi}\sqrt{\left(\frac{A_{in}}{A_{out}}\right)^2 - 1},\qquad(1)$$

where "MTT" is the mean transit time of water (in years), and A_{in} and A_{out} are the amplitudes of input and output isotope signals. The assumption about exponential distribution of transit times of water seems to be adequate to describe transport of precipitation input through a watershed. If the observed amplitudes of the input (precipitation) and output (stream) $\delta^{18}O$ curves are inserted into Eq. (1) the resulting mean transit time of water through the catchment of Dluga Woda stream, relatively to the river section used for sampling water, is 3.2 months.

Figure 9b shows the tritium content in local precipitation and in the Dluga Woda stream during the observation period.

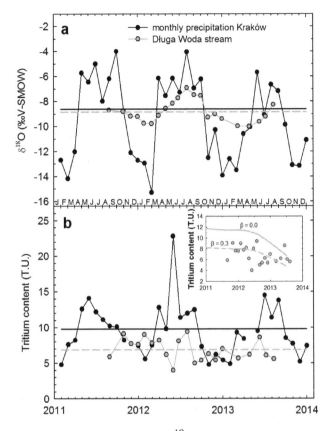

Figure 9. Seasonal variations of $\delta^{18}O$ (**a**) and tritium content (**b**) in the Dluga Woda stream during the period 2011–2013. The inset in (**b**) shows the comparison of modelled and measured tritium concentrations in the Dluga Woda stream. β is the fraction of tritium-free component in the total flow of the stream (see text for details).

The weighted mean tritium concentration in precipitation (9.8 TU) appears to be significantly higher than that of the Dluga Woda stream water (6.9 TU). Assuming that the total discharge of Dluga Woda is composed of a fast (MTT ca. 3.2 months) and a slow component devoid of tritium, the contribution of this old component can be easily assessed from first-order calculations based on tritium balance and is equal to approximately 30 %. A more appropriate approach based on lumped-parameter modelling (Maloszewski and Zuber, 1996) of tritium transport through the watershed of Dluga Woda stream confirms this rough assessment. The inset of Fig. 9b shows the results of lumped-parameter modelling of tritium record in the Dluga Woda stream using the following prescribed parameters: (i) the mean transit time of water containing tritium in the catchment equal to 3.2 months, (ii) exponential distribution of transit times, and (iii) the contribution of tritium-free component in the Dluga Woda discharge equal to zero and 30 %, respectively ($\beta = 0.0$ and 0.3 in the inset figure, where β is the fraction of tritium-free component in the total flow of the stream). It is obvious that the assumption of 30 % contribution of tritium-free component in the total discharge of Dluga Woda stream fits the experimen-

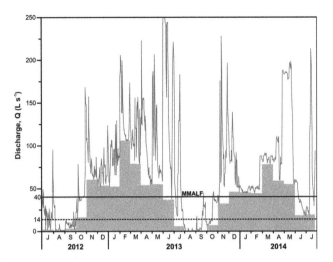

Figure 10. Hydrograph of Dluga Woda stream. Grey histogram reflects monthly low flows for the period July 2012–June 2014. The mean monthly annually low flow (MMALF) equal to $40\,L\,s^{-1}$ corresponds to the baseflow of the stream. The characteristic discharge rate of $14\,L\,s^{-1}$ (see Fig. 11) is marked by dashed line (see text).

tal data much better than the case neglecting this component. The contribution of tritium-free component also explains the difference between weighted mean $\delta^{18}O$ in precipitation for the period January 2011–December 2013 (−8.61 ‰) and the mean ^{18}O content of the Dluga Woda stream (−8.84 ‰) seen in Fig. 9a. Mass-balance calculations based on $\delta^{18}O$ data yield the contribution of the old component of the order of 20 %, assuming that its ^{18}O content is represented by the arithmetic mean of $\delta^{18}O$ values available for "Anna Spring" and wells Nos. 16, 42 and 54Q, all characterized by Holocene ages of groundwater.

A hydrograph of the Dluga Woda stream constructed for the period July 2012–July 2014 is shown in Fig. 10. It reveals large variability of the flow rate. The measured values varied from ca. $1.5\,L\,s^{-1}$ (31 August 2012) to $180\,L\,s^{-1}$ (8 June 2013) (see Table 4), while the corresponding values derived from the rating curve were equal to $0.8\,L\,s^{-1}$ (26 August 2012) and over $250\,L\,s^{-1}$ (26 June 2013). Persisting low flows during August and September 2012 and 2013 resulted from lower than normal precipitation in the preceding months.

The hydrograph presented in Fig. 10 allows quantitative assessment of the Dluga Woda baseflow. It was derived as the mean monthly annual low flow (MMALF) according to Wundt (1953). "Low flow" defines the lowest flow during the given month. MMALF reflects the discharge of the aquifer and may represent the annual baseflow in the river catchment. The MMALF value for the period July 2012–July 2014 was equal to $40\,L\,s^{-1}$. For the Dluga Woda catchment, with surface area of $8.2\,km^2$, this flow rate corresponds to an annual baseflow equal to $154\,mm$ (ca. 21 % of annual precipitation rate).

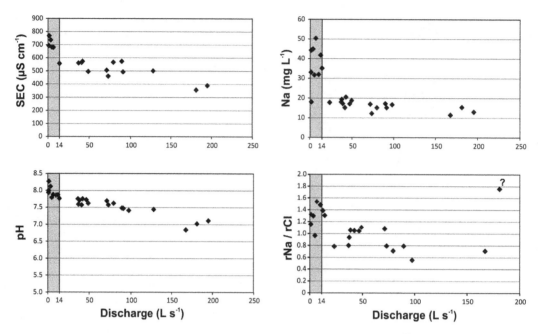

Figure 11. Electrical conductivity, pH, Na content and Na / Cl molar ratio in the Dluga Woda stream observed at monthly intervals during the period July 2011–June 2013, as a function of stream discharge rate measured at gauge station G (see Fig. 2).

Large fluctuations of Dluga Woda discharge rates are accompanied by substantial variability of the physico-chemical parameters of the stream water (see Table 4 and Fig. 11). The relationships between SEC, pH, Na content, Na / Cl molar ratio and the discharge rate of Dluga Woda shown in Fig. 11 clearly indicate that for the flow rates lower than ca. $14\,L\,s^{-1}$ the physico-chemical parameters of water attain distinct values (SEC $> 600\,\mu S\,cm^{-1}$; pH > 7.8; Na $> 30\,mg\,L^{-1}$, Na / Cl ratio higher than 1.3) not observed for higher flow rates. In addition, these low flow rates are accompanied by low tritium contents in the stream water. High pH values and high Na / Cl molar ratios in groundwater are typical for gradual freshening of sediments deposited in marine environment (Appelo and Postma, 2005). This strongly suggests that discharge of Dluga Woda stream at very low flow rates (ca. $< 14\,L\,s^{-1}$) carries significant contribution of waters seeping through clayey sediments separating water-bearing layers of the Neogene aquifer from the Quaternary shallow phreatic aquifer. Note that the flow rate of the order of $14\,L\,s^{-1}$ constitutes approximately 30 % of the MMALF value of $40\,L\,s^{-1}$, remarkably close to the percentage contribution of the tritium-free component in the total discharge of Dluga Woda derived from tritium data.

4.5   3-D flow and transport modelling of groundwater flow in the area of Wielkie Błoto fen

The 3-D flow and transport model of the entire Bogucice Sands (Neogene) aquifer was calibrated with the aid of environmental tracer data (Zuber et al., 2005; Witczak et al., 2008). In the framework of the presented study this model

was used to simulate the response of regional flow field to groundwater abstraction by the newly established Wola Batorska well field.

Figure 12 summarizes the measurements of the position of hydraulic head in well no. 32 located 1075 m north of the centre of Wola Batorska well field (see Fig. 2). The hydraulic head in this well changed radically after groundwater abstraction was initiated in July 2009. Initially slightly artesian, it stabilized at around 14 m below the surface after four years of operation of the new well field. Figure 12 also shows the changes of hydraulic head in well no. 32 simulated with the aid of a 3-D flow model forced by quarterly mean pumping rates of the entire well field. The agreement between modelled and observed evolution of the hydraulic head is satisfactory, particularly in the second part of the observation period.

The ratio of transmissivity to specific storage is a measure of the ability to transmit differences in hydraulic heads by groundwater systems (Alley et al., 2002, Sophocleous, 2012). The response of confined aquifers to changes in groundwater abstraction rates is relatively fast. The characteristic timescale of this response can be assessed using the approximate expression proposed by Alley et al. (2002):

$$T^* = \frac{S_S \cdot L_C^2}{K},$$ (2)

where T^* is the hydraulic response time for the basin (in days), S_S is specific storage (m^{-1}), L_C is characteristic length (m) of the basin and K is hydraulic conductivity $(m\,d^{-1})$. The response time of horizontal flow in the Bogucice Sands (Neogene) aquifer between the Wola Batorska

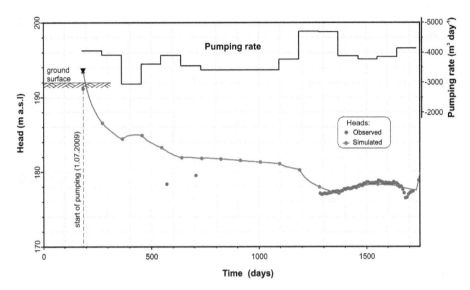

Figure 12. Changes of the hydraulic head in well no. 32 (see Fig. 2) after initialization of the operation of Wola Batorska well field in July 2009.

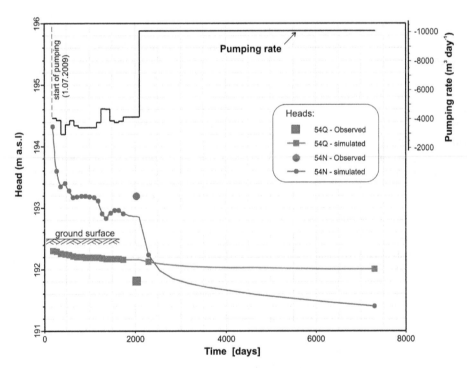

Figure 13. Changes of hydraulic heads in the shallow Quaternary and deeper Neogene aquifers in observation wells (54Q and 54N, respectively, see Fig. 2) simulated in the centre of the Wielkie Błoto fen. Pumping rate of Wola Batorska well field from the start in July 2009 till October 2014 was simulated as actual abstraction. The later part of the diagram shows future levels of groundwater abstraction with maximum permitted capacity (10 080 m^3 day^{-1}).

well field and the observation well (no. 32) was assessed using Eq. (2). With specific storage $S_S = 2.5 \times 10^{-5}$ m^{-1}, derived from fitting of the measurement data shown in Fig. 12, the characteristic length L_C equal to 1075 m and the hydraulic conductivity K set at 0.8 m d^{-1}, the first-order assessment of the hydraulic response time of Bogucice Sands

aquifer in the vicinity of Wola Batorska well field leads to a T^* value equal to approximately 36 days.

The assessment of the impact of groundwater abstraction by the Wola Batorska well field to hydraulic head changes in well no. 32 allowed us to calibrate the initially steady-state flow model to transient conditions during the first 4 years of well field operation. Further, such a calibrated transient model allowed us to assess the expected lowering of the hydraulic heads in the Wielkie Błoto fen area and to quantify changes of the upward seepage of groundwater from the deeper Neogene confined aquifer to the shallow Quaternary aquifer and the Dluga Woda stream draining the Wielkie Błoto area. Figure 13 shows the expected changes of hydraulic head in Quaternary and Neogene aquifers with respect to the observation wells (54Q and 54N, respectively), simulated in the centre of the Wielkie Błoto fen. The simulation takes into account both the actual discharge of the Wola Batorska well field (2009–2014) and the prognosis of maximum allowed pumping rate ($10\,080\,\text{m}^3\,\text{d}^{-1}$) for the next 15 years, from the end of 2014 to the end of 2029.

Lowering of the simulated hydraulic head in both aquifers was generally confirmed by the observations in two monitoring wells (54Q and 54N), starting from July 2014. The difference between simulated and observed heads is of the order of 30 cm. The monitoring of hydraulic heads in both aquifers in the years to come will provide the basis for refinement of modelling results.

According to model output, the expected shortage of groundwater flow to the fen and the Dluga Woda stream depends strongly on adopted scenarios of expected pumping rates. Before initialization of groundwater exploitation by the Wola Batorska well field, the simulated upward seepage from the deeper confined aquifer to the shallow Quaternary aquifer was of the order of $441\,\text{m}^3\,\text{d}^{-1}$. For the scenario with maximum permitted pumping capacity of the Wola Batorska well field ($10\,080\,\text{m}^3\,\text{d}^{-1}$) maintained from the end of 2014 to 2029, this upward seepage will reverse to downward flow of approximately $465\,\text{m}^3\,\text{d}^{-1}$ by the end of the simulation period. This means that the overall change will reach $906\,\text{m}^3\,\text{d}^{-1}$ ($10.5\,\text{L}\,\text{s}^{-1}$) in Dluga Woda outflow. It should be noted that simulations were run for mean yearly conditions. During low flow conditions occurring in the summer months (see Fig. 10) such drop of Dluga Woda discharge may lead to temporal disappearance of stream flow. By the end of the simulation period, the expected drop of water table in the centre of Wielkie Błoto fen (see Fig. 2) will be approximately 30 cm (from 192.31 to 192.00 m a.s.l.; see Fig. 13). It may be different for the wetland area. The climatic changes envisaged up to the end of 2029 were not considered in the simulation runs.

The impact of groundwater abstraction by the Wola Batorska well field, with the current mean pumping rate of ca. $3800\,\text{m}^3\,\text{d}^{-1}$, is already seen in the flow field of the deeper Neogene aquifer. Although the hydraulic head has dropped by about 1.5 m (from 194.33 to 192.88 m a.s.l.) since the beginning of pumping in July 2009 up to July 2014 when monitoring wells Nos. 54Q and 54N were drilled, the artesian conditions in the deeper confined aquifer have been maintained (Fig. 13). This will, however, change in future when more intense abstraction of groundwater takes place. As shown in Fig. 13, the simulated hydraulic heads of both aquifers will

be equal to approximately 310 days (0.85 years) after beginning of the exploitation of Wola Batorska well field with the maximum permitted capacity of $10\,080\,\text{m}^3\,\text{d}^{-1}$. At the end of the simulation period (end of 2029) the hydraulic head in the deeper aquifer will be about 0.6 m lower than in the shallow aquifer.

5   Conclusions

Steadily growing anthropogenic pressure on groundwater resources, with respect to both their quality and quantity, creates a conflict situation between nature and man in their competition for clean and safe sources of water. It is often forgotten that groundwater-dependent ecosystems have important functions in all climatic zones as they contribute to biological and landscape diversity and provide important economic and social services. The presented study has demonstrated that isotope and geochemical tools combined with 3-D flow and transport modelling may help to answer important questions related to the functioning of groundwater-dependent ecosystems and their interaction with the associated aquifers.

In the context of the presented study environmental tracers appeared to be particularly useful in quantifying timescales of groundwater flow through various parts of the Bogucice Sands aquifer, including its Quaternary cover. Environmental tracer data (tritium, stable isotopes of water) and physico-chemical parameters of groundwater and surface water in the study area provide strong collective evidence for upward seepage of groundwater from the deeper Neogene aquifer to the shallow Quaternary aquifer supporting the studied GDTE (Niepolomice forest and Wielkie Błoto fen).

Simulations of groundwater flow field with the aid of 3-D flow and the transport model developed for the studied aquifer and calibrated using environmental tracer data, strongly suggest that prolonged groundwater abstraction through the newly established cluster of water-supply wells at maximum permitted capacity (ca. $10\,000\,\text{m}^3\text{d}^{-1}$) represents a significant risk to the studied GDTE. It may lead to reorganization of groundwater flow field in the study area and a significant drop in the water table, leading to degradation of this valuable groundwater-dependent ecosystem in the near future.

Acknowledgements. The study was supported by the GENESIS project funded by the European Commission 7FP (project contract 226536) and by statutory funds of the AGH University of Science and Technology (project nos. 11.11.220.01 and 11.11.140.026).

Edited by: C. Stumpp

References

Alley, W. M., Healy, R. W., LaBaugh, J. W., and Reilly, T. E.: Flow and storage in groundwater systems, Science, 296, 1985–1990, 2002.

Appelo, C. A. J. and Postma, D.: Geochemistry, Groundwater and Pollution, 2nd Edn., A. A. Balkema Publishers, Amsterdam, the Netherlands, 649 pp., 2005.

Bertrand, G., Goldscheider, N., Gobat, J.-M. and Hunkeler, D.: Review: From multi-scale conceptualization to a classification system for inland groundwater-dependent ecosystems, Hydrogeol. J., 20, 5–25, 2012.

Bobachev, A.: Resistivity Sounding Interpretation – IPI2Win, Moscow State University, 2010.

Boulton, A. J.: Chances and challenges in the conservation of groundwaters and their dependent ecosystems, Aquat. Conserv., 15, 319–323, 2005.

Chełmicki, W., Ciszewski, S., and Żelazny, M.: Reconstructing groundwater level fluctuations in the 20th century in the forested catchment of Drwinka (Niepołomice Forest, S. Poland), in: Interdisciplinary Approaches in Small Catchment Hydrology: Monitoring and Research, edited by: Holko, L. and Miklanek, P., Proceedings of the 9th ERB Conference, Demanovska dolina, Slovakia, 25–28 September 2002, IHP Technical Documents in Hydrology, UNESCO, Paris, 67, 203–208, 2003.

Clark, I. D. and Fritz, P.: Environmental Isotopes in Hydrogeology, Lewis Publishers, New York, USA, 331 pp., 1997.

Coplen, T.: New guidelines for reporting stable hydrogen, carbon and oxygen isotope-ratio data, Geochim. Cosmochim. Ac., 60, 3359–3360, 1996.

Daniels, D. J.: Ground Penetrating Radar, 2nd Edn., The Institution of Electrical Engineers, London, UK, 726 pp., 2004.

Dresel, P. E., Clark, R., Cheng, X., Reid, M., Terry, A., Fawcett, J., and Cochrane, D.: Mapping Terrestrial Groundwater Dependent Ecosystems: Method Development and Example Output, Department of Primary Industries, Melbourne, Australia, 66 pp., 2010.

Dulinski, M., Rozanski, K., Kuc, T., Gorczyca, Z., Kania, J., and Kapusta, M.: Evolution of radiocarbon in a sandy aquifer across large temporal and spatial scales: case study from southern Poland, Radiocarbon, 55, 905–919, 2013.

EC: Directive 2000/60/EC of the European Parliament and of the Council establishing a framework for Community action in the field of water policy, OJ L 327, Office for Official Publications of the European Communities, Luxembourg, 2000.

EC: Common Implementation Strategy for the Water Framework Directive (2000/60/EC). The role of wetlands in the Water Framework Directive, Guidance document No. 12, Office for Official Publications of the European Communities, Luxembourg, 2003.

EC: Directive 2006/118/EC of the European Parliament and of the Council on the protection of groundwater against pollution and deterioration, OJ L 372, Office for Official Publications of the European Communities, Luxembourg, 2006.

EC: Common Implementation Strategy for the Water Framework Directive, Guidance on Risk Assessment and the Use of Conceptual Models for Groundwater, Guidance document No. 26, Office for Official Publications of the European Communities, Luxembourg, 2010.

EC: Common Implementation Strategy for the Water Framework Directive, Technical Report on Groundwater Dependent Ecosystems, Technical Report 6, Office for Official Publications of the European Communities, Luxembourg, 2011.

Florkowski, T., Grabczak, J., Kuc, T., and Rozanski, K.: Determination of radiocarbon in water by gas or liquid scintillation counting, Nukleonika, 20, 1053–1062, 1975.

Fontes, J. C. and Garnier, J. M.: Determination of the initial ^{14}C activity of total dissolved carbon: a review of existing models and a new approach, Water Resour. Res., 15, 399–413, 1979.

Forest Management Manual, The State Forests National Forest Holding, Warsaw, Poland, 655pp., 2012 (in Polish).

GENESIS: Deliverable 4.3: New indicators for assessing GDE vulnerability, available at:www.thegenesisproject.eu (last access: 15 December 2014), 2012.

Górka, J., Reczek, D., Gontarz, Ż, and Szklarz, K.: Annex to the Project Documenting Disposable Reserves of Groundwater and Delineating Protection Zones of Bogucice Sands Aquifer (GZWP 451), SEGI-AT Sp. z o.o., Warszawa, Poland, 2010 (in Polish).

Harbaugh, A. W., Banta, E. R., Hill, M. C., and McDonald, M. G.: MODFLOW-2000, the US Geological Survey Modular Ground-Water Model – User Guide to Modularization Concepts and the Ground-Water Flow Process: US Geological Survey Open-File Report 00-92, Reston, Virginia, USA, 2000.

Hatton, T. and Evans, R.: Dependence of Ecosystems on Groundwater and its Significance to Australia, LWRRDC Occasional Paper No 12/98, Canberra, Australia, 1998.

Hill, M. C. and Tiedeman, C. R.: Effective Groundwater Model Calibration With Analysis of Data, Sensitivities, Predictions, and Uncertainty, John Wiley and Sons Inc., Hoboken, NJ, USA, 480pp., 2007.

Hose, G. C., Bailey, J., Stumpp, C., and Fryirs, K.: Groundwater depth and topography correlate with vegetation structure of an upland peat swamp, Budderoo Plateau, NSW, Australia, Ecohydrol., 7, 1392–1402, 2014.

Kinzelbach, W.: Groundwater Modeling: An Introduction with Sample Programs in BASIC. Elsevier Science Publishers B. V., Amsterdam, the Netherlands, 334 pp., 1986.

Kleczkowski, A. S. (Ed.): The Map of the Critical Protection Areas (CPA) of the Major Groundwater Basins (MGWB) in Poland, Institute of Hydrogeology and Engineering Geology, Academy of Mining and Metallurgy, Krakow, Poland, 44 pp., 1990.

Kløve, B., Ala-aho, P., Bertrand, G., Boukalova, Z., Ertürk, A., Goldscheider, N., Ilmonen, J., Karakaya, N., Kupfersberger, H., Kværner, J., Lundberg, A., Mileusnic' , M., Moszczynska, A., Muotka, T., Preda, E., Rossi, P., Siergieiev, D., Šimek, J., Wachniew, P., and Widerlund, A.: Groundwater dependent ecosystems: Part I – Hydroecology, threats and status of ecosystems, Environ. Sci. Pollut. R., 14, 770–781, 2011a.

Kløve, B., Ala-aho, P., Allan, A., Bertrand, G., Druzynska, E., Ertürk, A., Goldscheider, N., Henry, S., Karakaya, N., Karjalainen, T.P., Koundouri, P., Kværner, J., Lundberg, A. Muotka, T., Preda, E., Pulido Velázquez, M., and Schipper, P.: Groundwater dependent ecosystems: Part II – ecosystem services and management under risk of climate change and land-use management, Environ. Sci. Pollut. R., 14, 782–793, 2011b.

Kløve, B., Ala-Aho, P., Bertrand, G., Gurdak, J. J., Kupfersberger, H., Kværner, J., Muotka T., Mykrä, H., Preda, E., Rossi, P., Bertacchi Uvo, C., Velasco, E., and Pulido-Velazquez, M.: Climate change impacts on groundwater and dependent ecosystems, J. Hydrol., 518, 250–266, 2014.

Koefoed, O.: Geosounding Principles, 1: Resistivity Sounding Measurements, Elsevier, Amsterdam Oxford New York, 276 pp., 1979.

Lipka, K.: The Wielkie Błoto Peat Bog in the Niepołomice Forest Near Szarów, Przewodnik LX Zjazdu PTG, Kraków, Poland, 143–146, 1989 (in Polish).

Lipka, K., Zając, E., and Zarzycki, J.: Course of plant succession in the post-harvest and post-fire areas of the Wielkie Bloto fen in the Niepolomice Primeveal Forest, Acta Agrophysica 7, 433–438, 2006.

Łajczak, A.: Geomorphological and hydrographic characterization of the "Royal Fern" nature reserve in the Niepolomice Forest, Ochrona Przyrody, 54, 81–90, 1997 (in Polish).

Małoszewski, P. and Zuber, A.: Lumped parameter models for the interpretation of environmental tracer data, in: Manual on Mathematical Models in Isotope Hydrogeology, IAEA-TECDOC-910, International Atomic Energy Agency, Vienna, Austria, 9–58, 1996.

Małoszewski, P., Rauert, W., Stichler, W., and Herrmann, A.: Application of flow models in an Alpine catchment area using tritium and deuterium data, J. Hydrol., 66, 319–330, 1983.

Mościcki, J.: Characterization of near-surface sediments based on DC resistivity soundings in the Starunia area, fore-Carpathian region, in: Ukraine: Polish and Ukrainian Geological Studies (2004–2005) at Starunia – the Area of Discoveries of Woolly Rhinoceroses, Polish Geological Institute and Society of Research on Environmental Changes "Geosphere", Warszawa-Kraków, Poland, 103–114, 2005.

Mook, W. G. and van der Plicht, J.: Reporting ^{14}C activities and concentrations, Radiocarbon, 41, 227–239, 1999.

Parkhurst, D. L. and Appelo, C. A. J.: User's Guide to PHREEQC (Version 2) – A Computer Program for Speciation, Batch-Reaction, One-Dimensional Transport, and Inverse Geochemical Calculations, Water-Resources Investigations Report no. 99-4259, USGS, Reston, Virginia, USA, 1999.

Pettit, N. E., Edwards, T., Boyd, T. C., and Froend, R. H.: Ecological Water Requirement (Interim) Framework Development, A conceptual framework for the maintenance of groundwater dependent ecosystems using state and transition modelling, Centre for Ecosystem Management, Report 2007-14, ECU Joondalup, Australia, 2007.

Plado, J., Sibul, I., Mustasaar, M., and Jõeleht, A.: Ground-penetrating radar study of the Rahivere peat bog, eastern Estonia, Est. J. Earth Sci., 60, 31–42, 2011.

Plummer, L. N., Prestemon, E. C., and Parkhurst, D. L.: An Interactive Code (NETPATH) for Modeling NET Geochemical Reactions Along a Flow PATH, Version 2.0, Water-Resources Investigations Report no. 94-4169, USGS, Reston, Virginia, USA, 1994.

Porebski, S. and Oszczypko, N.: Lithofacies and origin of the Bogucice Sands (Upper Badenian), Carpathian Foredeep, Proceedings of Polish Geological Institute, CLXVIII, 57–82, 1999 (in Polish).

Rozanski, K.: Deuterium and oxygen-18 in European groundwaters – link to atmospheric circulation in the past, Chem. Geol., 52, 349–363, 1985.

Salminen, R. (Ed.): Geochemical Atlas of Europe – Part 1: Background Information, Methodology and Maps, Geological Survey of Finland, Espoo, Finland, 2005.

Schaffers, A. P. and Sýkora, K. V.: Reliability of Ellenberg indicator values for moisture, nitrogen and soil reaction: a comparison with field measurements, J. Veg. Sci., 11, 225–244, 2000.

Schlumberger Water Services: Visual MODFLOW 2011.1 User's Manual; For Professional Applications in Three-Dimensional Groundwater Flow and Contaminant Transport Modeling, Schlumberger Water Services, Kitchener, Ontario, Canada, 2011.

Sinclair Knight Merz Pty Ltd.: Environmental Water Requirements of Groundwater Dependent Ecosystems, Environmental Flows Initiative Technical Report Number 2, Commonwealth of Australia, Canberra, 2001.

Slater, L. D. and Reeve, A.: Investigating peat stratigraphy and hydrology using integrated electrical geophysics, Geophysics, 67, 365–378, 2002.

Sophocleous, M.: On understanding and predicting groundwater response time, Ground Water, 50, 528–540, 2012.

Stuiver, M. and Polach, H.: Discussion: reporting of 14C data. Radiocarbon, 22, 355–363, 1977.

Wachniew, P., Witczak, S., Postawa, A., Kania, J., Żurek, A., Różański, K., and Duliński, M.: Groundwater dependent ecosystems and man: conflicting groundwater uses, Geol. Q., 58, 595–706, 2014.

Warner, B. G., Nobes, D. C., and Theimer, B. D.: An application of ground penetrating radar to peat stratigraphy of Ellice Swamp, southwestern Ontario, Can. J. Earth Sci., 27, 932–938, 1990.

Witczak, S., Zuber, A., Kmiecik, E., Kania, J., Szczepańska, J., and Różański, K.: Tracer based study of the Badenian Bogucice Sands aquifer, Poland, in: Natural Groundwater Quality, edited by: Edmunds, W. M. and Shand, P., Blackwell Publishing, Malden, MA, USA, 335–352, 2008.

Wundt, W.: Gewässerkunde, Springer-Verlag, Berlin Göttingen Heidelberg, 326 pp., 1953.

Zheng, C. and Wang, P. P.: MT3DMS, a Modular Three-Dimensional Multi-Species Transport Model for Simulations of Advection, Dispersion and Chemical Reactions of Contaminants in Groundwater Systems, Documentation and User's Guide, US Army Engineer Research and Development Center Contact Report SERDP-99-1, Vicksburg, MS, USA, 1999.

Zuber, A., Weise, S. M., Motyka, J., Osenbrück, K., and Rozanski, K.: Age and flow patter of groundwater in a Jurassic limestone aquifer and related Tertiary sands derived from combined isotope, noble gas and chemical data, J. Hydrol., 286, 87–112, 2004.

Zuber, A., Witczak, S., Rozanski, K., Sliwka, I., Opoka, M., Mochalski, P., Kuc, T., Karlikowska, J., Kania, J., Jackowicz-Korczynski, M., and Dulinski, M.: Groundwater dating with ^{3}H and SF_6 in relation to mixing pattern, transport modelling and hydrochemistry, Hydrol. Process., 19, 2247–2275, 2005.

Exploring water cycle dynamics by sampling multiple stable water isotope pools in a developed landscape in Germany

Natalie Orlowski[1,2], Philipp Kraft[1], Jakob Pferdmenges[1], and Lutz Breuer[1,3]

[1]Institute for Landscape Ecology and Resources Management (ILR), Research Centre for BioSystems, Land Use and Nutrition (iFZ), Justus Liebig University Gießen, Gießen, Germany
[2]Global Institute for Water Security, University of Saskatchewan, Saskatoon, Canada
[3]Centre for International Development and Environmental Research, Justus Liebig University Gießen, Gießen, Germany

Correspondence to: N. Orlowski (natalie.orlowski@umwelt.uni-giessen.de)

Abstract. A dual stable water isotope (δ^2H and δ^{18}O) study was conducted in the developed (managed) landscape of the Schwingbach catchment (Germany). The 2-year weekly to biweekly measurements of precipitation, stream, and groundwater isotopes revealed that surface and groundwater are isotopically disconnected from the annual precipitation cycle but showed bidirectional interactions between each other. Apparently, snowmelt played a fundamental role for groundwater recharge explaining the observed differences to precipitation δ values.

A spatially distributed snapshot sampling of soil water isotopes at two soil depths at 52 sampling points across different land uses (arable land, forest, and grassland) revealed that topsoil isotopic signatures were similar to the precipitation input signal. Preferential water flow paths occurred under forested soils, explaining the isotopic similarities between top- and subsoil isotopic signatures. Due to human-impacted agricultural land use (tilling and compression) of arable and grassland soils, water delivery to the deeper soil layers was reduced, resulting in significant different isotopic signatures. However, the land use influence became less pronounced with depth and soil water approached groundwater δ values. Seasonally tracing stable water isotopes through soil profiles showed that the influence of new percolating soil water decreased with depth as no remarkable seasonality in soil isotopic signatures was obvious at depths > 0.9 m and constant values were observed through space and time. Since classic isotope evaluation methods such as transfer-function-based mean transit time calculations did not provide a good fit between the observed and calculated data, we established a hydrological model to estimate spatially distributed groundwater ages and flow directions within the Vollnkirchener Bach subcatchment. Our model revealed that complex age dynamics exist within the subcatchment and that much of the runoff must has been stored for much longer than event water (average water age is 16 years). Tracing stable water isotopes through the water cycle in combination with our hydrological model was valuable for determining interactions between different water cycle components and unravelling age dynamics within the study area. This knowledge can further improve catchment-specific process understanding of developed, human-impacted landscapes.

1 Introduction

The application of stable water isotopes as natural tracers in combination with hydrodynamic methods has been proven to be a valuable tool for studying the origin and formation of recharged water as well as the interrelationship between surface water and groundwater (Blasch and Bryson, 2007), partitioning evaporation and transpiration (Wang and Yakir, 2000), and mixing processes between various water sources (Clark and Fritz, 1997c). Particularly in catchment hydrology, stable water isotopes play a major role since they can be utilized for hydrograph separations (Buttle, 2006), to calculate the mean transit time (McGuire and McDonnell, 2006), to investigate water flow paths (Barthold et al., 2011), or to improve hydrological model simulations (Windhorst et al., 2014). However, most of our current understanding results from studies in forested catchments. Spatio-temporal studies of stream water in developed, agriculturally dominated, and managed catchments are less abundant. This is partly caused by damped stream water isotopic signatures excluding traditional hydrograph separations in low-relief catchments (Klaus et al., 2015). Unlike the distinct watershed components found in steeper headwater counterparts, lowland areas often exhibit a complex groundwater–surface-water interaction (Klaus et al., 2015). Sklash and Farvolden (1979) showed that groundwater plays an important role as a generating factor for storm and snowmelt runoff processes. In many catchments, streamflow responds promptly to rainfall inputs but variations in passive tracers such as water isotopes are often strongly damped (Kirchner, 2003). This indicates that storm runoff in these catchments is dominated mostly by "old water" (Buttle, 1994; Neal and Rosier, 1990; Sklash, 1990). However, not all old water is the same (Kirchner, 2003). This catchment behaviour was described by Kirchner (2003) as the old-water paradox. Thus, there is evidence of complex age dynamics within catchments and much of the runoff is stored in the catchment for much longer than event water (Rinaldo et al., 2015). Still, some of the physical processes controlling the release of old water from catchments are poorly understood and roughly modelled, and the observed data do not suggest a common catchment behaviour (Botter et al., 2010). However, old-water paradox behaviour was observed in many catchments worldwide, but it may have the strongest effect in agriculturally managed catchments, where surprisingly only small changes in stream chemistry have been observed (Hrachowitz et al., 2016).

Moreover, almost all European river systems were already substantially modified by humans before river ecology research developed (Allan, 2004). Through changes in land use, land cover, irrigation, and draining, agriculture has substantially modified the water cycle in terms of both quality and quantity (Gordon et al., 2010) as well as hydrological functioning (Pierce et al., 2012). Hrachowitz et al. (2016) recently stated the need for a stronger linkage between catchment-scale hydrological and water quality communities. Further, McDonnell et al. (2007) concluded that we need to figure out a way to embed landscape heterogeneity or the consequence of the heterogeneity (i.e. of agriculturally dominated and managed catchments) into models as current-generation catchment-scale hydrological and water quality models are poorly linked (Hrachowitz et al., 2016).

One way to better understand catchment behaviour and the interaction among the various water sources (surface, subsurface, and groundwater) and their variation in space and time is a detailed knowledge about their isotopic composition. In principal, isotopic signatures of precipitation are altered by temperature, amount (or rainout), continental, altitudinal, and seasonal effects. Stream water isotopic signatures can reflect precipitation isotopic composition and, moreover, dependent on discharge variations, be affected by seasonally variable contributions of different water sources such as

bidirectional water exchange with the groundwater body during baseflow or high event-water contributions during storm flow (Genereux and Hooper, 1998; Koeniger et al., 2009). Precipitation falling on vegetated areas is partly intercepted by plants and re-evaporated isotopically fractionated. The remaining throughfall infiltrates slower and can be affected by evaporation resulting in an enrichment of heavy isotopes, particularly in the upper soil layers (Gonfiantini et al., 1998; Kendall and Caldwell, 1998). In the soil, specific isotopic profiles develop, characterized by an evaporative layer near the surface. The isotopic enrichment decreases exponentially with depth, representing a balance between the upward convective flux and the downward diffusion of the evaporative signature (Barnes and Allison, 1988). In humid and semi-humid areas, this exponential decrease is generally interrupted by the precipitation isotopic signal. Hence, the combination of the evaporation effect and the precipitation isotopic signature determine the isotope profile in the soil (Song et al., 2011). Once soil water reaches the saturated zone, this isotope information is finally transferred to the groundwater (Song et al., 2011). Soil water can therefore be seen as a link between precipitation and groundwater, and the dynamics of isotopic composition in soil water are indicative of the processes of precipitation infiltration, evaporation of soil water, and recharge to groundwater (Blasch and Bryson, 2007; Song et al., 2011).

We started our research with results obtained through an earlier study in the managed Schwingbach catchment that implied a high responsiveness of the system to precipitation inputs indicated by very fast rises in discharge and groundwater head levels (Orlowski et al., 2014). However, as there was only a negligible influence of the precipitation input signal on the stable water isotopic composition in streams, our initial data set showed evidence of complex age dynamics within the catchment. Nevertheless, a rapid flow response to a precipitation input may also be mistaken (as conceptualized in the vast majority of catchment-scale conceptual hydrological models) for the actual input signal already reaching the stream, while in reality it is the remainder of past input signals that have slowly travelled through the system (Hrachowitz et al., 2016). The observable hydrological response therefore acts on different timescales from the tracer response (Hrachowitz et al., 2016) as described by the celerity vs. velocity concept (McDonnell and Beven, 2014). The observed patterns in our catchment therefore inspired us to use a combined approach of hydrodynamic data analyses, stable water isotope investigations, and data-driven hydrological modelling to determine catchment dynamics (response times and groundwater age patterns) and unravel water flow paths on multiple spatial scales. This work should further improve our knowledge of hydrological flow paths in developed, human-impacted catchments.

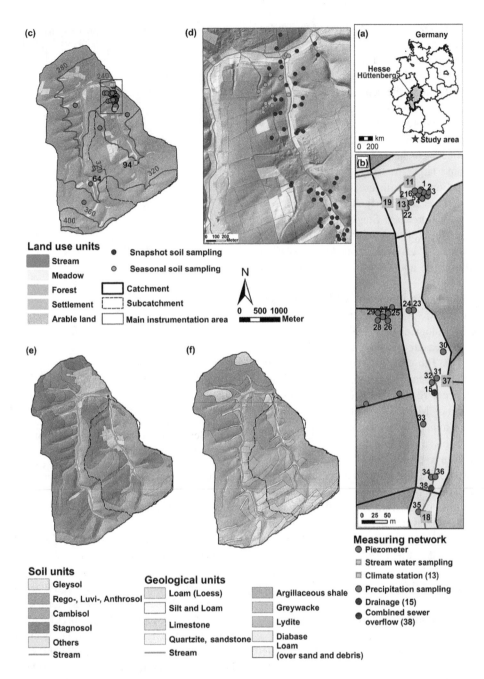

Figure 1. Maps show **(a)** the location of the Schwingbach catchment in Germany, **(b)** the main monitoring area, **(c)** the land use, elevation, and instrumentation, **(d)** the locations of the snapshot as well as the seasonal soil samplings, **(e)** soil types, and **(f)** geology of the Schwingbach catchment including the Vollnkirchener Bach subcatchment boundaries.

2   Materials and methods

2.1   Study area

The research was carried out in the Schwingbach catchment (50°30′4.23″ N, 8°33′2.82″ E) (Germany) (Fig. 1a). The Schwingbach and its main tributary the Vollnkirchener Bach are low mountainous creeks and have an altitudinal difference of 50–100 m over a 5 km distance (Perry and Tay-

lor, 2009) (Fig. 1c) with an altered physical structure of the stream system (channelled stream reaches, pipes, drainage systems, fishponds). The Schwingbach catchment (9.6 km^2) ranges from 233–415 m a.s.l. with an average slope of 8.0 %. The Vollnkirchener Bach tributary is 4.7 km in length and drains a 3.7 km^2 subcatchment area (Fig. 1c), with elevations from 235 to 351 m a.s.l. Almost 46 % of the overall Schwingbach catchment is forested, which slightly exceeds agricultural land use (35 %) (Fig. 1c). Grassland (10 %) is mainly distributed along streams, and smaller meadow orchards are located around the villages.

The Schwingbach main catchment is underlain by argillaceous shale in the northern parts, serving as aquicludes. Greywacke zones with lydite in the central, as well as limestone, quartzite, and sandstone regions in the headwater area provide aquifers with large storage capacities (Fig. 1f). Loess covers Paleozoic bedrock at north- and east-bounded hillsides (Fig. 1f). Streambeds consists of sand and debris covered by loam and some larger rocks (Lauer et al., 2013). Many downstream sections of both creeks are framed by armour stones (Orlowski et al., 2014). The dominant soil types in the overall study area are Stagnosols (41 %) and mostly forested Cambisols (38 %). Stagnic Luvisols with thick loess layers, Regosol, Luvisols, and Anthrosols are found under agricultural use and Gleysols under grassland along the creeks.

The climate is classified as temperate with a mean annual temperature of 8.2 °C. An annual precipitation sum of 633 mm (for the hydrological year 1 November 2012 to 31 October 2013) was measured at the catchment's climate station (site 13, Fig. 1b). The year 2012 to 2013 was an average hydrometeorological year. For comparison, the climate station Gießen/Wettenberg (25 km north of the catchment) operated by the German Meteorological Service (DWD, 2014) records a mean annual temperature of 9.6 °C and a mean annual precipitation sum of 666 ± 103 mm for the period 1980–2010. Discharge peaks from December to April (measured by the use of RBC flumes with a maximum peak flow of 114 L s^{-1}, Eijkelkamp Agrisearch Equipment, Giesbeek, NL), and low flows occur from July until November. Substantial snowmelt peaks were observed during December 2012 and February 2013. Furthermore, May 2013 was an exceptionally wet month characterized by discharge of 2–3 mm day^{-1}. A detailed description of runoff characteristics is given by Orlowski et al. (2014).

2.2 Monitoring network and water isotope sampling

The monitoring network consists of an automated climate station (site 13, Fig. 1b–c) (Campbell Scientific Inc., AQ5, UK; equipped with a CR1000 data logger), 3 tipping buckets, and 15 precipitation collectors, 6 stream water sampling points, and 22 piezometers (Fig. 1b–c). Precipitation data were corrected according to Xia (2006).

Two stream water sampling points (sites 13 and 18) in the Vollnkirchener Bach are installed with trapezium-shaped RBC flumes for gauging discharge (Eijkelkamp Agrisearch Equipment, Giesbeek, NL), and a V-notch weir is located at sampling point 64. RBC flumes and V-notch weir are equipped with Mini-Divers® (Eigenbrodt Inc. & Co. KG, Königsmoor, DE) for automatically recording water levels. Discharge at the remaining stream sampling points was manually measured applying the salt dilution method (WTW-cond340i, WTW, Weilheim, DE). The 22 piezome-

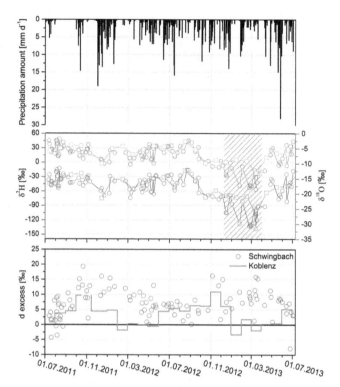

Figure 2. Temporal variation of precipitation amount, isotopic signatures (δ^2H and δ^{18}O) including snow samples (grey striped box) of the Schwingbach and GNIP station Koblenz, and d-excess values for the study area compared to monthly d-excess values (July 2011 to July 2013) of GNIP station Koblenz with reference d excess of GMWL (d = 10; solid black line).

ters (Fig. 1b) are made from perforated PVC tubes sealed with bentonite at the upper part of the tube to prevent contamination by surface water. For monitoring shallow groundwater levels, either combined water level and temperature loggers (Odyssey Data Flow System, Christchurch, NZ) or Mini-Diver® water level loggers (Eigenbrodt Inc. & Co. KG, Königsmoor, DE) are installed. The accuracy of Mini-Diver® is ± 5 mm and it is ± 1 mm for the Odyssey data logger. For calibration purposes, groundwater levels are additionally measured manually via an electric contact gauge.

Stable water isotope samples of rainfall and stream- and groundwater were taken from July 2011 to July 2013 at weekly intervals. In winter 2012–2013, snow core samples over the entire snow depth of < 0.15 m were collected in tightly sealed jars at the same sites as open rainfall was sampled. We sampled shortly after snowfall because sublimation, recrystallization, partial melting, rainfall on snow, and redistribution by wind can alter the isotopic composition (Clark and Fritz, 1997b). Samples were melted overnight following the method of Kendall and Caldwell (1998) and analysed for their isotopic composition. Open rainfall was collected in self-constructed samplers as in Windhorst et al. (2013). Grab samples of stream water were taken at six locations,

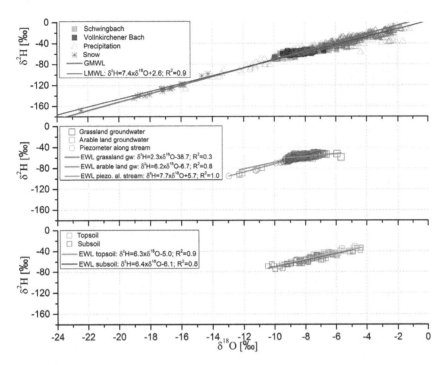

Figure 3. Local meteoric water line for the Schwingbach catchment (LMWL) in comparison to GMWL, including comparisons between precipitation, stream water, groundwater, and soil water isotopic signatures and the respective EWLs.

with three sampling points at each stream (Fig. 1b–c). Since spatial isotopic variations of groundwater among piezometers under the meadow were small, samples were collected at three out of eight sampling points under the meadow (sites 3, 6, and 21), five under the arable field (sites 25–29), and four next to the Vollnkirchener Bach (sites 24, 31, 32, and 35) (Fig. 1b). Additionally, a drainage pipe (site 15) located ~ 226 m downstream of site 18 was sampled. According to IAEA standard procedures, all samples were filled and stored in 2 mL brown glass vials, sealed with a solid lid, and wrapped up with Parafilm®.

2.3　Isotopic soil sampling

2.3.1　Spatial variability

In order to analyse the effect of small-scale characteristics such as distance to stream, topographic wetness index (TWI), and land use on soil isotopic signatures, we sampled a snapshot of 52 points evenly distributed over a 200 m grid around the Vollnkirchener Bach (Fig. 1d). Soil samples were taken on four consecutive rainless days (1 to 4 November 2011) at elevations of 235–294 m a.s.l. Sampling sites were selected via a stratified, GIS-based sampling plan (ArcGIS, Arc Map 10.2.1, Esri, California, USA), including three classes of TWIs (4.4–6.5; 6.5–7.7; 7.7–18.4), two different distances to the stream (0–121 and 121–250 m), and three land uses (arable land, forest, and grassland), with each class containing the same number of sampling points. Samples were col-

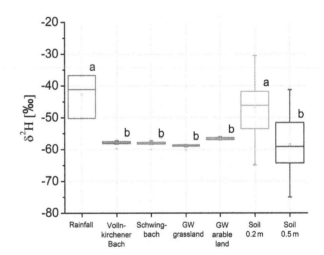

Figure 4. Box plots of $\delta^2\mathrm{H}$ values comparing precipitation, stream, groundwater, and soil isotopic composition at 0.2 and 0.5 m depth ($N = 52$ per depth). Different letters indicate significant differences ($p \leq 0.05$).

lected at depths of 0.2 and 0.5 m. Gravimetric water content was measured according to DIN-ISO 11465 by drying soils for 24 h at 110 °C. Soil pH was analysed following DIN-ISO 10390 on 1 : 1 soil–water mixture with a handheld pH meter (WTW cond340i, WTW Inc., DE). Bulk density was determined according to DIN-ISO 11272 and soil texture by finger testing.

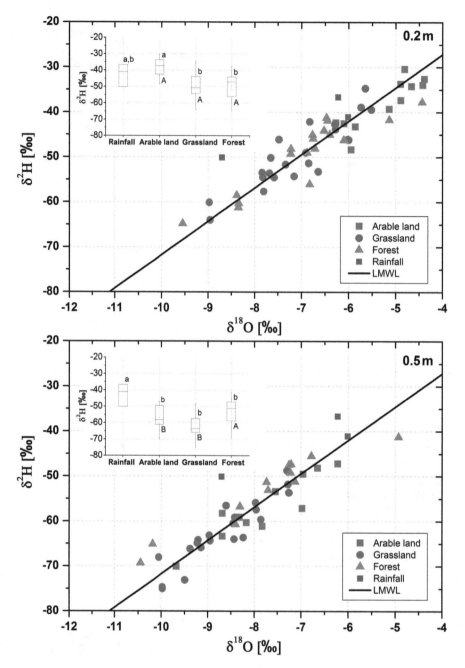

Figure 5. Dual isotope plot of soil water isotopic signatures at 0.2 and 0.5 m depth compared by land use including precipitation isotope data from 19, 21, and 28 October 2011. Insets: box plots comparing δ^2H isotopic signatures between different land use units and precipitation (small letters) in top- and subsoil (capital letters). Different letters indicate significant differences ($p \leq 0.05$).

2.3.2   Seasonal isotope soil profiling and isotope analysis

In order to trace the seasonal development of stable water isotopes from rainfall to groundwater, seven soil profiles were taken in the dry summer season (28 August 2011), seven in the wet winter period (28 March 2013), and two profiles in spring (24 April 2013) under different vegetation cover (arable land and grassland) (Fig. 1d). Soil was sampled from

the soil surface to 2 m depth utilizing a hand auger (Eijkelkamp Agrisearch Equipment BV, Giesbeek, DE). More samples were collected near the soil surface since this area is known to have the greatest isotopic variability (Barnes and Allison, 1988).

Soil samples were stored in amber glass tubes, sealed with Parafilm®, and kept frozen until water extraction. Soil water was extracted cryogenically with 180 min extraction duration, a vacuum threshold of 0.3 Pa, and an extraction tem-

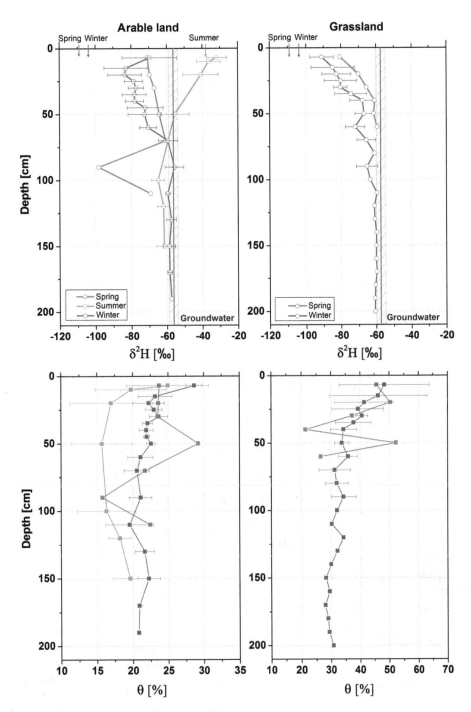

Figure 6. Seasonal δ^2H profiles of soil water (upper panels) and water content (lower panels) for winter (28 March 2013), summer (28 August 2011), and spring (24 April 2013). Error bars represent the natural isotopic variation of the replicates taken during each sampling campaign. For reference, mean groundwater (grey shaded) and mean seasonal precipitation δ^2H values are shown (coloured arrows at the top).

perature of 90°C following Orlowski et al. (2013). Isotopic signatures of $\delta^{18}O$ and δ^2H were analysed via off-axis integrated cavity output spectroscopy (OA-ICOS) (DLT-100, Los Gatos Research Inc., Mountain View, USA). Within each isotope analysis three calibrated stable water isotope stan-

dards of different water isotope ratios were included (Los Gatos Research (LGR) working standard number 1, 3, and 5; Los Gatos Research Inc., CA, US). After every fifth sample the LGR working standards are measured. For each sample, six sequential 900 µL aliquots of a water sample are injected

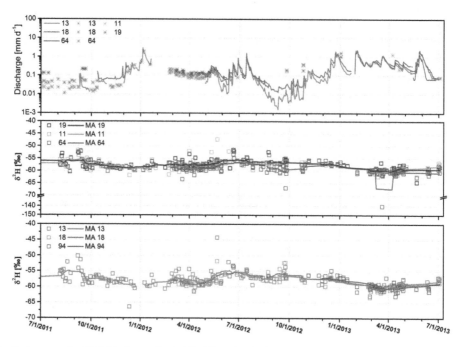

Figure 7. Mean daily discharge at the Vollnkirchener Bach (13, 18) and Schwingbach (site 11, 19, and 64) with automatically recorded data (solid lines) and manual discharge measurements (asterisks), temporal variation of δ^2H of stream water in the Schwingbach (site 11, 19, and 64) and Vollnkirchener Bach (site 13, 18, and 94) including moving averages (MAs) for streamflow isotopes.

into the analyser. Then, the first three measurements are discarded. The remaining are averaged and corrected for per mil scale linearity following the IAEA laser spreadsheet template (Newman et al., 2009). Following this IAEA standard procedure allows for drift and memory corrections. Isotopic ratios are reported in per mil (‰) relative to Vienna Standard Mean Ocean Water (VSMOW) (Craig, 1961b). The accuracy of analyses was 0.6‰ for δ^2H and 0.2‰ for $\delta^{18}O$ (LGR, 2013). Leaf water extracts typically contain a high fraction of organic contaminations, which might lead to spectral interferences when using isotope ratio infrared absorption spectroscopy, causing erroneous isotope values (Schultz et al., 2011). However, for soil water extracts there exists no need to check or correct such data (Schultz et al., 2011; Zhao et al., 2011).

2.4 Mean transit time estimation

To understand the connection between the different water cycle components in the Schwingbach catchment, mean transit times (MTTs) for both streams as well as from precipitation to groundwater were calculated using FlowPC (Maloszewski and Zuber, 2002). See Appendix A for details about the applied method.

2.5 Model-based groundwater age dynamics

To estimate the age dynamics of the groundwater body in the Vollnkirchener Bach subcatchment, a hydrological model was established on the basis of the conceptual model pre-

sented by Orlowski et al. (2014) and the isotopic measurements presented here. Appendix B outlines the modelling concept, model set-up, and its parameterization.

2.6 Statistical analyses

For statistical analyses, we used IBM SPSS Statistics (Version 22, SPSS Inc., Chicago, IL, US) and R (version Rx64 3.2.2). The R package igraph was utilized for plotting (Csardi and Nepusz, 2006). In order to study temporal and spatial variations in meteoric and groundwater, isotope data were tested for normal distribution. Subsequently, t tests or multivariate analyses of variance (MANOVAs) were applied and Tukey's honest significant difference (HSD) tests were run to determine which groups were significantly different ($p \le 0.05$). Event mean values of isotopes in precipitation, stream, and groundwater were calculated when no spatial variation was observed. Regression analyses were run to determine the effect of small-scale characteristics such as distance to stream, TWI, and land use on soil isotopic signatures.

We used a topology inference network map (Kolaczyk, 2014) in combination with a principal component analysis to show $\delta^{18}O$ isotope relationships between surface and groundwater sampling points. To explore the sensitivity of missing data, we used both the complete isotope time series and randomly selected 80 % of the whole data sets. Overall, the cluster relationships of the surface and groundwater sampling points are largely similar for both entire and subsets of

Table 1. Descriptive statistics of δ^2H, δ^{18}O, and d-excess values for precipitation, stream, and groundwater over the 2-year observation period including all sampling points.

Sample type	Mean ± SD		Min		Max		D-excess mean ± SD	N
	δ^2H (‰)	δ^{18}O (‰)	δ^2H (‰)	δ^{18}O (‰)	δ^2H (‰)	δ^{18}O (‰)		
Precipitation	−43.9 ± 23.4	−6.2 ± 3.1	−167.6	−22.4	−8.3	−1.2	5.9 ± 5.7	592
Vollnkirchener Bach	−58.0 ± 2.8	−8.4 ± 0.4	−66.3	−10.0	−26.9	−6.7	9.0 ± 2.3	332
Schwingbach	−58.2 ± 4.3	−8.4 ± 0.6	−139.7	−18.3	−47.2	−5.9	9.0 ± 2.2	463
Groundwater meadow	−57.6 ± 1.6	−8.2 ± 0.4	−64.9	−9.2	−50.8	−5.7	7.9 ± 5.5	375
Groundwater arable land	−56.2 ± 3.7	−8.0 ± 0.5	−91.6	−12.3	−49.5	−6.8	1.7 ± 5.0	338
Groundwater along stream	−59.9 ± 6.8	−8.5 ± 0.9	−94.5	−13.0	−49.5	−7.0	8.2 ± 1.5	108

isotope data sets, despite some differences in the exact cluster centroid locations. We therefore decided to use randomly selected 80 % of the isotope time series to illustrate our results. In the network map, each node of the network represents an isotope sampling point. The locations of the nodes are based on the first two components (PC1 and PC2). The correlations between isotope time series are represented by the edges connecting nodes. The thickness of edges characterizes the strength of the correlations. The p values of correlations are approximated by using the F distributions and mid-ranks are used for the ties (Hollander et al., 2013). Only statistically significant connections ($p < 0.05$) are shown.

To compare different water sources on the catchment scale, a local meteoric water line (LMWL) was developed and evaporation water lines (EWLs) were used. They represent the linear relationship between δ^2H and δ^{18}O of meteoric waters (Cooper, 1998) in contrast to the global meteoric water line (GMWL), which describes the worldwide average stable isotopic composition in precipitation (Craig, 1961a). Identifying the origin of water vapour sources and moisture recycling (Gat et al., 2001; Lai and Ehleringer, 2011), the deuterium excess (d excess), defined by Dansgaard (1964) as d= δ^2H–8 × δ^{18}O was used.

For comparisons, precipitation isotope data from the closest GNIP (Global Network of Isotopes in Precipitation) station Koblenz (DE; 74 km SW of the study area, 97 m a.s.l.) were used (IAEA, 2014; Stumpp et al., 2014). For monthly comparisons with Schwingbach d-excess values, we used a data set from the GNIP station Koblenz that includes 24 values starting from July 2011 to July 2013.

3   Results

3.1   Variations of precipitation isotopes and d excess

The δ^2H values of all precipitation isotope samples ranged from −167.6 to −8.3 ‰ (Table 1). To examine the spatial isotopic variations, rainfall was collected at 15 open-field site locations throughout the Schwingbach main catchment (Fig. 1b–c) for a 7-month period, but no spatial variation could be observed. Thus, rainfall was collected at the catchment outlet (site 13) from 23 October 2014 onward. We could

neither identify an amount effect nor an altitude effect in our precipitation isotope data. The greatest altitudinal difference between sampling points was also only 101 m. Nevertheless, a slight temperature effect ($R^2 = 0.5$ for δ^2H and $R^2 = 0.6$ for δ^{18}O) was observed showing enriched isotopic signatures at higher temperatures.

Strong temporal variations in precipitation isotopic signatures as well as pronounced seasonal isotopic effects were measured, with greatest isotopic differences occurring between summer and winter. Samples taken in the fall and spring were isotopically similar but differed from winter isotopic signature, which were somewhat lighter (Fig. 2). Furthermore, in the winter of 2012–2013 snow was sampled, which decreased the mean winter isotopic values for this period in comparison to the previous winter period (2011–2012) where no snow sampling could be conducted. The mean δ^2H isotope values of snow samples were approximately 84 ‰ lighter than mean precipitation isotopic signatures (Fig. 3). Furthermore, no statistically significant ($p > 0.05$) interannual variation was detected between the summer periods of 2011 and 2012 (Fig. 2).

Examining the influence of moisture recycling on the isotopic compositions of precipitation, the d excess was calculated for each individual rain event at the Schwingbach catchment. D excess values ranged from −7.8 to +19.4 ‰ and averaged +7.1 ‰ (Fig. 2). In general, 37 % of all events were sampled in summer periods (21 June–21/22 September). These summer events showed lower d-excess values in comparison to the 19 % winter precipitation events (21/22 December to 19/20 March) (Fig. 2). D excess greater than +10 ‰ was determined for 22 % of all events. Lowest values corresponded to summer precipitation events where evaporation of the raindrops below the cloud base may occur. Most of the higher values (> +10 ‰) appeared in cold seasons (fall or winter) and winter snow samples of the Schwingbach catchment with much depleted δ values showed highest d excess (Fig. 2).

In comparison with the GNIP station Koblenz (2011–2013), the mean annual d excess at the Schwingbach catchment was on average 3.9 ‰ higher, showing a greater impact of oceanic moisture sources than the station Koblenz, located further south-west. The long-term mean d excess was 4.4 ‰

for the Koblenz station (1978–2009) (Stumpp et al., 2014). Highest d excesses at the GNIP station matched the highest values in the Schwingbach catchment, both occurring in the cold seasons (October to December 2011 and November to December 2012).

The linear relationship of δ^2H and $\delta^{18}O$ content in local precipitation, results in a local meteoric water line (LMWL) (Fig. 3). The slope of the Schwingbach LMWL is in good agreement with the one from the GNIP station Koblenz ($\delta^2H = 7.66 \times \delta^{18}O + 2.0\%_o$; $R^2 = 0.97$; 1978–2009; Stumpp et al., 2014) but is slightly lower in comparison to the GMWL, showing stronger local evaporation conditions. Since evaporation causes a differential increase in δ^2H and $\delta^{18}O$ values of the remaining water, the slope for the linear relationship between δ^2H and $\delta^{18}O$ is lower in comparison to the GMWL (Rozanski et al., 2001; Wu et al., 2012).

3.2 Isotopes of soil water

3.2.1 Spatial variability

Determining the impact of landscape characteristics on soil water isotopic signatures, we found no statistically significant connection between the parameters' distance to stream, TWI, soil water content, soil texture, pH, and bulk density with the soil isotopic signatures at both soil depths, except for land use.

The mean δ values in the top 0.2 m of the soil profile are higher than in the subsoil, reflecting a stronger impact of precipitation in the topsoil (Table 2, Fig. 4). While the δ values for subsoil and precipitation differed significantly ($p \leq 0.05$), they did not do so for topsoil (Fig. 4). Subsoil isotopic values were statistically equal to stream water and groundwater (Fig. 4).

Generally, all soil water isotopic values fell on the LMWL, indicating no evaporative enrichment (Fig. 5). Comparing soil isotopic signatures between different land covers showed generally higher and statistically significantly different δ values ($p \leq 0.05$) at 0.2 m soil depth under arable land as compared to forests and grasslands. For the lower 0.5 m of the soil column, isotopic signatures under all land uses showed statistically similar values. Comparing soil water δ^2H values between top- and subsoil under different land use units showed significant differences ($p \leq 0.05$) under arable and grassland but not under forested sites (Fig. 5).

3.2.2 Seasonal isotope soil profiling

Isotope compositions of soil water varied seasonally: more depleted soil water was found in the winter and spring (Fig. 6); by contrast, soil water was enriched in summer due to evaporation during warmer and drier periods (Darling, 2004). For summer soil profiles in the Vollnkirchener subcatchment, no evidence for evaporation was obvious below 0.4 m soil depth. However, snowmelt isotopic signatures could be traced down to a soil depth of 0.9 m during spring rather than winter, pointing to a depth translocation of meltwater in the soil, more remarkable for the deeper profile under arable land (Fig. 6, upper left panel). Furthermore, shallow soil water (< 0.4 m) showed larger standard deviations with values closer to mean seasonal precipitation inputs (Fig. 6, upper panels). Winter profiles exhibited somewhat greater standard deviations in comparison to summer isotopic soil profiles. The observed seasonal amplitude became less pronounced with depth as soil water isotope signals approached a groundwater average at > 0.9 m depth. Generally, deeper soil water isotope values were relatively constant through time and space.

3.3 Isotopes of stream water

No statistically significant differences were found between the Schwingbach and Vollnkirchener Bach stream water (Fig. 7). All stream water isotope samples fell on the LMWL except for a few evaporatively enriched samples (Fig. 3). $\delta^{18}O$ values varied for the Vollnkirchener Bach by $-8.4 \pm 0.4\%_o$ and for the Schwingbach by $-8.4 \pm 0.6\%_o$ (Table 1). Stream water isotopic signatures were by approximately $-15\%_o$ in δ^2H more depleted than precipitation signatures and were similar to groundwater (Table 1).

A damped seasonality of the isotope concentration in stream water versus precipitation occurred between summer and winter (Fig. 7). Most outlying depleted stream water isotopic signatures (e.g. in March 2012 and 2013) can be explained by snowmelt (Fig. 7). However, the outlier at the Schwingbach stream water sampling site 64 ($-66.7\%_o$ for δ^2H) is $8.5\%_o$ more depleted than the 2-year average of Schwingbach stream water (Table 1). Rainfall falling on 24 September 2012 was $-31.9\%_o$ for δ^2H. This period in September was generally characterized by low flow and little rainfall. Thus, little contribution of new water was observed and stream water isotopic signatures were groundwater-dominated. For site 13, the outlier in May 2012 ($-44.2\%_o$ for δ^2H) was $13.8\%_o$ more enriched than the average stream water isotopic composition of the Vollnkirchener Bach over the 2-year observation period (Table 1). A runoff peak at site 13 of 0.15 mm day^{-1} and a 2.9 mm rainfall event were recorded on 23 May 2012. Thus, this outlier could be explained by precipitation contributing to stream flow causing more enriched isotopic values in stream water, which approached average precipitation δ values (-43.9 ± 23.4).

MTT calculations for the Schwingbach and the Vollnkirchener Bach did not provide a good fit in terms of the quality criteria sigma and model efficiency (Timbe et al., 2014) ($ME_{Schwingbach} - 0.1-0.0$, $ME_{Vollnkirchener\ Bach}$ 0.0–0.4; sigma for all sampling points 0.1). Bias correction of the input data did not improve the model outputs (sigma = 0.1).

Table 2. Mean and standard deviation (SD) for isotopic signatures and soil physical properties at 0.2 m and 0.5 m soil depth ($N = 52$ per depth).

	δ^2H (‰)		δ^{18}O (‰)		Water content (% w/w)		pH		Bulk density (g cm^{-3})	
	0.2 m	0.5 m	0.2 m	0.5 m	0.2 m	0.5 m	0.2 m	0.5 m	0.2 m	0.5 m
Mean ± SD	−46.9 ± 8.4	−58.5 ± 8.3	−6.6 ± 1.2	−8.2 ± 1.2	16.8 ± 7.2	16.1 ± 8.3	5.0 ± 1.0	5.3 ± 1.0	1.3 ± 0.2	1.3 ± 0.2

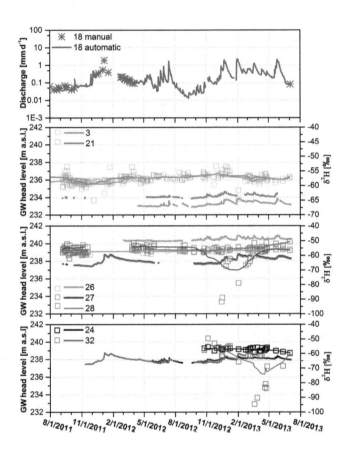

Figure 8. Temporal variation of discharge at the Vollnkirchener Bach with automatically recorded data (solid line) and manual discharge measurements (asterisks) (site 18), groundwater head levels, and δ^2H values (coloured dots) for selected piezometers under the meadow (sites 3 and 21), arable land (sites 26, 27, and 28), and beside the Vollnkirchener Bach (sites 24 and 32) including moving averages for groundwater isotopes.

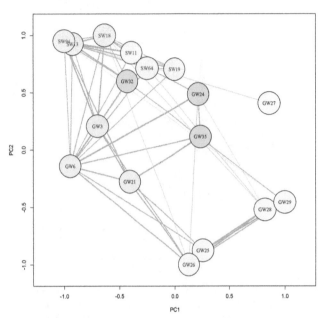

Figure 9. Network map of δ^{18}O relationships between surface water (SW) and groundwater (GW) sampling points. Yellow circles represent groundwater sampling points on the arable field, light green circles are piezometers located on the grassland close to the conjunction of the Schwingbach with the Vollnkirchener Bach, and dark green circles represent piezometers along the Vollnkirchener Bach. Light blue circles stand for Schwingbach and darker blue circles for Vollnkirchener Bach surface water sampling points. See Fig. 1 for an overview of all sampling points. Only statistically significant connections between δ^{18}O time series ($p < 0.05$) are shown in the network diagram.

3.4 Isotopes of groundwater

For the piezometers under the meadow, almost constant isotopic values (Fig. 8, Table 1) were observed (δ^2H: −57.6 ± 1.6‰). Most depleted groundwater isotopic values (< −80‰ for δ^2H) were measured for piezometer 32 during snowmelt events in March and April 2013 as well as for piezometer 27 from December 2012 to February 2013. Piezometer 32 is highly responsive to rainfall–runoff events and groundwater head elevations showed significant correla-

tions with mean daily discharge at this site (Orlowski et al., 2014).

Groundwater under the meadow differed from mean precipitation values by about −14‰ for δ^2H, showing no evidence of a rapid transfer of rainfall isotopic signatures to the groundwater (Fig. 8). For the MTT estimations of the 13 piezometers, the calculated output data did not fit the observed values, showing very low MEs (ME: −0.62 – −0.09 for δ^{18}O and −0.49–0.16 for δ^2H; sigma: 0.08–0.15 for δ^{18}O and 0.62–1.11 for δ^2H).

Due to different water flow paths of groundwater along the studied stream, we expected to find distinct groundwater isotopic signatures. In fact, we could identify spatial statistical differences between grassland and arable land groundwater isotopic signatures (Fig. 9). Groundwater isotopic signa-

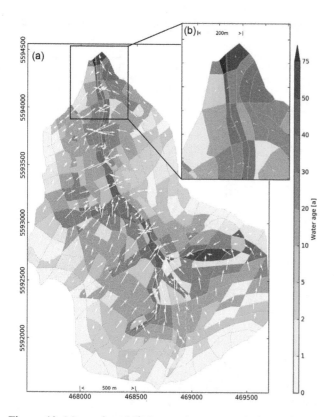

Figure 10. Maps of modelled groundwater ages (colour scheme) and flow directions (white arrows) of **(a)** the Vollnkirchener Bach subcatchment and **(b)** a detailed view of the northern part of the subcatchment. The length of the white arrows depicts the intensity of flow. UTM-32N (WGS84) coordinates on both axes.

tures under arable land (sites: 25–29, Fig. 1b) showed more enriched values (Fig. 8) and showed significant correlations ($p < 0.05$) among each other (Fig. 9). Arable land groundwater plotted furthest away from surface water sampling points in our network map, showing no significant correlations to either the Schwingbach or the Vollnkirchener Bach. $\delta^{18}O$ time series of piezometers along the stream and under the meadow showed the closest relationship to surface water sampling points (Fig. 9). We further found high correlations ($R^2 > 0.6$) of $\delta^{18}O$ time series of piezometers located under the meadow with each other. Additionally, $\delta^{18}O$ values of piezometer 3 correlated significantly ($p < 0.05$) with surface water sampling points 18 and 94 ($R^2 = 0.6$ and 0.8, respectively) and those of piezometer 32 with sampling points 13 and 64 ($R^2 = 0.8$ and 0.6, respectively).

We further observed a close relationship ($p < 0.05$) among $\delta^{18}O$ values of Vollnkirchener Bach sampling sites 13, 18, and 94 as well as of Schwingbach sites 11, 19, and 64 along with significant correlations between each other.

3.5   Groundwater age dynamics

Since MTT calculations did not provide a good fit between

the observed and calculated output data, we modelled the groundwater age in the Vollnkirchener Bach subcatchment using catchment modelling framework (CMF; Appendix B), applying observed hydrometric as well as stable water isotope data (Fig. 10).

The maximum age of water is highly variable throughout the subcatchment, which results in a heterogeneous spatial age distribution. The groundwater in most of the outer cells is young (0–10 years), whereas the inner cells, which incorporate the Vollnkirchener Bach, contain older water (>30 years). The oldest water (≥ 55 years) can be found in the northern part of the catchment (Fig. 10, detail view), where the Vollnkirchener Bach drains into the Schwingbach. The main outlets of the subcatchment (dark red coloured cell and green cell) even reach an age of 100 and 55 years, respectively. This can be explained by the fact that it is the lowest cell within the subcatchment and that water accumulates here. The overall flow path to this cell is the longest and as a consequence the groundwater age in this cell is the highest.

In general, 2 % of cells contain groundwater that is older than 50 years, <1 % reveal ages >70 years, 13 % contain water with an age of less than 1 year, and 52 % with an age <15 years. Thus, most of the cells contain young to moderately old water (<15 years), while few cells comprise old water (>50 years). The average groundwater age in the Vollnkirchener Bach subcatchment is 16 years. Correlating the groundwater age with the distance to the stream, we found a linear correlation ($R^2 = 0.3$) with a distinct trend. The water tends to be younger with greater distance to the stream.

The amount of flowing water depicted by the length of the arrows is generally higher near the stream, whereas in most of the outer cells the amount is very low (Fig. 10). The modelled main flow direction is towards the Vollnkirchener Bach, but many arrows show a flow direction across the stream, indicating bidirectional water exchange between the stream and the groundwater body.

4   Discussion

4.1   Variations of precipitation isotopes and d excess

We found no spatial variation in precipitation isotopes throughout the Schwingbach catchment. For north-western Europe, Mook et al. (1974) also observed that precipitation collected over periods of 8 and 24 h from three different locations within 6 km^2 at the same elevation were consistent within 0.3 ‰ for $\delta^{18}O$. Further, we detected no amount or altitude effect on isotopes in precipitation. Amount effects are generally most likely to occur in the tropics or for intense convective rain events and are not a key factor for explaining isotope distributions in German precipitation (Stumpp et al., 2014).

The observed linear relationship ($\delta^{18}O = 0.44T - 12.05‰$) between air temperature and precipitation $\delta^{18}O$ values compares reasonably well with a correlation reported by Yurtsever (1975) based on North Atlantic and European stations from the GNIP network $\delta^{18}O = (0.521 \pm 0.014)T - (14.96 \pm 0.21)‰$. The same is true for a correlation found by Rozanski et al. (1982) for the GNIP station Stuttgart, 196 km south of the Schwingbach. Stumpp et al. (2014) analysed long-term precipitation data from meteorological stations across Germany and found that 23 out of 24 tested stations showed a positive long-term temperature trend over time. The observed correspondence between the degree of isotope depletion and the temperature reflects the influence of the temperature effect in the Schwingbach catchment, which mainly appears in continental, middle–high latitudes (Jouzel et al., 1997). Furthermore, the correlation between δ^2H in monthly precipitations and local surface air temperature becomes increasingly stronger towards the centre of the continent (Rozanski et al., 1982). Thus, the observed seasonal differences in precipitation δ values in the Schwingbach catchment could mainly be attributed to seasonal differences in air temperature and the presence of snow in the winter of 2012–2013 (Fig. 2).

Precipitation events originating from oceanic moisture show d-excess values close to +10‰ (Craig, 1961a; Dansgaard, 1964; Wu et al., 2012), and one of the main sources for precipitation in Germany is moisture from the Atlantic Ocean (Stumpp et al., 2014). Lowest values corresponded to summer precipitation events where the evaporation of the falling raindrops below the cloud base occurs. The same observations were made by Rozanski et al. (1982) for European GNIP stations. Winter snow samples of the Schwingbach catchment with very depleted δ values showed the highest d-excess values (>+10‰), in good agreement with results of Rozanski et al. (1982) for European GNIP stations. The observed differences in d-excess values between the Schwingbach catchment and the GNIP station Koblenz can be attributed to differences in elevation range and the different regional climatic settings at both sites (Koblenz is located in the relatively warmer Rhine River valley).

4.2 Isotopes of soil water

4.2.1 Spatial variability

We found no statistically significant connection between the parameters' distance to stream, TWI, soil water content, soil texture, pH, and bulk density with the soil isotopic signatures in both soil depths. This was potentially attributed to the small variation in soil textures (mainly clayey silts and loamy sandy silts), bulk densities, and pH values for both soil depths (Table 2). Garvelmann et al. (2012) obtained high-resolution

δ^2H vertical depth profiles of pore water at various points along two fall lines of a pasture hillslope in the Black Forest (Germany) by applying the H_2O(liquid)–H_2O(vapour) equilibration laser spectroscopy method. The authors showed that groundwater was flowing through the soil in the riparian zone (downslope profiles) and dominated streamflow during baseflow conditions. Their comparison indicated that the percentage of pore water soil samples with a very similar stream water δ^2H signature increases towards the stream channel (Garvelmann et al., 2012). In contrast, we found no such relationship between the distance to stream or TWI and soil isotopic values in the Vollnkirchener Bach subcatchment over various elevations (235–294 m a.s.l.) and locations. We attributed this to the gentle hillslopes and the low subsurface flow contribution in large parts of the catchment.

In our study, the δ values of topsoil and precipitation did not differ statistically (Fig. 4), but for precipitation and subsoil they did. The latter indicates either the influence of evaporation in the topsoil or the mixing with groundwater in the subsoil. However, a mixing and homogenization of new and old soil water with depth could not be seen clearly at 0.5 m soil depth, which would have resulted in a lower standard deviation (Song et al., 2011), but standard deviations of isotopic signatures in top- and subsoil were similar (Table 2). Subsoil isotopic values were statistically equal to stream water and groundwater (Fig. 4), implying that capillary rise of groundwater occurred. Overall, the rainfall isotopic signal was not directly transferred through the soil to the groundwater; even so the groundwater head level rose promptly after rainfall events. This behaviour reflects the differences of celerity and velocity in the catchment's rainfall–runoff response (McDonnell and Beven, 2014).

Soil water δ^2H between top- and subsoil showed significant differences ($p \leq 0.05$) under arable land and grassland but not under forested sites (Fig. 5). This could be explained through the occurrence of vertical preferential flow paths and interconnected macropore flow (Buttle and McDonald, 2002) characteristic of forested soils. Alaoui et al. (2011) showed that macropore flow with high interaction with the surrounding soil matrix occurred in forest soils, while macropore flow with low to mixed interaction with the surrounding soil matrix dominates in grassland soils. Seasonal tilling prevents the establishment of preferential flow paths under agricultural sites and is regularly done in the Schwingbach catchment, whereas the structure of forest soils, may remain uninterrupted throughout the entire soil profile for years (in particular the macropores and biopores) (Alaoui et al., 2011). This is reflected in the bulk density of the soils in the Schwingbach catchment, which increases from forests (1.10 g cm^{-3}) to grassland (1.25 g cm^{-3}) to arable land (1.41 g cm^{-3}) in the topsoil. We infer that reduced hydrological connection between top- and subsoil under arable and grassland led to different isotopic signatures (Fig. 5).

Although vegetation cover has often shown an impact on soil water isotopes (Gat, 1996), only few data are available for Central Europe (Darling, 2004). Burger and Seiler (1992) found that soil water isotopic enrichment under spruce forest in Upper Bavaria was double that beneath neighbouring arable land, but soil isotope values were not comparable to groundwater (Burger and Seiler, 1992). Gehrels et al. (1998) also detected (though only slightly) heavier isotopic signatures under forested sites in the Netherlands in comparison to non-forested sites (grassland and heathland). By contrast, in southern Germany, Brodersen et al. (2000) observed only a negligible effect of throughfall isotopic signatures (of spruce and beech) on soil water isotopes, since soil water in the upper layers followed the seasonal trend in the precipitation input and had a very constant signature at greater depth. In a study by Sprenger et al. (2016b) the differences between the investigated soil profiles across the Attert catchment (LU) were mostly driven by soil types, which was also seen in the pore water stable isotope dynamics reported for soils in the Scottish Highlands (Geris et al., 2015). However, for the Schwingbach catchment, we conclude that the observed land use effect in the upper soil column is mainly attributed to different preservation and transmission of the precipitation input signal. It is most likely not attributable to distinguished throughfall isotopic signatures, the impact of evaporation or interception losses, since topsoil water isotopic signals followed the precipitation input signal under all land use units.

4.2.2 Seasonal isotope soil profiling

Soil water was enriched in summer due to evaporation during warmer and drier periods. The depth to which soil water isotopes are significantly affected by evaporation is rarely more than 1–2 m below ground and often less under temperate climates (Darling, 2004). In contrast, winter profiles exhibited somewhat greater standard deviations in comparison to summer isotopic soil profiles, indicative of wetter soils (Fig. 6, lower panels) and shorter residence times (Thomas et al., 2013). Isotope profiles taken during or after snowmelt in a study by Sprenger et al. (2016b) did not show an isotopic depletion at a certain depth as observed for example by Stumpp and Hendry (2012) and Peralta-Tapia et al. (2015). Generally, deeper soil water isotope values in our study were relatively constant through time and space. Similar findings were made by Foerstel et al. (1991) on a sandy soil in western Germany, by McConville et al. (2001) under predominately agriculturally used gley and till soils in Northern Ireland, Thomas et al. (2013) in a forested catchment in central Pennsylvania, USA, and by Bertrand et al. (2014) on the Pfyn alluvial forest (CH). Furthermore, Tang and Feng (2001) showed, for a sandy loam in New Hampshire (USA), that the influence of summer precipitation decreased with increasing depth, and soils at 0.5 m only received water from large storms. Pore water δ^2H profiles taken at the catchment of the groundwater aquifer Freiburger Bucht (DE) in a study by Sprenger et al. (2016a) showed how the isotopic signal of rain water over time is preserved in the unsaturated soil profile.

However, the input signal was dampened due to mixing processes. In our summer soil profiles under arable land, precipitation input signals decreased with depth (Fig. 6, upper left panel). Dampening of precipitation's isotopic fluctuations with increasing soil depth was in line with other studies (e.g. Muñoz-Villers and McDonnell, 2012; Timbe et al., 2014; Wang et al., 2010). Generally, the replacement of old soil water with new infiltrating water is dependent on the frequency and intensity of precipitation and the soil texture, structure, wetness, and water potential of the soil (Li et al., 2007; Tang and Feng, 2001). As a result, the amount of percolating water decreases with depth and consequently, deeper soil layers have less chance to obtain new water (Tang and Feng, 2001). In the growing season, the percolation depth is additionally limited by plants' transpiration (Tang and Feng, 2001). For the Schwingbach catchment we conclude that the percolation of new soil water is low as no remarkable seasonality in soil isotopic signatures was obvious at > 0.9 m and constant values were observed through space and time. Although replications over several years are missing, this result indicates a transit time through the rooting zone (1m) of approximately 1 year.

4.3   Linkages between water cycle components

Stream water isotopic time series of the Vollnkirchener Bach and Schwingbach showed little deflections through time. Due to the observed isotopic similarities of stream and groundwater, we conclude that groundwater predominantly feeds baseflow (discharge $< 10\,\mathrm{L\,s^{-1}}$). Even during peak flow occurring in January 2012 and December to April or May 2013, rainfall input did not play a major role for stream water isotopic composition although fast rainfall–runoff behaviours were observed by Orlowski et al. (2014). The damped groundwater isotopic signatures seemed to be a mixture of former lighter precipitation events and snowmelt, since meltwater is known to be depleted in stable isotopes as compared to precipitation or groundwater (Rohde, 1998) (Fig. 3). However, differences in the snow sampling method (new snow, snow pit layers, meltwater) can affect the isotopic composition (Penna et al., 2014; Taylor et al., 2001). As groundwater at the observed piezometers in the Vollnkirchener subcatchment is shallow (Orlowski et al., 2014), the snowmelt signal is able to move rapidly through the soil. Pulses of snowmelt water causing a depletion in spring and early summer were also observed by other studies (Darling, 2004; Kortelainen and Karhu, 2004). We therefore conclude that groundwater is mainly recharged throughout the winter. During spring runoff when soils are saturated, temperatures are low, and vegetation is inactive, recharge rates are generally highest. In contrast, recharge is very low during summer when most precipitation is transpired back to the atmosphere (Clark and Fritz, 1997a). Similarly, O'Driscoll et al. (2005) showed that summer precipitation does not significantly contribute to recharge in the Spring Creek watershed (Pennsylvania, USA) since δ^{18}O values in summer precipitation were enriched compared to mean annual groundwater composition.

Further, Orlowski et al. (2014) showed that influent and effluent conditions (bidirectional water exchange) occurred simultaneously in different stream sections of the Vollnkirchener Bach, affecting stream and groundwater isotopic compositions equally. Our network map supported this assumption (Fig. 9) as surface water sampling points plotted close to groundwater sampling points (especially to the sampling points under the meadow and along the stream). This was also underlined by our groundwater model showing flow directions across the Vollnkirchener Bach. Nevertheless, both stream and groundwater differed significantly from rainfall isotopic signatures (Table 1). Thus, our catchment showed double water paradox behaviour as per Kirchner (2003), with fast release of very old water but little variation in tracer concentration.

4.4　Water age dynamics

Our MTT calculations did not provide a good fit between the observed and calculated data. Just by comparing mean precipitation, stream, and groundwater isotopic signatures (Table 1), one could expect that simple mixing calculations would not work to derive MTTs, i.e. showing predominant groundwater contribution. The same observations were made by Jin et al. (2012), indicating good hydraulic connectivity between surface water and shallow groundwater. Just as in the results presented here, Klaus et al. (2015) had difficulties to apply traditional methods of isotope hydrology (MTT estimation, hydrograph separation) to their data set due to the lack of temporal isotopic variation in stream water of a forested low mountainous catchment in South Carolina (USA). Furthermore, stable water isotopes can only be utilised for estimations of younger water (< 5 years) (Stewart et al., 2010) as they are blind to older contributions (Duvert et al., 2016). In our catchment, transit times are orders of magnitudes longer than the timescale of hydrologic response (prompt discharge of old water) (McDonnell et al., 2010) and the range used for stable water isotopes.

Accurately capturing the transit time of the old water fraction is essential (Duvert et al., 2016) and could previously only be determined via other tracers such as tritium (e.g. Michel, 1992). Current studies on mixing assumptions either consider spatial or time-varying MTTs. Heidbüchel et al. (2012) proposed the concept of the master transit time distribution that accounts for the temporal variability of MTT. The time-varying transit time concept of Botter et al. (2011) and van der Velde et al. (2012) was recently reformulated by Harman (2015) so that the storage selection function became a function of the watershed storage and actual time. Instead of quantifying time-variant travel times, our model facilitates the estimation of spatially distributed groundwater ages,

which opens up new opportunities to compare groundwater ages from over a range of scales within catchments. Furthermore, it gives a deeper understanding of the groundwater–surface water connection across the landscape than a classical MTT calculation could provide. Our work complements recent advances in spatially distributed modelling of age distributions through transient groundwater flows (e.g. Gomez and Wilson, 2013; Woolfenden and Ginn, 2009). The results of our model reveal a spatially highly heterogeneous age distribution of groundwater throughout the Vollnkirchener Bach subcatchment (ages of 2 days–100 years), with the oldest water near the stream. Thus, our model provides the opportunity to make use of stable water isotope information along with climate, land use, and soil type data, in combination with a digital elevation map to estimate residence times > 5 years. If stable water isotope information is used alone, it is known to cause a truncation of stream residence time distributions (Stewart et al., 2010). Further, our groundwater model suggests that the main groundwater flow direction is towards and across the stream and the quantity of flowing water is highest near the stream (Fig. 10). This further supports the assumption that stream water is mainly fed by older groundwater. Moreover, the simulation underlines the conclusion that the groundwater body and stream water are isotopically disconnected from the precipitation cycle, since only 13 % of cells contained water with an age < 1 year.

However, our semi-conceptual model approach also has some limitations. During model set-up a series of assumptions and simplifications were made to develop a realistic hydrologic model without a severe loss in performance. Due to the assumption of a constant groundwater recharge over the course of a year, no seasonality was simulated. Moreover, no spatial differences in soil properties of the groundwater layer were considered. Further, several parameters such as the depth of the groundwater body are only rough estimations, while others like evapotranspiration are based on simulations. Moreover, the groundwater body is highly simplified since, e.g., properties of the simulated aquifer are assumed to be constant over the subcatchment. Nevertheless, as shown by the diverse ages of water in the stream cells and the assumption of spatially gaining conditions, the model confirms that the stream contains water with different transit times and supports the assumption that surface and groundwater are isotopically disconnected from precipitation. Therefore, the stream water does not have a discrete age, but a distribution of ages due to variable flow paths (Stewart et al., 2010). In future models a more diverse groundwater body based on small-scale measurements of aquifer parameters should be implemented. Especially data of saturated hydraulic conductivity with a high spatial resolution, as well as the implementation of a temporal dynamic groundwater recharge could lead to an enhanced model performance.

5   Conclusions

Conducting a stable water isotope study in the Schwingbach catchment helped to identify relationships between precipitation, stream, soil, and groundwater in a developed (managed) catchment. The close isotopic link between groundwater and the streams revealed that groundwater controls streamflow. Moreover, it could be shown that groundwater was predominately recharged during winter but was decoupled from the annual precipitation cycle. Although streamflow and groundwater head levels promptly responded to precipitation inputs, there was no obvious change in their isotopic composition due to rain events.

Nevertheless, the lack of temporal variation in stable isotope time series of stream and groundwater limited the application of classical methods of isotope hydrology, i.e. transfer-function-based MTT estimations. By splitting the flow path into different compartments (upper and lower vadose zone, groundwater, stream), we were able to determine, where the water age passes the limit of using stable isotopes for age calculations. This limit is in the lower vadose zone, approximately 1–2 m below ground. To estimate the total transit time to the stream, we set up a hydrological model calculating spatially distributed groundwater ages and flow directions in the Vollnkirchener Bach subcatchment. Our model results supported the finding that the water in the catchment is > 5 years old (on average 16 years) and that stream water is mainly fed by groundwater. Our modelling approach was valuable to overcome the limitations of MTT calculations with traditional methods and/or models. Further, our dual isotope study in combination with the hydrological model approach enabled the determination of connection and disconnection between different water cycle components.

Appendix A: Mean transit time estimation

We applied a set of five different models to estimate the MTT using the FlowPC software (Maloszewski and Zuber, 2002): a dispersion model (with different dispersion parameters $D_p = 0.05$, 0.4, and 0.8), an exponential model, an exponential-piston-flow model, a linear model, and a linear-piston-flow model. We evaluated these results using two goodness-of-fit criteria, i.e. sigma (σ) and model efficiency (ME) following Maloszewski and Zuber (2002):

$$\sigma = \frac{\sqrt{\sum (c_{mi} - c_{oi})^2}}{m}, \tag{A1}$$

$$ME = 1 - \frac{\sum (c_{mi} - c_{oi})^2}{\sum (c_{oi} - \bar{c}_o)^2}, \tag{A2}$$

where c_{mi} is the i-th model result, c_{oi} is the i-th observed result, and $\bar{c}_o$ is the arithmetic mean of all observations.

A model efficiency ME = 1 indicates an ideal fit of the model to the concentrations observed, while ME = 0 indicates that the model fits the data no better than a horizontal

line through the mean observed concentration (Maloszewski and Zuber, 2002). The same is true for sigma. For calculations with FlowPC, weekly averages of precipitation and stream water isotopic signatures are calculated. We firstly calculated the MTT from precipitation to the streams for three sampling points in the Vollnkirchener Bach (sites 13, 18, and 94) and three points in the Schwingbach (sites 11, 19, and 64). For the second set of simulations, the mean residence time from precipitation to groundwater comprising 13 groundwater sampling points was determined. We also bias-corrected the precipitation input data with two different approaches. The mean precipitation value is subtracted from every single precipitation value and then divided by the standard deviation of precipitation isotopic signatures. Afterwards, this value is subtracted from the weekly precipitation values (bias1). For the second approach, the difference in the mean stream water isotopic value and the mean precipitation value is calculated and also subtracted from the weekly precipitation values (bias2).

Appendix B: Model-based groundwater age dynamics

B1   Objective

Stable water isotopes are only a tool to determine the residence time for a few years (McDonnell et al., 2010). In cases of longer residence times and a strong mixing effect, seasonal variation of isotopes vanishes and results in barely varying isotopic signals. To get a rough estimate of residence times greater than the limit of stable water isotopes (> 5 years), we split the water flow path in our catchment in two parts: the flow from precipitation to groundwater, which was calculated via FlowPC and the longer groundwater transport. The simplest method to estimate the residence time of groundwater transport is via the storage-to-input relation, with the storage as the aquifer size and the input as the groundwater recharge time. However, this method ignores the topographic setting and water input heterogeneity. In our study we used a simplified groundwater flow model with tracer transport to calculate the groundwater age dynamics. The numerical output of water ages cannot be validated with the given isotope data, since the model is used to fill a residence time gap, where it is not feasible to apply stable water isotopes. The model is falsified, however, if the residence time is short enough (< 5 years) to be calculable via FlowPC. Hence, the results of the groundwater age model should be handled with care and only seen as the order of magnitude of flow timescales.

B2   Model setup

We set up a tailored hydrological model for the Vollnkirchener Bach subcatchment using the CMF by Kraft et al. (2011). CMF is a modular framework for hydrological modelling based on the concept of finite volume method by Qu and Duffy (2007). CMF is applicable for simulating one-to three-dimensional water fluxes but also advective transport of stable water isotopes (^{18}O and ^{2}H). Thus, it is especially suitable for our tracer study and can be used to study the origin (Windhorst et al., 2014) and age of water. To avoid errors in transit time calculations from small differences between the isotopic signal in groundwater and stream water, we are tracing the transit time of groundwater and not the real isotopic values in this study. The generated model is a highly simplified representation of the Vollnkirchener Bach subcatchment's groundwater body. The subcatchment is divided into 353 polygonal-shaped cells ranging from 100 to 40 000 m^2 in size based on land use, soil type, and topography. The model is vertically divided into two compartments, the upper soft rock aquifer, and the lower bedrock aquifer, referred to as upper and lower layer from now onwards.

The layers of each cell are connected using a mass conservative Darcy approach with a finite volume discretization. The water storage dynamic of one layer in one cell i of the groundwater model is given as

$$\frac{dV_{i,s}}{dt} = R_i - S_i - \sum_{j=1}^{N_i} \left(K_s \frac{\Psi_{i,s} - \Psi_{j,s}}{d_{ij}} A_{ij,s} \right), \qquad (B1)$$

$$\frac{dV_{i,b}}{dt} = S_i - \sum_{j=1}^{N_i} \left(K_b \frac{\Psi_{i,b} - \Psi_{j,b}}{d_{ij}} A_{ij,b} \right), \qquad (B2)$$

where V_i is the water volume stored by the layer in m^{-3} in cell I for soft rock (s) and bedrock (b); R_i is the groundwater recharge rate in m^2 day^{-1}; S_i is the percolation from the soft rock to the bedrock aquifer, calculated by the gradient and geometric mean conductivity between the layers: $S_i = \sqrt{K_s K_b} \frac{\Psi_{i,s} - \Psi_{i,b}}{d_{sb}} A_i$, where d_{sb} is the distance between the layers and A_i is the cell area; N_i is the number of adjacent cells to cell i; K is the saturated hydraulic conductivity in m day^{-1} for soft rock (s) and bedrock (b), respectively; Ψ is the water head in the current cell i and the neighbour cell j in metres for soft rock (s) and bedrock (b); d_{ij} is the distance between the current cell i and the neighbour cell j in metres; and $A_{i,j,x}$ is the wetted area of the joint layer boundary in m^2 between cells i and j in layer x.

The volume head relation is linearized as $\Psi = \phi \frac{V}{A}$, with ϕ being the fillable porosity and A the cell area. The resulting ordinary differential equation system is integrated using the CVODE solver by Hindmarsh et al. (2005), an error-controlled Krylov–Newton multistep implicit solver with an adaptive order of 1–5 according to stability constraints.

B3 Boundary conditions

The upper boundary condition of the groundwater system – the mean groundwater recharge – is modelled applying a Richard's equation based model using measured rainfall data (2011–2013) and calculated evapotranspiration with the Shuttleworth–Wallace method (Shuttleworth and Wallace, 1985) including land cover and climate data. To retrieve long-term steady-state conditions, the groundwater recharge is averaged and used as a constant-flow Neumann bound-

ary condition. The total outflow is calibrated against measured outflow data; hence, the unsaturated model's role is mainly to account for spatial heterogeneity of groundwater recharge. As an additional input, a combined sewer overflow (site 38, Fig. 1b) is considered based on findings of Orlowski et al. (2014). Moreover, there are two water outlets in the two lowest cells for efficient draining, reflecting measured groundwater flow directions throughout most of the year at piezometers 1–6 (Fig. 1b). Both cells are located in the very north of the subcatchment and their outlets are modelled as constant head Dirichlet boundary condition.

B4 Parameters

The saturated hydraulic conductivity of the groundwater body is set to 0.1007 m day^{-1}, as measured in the study area. For the lower bedrock compartment there are no data available. However, expecting a high rate of joints, preliminary testing revealed that a saturated hydraulic conductivity of 0.25 m day^{-1} seemed to be a realistic estimation (based on field measurements).

B5 Water age

To calculate the water age in each cell, a virtual tracer flows through the system using advective transport. To calculate the water age from the tracer that enters the system with a unity concentration by groundwater recharge, a linear decay is used to reduce the tracer concentration with time:

$$\frac{dX_{i,s}}{dt} = 1 \frac{u}{m^3} R_i - S_i [X]_{i,s}$$
$$- \sum_{j=1}^{N_i} \left([X]_{i,s} K_s \frac{\Psi_{i,s} - \Psi_{j,s}}{d_{ij}} A_{ij,s} \right) - r X_{i,s}, \qquad (B3)$$

$$\frac{dX_{i,b}}{dt} = S_i [X]_{i,s} - \sum_{j=1}^{N_i} \left([X]_{i,b} K_b \frac{\Psi_{i,b} - \Psi_{j,b}}{d_{ij}} A_{ij,b} \right)$$
$$- r X_{i,b} t_{ix} = \frac{\ln [X]_{ix}}{r}, \qquad (B4)$$

where $X_{i,x}$ is the amount of virtual tracer in layer x in cell i in virtual unit u; $1\,u\,m^{-3} R_i$ is the tracer input with groundwater recharge R with unity concentration; $[X]_{i,x}$ is the concentration of tracer in layer x of cell i in $u\,m^{-3}$; r is the arbitrarily chosen decay constant, for water age calculation in day^{-1} – rounding errors occur due to low concentrations when r is set to a high value and we found a good numerical performance with values between 10^{-6} and 10^{-9} day^{-1}; and t_{ix}: water age in days in layer x in cell i.

To ensure long-term steady-state conditions, the model is run for 2000 years. However, after 300 years of model run time, steady state is reached.

Acknowledgements. The first author acknowledges financial support by the Friedrich-Ebert-Stiftung (Bonn, DE). Furthermore, this work was supported by the Deutsche Forschungsgemeinschaft under Grant BR2238/10-1. Thanks also to the student assistants and BSc and MSc students for their help during field sampling campaigns and hydrological modelling. In this context we would like to acknowledge especially Julia Mechsner, Julia Klöber, and Judith Henkel. We thank Christine Stumpp from the Helmholtz Zentrum München for providing us with the isotope data from GNIP station Koblenz. We thank Kwok Pan (Sun) Chun for statistical support and suggestions along the way.

Edited by: M. Weiler

References

Alaoui, A., Caduff, U., Gerke, H. H., and Weingartner, R.: Preferential Flow Effects on Infiltration and Runoff in Grassland and Forest Soils, Vadose Zone J., 10, 367–377, doi:10.2136/vzj2010.0076, 2011.

Allan, J. D.: Landscapes and Riverscapes: The Influence of Land Use on Stream Ecosystems, Annu. Rev. Ecol. Evol. Syst., 35, 257–284, doi:10.1146/annurev.ecolsys.35.120202.110122, 2004.

Barnes, C. J. and Allison, G. B.: Tracing of water movement in the unsaturated zone using stable isotopes of hydrogen and oxygen, J. Hydrol., 100, 143–176, doi:10.1016/0022-1694(88)90184-9, 1988.

Barthold, F. K., Tyralla, C., Schneider, K., Vaché, K. B., Frede, H.-G., and Breuer, L.: How many tracers do we need for end member mixing analysis (EMMA)? A sensitivity analysis, Water Resour. Res., 47, W08519, doi:10.1029/2011WR010604, 2011.

Bertrand, G., Masini, J., Goldscheider, N., Meeks, J., Lavastre, V., Celle-Jeanton, H., Gobat, J.-M., and Hunkeler, D.: Determination of spatiotemporal variability of tree water uptake using stable isotopes (δ^{18}O, δ^2H) in an alluvial system supplied by a high-altitude watershed, Pfyn forest, Switzerland, Ecohydrol., 7, 319–333, doi:10.1002/eco.1347, 2014.

Blasch, K. W. and Bryson, J. R.: Distinguishing Sources of Ground Water Recharge by Using δ^2H and δ^{18}O, Ground Water, 45, 294–308, doi:10.1111/j.1745-6584.2006.00289.x, 2007.

Botter, G., Bertuzzo, E., and Rinaldo, A.: Transport in the hydrologic response: Travel time distributions, soil moisture dynamics, and the old water paradox, Water Resour. Res., 46, W03514, doi:10.1029/2009WR008371, 2010.

Botter, G., Bertuzzo, E., and Rinaldo, A.: Catchment residence and travel time distributions: The master equation, Geophys. Res. Lett., 38, L11403, doi:10.1029/2011GL047666, 2011.

Brodersen, C., Pohl, S., Lindenlaub, M., Leibundgut, C., and Wilpert, K. V: Influence of vegetation structure on isotope content of throughfall and soil water, Hydrol. Process., 14, 1439–1448, doi:10.1002/1099-1085(20000615)14:8<1439::AID-HYP985>3.0.CO;2-3, 2000.

Burger, H. M. and Seiler, K. P.: Evaporation from soil water under humid climate conditions and its impact on deuterium and ^{18}O concentrations in groundwater, edited by: International Atomic Energy Agency, International Atomic Energy Agency, Vienna, Austria, 674–678, 1992.

Buttle, J. M.: Isotope hydrograph separations and rapid delivery of pre-event water from drainage basins, Prog. Phys. Geogr., 18, 16–41, doi:10.1177/030913339401800102, 1994.

Buttle, J. M.: Isotope Hydrograph Separation of Runoff Sources, in Encyclopedia of Hydrological Sciences, edited by: Anderson, M. G., p. 10, 116, John Wiley & Sons, Ltd, Chichester, Great Britain, 2006.

Buttle, J. M. and McDonald, D. J.: Coupled vertical and lateral preferential flow on a forested slope, Water Resour. Res., 38, 18-1–18-16, doi:10.1029/2001WR000773, 2002.

Clark, I. D. and Fritz, P.: Groundwater, in Environmental Isotopes in Hydrogeology, p. 80, CRC Press, Florida, FL, USA, 1997a.

Clark, I. D. and Fritz, P.: Methods for Field Sampling, in Environmental Isotopes in Hydrogeology, p. 283, CRC Press, Florida, FL, USA, 1997b.

Clark, I. D. and Fritz, P.: The Environmental Isotopes, in: Environmental Isotopes in Hydrogeology, 2–34, CRC Press, Florida, FL, USA, 1997c.

Cooper, L. W.: Isotopic Fractionation in Snow Cover, in Isotope Tracers in Catchment Hydrology, edited by: Kendall, C. and McDonnell, J. J., 119–136, Elsevier, Amsterdam, the Netherlands, 1998.

Craig, H.: Isotopic Variations in Meteoric Waters, Science, 133, 1702–1703, doi:10.1126/science.133.3465.1702, 1961a.

Craig, H.: Standard for reporting concentrations of deuterium and oxygen-18 in natural waters, Science, 133, 1833–1834, doi:10.1126/science.133.3467.1833, 1961b.

Csardi, G. and Nepusz, T.: The igraph software package for complex network research, Complex Systems 1695, available at: http://igraph.sf.net (last access: 24 December 2015), 2006.

Dansgaard, W.: Stable isotopes in precipitation, Tellus, 16, 436–468, doi:10.1111/j.2153-3490.1964.tb00181.x, 1964.

Darling, W. G.: Hydrological factors in the interpretation of stable isotopic proxy data present and past: a European perspective, Quat. Sci. Rev., 23, 743–770, doi:10.1016/j.quascirev.2003.06.016, 2004.

Duvert, C., Stewart, M. K., Cendón, D. I., and Raiber, M.: Time series of tritium, stable isotopes and chloride reveal short-term variations in groundwater contribution to a stream, Hydrol. Earth Syst. Sci., 20, 257–277, doi:10.5194/hess-20-257-2016, 2016.

DWD: Deutscher Wetterdienst – Wetter und Klima, Bundesministerium für Verkehr und digitale Infrastruktur, available at: http://dwd.de/ (last access: 17 February 2014), 2014.

Foerstel, H., Frinken, J., Huetzen, H., Lembrich, D., and Puetz, T.: Application of H_2^{18}O as a tracer of water flow in soil, International Atomic Energy Agency, Vienna, Austria, 523–532, 1991.

Garvelmann, J., Kuells, C., and Weiler, M.: A porewater-based stable isotope approach for the investigation of subsurface hydrological processes, Hydrol. Earth Syst. Sci., 16, 631–640, doi:10.5194/hess-16-631-2012, 2012.

Gat, J. R.: Oxygen and Hydrogen Isotopes in the Hydrologic Cycle, Annu. Rev. Earth Planet. Sci., 24, 225–262, doi:10.1146/annurev.earth.24.1.225, 1996.

Gat, J., R., Mook, W. G., and Meijer, H. A. J.: Environmental isotopes in the hydrological cycle: Principles and Applications, edited by: Mook, W. G., International Hydrological Programme, United Nations Educational, Scientific and Cultural Organization

Exploring water cycle dynamics by sampling multiple stable water isotope pools in a developed landscape...

77

and International Atomic Energy Agency, Paris, France, p. 73, 2001.

Gehrels, J. C., Peeters, J. E. M., de Vries, J. J., and Dekkers, M.: The mechanism of soil water movement as inferred from ^{18}O stable isotope studies, Hydrol. Sci. J., 43, 579–594, doi:10.1080/02626669809492154, 1998.

Genereux, D. P. and Hooper, R. P.: Oxygen and Hydrogen Isotopes in Rainfall-Runoff Studies, in Isotope Tracers in Catchment Hydrology, edited by: Kendall, C. and McDonnell, J. J., 319–346, Elsevier, Amsterdam, the Netherlands, 1998.

Geris, J., Tetzlaff, D., McDonnell, J., and Soulsby, C.: The relative role of soil type and tree cover on water storage and transmission in northern headwater catchments, Hydrol. Process., 29, 1844–1860, doi:10.1002/hyp.10289, 2015.

Gomez, J. D. and Wilson, J. L.: Age distributions and dynamically changing hydrologic systems: Exploring topography-driven flow, Water Resour. Res., 49, 1503–1522, doi:10.1002/wrcr.20127, 2013.

Gonfiantini, R., Fröhlich, K., Araguás-Araguás, L., and Rozanski, K.: Isotopes in Groundwater Hydrology, in Isotope Tracers in Catchment Hydrology, edited by: Kendall, C. and McDonnell, J. J., 203–246, Elsevier, Amsterdam, the Netherlands, 1998.

Gordon, L. J., Finlayson, C. M., and Falkenmark, M.: Managing water in agriculture for food production and other ecosystem services, Agr. Water Manag., 97, 512–519, doi:10.1016/j.agwat.2009.03.017, 2010.

Harman, C. J.: Time-variable transit time distributions and transport: Theory and application to storage-dependent transport of chloride in a watershed, Water Resour. Res., 51, 1–30, doi:10.1002/2014WR015707, 2015.

Heidbüchel, I., Troch, P. A., Lyon, S. W., and Weiler, M.: The master transit time distribution of variable flow systems, Water Resour. Res., 48, W06520, doi:10.1029/2011WR011293, 2012.

Hindmarsh, A. C., Brown, P. N., Grant, K. E., Lee, S. L., Serban, R., Shumaker, D. E., and Woodward, C. S.: SUNDIALS: Suite of Nonlinear and Differential/Algebraic Equation Solvers, ACM Trans. Math. Softw., 31, 363–396, doi:10.1145/1089014.1089020, 2005.

Hollander, M., Wolfe, D. A., and Chicken, E.: Nonparametric Statistical Methods, John Wiley & Sons, New York, NY, USA, 2013.

Hrachowitz, M., Benettin, P., van Breukelen, B. M., Fovet, O., Howden, N. J. K., Ruiz, L., van der Velde, Y., and Wade, A. J.: Transit times – the link between hydrology and water quality at the catchment scale, Wiley Interdiscip. Rev. Water, 3, 629–657, doi:10.1002/wat2.1155, 2016.

IAEA: International Atomic Energy Agency: Water Resources Programme – Global Network of Isotopes in Precipitation, available at: http://www-naweb.iaea.org/napc/ih/IHS_resources_gnip.html (last access: 11 August 2014), 2014.

Jin, L., Siegel, D. I., Lautz, L. K., and Lu, Z.: Identifying streamflow sources during spring snowmelt using water chemistry and isotopic composition in semi-arid mountain streams, J. Hydrol., 470–471, 289–301, doi:10.1016/j.jhydrol.2012.09.009, 2012.

Jouzel, J., Alley, R. B., Cuffey, K. M., Dansgaard, W., Grootes, P., Hoffmann, G., Johnsen, S. J., Koster, R. D., Peel, D., Shuman, C. A., Stievenard, M., Stuiver, M., and White, J.: Validity of the temperature reconstruction from water isotopes in ice cores, J. Geophys. Res., 102, 26471–26487, doi:10.1029/97JC01283, 1997.

Kendall, C. and Caldwell, E. A.: Fundamentals of Isotope Geochemistry, in Isotope Tracers in Catchment Hydrology, edited by: Kendall, C. and McDonnell, J. J., 51–86, Elsevier, Amsterdam, the Netherlands, 1998.

Kirchner, J. W.: A double paradox in catchment hydrology and geochemistry, Hydrol. Process., 17, 871–874, doi:10.1002/hyp.5108, 2003.

Klaus, J., McDonnell, J. J., Jackson, C. R., Du, E., and Griffiths, N. A.: Where does streamwater come from in low-relief forested watersheds? A dual-isotope approach, Hydrol. Earth Syst. Sci., 19, 125–135, doi:10.5194/hess-19-125-2015, 2015.

Koeniger, P., Leibundgut, C., and Stichler, W.: Spatial and temporal characterisation of stable isotopes in river water as indicators of groundwater contribution and confirmation of modelling results; a study of the Weser river, Germany, Isotopes Environ. Health Stud., 45, 289–302, doi:10.1080/10256010903356953, 2009.

Kolaczyk, E. D.: Statistical Analysis of Network Data with R, Springer Science & Business Media, New York, NY, USA, 187–188, 2014.

Kortelainen, N. M. and Karhu, J. A.: Regional and seasonal trends in the oxygen and hydrogen isotope ratios of Finnish groundwaters: a key for mean annual precipitation, J. Hydrol., 285, 143–157, doi:10.1016/j.jhydrol.2003.08.014, 2004.

Kraft, P., Vaché, K. B., Frede, H.-G., and Breuer, L.: CMF: A Hydrological Programming Language Extension For Integrated Catchment Models, Environ. Model. Softw., 26, 828–830, doi:10.1016/j.envsoft.2010.12.009, 2011.

Lai, C.-T. and Ehleringer, J. R.: Deuterium excess reveals diurnal sources of water vapor in forest air, Oecologia, 165, 213–223, doi:10.1007/s00442-010-1721-2, 2011.

Lauer, F., Frede, H.-G., and Breuer, L.: Uncertainty assessment of quantifying spatially concentrated groundwater discharge to small streams by distributed temperature sensing, Water Resour. Res., 49, 400–407, doi:10.1029/2012WR012537, 2013.

LGR: Los Gatos Research, Greenhouse Gas, isotope and trace gas analyzers, available at: http://www.lgrinc.com/ (last access: 5 February 2013), 2013.

Li, F., Song, X., Tang, C., Liu, C., Yu, J., and Zhang, W.: Tracing infiltration and recharge using stable isotope in Taihang Mt., North China, Environ. Geol., 53, 687–696, doi:10.1007/s00254-007-0683-0, 2007.

Maloszewski, P. and Zuber, A.: Manual on lumped parameter models used for the interpretation of environmental tracer data in groundwaters, in: Use of isotopes for analyses of flow and transport dynamics in groundwater systems, p. 50, International Atomic Energy Agency, Vienna, Austria, 2002.

McConville, C., Kalin, R. M., Johnston, H., and McNeill, G. W.: Evaluation of Recharge in a Small Temperate Catchment Using Natural and Applied δ^{18}O Profiles in the Unsaturated Zone, Ground Water, 39, 616–623, doi:10.1111/j.1745-6584.2001.tb02349.x, 2001.

McDonnell, J. J. and Beven, K.: Debates – The future of hydrological sciences: A (common) path forward? A call to action aimed at understanding velocities, celerities and residence time distributions of the headwater hydrograph, Water Resour. Res., 50, 5342–5350, doi:10.1002/2013WR015141, 2014.

McDonnell, J. J., Sivapalan, M., Vaché, K., Dunn, S., Grant, G., Haggerty, R., Hinz, C., Hooper, R., Kirchner, J., Roderick, M. L., McDonnell, J. J., Sivapalan, M., Vaché, K., Dunn, S., Grant,

G., Haggerty, R., Hinz, C., Hooper, R., Kirchner, J., Roderick, M. L., Selker, J., and Weiler, M.: Moving beyond heterogeneity and process complexity: a new vision for watershed hydrology, Water Resour. Res., 43, W07301, doi:10.1029/2006WR005467, 2007.

McDonnell, J. J., McGuire, K., Aggarwal, P., Beven, K. J., Biondi, D., Destouni, G., Dunn, S., James, A., Kirchner, J., Kraft, P., Lyon, S., Maloszewski, P., Newman, B., Pfister, L., Rinaldo, A., Rodhe, A., Sayama, T., Seibert, J., Solomon, K., Soulsby, C., Stewart, M., Tetzlaff, D., Tobin, C., Troch, P., Weiler, M., Western, A., Worman, A., and Wrede, S.: How old is streamwater? Open questions in catchment transit time conceptualization, modelling and analysis, Hydrol. Process., 24, 1745–1754, 2010.

McGuire, K. J. and McDonnell, J. J.: A review and evaluation of catchment transit time modeling, J. Hydrol., 330, 543–563, 2006.

Michel, R. L.: Residence times in river basins as determined by analysis of long-term tritium records, J. Hydrol., 130, 367–378, doi:10.1016/0022-1694(92)90117-E, 1992.

Mook, W. G., Groeneveld, D. J., Brouwn, A. E., and van Ganswijk, A. J.: Analysis of a run-off hydrograph by means of natural ^{18}O, in Isotope techniques in groundwater hydrology, 1, 159–169, International Atomic Energy Agency, Vienna, Austria, 1974.

Muñoz-Villers, L. E. and McDonnell, J. J.: Runoff generation in a steep, tropical montane cloud forest catchment on permeable volcanic substrate, Water Resour. Res., 48, W09528, doi:10.1029/2011WR011316, 2012.

Neal, C. and Rosier, P. T. W.: Chemical studies of chloride and stable oxygen isotopes in two conifer afforested and moorland sites in the British uplands, J. Hydrol., 115, 269–283, doi:10.1016/0022-1694(90)90209-G, 1990.

Newman, B., Tanweer, A., and Kurttas, T.: IAEA Standard Operating Procedure for the Liquid-Water Stable Isotope Analyser, Laser Proced, IAEA Water Resour. Programme, 2009.

O'Driscoll, M. A., DeWalle, D. R., McGuire, K. J., and Gburek, W. J.: Seasonal ^{18}O variations and groundwater recharge for three landscape types in central Pennsylvania, USA, J. Hydrol., 303, 108–124, doi:10.1016/j.jhydrol.2004.08.020, 2005.

Orlowski, N., Frede, H.-G., Brüggemann, N., and Breuer, L.: Validation and application of a cryogenic vacuum extraction system for soil and plant water extraction for isotope analysis, J. Sens. Sens. Syst., 2, 179–193, doi:10.5194/jsss-2-179-2013, 2013.

Orlowski, N., Lauer, F., Kraft, P., Frede, H.-G., and Breuer, L.: Linking Spatial Patterns of Groundwater Table Dynamics and Streamflow Generation Processes in a Small Developed Catchment, Water, 6, 3085–3117, doi:10.3390/w6103085, 2014.

Penna, D., Ahmad, M., Birks, S. J., Bouchaou, L., Brenčič, M., Butt, S., Holko, L., Jeelani, G., Martínez, D. E., Melikadze, G., Shanley, J. B., Sokratov, S. A., Stadnyk, T., Sugimoto, A., and Vreča, P.: A new method of snowmelt sampling for water stable isotopes, Hydrol. Process., 28, 5637–5644, doi:10.1002/hyp.10273, 2014.

Peralta-Tapia, A., Sponseller, R. A., Tetzlaff, D., Soulsby, C., and Laudon, H.: Connecting precipitation inputs and soil flow pathways to stream water in contrasting boreal catchments, Hydrol. Process., 29, 3546–3555, doi:10.1002/hyp.10300, 2015.

Perry, C. and Taylor, K.: Environmental Sedimentology, p. 36, Blackwell Publishing, Oxford, OX, UK, 2009.

Pierce, S. C., Kröger, R., and Pezeshki, R.: Managing Artificially Drained Low-Gradient Agricultural Headwaters

for Enhanced Ecosystem Functions, Biology, 1, 794–856, doi:10.3390/biology1030794, 2012.

Qu, Y. and Duffy, C. J.: A semidiscrete finite volume formulation for multiprocess watershed simulation, Water Resour. Res., 43, W08419, doi:10.1029/2006WR005752, 2007.

Rinaldo, A., Benettin, P., Harman, C. J., Hrachowitz, M., McGuire, K. J., van der Velde, Y., Bertuzzo, E., and Botter, G.: Storage selection functions: A coherent framework for quantifying how catchments store and release water and solutes, Water Resour. Res., 51, 4840–4847, doi:10.1002/2015WR017273, 2015.

Rohde, A.: Snowmelt-Dominated Systems, in Isotope Tracers in Catchment Hydrology, edited by: Kendall, C. and McDonnell, J. J., 391–433, Elsevier, Amsterdam, the Netherlands, 1998.

Rozanski, K., Sonntag, C., and Münnich, K. O.: Factors controlling stable isotope composition of European precipitation, Tellus, 34, 142–150, doi:10.1111/j.2153-3490.1982.tb01801.x, 1982.

Rozanski, K., Froehlich, K., and Mook, W. G.: in Environmental isotopes in the hydrological cycle: Principles and Applications, vol. 3, International Hydrological Programme, United Nations Educational, Scientific and Cultural Organization and International Atomic Energy Agency, Paris, Vienna, 2001.

Schultz, N. M., Griffis, T. J., Lee, X., and Baker, J. M.: Identification and correction of spectral contamination in ^{2}H/^{1}H and ^{18}O/^{16}O measured in leaf, stem, and soil water, Rapid Commun. Mass Spectrom., 25, 3360–3368, doi:10.1002/rcm.5236, 2011.

Shuttleworth, W. J. and Wallace, J. S.: Evaporation from sparse crops-an energy combination theory, Q. J. Roy. Meteor. Soc., 111, 839–855, doi:10.1002/qj.49711146910, 1985.

Sklash, M. G.: Environmental isotope studies of storm and snowmelt runoff generation, in Process studies in hillslope hydrology, edited by: Anderson, M. G., 410–435, Wiley, New York, NY, USA, 1990.

Sklash, M. G. and Farvolden, R. N.: The Role Of Groundwater In Storm Runoff, in Developments in Water Science, vol. 12, edited by: Back, W. and Stephenson, D. A., 45–65, Elsevier, Amsterdam, Netherlands., 1979.

Song, X., Wang, P., Yu, J., Liu, X., Liu, J., and Yuan, R.: Relationships between precipitation, soil water and groundwater at Chongling catchment with the typical vegetation cover in the Taihang mountainous region, China, Environ. Earth Sci., 62, 787–796, doi:10.1007/s12665-010-0566-7, 2011.

Sprenger, M., Erhardt, M., Riedel, M., and Weiler, M.: Historical tracking of nitrate in contrasting vineyards using water isotopes and nitrate depth profiles, Agric. Ecosyst. Environ., 222, 185–192, doi:10.1016/j.agee.2016.02.014, 2016a.

Sprenger, M., Seeger, S., Blume, T., and Weiler, M.: Travel times in the vadose zone: Variability in space and time, Water Resour. Res., doi:10.1002/2015WR018077, 2016b.

Stewart, M. K., Morgenstern, U., and McDonnell, J. J.: Truncation of stream residence time: how the use of stable isotopes has skewed our concept of streamwater age and origin, Hydrol. Process., 24, 1646–1659, doi:10.1002/hyp.7576, 2010.

Stumpp, C. and Hendry, M. J.: Spatial and temporal dynamics of water flow and solute transport in a heterogeneous glacial till: The application of high-resolution profiles of δ^{18}O and δ^2H in pore waters, J. Hydrol., 438–439, 203–214, doi:10.1016/j.jhydrol.2012.03.024, 2012.

Stumpp, C., Klaus, J., and Stichler, W.: Analysis of long-term stable isotopic composition in German precipitation, J. Hydrol., 517, 351–361, doi:10.1016/j.jhydrol.2014.05.034, 2014.

Tang, K. and Feng, X.: The effect of soil hydrology on the oxygen and hydrogen isotopic compositions of plants' source water, Earth Planet. Sci. Lett., 185, 355–367, 2001.

Taylor, S., Feng, X., Kirchner, J. W., Osterhuber, R., Klaue, B., and Renshaw, C. E.: Isotopic evolution of a seasonal snowpack and its melt, Water Resour. Res., 37, 759–769, doi:10.1029/2000WR900341, 2001.

Thomas, E. M., Lin, H., Duffy, C. J., Sullivan, P. L., Holmes, G. H., Brantley, S. L., and Jin, L.: Spatiotemporal Patterns of Water Stable Isotope Compositions at the Shale Hills Critical Zone Observatory: Linkages to Subsurface Hydrologic Processes, Vadose Zone J., 12, doi:10.2136/vzj2013.01.0029, 2013.

Timbe, E., Windhorst, D., Crespo, P., Frede, H.-G., Feyen, J., and Breuer, L.: Understanding uncertainties when inferring mean transit times of water trough tracer-based lumped-parameter models in Andean tropical montane cloud forest catchments, Hydrol. Earth Syst. Sci., 18, 1503–1523, doi:10.5194/hess-18-1503-2014, 2014.

van der Velde, Y., Torfs, P. J. J. F., van der Zee, S. E. A. T. M., and Uijlenhoet, R.: Quantifying catchment-scale mixing and its effect on time-varying travel time distributions, Water Resour. Res., 48, W06536, doi:10.1029/2011WR011310, 2012.

Wang, P., Song, X., Han, D., Zhang, Y., and Liu, X.: A study of root water uptake of crops indicated by hydrogen and oxygen stable isotopes: A case in Shanxi Province, China, Agric. Water Manag., 97, 475–482, 2010.

Wang, X. F. and Yakir, D.: Using stable isotopes of water in evapotranspiration studies, Hydrol. Process., 14, 1407–1421, doi:10.1002/1099-1085(20000615)14:8<1407::AID-HYP992>3.0.CO;2-K, 2000.

Windhorst, D., Waltz, T., Timbe, E., Frede, H.-G., and Breuer, L.: Impact of elevation and weather patterns on the isotopic composition of precipitation in a tropical montane rainforest, Hydrol. Earth Syst. Sci., 17, 409–419, doi:10.5194/hess-17-409-2013, 2013.

Windhorst, D., Kraft, P., Timbe, E., Frede, H.-G., and Breuer, L.: Stable water isotope tracing through hydrological models for disentangling runoff generation processes at the hillslope scale, Hydrol. Earth Syst. Sci., 18, 4113–4127, doi:10.5194/hess-18-4113-2014, 2014.

Woolfenden, L. R. and Ginn, T. R.: Modeled Ground Water Age Distributions, Ground Water, 47, 547–557, doi:10.1111/j.1745-6584.2008.00550.x, 2009.

Wu, J., Ding, Y., Ye, B., Yang, Q., Hou, D., and Xue, L.: Stable isotopes in precipitation in Xilin River Basin, northern China and their implications, Chin. Geogr. Sci., 22, 531–540, doi:10.1007/s11769-012-0543-z, 2012.

Xia, Y.: Optimization and uncertainty estimates of WMO regression models for the systematic bias adjustment of NLDAS precipitation in the United States, J. Geophys. Res.-Atmos., 111, D08102, doi:10.1029/2005JD006188, 2006.

Yurtsever, Y.: Worldwide survey of stable isotopes in precipitation, International Atomic Energy Agency, Vienna, Austria, 44 pp., 1975.

Zhao, L., Xiao, H., Zhou, J., Wang, L., Cheng, G., Zhou, M., Yin, L., and McCabe, M. F.: Detailed assessment of isotope ratio infrared spectroscopy and isotope ratio mass spectrometry for the stable isotope analysis of plant and soil waters, Rapid Commun. Mass Spectrom., 25, 3071–3082, doi:10.1002/rcm.5204, 2011.

5

Response of water vapour D-excess to land–atmosphere interactions in a semi-arid environment

Stephen D. Parkes[1], **Matthew F. McCabe**[1,2], **Alan D. Griffiths**[3], **Lixin Wang**[4], **Scott Chambers**[3], **Ali Ershadi**[1,2], **Alastair G. Williams**[3], **Josiah Strauss**[2], and **Adrian Element**[3]

[1]Water Desalination and Reuse Centre, King Abdullah University of Science and Technology (KAUST), Jeddah, Saudi Arabia
[2]Department of Civil and Environmental Engineering, University of New South Wales, Sydney, Australia
[3]Australian Nuclear Science and Technology Organization, Sydney, New South Wales, Australia
[4]Department of Earth Sciences, Indiana University – Purdue University Indianapolis (IUPUI), Indianapolis, IN, USA

Correspondence to: Stephen D. Parkes (stephen.parkes@kaust.edu.sa)

Abstract. The stable isotopic composition of water vapour provides information about moisture sources and processes difficult to obtain with traditional measurement techniques. Recently, it has been proposed that the D-excess of water vapour ($d_v = \delta^2 H - 8 \times \delta^{18} O$) can provide a diagnostic tracer of continental moisture recycling. However, D-excess exhibits a diurnal cycle that has been observed across a variety of ecosystems and may be influenced by a range of processes beyond regional-scale moisture recycling, including local evaporation (ET) fluxes. There is a lack of measurements of D-excess in evaporation (ET) fluxes, which has made it difficult to assess how ET fluxes modify the D-excess in water vapour (d_v). With this in mind, we employed a chamber-based approach to directly measure D-excess in ET (d_{ET}) fluxes. We show that ET fluxes imposed a negative forcing on the ambient vapour and could not explain the higher daytime d_v values. The low d_{ET} observed here was sourced from a soil water pool that had undergone an extended drying period, leading to low D-excess in the soil moisture pool. A strong correlation between daytime d_v and locally measured relative humidity was consistent with an oceanic moisture source, suggesting that remote hydrological processes were the major contributor to daytime d_v variability. During the early evening, ET fluxes into a shallow nocturnal inversion layer caused a lowering of d_v values near the surface. In addition, transient mixing of vapour with a higher D-excess from above the nocturnal inversion modified these values, causing large variability during the night. These results indicate d_{ET} can generally be expected to show

large spatial and temporal variability and to depend on the soil moisture state. For long periods between rain events, common in semi-arid environments, ET would be expected to impose negative forcing on the surface d_v. Spatial and temporal variability of D-excess in ET fluxes therefore needs to be considered when using d_v to study moisture recycling and during extended dry periods with weak moisture recycling may act as a tracer of the relative humidity at the oceanic moisture source.

1 Introduction

Climate change has the potential to significantly impact surface and atmospheric water budgets. Our best understanding of future exchanges between the atmospheric water cycle and the land surface for regional to global scales is likely to be gained through analysis of numerical simulations (Decker et al., 2015; Evans and McCabe, 2010; Harding and Snyder, 2012; Wei et al., 2012). Consequently, continual improvement of available models is essential, but this is contingent upon ongoing validation and evaluation of model performance over a broad range of landscapes and climate types (McCabe et al., 2016). To do this effectively, a diversity of datasets that directly quantify processes represented within these models are required (McCabe et al., 2005). Unfortunately, datasets that directly measure land–atmosphere exchange at the process level are limited (Jana et al., 2016).

Water is composed of a number of stable isotopologues that have sufficient abundance to be measured in atmospheric water vapour ($^1H_2^{16}O$, $^1H^2H^{16}O$, $^1H_2^{18}O$ and $^1H_2^{17}O$). Deviations of water isotope ratios are reported as

$$\delta = \left[\frac{R_{sample}}{R_{VSMOW}} - 1 \right] ‰, \qquad (1)$$

where R is the isotope ratio ($^2H/^1H$ or $^{18}O/^{16}O$), and VSMOW (Vienna Standard Mean Ocean Water) is the international standard for reporting water isotope ratios, and these ratios have the potential to evaluate land–atmosphere exchange by discriminating processes based on their isotopic signature (Berkelhammer et al., 2013; Lee et al., 2009; Noone et al., 2013; Risi et al., 2013). Isotopic ratios of water vapour (δ^2H and $\delta^{18}O$) can therefore provide information that is complimentary or even unobtainable when using conventional measurement techniques.

The utility of water isotope ratios for tracing sources of moisture derives from the characteristic equilibrium and kinetic isotopic fractionation that occurs when water undergoes a phase change, causing light water molecules to preferentially accumulate in the vapour phase. Soil moisture is typically enriched in heavy isotopes relative to the ocean (Gat, 1996), so water vapour derived from land surface evaporation is expected to have a different isotopic composition to moisture evaporated from the ocean. This has led to a number of studies using stable isotopes in precipitation to partition oceanic and land-derived sources (Froehlich et al., 2008; Tian et al., 2001). However, land–atmosphere exchange is not restricted to periods of precipitation, and there are relatively few studies examining the role of land–atmosphere exchange on ambient humidity budgets using stable isotope observations of vapour (e.g. Aemisegger et al., 2014; Risi et al., 2013).

In addition to the source of moisture, the magnitude of isotopic fractionation that occurs when water evaporates is related to the liquid surface temperature and humidity gradient between the evaporating surface and atmosphere (Craig and Gordon, 1965). The temperature-dependent equilibrium exchange between liquid and vapour is the largest contributor to isotopic fractionation during evaporation, with the fractionation for δ^2H approximately a factor of 8 greater than $\delta^{18}O$. The effect of kinetic fractionation associated with moisture diffusing from the thin laminar layer of vapour in equilibrium with the water surface to the turbulent atmosphere above is influenced by the relative humidity of the atmosphere and wind speed (Merlivat and Jouzel, 1979). The kinetic fractionation factors for δ^2H and $\delta^{18}O$ are similar, causing the ratio of δ^2H to $\delta^{18}O$ in the evaporating vapour to decrease as kinetic effects increase with decreasing relative humidity. This phenomenon has been observed for evaporative conditions over the Mediterranean Sea (Gat et al., 2003; Pfahl and Wernli, 2009) and the Great Lakes in northern USA (Gat et al., 1994; Vallet-Coulomb et al., 2008).

The D-excess (D-excess $= \delta^2H - 8 \times \delta^{18}O$) parameter (Dansgaard, 1964) quantifies the non-equilibrium isotopic fractionation. A reproducible relationship between the D-excess and relative humidity near the ocean surface has been observed across a wide range of locations (Kurita, 2011; Pfahl and Wernli, 2008; Steen-Larsen et al., 2015; Uemura et al., 2008). Therefore, it has been suggested that for precipitation, D-excess is a good tracer of sea surface evaporative conditions (Masson-Delmotte et al., 2005; Merlivat and Jouzel, 1979). However, this view has recently been challenged due to the role local and regional scale land–atmosphere coupling has in modifying the D-excess of atmospheric humidity over diurnal (Lai and Ehleringer, 2011; Simonin et al., 2014; Welp et al., 2012; Zhao et al., 2014) and synoptic timescales (Aemisegger et al., 2014). As evidence for the role ET plays in modifying the D-excess of water vapour (d_v), a diurnal cycle of d_v near the land surface across a range of land surface types has been observed (Berkelhammer et al., 2013; Simonin et al., 2014; Welp et al., 2012). The diurnal cycle shows higher values during the day, which has been proposed to be driven by entrainment (Lai and Ehleringer, 2011; Welp et al., 2012), local evapotranspiration sources (Simonin et al., 2014; Zhao et al., 2014) and meteorological conditions affecting the D-excess of the evaporative fluxes (d_{ET}) (Welp et al., 2012; Zhao et al., 2014), coupled with low nocturnal values resulting from equilibrium exchange between liquid and vapour pools (Simonin et al., 2014) and dewfall (Berkelhammer et al., 2013). For synoptic scales, Aemisegger et al. (2014) showed that moisture recycling from the land surface had a significant impact on d_v for in situ measurements in Switzerland. These studies have largely relied on isotopic models to assess the contribution of ET fluxes, but a lack of d_{ET} measurements makes it difficult to draw robust conclusions.

The evidence provided by these studies suggest d_v is a tracer of moisture recycling both on diurnal and synoptic timescales, and is influenced by the dynamics of surface moisture budgets in the atmospheric boundary layer (ABL). However, as noted by Welp et al. (2012), ET and entrainment fluxes both increase as the ABL grows through the previous-days residual layer, which can make interpreting the role of local moisture recycling on d_v difficult. To overcome this, Simonin et al. (2014) used a trajectory model to simulate the D-excess of vapour evaporated over the ocean. As the d_v was greater than the modelled oceanic moisture source, it was assumed that high daytime values were supported by local ET fluxes. Zhao et al. (2014) suggested that since, on cloudy days, no diurnal cycle was observed for the d_v, ET fluxes therefore played a dominant role. Whilst these studies provide compelling evidence for the role of ET driving the diurnal cycle of d_v, no measurements of d_{ET} were made. To date the only measurements of d_{ET} have been presented by Huang and Wen (2014) over a maize crop in Northwest China. Interestingly, their direct measurements conflicted with previous interpretations and showed that the d_{ET} invoked a negative

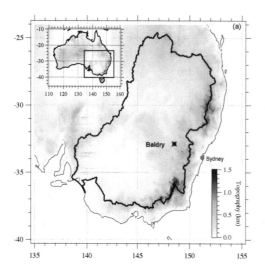
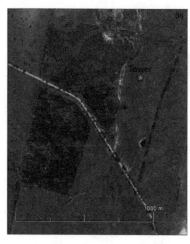

Figure 1. (a) Location of the Baldry Hydrological Observatory, with the heavy black border outlining the extent of the Murray–Darling Basin, **(b)** location of the field site used for the campaign, illustrating the semi-arid grassland and adjacent reforested site.

forcing on d_v, even though a strong diurnal cycle of high values during the day and low values at night were observed. In order to better interpret the role of local moisture recycling on the diurnal cycle of d_v, measurements of d_{ET} are required to assess if the negative forcing is consistent across different ecosystems.

The aim of this work is to provide much needed d_{ET} measurements to investigate how ET fluxes modulate the d_v diurnal cycle. To do this, chamber-based measurements of the ET flux isotopic compositions were combined with in situ measurements of water vapour isotope ratios, meteorological and radon concentration observations. The data were collected in a region of the semi-arid Murray–Darling basin in southeastern Australia. These data represent the first such collection of the δ^2H, $\delta^{18}O$ and D-excess in water vapour from this region of Australia. The augmentation of the chamber-based measurements with in situ observations provides a framework to directly assess the role local ET fluxes have on ambient vapour D-excess.

2 Methods

2.1 Site description

During the austral autumn of 2011, a field campaign covering the period 27 April to 11 May was conducted at the Baldry Hydrological Observatory (BHO) (−32.87, 148.54, 460 m a.s.l. – above sea level) located in the central-west of New South Wales, Australia (Fig. 1). The climate of the region is characterised as semi-arid with no clear wet season, a mean annual rainfall of 600 mm, and a mean annual temperature of 24.2 °C (source Australian Bureau of Meteorology, 2016; http://www.bom.gov.au/). The BHO grassland eddy covariance flux tower was the central site of measurements

and was located in a natural grassland paddock of dimensions approximately 900 m (north–south) by 300 m (west–east), with a gentle slope decreasing in elevation by approximately 20 m from south-east to north-west. The flux tower was located 650 m from the road to the south and 200 m from a reforested paddock to the west. The forest site to the west and south-west was reforested in 2001 with *Eucalyptus camaldulensis*, *Eucalyptus crebra* and *Corymbia maculata*. At the time of the campaign these trees were approximately 10 m tall. All other adjacent paddocks and most of the surrounding region had similar surface characteristics to the grassland measurement site.

2.2 Water stable isotope analyses

2.2.1 In situ water vapour calibration and sampling

In situ water vapour isotope ratios were monitored using a Wavelength Scanning Cavity Ring Down Spectrometer (WS-CRDS L115-I, Picarro Inc., Sunnyvale, CA, USA), while flux chambers were interfaced to an Off-Axis Integrated Cavity Output Spectrometer (OA-ICOS, DLT100, Los Gatos Research – LGR, Mountain View, CA, USA) to determine the isotopic composition of ET fluxes. Using an automated continuous flow calibration system (built in-house), we simultaneously determined calibration coefficients for both analysers. Calibration experiments were designed to determine the water vapour mixing ratio cross-sensitivity of isotope ratios and linearity of the δ^2H and $\delta^{18}O$ measurements. More details on the calibration procedure are found in the Supplement. Due to logistical constraints, the calibration system was not transported into the field, so corrections were determined by compositing multiple calibration experiments run before and after the campaign.

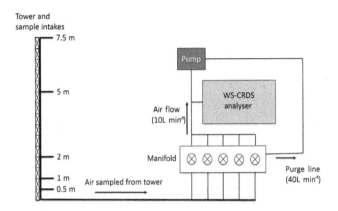

Figure 2. Sampling system for the automated in situ collection and measurement of water vapour isotopes from the tower.

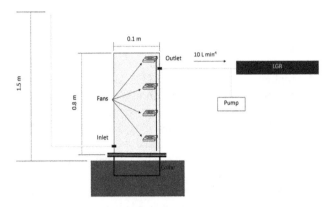

Figure 3. Chamber design used for determining the isotopic compositions of ET fluxes.

During the campaign, a secondary portable calibration system was employed to monitor time-dependent drift of the Picarro analyser (CTC HTC Pal liquid autosampler; LEAP Technologies, Carrboro, NC, USA). Two standards spanning expected water vapour δ^2H (-49.1 and $-221.9\,\permil$) and $\delta^{18}O$ (-9.17 and $-27.57\,\permil$) ranges were injected at approximately $18\,mmol\,mol^{-1}$ on three occasions during the campaign.

The uncertainty of measurements from both isotopic analysers was estimated by applying mixing ratio cross-sensitivity and linearity corrections to all calibration measurements collected prior to, during and after the campaign. For the Picarro instrument, measurement uncertainty was 0.8, 0.2 and $1.9\,\permil$ for δ^2H_v, $\delta^{18}O_v$ and d_v, respectively. No calibrations were performed for the LGR in field, so the measurement uncertainty was estimated by compositing calibration measurements made before and after the campaign, which were 0.9, 0.4 and $3.3\,\permil$ for δ^2H, $\delta^{18}O$ and d_v, respectively.

Although no calibration experiments were run on the LGR during the campaign, simultaneous in situ measurements were made with the Picarro when chamber measurements were not operated. During the day, average differences were -0.06 (± 2.0), 0.13 (± 0.5) and 0.4 (± 3.3)$\,\permil$ for δ^2H, $\delta^{18}O$ and d_v, respectively. A comparison of the analysers is shown in Fig. S1 in the Supplement. At night, while the Picarro was able to maintain a steady cavity and optical housing temperature, the LGR cavity temperature dropped by up to 8 °C. In response to the drop in cavity temperature, nighttime LGR measurements of $\delta^{18}O$ and d_v, and to a lesser extent the δ^2H, were physically unrealistic and discarded from subsequent analyses. Chamber measurements were therefore restricted to between 09:00 LST (when the LGR cavity temperature had stabilised and in situ measurements were again in agreement with the Picarro) and 17:00 LST (before LGR cavity temperatures began dropping).

A schematic diagram illustrating the sampling design for water vapour is shown in Fig. 2. Half-hourly vertical pro-

files of humidity and isotopes were sampled by drawing air to the in situ analyser through 10 mm (outside diameter) PTFE tubing, located at five heights on a 7.5 m tower (0.5, 1, 2, 5 and 7.5 m a.g.l.). The instrument was interfaced to a five-inlet manifold that enabled sequential sampling of the different heights. Approximately 20 m of tubing was required to connect the tower inlet to the analyser. A vacuum pump (MV 2 NT, Vacuubrand, Wertheim, Germany) was used to draw air through all inlets to the analyser at a flow rate of $10\,L\,min^{-1}$, with the Picarro bleeding off the $0.03\,L\,min^{-1}$ through its measurement cavity. To avoid condensation, sample tubes and intakes were wrapped in $15\,W\,m^{-1}$ heat tape, insulated by Thermobreak pipe and placed inside 100 mm PVC pipe. The sample tube temperature was controlled using a resistance thermometer detector coupled to a CAL3300 temperature controller (CAL controls Ltd., Grayslake, IL, USA). The inlets at each height were constructed from inverted funnels with mesh filters. In this study we present block hourly averages of all measurements collected at all heights.

2.2.2 Flux chambers

To separate the isotopic signatures of the ET flux components, flux chambers were deployed on both bare soil and vegetated plots to determine the isotopic signature of the evaporative fluxes. An open chamber was designed with a high volume-to-footprint ratio to avoid the chamber mixing ratio rapidly reaching the dew point temperature (causing condensation) and to minimise impacts on the evaporation environment. A schematic of the chamber design is shown in Fig. 3. Four flanged metal collars were inserted $\sim 10\,cm$ into the soil column 2 days before the beginning of the campaign. While this was a short settling time for chamber bases, shallow roots of grass cover within the chamber were largely unaffected. All vegetation was removed from bare soil plots when the metal collars were inserted into the soil. A single chamber cover was constructed out of 4 mm G-UVT Plexiglas (Image Plastics, Padstow, Australia), selected for its higher transmittance of UV (ultraviolet) and blue light. The dimensions of the chamber were $0.1 \times 0.1 \times 0.8$ m (width $\times$ length $\times$ height), with the inlets and outlets at 0.1 and 0.7 m above the surface, respectively. All sampling tubes were 10 mm PTFE. The inlet to the chamber was connected to tubing that drew in air from 1.5 m above the ground surface. The outlet was connected to a flowmeter (VFA-25, Dwyers, Michigan City, IN, USA) that regulated the airflow at $10 \, \text{L min}^{-1}$ and was driven by a two-stage diaphragm pump. A T-piece was connected to the LGR, which bled off approximately $0.8 \, \text{L min}^{-1}$. All tubing between the chamber and the analyser were wrapped in heating tape ($15 \, \text{W m}^{-2}$) and foam insulation. High flow rates were used to combat memory effects modifying the isotopic composition of the vapour within the chamber. Analysis of chamber measurements was conducted on 2–5 min of data, so 2.5–6.25 chamber volumes were exchanged.

To monitor the internal chamber environment, an air temperature and humidity probe (HMP155, Vaisala, Vantaa, Finland) was mounted inside the chamber. To monitor the attenuation of the incoming radiation by the chamber, the photosynthetic flux density was measured (LI-190R, Licor, Lincoln, NE, USA) inside and outside the chamber; 10 s averages of the temperature, relative humidity and photosynthetic flux density were stored in a datalogger (CR1000, Campbell Scientific, Logan, UT, USA). In the Supplement we use these ancillary measurements to assess the impact of observed changes in the chamber environment on the isotopic composition of ET fluxes. The largest contributor to uncertainty caused by changing the evaporative environment was the temperature, although these affects were small compared to the overall variability of the chamber-derived ET isotopic compositions.

2.2.3 Isotopic composition of ET flux from chamber measurements

Mass balance or Keeling-mixing (Keeling, 1958; Wang et al., 2013a) models have been applied to determine the isotopic composition of ET fluxes from chamber measurements (Lu et al., 2017; Wang et al., 2013b). The focus of this work was not to evaluate chamber measurement techniques. Considering that it has been shown that Keeling and mass balance methods give very similar results (Lu et al., 2017; Wang et al., 2013b) we focus on using the Keeling-mixing model, given by

$$\delta_{\text{chamber}} = q_{\text{BG}} \frac{(\delta_{\text{BG}} - \delta_{\text{ET}})}{q_{\text{chamber}}} + \delta_{\text{ET}} \qquad (2)$$

where q_{BG} is the water vapour mixing ratio entering the chamber through the inlet and δ_{BG} its isotopic composition, q_{chamber} is the mixing ratio in the chamber and δ_{ET} is the isotopic composition of the ET flux. The δ_{ET} is determined from the intercept of δ_{chamber} against $1/q_{\text{chamber}}$. A key assumption of the Keeling method is that the isotopic composition of

the background vapour and the evaporation flux remain constant during the chamber measurements. For chamber measurements longer than 5 min, non-linear Keeling plots were commonly observed, indicating a change in isotopic composition of one of the sources of vapour. We therefore restricted the Keeling analysis to a maximum of 5 min after an increase in the concentration was observed by the analyser. Ensuring the linearity of Keeling plots also ensured that the influence of memory effects was minimised. Memory effects would constitute an additional moisture source, violating the two-source assumption of the Keeling methods and reducing Keeling plot linearity. The analysis was also restricted to periods where the H_2O mixing ratio was increasing, so analysis was generally performed on 2–5 min of data. In addition, only chamber measurements where the correlation between δ_{chamber} and $1/q_{\text{chamber}}$ was significant ($p < 0.001$) were included in this analysis. A few chamber measurements where obvious non-linearity or very small changes in q_{chamber} occurred were also subjectively removed. Of a total of 105 chamber measurements made from the 4 vegetation plots during the campaign, 99 measurements of the $\delta^2 H_{\text{ET}}$, and 97 measurements of $\delta^{18}O_{\text{ET}}$ and d_{ET} were retained. For the bare soil plots, 84 of the 86 chamber measurements were retained for the $\delta^2 H_{\text{ET}}$, and 77 of the $\delta^{18}O_{\text{ET}}$ and d_{ET}. The eight plots were sampled 2 to 4 times each day on all days except the first two days of the campaign, and 2 and 5 May. Sampling was restricted to between 09:00 and 17:00 LST (local solar time) as the large temperature dependence of the LGR at low ambient temperatures limited the accuracy of the chamber measurements.

Results from vegetated plots were used to determine ET flux isotopic compositions and determine how ET influences d_{v}. The bare soil plots were used to determine the isotopic composition of soil evaporation fluxes and to provide an estimate of the isotopic composition of water at the evaporation front. The isotopic composition of the water at the evaporation front (δ_{L}) was determined by rearranging the Craig and Gordon model:

$$\delta_{\text{L}} = \frac{\delta_{\text{E}}(1 - \text{RH}) + \text{RH}\delta_{\text{A}} + \varepsilon + \varepsilon_{\text{k}}}{\alpha} \qquad (3)$$

where the isotopic composition of the evaporation flux (δ_{E}) is taken from the bare soil chamber measurements, relative humidity (RH) normalised to the surface temperature determined from infrared surface temperature measurements (Sect. 2.3), and the ambient vapour isotope composition (δ_{A}) determined from Picarro in situ measurements. Equilibrium fractionation and enrichment factors (α, $\varepsilon = (\alpha - 1) \%o$) were calculated from the surface temperature measurements using the equations of Horita and Wesolowski (1994), while the kinetic enrichment factors (ε_{k}) were determined as in Gat (1996), but using the parameterisation of the exponent of the diffusion coefficients described by Mathieu and Bariac (1996) and the diffusion coefficients determined by Merlivat (1978).

2.2.4 Isoforcing of ET

The isotopic composition of the near-surface atmospheric water vapour is modified by surface ET fluxes. The impact of ET fluxes on surface vapour isotopes varies over diurnal timescales with the strength of vertical mixing in the ABL or over synoptic timescales as background moisture conditions change. The magnitude and isotopic composition of the ET flux as well as the amount of water vapour in the atmosphere also have an influence. The ET isoforcing (I_{ET}) represents a useful quantity to study the influence of ET fluxes on the surface vapour and is defined as

$$I_{ET} = \frac{F_{ET}}{H_2O} \left(\delta_{ET} - \delta_A \right), \tag{4}$$

where F_{ET} is the ET flux in $mol\, m^{-2}\, s^{-1}$, H_2O is the ambient mixing ratio in $mol\text{-}air\, mol\text{-}H_2O^{-1}$ measured by the local meteorological tower, and δ_{ET} and δ_A are the isotopic compositions of the evaporation flux and ambient water vapour, respectively (Lee et al., 2009).

For each chamber measurement, a surface isoforcing was calculated for δ^2H, $\delta^{18}O$ and D-excess from the determined ET isotopic composition, as well as the hourly averaged ET flux, mixing ratio and δ_A values. The importance of surface fluxes modifying surface vapour isotope composition was investigated for diurnal and synoptic timescales.

2.2.5 Plant and soil sampling

Grass samples were collected three times a day for the duration of the campaign. They were sampled randomly within 100 m of the instrumentation. Each sample consisted of approximately 10 grass leaves, which were placed in 12 mL Exetainer vials (Labco, Ceredigion, UK). The grass samples were assumed to represent bulk leaf water. Soil samples were collected every 2 days throughout the campaign by sampling from the top 5 cm of the soil column. They were collected in 50 mL glass bottles. Soil and plant samples were stored in a fridge (4 °C), before using the distillation method of West et al. (2006) to extract liquid water samples that were analysed on a Delta V Advantage Isotope Ratio Mass Spectrometer (Thermo Fisher Scientific Corporation, Massachusetts, USA). For δ^2H analysis, water samples were introduced into a H-device containing a chromium reactor, while for the $\delta^{18}O$ analysis, water samples were equilibrated with CO_2 on a Gas Bench II chromatography column (Thermo Fisher Scientific Corporation, Massachusetts, USA) before being transferred to the Isotope Ratio Mass Spectrometer (IRMS) for analysis.

2.3 ET fluxes and meteorological measurements

To measure ET fluxes, an eddy covariance system comprising a Campbell Scientific 3-D sonic anemometer (CSAT-3, Campbell Scientific, Logan, UT, USA) along with a Licor 7500 (Li-7500, Licor Biosciences, Lincoln, NB, USA) analyser was installed at an elevation of 2.5 m. The system was located approximately 10 m from the stable isotope observation tower and sampled at 10 Hz, with flux averages output at 30 min intervals. The ET fluxes from the eddy covariance tower are used to quantify the isoforcing of ET on the overlying atmosphere.

A meteorological tower was co-located with the eddy covariance system, providing complementary data to aid in the interpretation of measurements. The tower comprised a Kipp and Zonen CNR4 radiometer, Apogee infrared surface temperature, RIMCO rain gauge, Vaisala HMP75C temperature and humidity probe, RM Young wind sentry (wind speed and direction), Huskeflux ground heat flux plate and Vaisala BaroCap barometric pressure sensor. Both meteorological tower data and eddy covariance data were inspected visually to detect and remove spikes. The low-frequency eddy covariance data (30 min resolution) were corrected for coordinate rotation (Finnigan et al., 2003) and density effects (Leuning, 2007) using the PyQC software tool (available from http://code.google.com/p/eddy).

2.4 Radon-222 measurements

The naturally occurring radioactive gas radon (^{222}Rn) is predominantly of terrestrial origin and its only atmospheric sink is radioactive decay (Zahorowski et al., 2004). The surface flux density of radon is relatively constant in space and time, and since the half-life is much greater than ABL-mixing timescales, it is an ideal tracer of vertical mixing strength within the ABL (Chambers et al., 2015c; Griffiths et al., 2013; Williams et al., 2010). Hourly radon concentrations were measured by an Alpha Guard (Saphymo GmbH, Frankfurt, Germany) placed in a $\sim 20\,L$ enclosure. The enclosure was purged at $\sim 15\,L\, min^{-1}$ with a vacuum pump (2107 Series, Thomas, Wisconsin, USA) that sampled from a height of 2 m through 10 mm (outside diameter) PTFE tubing. Radon measurements were used to aid the interpretation of the diurnal variations in vertical mixing (see Griffiths et al., 2013).

3 Results

3.1 Meteorological observations

The last rain event was 10 days prior to the campaign, after which clear skies saw the soil dry to a moisture content close to minimum values observed for the site (Fig. 4). The meteorological and radon measurements shown in Fig. 5 indicate the 2-week campaign was conducted under predominantly calm meteorological conditions. In the middle of the campaign (2 May), a cold front moved across south-eastern Australia, producing cloudy conditions and 1.4 mm of precipitation at the site. No change in soil moisture was observed over the 0–10 cm soil layer following the rain event.

Wind directions were variable during the campaign (see Fig. S2a and b). Figure S3 shows that from 27 to 30 April,

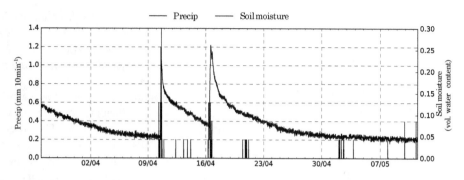

Figure 4. Precipitation and 0–10 cm soil moisture for the month leading up to and including the field campaign.

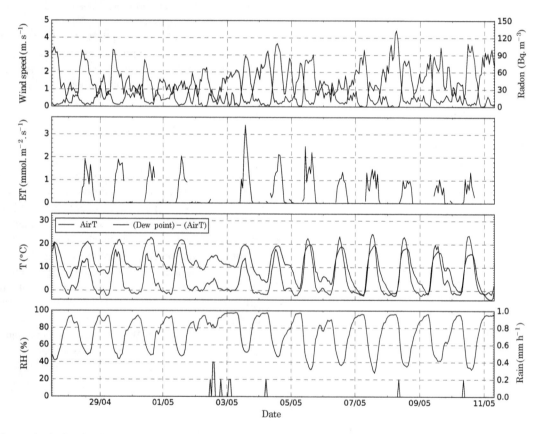

Figure 5. Meteorological and radon measurements collected throughout the field campaign. Meteorological measurements are block hourly averages calculated from 15 min observations. Small rain events on 4, 8 and 10 May were most likely dewfall rather than precipitation.

dominant daytime wind directions were mainly from the east. After 3 May winds were from the south, except on 7 and 8 May when the wind was from the west and had a fetch from the adjacent forest. At other times the fetch did not overlap the forested site. Daily maximum temperatures on clear days ranged from 16 to 23 °C, whilst nighttime minimum temperatures fell to between 8 and −4 °C. From 7 May onwards nocturnal temperatures fell below zero. On clear nights the surface temperature fell below dew point temperature, indicating dewfall. Apart from the night of 27–28 April and the cloudy nights between 1 and 3 May, the surface tempera-

ture fell below dew point temperature and dew or frost was observed in the morning, although heavier from 7 May onwards.

Radon concentrations were low during the day, when the convective boundary layer reached its maximum height, and high at night, when radon emissions were confined within the shallow nocturnal boundary layer. The accumulation at night was variable indicating a varying degree of nocturnal stability, mixing depth and occurrence of transient mixing events (Griffiths et al., 2013). There was general agreement between high nocturnal radon concentrations and low wind speeds, but no direct relationship. The lack of a direct relationship indicates that radon can provide additional information about nocturnal mixing and surface exchange that compliments standard meteorological measurements (Chambers et al., 2015a, b; Williams et al., 2013).

ET fluxes were in general quite low, reflecting the low soil moisture content. The ET flux did show a marked increase the day after the small rain event on 2 May and noticeably smaller fluxes were observed after the first night frost was observed. The health of the grass visibly deteriorated from 7 May, coinciding with frost formation.

3.2 Relationship between $\delta^2 H$ and $\delta^{18} O$ of the different water pools

A summary of the isotopic composition of all observed and modelled water pools are presented in Fig. 6. The local Meteoric Water Line (MWL) (Hughes and Crawford, 2013) is to the left of the global MWL (Craig, 1961), illustrating the characteristically high D-excess of precipitation in the region (Crawford et al., 2013). Ambient vapour observations aligned closely with the local MWL, but with a distribution that fell both to the left and right of the local MWL. Alignment between observations and the MWL show that equilibrium fractionation was the dominant process modifying $\delta^2 H$ and $\delta^{18} O$ in water vapour, while non-equilibrium kinetic processes shift observations away from the MWL and are more easily observed for d_v measurements.

Plant and soil water pools were enriched relative to the vapour and distributed to the right of the MWL, indicating evaporative enrichment. Soil water isotopes at the evaporation front (δ_L) were very enriched and had lower D-excess values (50 ± 12, 31 ± 3.8 and -131 ± 22‰ for $\delta^2 H$, $\delta^{18} O$ and D-excess) relative to the average soil moisture between 0 and 5 cm (-15 ± 4.2, 2.6 ± 2.5 and -36 ± 17‰ for $\delta^2 H$, $\delta^{18} O$ and D-excess). Low D-excess and enriched isotopes indicated large evaporative enrichment under non-equilibrium conditions consistent with the $\delta^{18} O$ soil profile measurements of Dubbert et al. (2013) and the $\delta^2 H$ profiles of Allison et al. (1983). The uncertainty of modelled isotope values was the most sensitive to parameterisation of the Craig–Gordon model. Changing the diffusion coefficient exponent (n) had the greatest impact on modelled soil water ($n = 0.66$, 42.7 ± 12, 21.8 ± 3.8 and -130.8 ± 22‰). However, changing parameterisation did not change the conclusion that soil moisture at the evaporation front was heavily enriched with very low D-excess values.

ET flux isotopic compositions from vegetated chambers were enriched relative to vapour and distributed to the right of the MWL (slope = 3.2). Similar isotopic compositions were measured from bare soil and vegetated chambers. Mean and standard deviations (1σ) for vegetated and soil chambers were -47.1 (± 13) and -50.2 (± 11) for $\delta^2 H$, -5.03 (± 3.8) and -6.3 (± 2.7)‰ for $\delta^{18} O$, and -6.3 (± 23) and -0.12 (± 15)‰ for D-excess, respectively. The similar ET isotopic

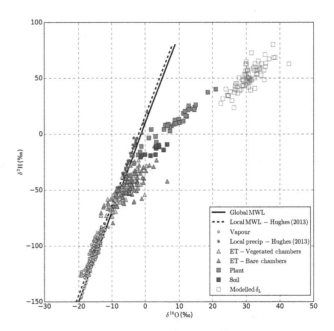

Figure 6. Relationship between $\delta^2 H$ and $\delta^{18} O$ for observed and modelled water pools. Linear regressions are shown for local and global meteoric water lines (MWL). Data from Hughes and Crawford (2013) are for monthly cumulative rainfall samples between 2005 and 2008.

composition from bare soil and vegetated chambers could indicate soil evaporation was the dominant process contributing to total ET. However, as pointed out in the discussion (Sect. 4.3), convergence of soil evaporation and transpiration isotope compositions as the soil evaporation source becomes progressively enriched (and D-excess lower), probably makes it difficult to identify the dominant process from these observations. Nevertheless, since the last significant rain event prior to the campaign, progressive reduction of D-excess of moisture at the evaporation front, and to a lesser extent in the 0–5 cm layer, caused low D-excess of overall ET fluxes compared to d_v. This would indicate that ET imposes a negative forcing on d_v.

Temporally, a clear trend was not observed for ET isotopic compositions over the measured portion of the diurnal cycle or over the campaign. No measurements were made at night or during the rapidly changing conditions of the morning transition, which may have led to our data missing some observed changes in ET isotope compositions.

3.3 In situ water vapour isotopes and ET isoforcing

Observed water vapour mixing ratios and stable isotope compositions are shown in Fig. 7. $\delta^2 H$ and $\delta^{18} O$ variability was similar, reflecting changes in both the synoptic and local meteorology. Prior to the rain event (2 May), relatively moist conditions (higher $H_2 O$ mixing ratios) were observed as air was transported from the warmer ocean off the east coast of Australia (see wind direction in Fig. S3). After 5 May, trans-

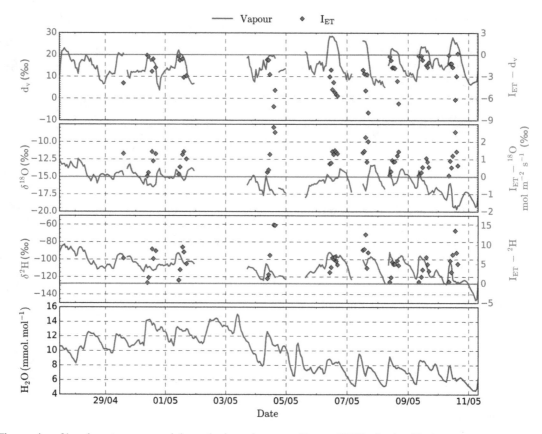

Figure 7. Time series of hourly water vapour mixing ratio, isotopic composition and ET isoforcing (I_{ET}).

port of air masses from the colder sea surface south of continental Australia brought drier conditions to the site (lower H_2O mixing ratios). Moisture source regions were confirmed by backward air trajectories calculated using the Stochastic Time-Inverted Lagrangian Transport Model (STILT; Lin et al., 2003, not shown). These two time periods are hereinafter referred as the "wet period" (before 2 May) and the "dry period" (after 5 May). The wet period coincided with more enriched isotopes and less diurnal variability. In the later part of the campaign, a reproducible diurnal cycle for δ^2H and $\delta^{18}O$ was observed (see Fig. 8 for diurnal composites), presenting a sharp increase at sunrise before decreasing from mid-morning (when vertical mixing increased) until the next sunrise. These observations emphasise the complex relationship between stable isotope observations in water vapour and both local- and synoptic-scale meteorology.

The d_v dataset showed a robust diurnal cycle of high values during the day and low values at night, consistent with what has been observed across a growing number of locations (Bastrikov et al., 2014; Berkelhammer et al., 2013; Simonin et al., 2014; Welp et al., 2012; Zhao et al., 2014). Wet period daytime d_v values were on average lower than those observed for the dry period. Nocturnal d_v was consistently lower during the night, but variable from night to night and across individual nights, with no clear difference observed

between wet and dry periods. Contrasting daytime measurements of wet and dry periods indicate a role of large-scale processes, whilst the lack of contrast for nocturnal observations shows the importance of local processes.

The I_{ET} was always positive for δ^2H and $\delta^{18}O$ and mostly negative for D-excess, but showed large variability across individual days (Fig. 7). I_{ET} was most sensitive to the magnitude of the ET fluxes, producing the greatest forcing on ambient vapour in the middle of the day. The I_{ET} time series did not correspond to temporal variability of vapour δ^2H, $\delta^{18}O$ or D-excess. δ^2H and $\delta^{18}O$ often decreased during the day while I_{ET} was positive. Whilst the high d_v values observed during the day were associated with negative isoforcing, over the course of the campaign the highest daytime d_v values did not correspond to the least negative I_{ET}. These observations illustrate that local ET fluxes were not overly important for day-to-day and diurnal d_v trends.

The level of agreement between the analysers presented some uncertainty in calculating the D-excess isoforcing. The sign of the isoforcing is dependent on the difference between d_v and d_{ET} (Eq. 8). In some cases this difference was small and within the range of agreement between the two analysers. While this caused problems for accurate calculation of the absolute values of D-excess isoforcing, for all chamber measurements passing our quality control requirements, D-

Table 1. Correlation between meteorological variables and the isotopic composition of water vapour. Values outside the brackets are statistics for the hourly observations. Inside the brackets are correlation statistics for average values calculated between 11:00 and 15:00 LST, hence representing activity during a convective boundary layer. Significant correlation are shown in bold; $p < 0.001$ for hourly observations and $p < 0.05$ for the daytime averages (due to the smaller number of points).

		T	RH	ET	H_2O	I_{ET}^*
δ^2H	Slope	**0.83** (0.51)	**−0.17** (0.23)	1.4 (6.1)	**2.1** (0.85)	−1.1 (**−3.0**)
	Intercept	**−120** (−140)	**−95** (−110)	−110 (−110)	**−110** (−130)	−99 (**−83**)
	R^2	**0.24** (0.13)	**0.09** (0.02)	0.001 (0.04)	**0.2** (0.04)	0.2 (**0.45**)
	p	**< 0.001** (0.3)	**< 0.001** (0.7)	0.32 (0.6)	**< 0.001** (0.5)	0.002 (**0.05**)
$\delta^{18}O$	Slope	**0.046** (0.44)	−0.01 (−0.01)	−0.37 (1.8)	**0.27** (0.29)	−0.7 (−1.9)
	Intercept	**−16** (−24)	−16 (−20)	−15 (−18)	**−18** (−19)	−15 (−14)
	R^2	**0.04** (0.30)	0.004 (0.2)	0.02 (0.16)	**0.2** (0.2)	0.14 (0.32)
	p	**< 0.001** (0.08)	0.26 (0.19)	0.05 (0.26)	**< 0.001** (0.15)	0.008 (0.11)
d_v	Slope	**0.51** (**−1.4**)	**−0.21** (**−0.52**)	0.01 (**−0.16**)	0.15 (**−1.3**)	−1.4 (−2.4)
	Intercept	**−9.9** (**48**)	**31** (**44**)	−15 (**−18**)	14 (**35**)	21 (20)
	R^2	**0.40** (**0.48**)	**0.62** (**0.74**)	0.22 (**0.30**)	0.004 (**0.71**)	0.06 (0.08)
	p	**< 0.001** (**0.02**)	**< 0.001** (**< 0.01**)	0.05 (**0.01**)	0.26 (**< 0.01**)	0.01 (0.44)

* Isoforcing correlations were calculated for simultaneous vapour and chamber measurements. Hourly averaged values were used for both.

excess decreased with concentration. This indicates that for all measurements, the D-excess isoforcing was negative.

3.4 Relationship between water vapour isotopes and local meteorology

The relationships between local meteorological variables and water vapour isotopes were examined to interpret the role of local processes (Table 1). Regression statistics are shown for both hourly observations and average daytime values (between 11:00 and 16:00 LST). Selecting daytime measurements removes variability associated with transition between the stable nocturnal and daytime convective boundary layer, as well as nocturnal periods when local surface equilibrium exchange and dewfall affect vapour isotope compositions. Correlations determined using only measurements in the middle of the day therefore provide a better indicator of how local meteorology and ET isotopic composition modified ambient water vapour isotope ratios from day to day.

Correlations calculated with hourly data were weak for $\delta^{18}O$ and δ^2H. Only correlations with air temperature ($R^2 = 0.24$ and 0.04, respectively) and mixing ratio ($R^2 = 0.2$ for both isotopes) were significant, and δ^2H also showed a weak correlation with RH ($R^2 = 0.09$). For daytime observations, only δ^2H showed a significant correlation with daytime I_{ET} ($R^2 = 0.45$, $p < 0.05$), but the slope was negative in contrast to positive isoforcing. The weak relationships with local meteorology indicate the importance of larger-scale precipitation processes and atmospheric mixing occurring as moisture was transported to the site.

As the diurnal cycle for d_v was consistent with growth and decay of the ABL, strong relationships were observed with air temperature and RH for the hourly observations. While

the local air temperature and RH could modify d_{ET} on diurnal timescales and in turn local d_v, the chamber measurements showed relatively constant d_{ET}. These correlations therefore result from the coincident diurnal variation of the d_v, RH and air temperature.

Daytime average d_v showed significant correlations with the air temperature, RH, ET flux and mixing ratio. The relationship with ET fluxes was weak ($R^2 = 0.3$) and positive, but as negative D-excess isoforcing was observed, a negative relationship would be expected. Likewise, the slope between air temperature and d_v was negative, counter to what theory would predict for local or remote moisture sources. The strongest relationship was observed with daytime RH ($R^2 = 0.74$), which had a negative slope ($-0.52\,‰\,\%^{-1}$) consistent with an inverse relationship between d_v and RH for a large unchanging evaporation source. The strong relationship of d_v with the daytime RH could indicate an important role for the evaporation conditions at remote moisture sources, as is discussed below in Sect. 4.2.

3.5 Diurnal variability of vapour isotopes

Diurnal composites were divided into dry and wet periods and are shown in Fig. 8. At sunrise (approximately 06:30 LST) surface heating initiated vertical mixing, shown by the radon concentration maximum, causing temperature and ET flux to increase and RH to decrease. Weak vertical mixing immediately after sunrise and injection of ET into the still shallow surface layer caused near-surface humidity to increase. Similarly, for δ^2H and $\delta^{18}O$, the observed spike immediately after sunrise was likely caused by ET fluxes with an enriched heavy isotope composition, possibly from re-evaporation of dewfall. During the dry period, vapour δ^2H

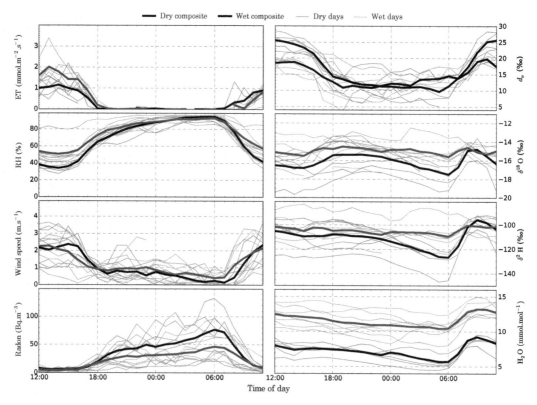

Figure 8. Data plotted by time of day and divided into dry and wet periods (see text in Sect. 3.5). Diurnal composites are shown for dry (red) and wet (blue) periods.

and $\delta^{18}O$ increased more steeply, caused by the combination of a shallower surface layer observed at the start of the morning transition, shown by higher radon concentrations, and more dewfall on the surface providing a greater initial evaporation source. Rapidly decreasing radon concentrations during this morning ABL transition caused by vigorous vertical mixing entraining air from the residual layer of the previous day diluted ET fluxes and caused the δ^2H, $\delta^{18}O$ and the mixing ratio to first stabilise and then decrease. ET fluxes rapidly increased as the ABL grew, but were not large enough to offset the dilution by dry air being mixed down from above or stop depletion of surface δ^2H and $\delta^{18}O$.

The d_v also increased after sunrise, but aligned more closely to when strong vertical mixing commenced, as shown by the close agreement with radon concentrations. The D-excess isoforcing was negative, evidence that d_v increased from encroachment mixing as the new mixed layer grew in depth and not ET fluxes. The dry period showed a greater increase in d_v during the morning transition, likely the result of higher d_v in background water vapour and greater differences between the d_v of the residual and nocturnal layer.

In the afternoon, d_v decreased back to values similar to those observed prior to sunrise, with a simultaneous decrease in solar insolation, ET and a decay of convective mixing. Radon shows how reduction in vertical mixing causes the concentration of tracers emitted from the surface to increase.

So while ET decreased, small fluxes were still observed well after 18:00 LST, when large changes in $\delta^{18}O$ and d_v were observed. Hence, as the I_{ET} was positive and negative for $\delta^{18}O$ and D-excess, respectively, small ET fluxes into a poorly mixed surface layer may have led to observed changes.

During the night, dewfall caused δ^2H and $\delta^{18}O$ to decrease as heavy isotopes were removed in condensation, especially during the dry period when greater surface cooling was observed. However, dew formation is an equilibrium processes and therefore did not affect d_v. Composites of dry and wet period nocturnal d_v measurements do not show clear nocturnal trends, but individual nights showed considerable variability. A regression of nocturnal d_v with radon concentrations produced a significant negative relationship ($p < 0.001$, $R^2 = 0.31$), indicating that atmospheric stability has some control over nocturnal d_v. High radon is associated with the most stable atmospheres, enhancing the effect of surface exchange in the early evening. Low radon, on the other hand, is associated with periods of atmospheric turbulence in which moisture above the nocturnal inversion with a high d_v is mixed down towards the surface.

4 Discussion

As has been previously observed (Steen-Larsen et al., 2013; Welp et al., 2012) and predicted by isotopic models (Gat, 1996), our observations showed water vapour $\delta^2 H$ and $\delta^{18}O$ are controlled by different atmospheric and hydrological processes than d_v. The diurnal cycle was the dominant mode of variability for d_v, consistent with previous studies for a range of ecosystems (Simonin et al., 2014; Welp et al., 2012; Zhao et al., 2014). However, results also showed that D-excess variability was controlled by local meteorological conditions and surface exchange at night, ABL growth and decay during transitional periods between the nocturnal and convective ABL, and larger-scale processes in the middle of the day.

4.1 Entrainment and the d_v diurnal cycle

The radon measurements showed that when the depth of the ABL was rapidly changing through the morning and evening transitions, entrainment from the residual layer and ET fluxes into a rapidly decaying convective boundary layer caused the observed d_v diurnal cycle. Between these transitions, when mixing extends to the capping inversion, entrainment fluxes introduce an additional moisture source from the free troposphere that could modify surface vapour isotopic compositions. Air above the ABL is drier and moisture is more depleted than at the surface. Drying and depleting trends for water vapour, $\delta^2 H$ and $\delta^{18}O$ throughout the day, particularly during the dry period (Fig. 8), indicate an important role for entrainment from the free troposphere. Whether this moisture flux impacts on d_v is less clear, as it remained reasonably stable once a maximum was reached after the morning transition period. The sign of the isoforcing of moisture entrained from the free troposphere is uncertain, as few free tropospheric d_v measurements exist (He and Smith, 1999; Samuels-Crow et al., 2014). Nevertheless, d_v values did not show a clear trend until vertical mixing began decaying later in the afternoon; therefore, free tropospheric d_v probably has a similar value to moisture already residing in the ABL.

4.2 Remote hydrometeorological processes

While the main focus of this study was to examine the role of local land–atmosphere exchange for the diurnal variability of d_v, the synoptic context of measurements warrants further examination for comparison against previous studies of d_v diurnal cycles. The slope between daytime RH and d_v ($-0.52\,‰\,\%^{-1}$, Table 1) was similar to those determined for measurements over the Mediterranean Sea and different ocean basins (between -0.43 and $-0.53\,‰\,\%^{-1}$) (Kurita, 2011; Pfahl and Wernli, 2008; Steen-Larsen et al., 2014, 2015; Uemura et al., 2008). Aemisegger et al. (2014) showed that this robust relationship is not restricted to coastal locations or measurements over the ocean surface. Using a trajectory model to investigate continental moisture recy-

cling in Europe, they found a similar relationship between d_v and RH of remote moisture sources during the cold season ($-0.57\,‰\,\%^{-1}$), but not for warm season observations. They concluded moisture recycling is weakest during winter, causing d_v to retain the signature of the RH of oceanic moisture sources, while in summer moisture recycling increased and attenuated the relationship. Similarities with their winter data indicate that our daytime d_v measurements were at least partly determined by RH at the oceanic moisture source.

Along an air masses back trajectory, entrainment fluxes from the free troposphere could be a major driver of daytime d_v variability. Mixing of warm dry air down to the surface, presumably with a relatively high D-excess (He and Smith, 1999; Samuels-Crow et al., 2014), would give the same negative relationship between d_v and RH observed here. However, for a strong relationship between d_v and RH, there must be a dominant moisture source. For the fraction of entrained air in the ABL to cause the strong linear relationship, the D-excess of vapour and RH in both the ABL and free troposphere must be reasonably constant, as in a two-source mixing model. Considering the variability of synoptic-scale weather patterns observed (Sect. 3.1), this seems unlikely. Thus, while we cannot definitively rule out the importance of entrainment along back trajectories, it seems more likely that the d_v–RH relationship was derived from a large unchanging moisture source such as the ocean.

A practical application of the d_v–RH relationship introduced by Aemisegger et al. (2014) was to determine the D-excess of the liquid moisture source. Based on the closure assumption of Merlivat and Jouzel (1979), it was shown when RH is 100 %, d_v is equal to the D-excess of the liquid moisture source. If no further kinetic fractionation or mixing of vapour with a different d_v–RH occurred between the point of evaporation and measurement location, extrapolating the regression between d_v and RH to 100 % RH gives an estimate of moisture source D-excess. For our measurements, a value of $-8\,‰$ was determined, remarkably similar to the D-excess determined for ocean water off the east coast of Australia by Xu et al. (2012) using a global ocean model. In contrast to recent literature (Simonin et al., 2014; Welp et al., 2012; Zhao et al., 2014), this suggests that although the common diurnal cycle was observed, daytime observations are potentially a tracer of RH at the oceanic moisture source, but it is likely restricted to periods when moisture recycling is weak.

Whilst we have shown a relationship between the RH and d_v consistent with an oceanic vapour source, the consistency of the relationship over longer time periods is uncertain. Indeed, it may be the reason why we show a strong relationship, whereas the study of Welp et al. (2012) did not for six mid-latitude sites in China and the USA, where longer datasets were available. As pointed out earlier, lowers slopes and weaker relationships result from stronger moisture recycling, which indicates moisture recycling and soil moisture state may be the most important variable controlling the relationship between d_v and RH. Here we present data from after an extended dry period, where the dominant moisture source is the ocean surrounding the Australian continent. So during wetter periods, an increase in the local and remote moisture recycling probably weaken the relationship between local d_v and RH (Aemisegger et al., 2014). However, for locations such as semi-arid Australia where extended dry periods prevail, the relationship between d_v and RH may be reasonably robust and prevail as a tracer of oceanic evaporative environments.

4.3   Controls of d_{ET}

The chamber d_{ET} measurements showed ET fluxes imposed a negative isoforcing on d_v, in contrast to interpretations in previous studies investigating d_v variability on diurnal timescales (Simonin et al., 2014; Zhao et al., 2014). However, it is expected that the sign and magnitude of the D-excess isoforcing would vary both spatially and temporally, in particular with the soil moisture state. After a rain event, soil moisture D-excess would decrease following a pseudo-Rayleigh process (Barnes and Allison, 1988). Therefore, immediately after a rainfall event, d_{ET} would be higher and probably impose a positive isoforcing. Here the negative d_{ET} caused the d_v to decrease rapidly as convective mixing shut down. When isoforcing is positive after a rain event, the diurnal cycle observed here and elsewhere may therefore not be observed. Although equilibration between liquid and vapour pools, as eluded to by Simonin et al. (2014), may still help maintain observed trends. As soil dries, a tipping point when the ET fluxes switch from positive to negative isoforcing will be observed. This has implications for studies attempting to use d_v as a tracer of continental moisture recycling, as the large spatial variability of rainfall and the associated soil moisture state would lead to large spatial and temporal variability for d_{ET}. Although, the strongest moisture recycling is expected for wet soils when d_{ET} is higher, variability in d_{ET} may still be important.

Relative magnitudes of evaporation and transpiration fluxes are important for d_{ET}, as the two processes could have different D-excess values and could vary strongly between precipitation or irrigation events. The classical view of ET isotope fluxes is that transpiration has an isotopic composition closer to the source moisture than evaporation, and therefore a higher D-excess. However, greater fractionation of the evaporation source pool causes its D-excess value to decrease over time, causing the D-excess of the fluxes to converge. The impact of converging isotopic signatures of ET component fluxes on moisture recycling would depend on the land surface type, but would constitute an important variable influencing the D-excess of local and remote moisture recycling. Further studies investigating how ET partitioning and drying of soil moisture reservoirs following irrigation or precipitation events would lead to a better understanding of how moisture recycling influences the ambient d_v on continental and local scales.

5   Conclusions

To determine how local ET fluxes modified water vapour D-excess, in situ observations were collected in a semi-arid region of south-eastern Australia. The diurnal cycle exhibited high values during the day and low values at night, reflected findings from previous studies. With chamber-based measurements of isotopic compositions in evaporative fluxes, it was shown that local ET fluxes exhibited a negative forcing on the ambient water vapour D-excess that could not explain the high daytime values. A strong negative relationship was observed between the locally measured relative humidity and vapour D-excess during the daytime, consistent with relationships observed for oceanic moisture sources. During the evening transition, collapse of the convective boundary layer and small ET fluxes with negative D-excess isoforcing were responsible for lowering the D-excess of water vapour near the surface. In addition, a negative nocturnal correlation between D-excess in water vapour and radon concentrations indicated transient nocturnal mixing events shifted the D-excess back towards the higher values observed during the day, with the most stable (least turbulent) nights producing the lowest D-excess values. In the morning, encroachment and entrainment of high D-excess air from above caused D-excess of surface vapour to increase back to the synoptic values.

Overall, it was found that the magnitude of the D-excess diurnal cycle was controlled predominantly by interplay between synoptic forcing and local ABL processes and was modified further by nocturnal surface exchange processes and turbulent mixing. The low D-excess of the ET fluxes determined from flux chambers in this study illustrated that the impact of large-scale moisture recycling may be both spatially and temporally variable, depending on the soil moisture state. This has implications for studies using D-excess to investigate moisture recycling.

Acknowledgements. Stephen Parkes was supported by the Atmospheric Mixing and Pollution Transport (AMPT) project at the Australian Nuclear Science and Technology Organization (ANSTO) and the King Abdullah University of Science and Technology. The Baldry Hydrological Observatory field campaign was supported by Australian Research Council Discovery grants DP0987478 and DP120104718. Matthew McCabe acknowledges the support of the King Abdullah University of Science and

Technology. We thank Peter Graham, Cecilia Azcurra, Jin Wang and Yingzhe Cai for their assistance during the campaign. We also appreciate the support of Diana and Jason Tremain for access to the Baldry Hydrological Observatory and surrounding farmland, Chris Dimovski for performing plant and soil water extractions and Barbara Neklapilova analysis of plant and soil water samples.

Edited by: C. Stumpp

References

Aemisegger, F., Pfahl, S., Sodemann, H., Lehner, I., Seneviratne, S. I., and Wernli, H.: Deuterium excess as a proxy for continental moisture recycling and plant transpiration, Atmos. Chem. Phys., 14, 4029–4054, doi:10.5194/acp-14-4029-2014, 2014.

Allison, G. B., Barnes, C. J., and Hughes, M. W.: The distribution of deuterium and ^{18}O in dry soils 2. Experimental, J. Hydrol., 64, 377–397, doi:10.1016/0022-1694(83)90078-1, 1983.

Australian Bureau of Meteorology: http://www.bom.gov.au, last access: 2016.

Barnes, C. J. and Allison, G. B.: Tracing of water movement in the unsaturated zone using stable isotopes of hydrogen and oxygen, J. Hydrol., 100, 143–176, doi:10.1016/0022-1694(88)90184-9, 1988.

Bastrikov, V., Steen-Larsen, H. C., Masson-Delmotte, V., Gribanov, K., Cattani, O., Jouzel, J., and Zakharov, V.: Continuous measurements of atmospheric water vapour isotopes in western Siberia (Kourovka), Atmos. Meas. Tech., 7, 1763–1776, doi:10.5194/amt-7-1763-2014, 2014.

Berkelhammer, M., Hu, J., Bailey, A., Noone, D. C., Still, C. J., Barnard, H., Gochis, D., Hsiao, G. S., Rahn, T., and Turnipseed, A.: The nocturnal water cycle in an open-canopy forest, J. Geophys. Res.-Atmos., 118, 10225–10242, doi:10.1002/jgrd.50701, 2013.

Chambers, S. D., Wang, F., Williams, A. G., Xiaodong, D., Zhang, H., Lonati, G., Crawford, J., Griffiths, A. D., Ianniello, A., and Allegrini, I.: Quantifying the influences of atmospheric stability on air pollution in Lanzhou, China, using a radon-based stability monitor, Atmos. Environ., 107, 233–243, doi:10.1016/j.atmosenv.2015.02.016, 2015a.

Chambers, S. D., Williams, A. G., Crawford, J., and Griffiths, A. D.: On the use of radon for quantifying the effects of atmospheric stability on urban emissions, Atmos. Chem. Phys., 15, 1175–1190, doi:10.5194/acp-15-1175-2015, 2015b.

Chambers, S. D., Williams, A. G., Crawford, J., and Griffiths, A. D.: On the use of radon for quantifying the effects of atmospheric stability on urban emissions, Atmos. Chem. Phys., 15, 1175–1190, doi:10.5194/acp-15-1175-2015, 2015c.

Craig, H.: Isotopic Variations in Meteoric Waters, Science, 133, 1702–1703, doi:10.1126/science.133.3465.1702, 1961.

Craig, H. and Gordon, L. I.: Deuterium and oxygen-18 variations in the ocean and marine atmosphere, in: Stable isotopes in oceanographic studies and paleotemperatures, Proceedings, Spoleto, Italy, edited by: Tongiogi, E., Pisa, Italy, 9–130, 1965.

Crawford, J., Hughes, C. E., and Parkes, S. D.: Is the isotopic composition of event based precipitation driven by moisture source or synoptic scale weather in the Sydney Basin, Australia?, J. Hydrol., 507, 213–226, doi:10.1016/j.jhydrol.2013.10.031, 2013.

Dansgaard, W.: Stable isotopes in precipitation, Tellus, 16, 436–468, doi:10.1111/j.2153-3490.1964.tb00181.x, 1964.

Decker, M., Pitman, A., and Evans, J.: Diagnosing the seasonal land–atmosphere correspondence over northern Australia: dependence on soil moisture state and correspondence strength definition, Hydrol. Earth Syst. Sci., 19, 3433–3447, doi:10.5194/hess-19-3433-2015, 2015.

Dubbert, M., Cuntz, M., Piayda, A., Maguás, C., and Werner, C.: Partitioning evapotranspiration – Testing the Craig and Gordon model with field measurements of oxygen isotope ratios of evaporative fluxes, J. Hydrol., 496, 142–153, doi:10.1016/j.jhydrol.2013.05.033, 2013.

Evans, J. P. and McCabe, M. F.: Regional climate simulation over Australia's Murray-Darling basin: A multitemporal assessment, J. Geophys. Res.-Atmos., 115, D14114, doi:10.1029/2010JD013816, 2010.

Finnigan, J. J., Clement, R., Malhi, Y., Leuning, R., and Cleugh, H. A.: A Re-Evaluation of Long-Term Flux Measurement Techniques Part I: Averaging and Coordinate Rotation, Bound.-Lay. Meteorol., 107, 1–48, doi:10.1023/A:1021554900225, 2003.

Froehlich, K., Kralik, M., Papesch, W., Rank, D., Scheifinger, H., and Stichler, W.: Deuterium excess in precipitation of Alpine regions – moisture recycling, Isotopes Environ. Health Stud., 44, 61–70, doi:10.1080/10256010801887208, 2008.

Gat, J. R.: Oxygen and hydrogen isotopes in the hydrologic cycle, Annu. Rev. Earth Planet. Sci., 24, 225–262, doi:10.1146/annurev.earth.24.1.225, 1996.

Gat, J. R., Bowser, C. J., and Kendall, C.: The contribution of evaporation from the Great Lakes to the continental atmosphere: estimate based on stable isotope data, Geophys. Res. Lett., 21, 557–560, doi:10.1029/94GL00069, 1994.

Gat, J. R., Klein, B., Kushnir, Y., Roether, W., Wernli, H., Yam, R., and Shemesh, A.: Isotope composition of air moisture over the Mediterranean Sea: an index of the air–sea interaction pattern, Tellus B, 55, 953–965, doi:10.1034/j.1600-0889.2003.00081.x, 2003.

Griffiths, A. D., Parkes, S. D., Chambers, S. D., McCabe, M. F., and Williams, A. G.: Improved mixing height monitoring through a combination of lidar and radon measurements, Atmos. Meas. Tech., 6, 207–218, doi:10.5194/amt-6-207-2013, 2013.

Harding, K. J. and Snyder, P. K.: Modeling the Atmospheric Response to Irrigation in the Great Plains. Part I: General Impacts on Precipitation and the Energy Budget, J. Hydrometeorol., 13, 1667–1686, doi:10.1175/JHM-D-11-098.1, 2012.

He, H. and Smith, R. B.: Stable isotope composition of water vapor in the atmospheric boundary layer above the forests of New England, J. Geophys. Res.-Atmos., 104, 11657–11673, doi:10.1029/1999JD900080, 1999.

Horita, J. and Wesolowski, D. J.: Liquid-vapor fractionation of oxygen and hydrogen isotopes of water from the freezing to the critical temperature, Geochim. Cosmochim. Ac., 58, 3425–3437, doi:10.1016/0016-7037(94)90096-5, 1994.

Huang, L. and Wen, X.: Temporal variations of atmospheric water vapor δD and δ^{18}O above an arid artificial oasis cropland in the Heihe River Basin, J. Geophys. Res.-Atmos., 119, 11456–11476, doi:10.1002/2014JD021891, 2014.

Hughes, C. E. and Crawford, J.: Spatial and temporal variation in precipitation isotopes in the Sydney Basin, Australia, J. Hydrol., 489, 42–55, doi:10.1016/j.jhydrol.2013.02.036, 2013.

Jana, R. B., Ershadi, A., and McCabe, M. F.: Examining the relationship between intermediate-scale soil moisture and terrestrial evaporation within a semi-arid grassland, Hydrol. Earth Syst. Sci., 20, 3987–4004, doi:10.5194/hess-20-3987-2016, 2016.

Keeling, C. D.: The concentration and isotopic abundances of atmospheric carbon dioxide in rural areas, Geochim. Cosmochim. Ac., 13, 322–334, doi:10.1016/0016-7037(58)90033-4, 1958.

Kurita, N.: Origin of Arctic water vapor during the ice-growth season, Geophys. Res. Lett., 38, L02709, doi:10.1029/2010GL046064, 2011.

Lai, C.-T. and Ehleringer, J.: Deuterium excess reveals diurnal sources of water vapor in forest air, Oecologia, 165, 213–223, doi:10.1007/s00442-010-1721-2, 2011.

Lee, X., Griffis, T. J., Baker, J. M., Billmark, K. A., Kim, K., and Welp, L. R.: Canopy-scale kinetic fractionation of atmospheric carbon dioxide and water vapor isotopes, Global Biogeochem. Cy., 23, GB1002, doi:10.1029/2008GB003331, 2009.

Leuning, R.: The correct form of the Webb, Pearman and Leuning equation for eddy fluxes of trace gases in steady and non-steady state, horizontally homogeneous flows, Bound.-Lay. Meteorol., 123, 263–267, doi:10.1007/s10546-006-9138-5, 2007.

Lin, J. C., Gerbig, C., Wofsy, S. C., Andrews, A. E., Daube, B. C., Davis, K. J., and Grainger, C. A.: A near-field tool for simulating the upstream influence of atmospheric observations: The Stochastic Time-Inverted Lagrangian Transport (STILT) model, J. Geophys. Res., 108, 4493, doi:10.1029/2002JD003161, 2003.

Lu, X., Liang, L. L., Wang, L., Jenerette, G. D., McCabe, M. F., and Grantz, D. A.: Partitioning of evapotranspiration using a stable isotope technique in an arid and high temperature agricultural production system, Agr. Water Manage., 179, 103–109, doi:10.1016/j.agwat.2016.08.012, 2017.

Masson-Delmotte, V., Jouzel, J., Landais, A., Stievenard, M., Johnsen, S. J., White, J. W. C., Werner, M., Sveinbjornsdottir, A., and Fuhrer, K.: GRIP Deuterium Excess Reveals Rapid and Orbital-Scale Changes in Greenland Moisture Origin, Science, 309, 118–121, doi:10.1126/science.1108575, 2005.

Mathieu, R. and Bariac, T.: A numerical model for the simulation of stable isotope profiles in drying soils, J. Geophys. Res.-Atmos., 101, 12685–12696, doi:10.1029/96JD00223, 1996.

McCabe, M. F., Franks, S. W., and Kalma, J. D.: Calibration of a land surface model using multiple data sets, J. Hydrol., 302, 209–222, doi:10.1016/j.jhydrol.2004.07.002, 2005.

McCabe, M. F., Ershadi, A., Jimenez, C., Miralles, D. G., Michel, D., and Wood, E. F.: The GEWEX LandFlux project: evaluation of model evaporation using tower-based and globally gridded forcing data, Geosci. Model Dev., 9, 283–305, doi:10.5194/gmd-9-283-2016, 2016.

Merlivat, L.: Molecular diffusivities of $H_2O^{16}O$, $HD^{16}O$, and $H_2^{18}O$ in gases, J. Chem. Phys., 69, 2864, doi:10.1063/1.436884, 1978.

Merlivat, L. and Jouzel, J.: Global climatic interpretation of the deuterium-oxygen 18 relationship for precipitation, J. Geophys. Res.-Oceans, 84, 5029–5033, doi:10.1029/JC084iC08p05029, 1979.

Noone, D., Risi, C., Bailey, A., Berkelhammer, M., Brown, D. P., Buenning, N., Gregory, S., Nusbaumer, J., Schneider, D., Sykes, J., Vanderwende, B., Wong, J., Meillier, Y., and Wolfe, D.: Determining water sources in the boundary layer from tall tower profiles of water vapor and surface water isotope ratios after a snowstorm in Colorado, Atmos. Chem. Phys., 13, 1607–1623, doi:10.5194/acp-13-1607-2013, 2013.

Pfahl, S. and Wernli, H.: Air parcel trajectory analysis of stable isotopes in water vapor in the eastern Mediterranean, J. Geophys. Res.-Atmos., 113, D20104, doi:10.1029/2008JD009839, 2008.

Pfahl, S. and Wernli, H.: Lagrangian simulations of stable isotopes in water vapor: An evaluation of nonequilibrium fractionation in the Craig-Gordon model, J. Geophys. Res.-Atmos., 114, D20108, doi:10.1029/2009JD012054, 2009.

Risi, C., Noone, D., Frankenberg, C., and Worden, J.: Role of continental recycling in intraseasonal variations of continental moisture as deduced from model simulations and water vapor isotopic measurements, Water Resour. Res., 49, 4136–4156, doi:10.1002/wrcr.20312, 2013.

Samuels-Crow, K. E., Galewsky, J., Sharp, Z. D., and Dennis, K. J.: Deuterium-excess in subtropical free troposphere water vapor: continuous measurements from the Chajnantor Plateau, northern Chile, Geophys. Res. Lett., 41, 8652–8659, doi:10.1002/2014GL062302, 2014.

Simonin, K. A., Link, P., Rempe, D., Miller, S., Oshun, J., Bode, C., Dietrich, W. E., Fung, I., and Dawson, T. E.: Vegetation induced changes in the stable isotope composition of near surface humidity, Ecohydrology, 7, 936–949, doi:10.1002/eco.1420, 2014.

Steen-Larsen, H. C., Johnsen, S. J., Masson-Delmotte, V., Stenni, B., Risi, C., Sodemann, H., Balslev-Clausen, D., Blunier, T., Dahl-Jensen, D., Ellehøj, M. D., Falourd, S., Grindsted, A., Gkinis, V., Jouzel, J., Popp, T., Sheldon, S., Simonsen, S. B., Sjolte, J., Steffensen, J. P., Sperlich, P., Sveinbjörnsdóttir, A. E., Vinther, B. M., and White, J. W. C.: Continuous monitoring of summer surface water vapor isotopic composition above the Greenland Ice Sheet, Atmos. Chem. Phys., 13, 4815–4828, doi:10.5194/acp-13-4815-2013, 2013.

Steen-Larsen, H. C., Sveinbjörnsdottir, A. E., Peters, A. J., Masson-Delmotte, V., Guishard, M. P., Hsiao, G., Jouzel, J., Noone, D., Warren, J. K., and White, J. W. C.: Climatic controls on water vapor deuterium excess in the marine boundary layer of the North Atlantic based on 500 days of in situ, continuous measurements, Atmos. Chem. Phys., 14, 7741–7756, doi:10.5194/acp-14-7741-2014, 2014.

Steen-Larsen, H. C., Sveinbjörnsdottir, A. E., Jonsson, T., Ritter, F., Bonne, J.-L., Masson-Delmotte, V., Sodemann, H., Blunier, T., Dahl-Jensen, D., and Vinther, B. M.: Moisture sources and synoptic to seasonal variability of North Atlantic water vapor isotopic composition, J. Geophys. Res.-Atmos., 120, 5757–5774, doi:10.1002/2015JD023234, 2015.

Tian, L., Masson-Delmotte, V., Stievenard, M., Yao, T., and Jouzel, J.: Tibetan Plateau summer monsoon northward extent revealed by measurements of water stable isotopes, J. Geophys. Res.-Atmos., 106, 28081–28088, doi:10.1029/2001JD900186, 2001.

Uemura, R., Matsui, Y., Yoshimura, K., Motoyama, H., and Yoshida, N.: Evidence of deuterium excess in water vapor as an indicator of ocean surface conditions, J. Geophys. Res.-Atmos., 113, D19114, doi:10.1029/2008JD010209, 2008.

Vallet-Coulomb, C., Gasse, F., and Sonzogni, C.: Seasonal evolution of the isotopic composition of atmospheric water vapour above a tropical lake: Deuterium excess and implication for water recycling, Geochim. Cosmochim. Ac., 72, 4661–4674, doi:10.1016/j.gca.2008.06.025, 2008.

Wang, L., Niu, S., Good, S. P., Soderberg, K., McCabe, M. F., Sherry, R. A., Luo, Y., Zhou, X., Xia, J., and Caylor, K. K.: The effect of warming on grassland evapotranspiration partitioning using laser-based isotope monitoring techniques, Geochim. Cosmochim. Ac., 111, 28–38, doi:10.1016/j.gca.2012.12.047, 2013a.

Wang, L., Niu, S., Good, S. P., Soderberg, K., McCabe, M. F., Sherry, R. A., Luo, Y., Zhou, X., Xia, J., and Caylor, K. K.: The effect of warming on grassland evapotranspiration partitioning using laser-based isotope monitoring techniques, Geochim. Cosmochim. Ac., 111, 28–38, doi:10.1016/j.gca.2012.12.047 2013b.

Wei, J., Dirmeyer, P. A., Wisser, D., Bosilovich, M. G., and Mocko, D. M.: Where Does the Irrigation Water Go? An Estimate of the Contribution of Irrigation to Precipitation Using MERRA, J. Hydrometeorol., 14, 275–289, doi:10.1175/JHM-D-12-079.1, 2012.

Welp, L. R., Lee, X., Griffis, T. J., Wen, X.-F., Xiao, W., Li, S., Sun, X., Hu, Z., Val Martin, M., and Huang, J.: A meta-analysis of water vapor deuterium-excess in the midlatitude atmospheric surface layer, Global Biogeochem. Cy., 26, GB3021, doi:10.1029/2011GB004246, 2012.

West, A. G., Patrickson, S. J., and Ehleringer, J. R.: Water extraction times for plant and soil materials used in stable isotope analysis, Rapid Commun. Mass Spectrom., 20, 1317–1321, doi:10.1002/rcm.2456, 2006.

Williams, A. G., Zahorowski, W., Chambers, S., Griffiths, A., Hacker, J. M., Element, A., and Werczynski, S.: The Vertical Distribution of Radon in Clear and Cloudy Daytime Terrestrial Boundary Layers, J. Atmos. Sci., 68, 155–174, doi:10.1175/2010JAS3576.1, 2010.

Williams, A. G., Chambers, S., and Griffiths, A.: Bulk Mixing and Decoupling of the Nocturnal Stable Boundary Layer Characterized Using a Ubiquitous Natural Tracer, Bound.-Lay. Meteorol., 149, 381–402, doi:10.1007/s10546-013-9849-3, 2013.

Xu, X., Werner, M., Butzin, M., and Lohmann, G.: Water isotope variations in the global ocean model MPI-OM, Geosci. Model Dev., 5, 809–818, doi:10.5194/gmd-5-809-2012, 2012.

Zahorowski, W., Chambers, S. D., and Henderson-Sellers, A.: Ground based radon-222 observations and their application to atmospheric studies, J. Environ. Radioact., 76, 3–33, doi:10.1016/j.jenvrad.2004.03.033, 2004.

Zhao, L., Wang, L., Liu, X., Xiao, H., Ruan, Y., and Zhou, M.: The patterns and implications of diurnal variations in the d-excess of plant water, shallow soil water and air moisture, Hydrol. Earth Syst. Sci., 18, 4129–4151, doi:10.5194/hess-18-4129-2014, 2014.

6

Identifying the origin and geochemical evolution of groundwater using hydrochemistry and stable isotopes in the Subei Lake basin, Ordos energy base, Northwestern China

F. Liu[1,2]**, X. Song**[1]**, L. Yang**[1]**, Y. Zhang**[1]**, D. Han**[1]**, Y. Ma**[1]**, and H. Bu**[1]

[1]Key Laboratory of Water Cycle and Related Land Surface Processes, Institute of Geographic Sciences and Natural Resources Research, Chinese Academy of Sciences, 11 A, Datun Road, Chaoyang District, Beijing, 100101, China
[2]University of Chinese Academy of Sciences, Beijing, 100049, China

Correspondence to: X. Song (songxf@igsnrr.ac.cn)

Abstract. A series of changes in groundwater systems caused by groundwater exploitation in energy base have been of great concern to hydrogeologists. The research aims to identify the origin and geochemical evolution of groundwater in the Subei Lake basin under the influence of human activities. Water samples were collected, and major ions and stable isotopes (δ^{18}O, δD) were analyzed. In terms of hydrogeological conditions and the analytical results of hydrochemical data, groundwater can be classified into three types: the Quaternary groundwater, the shallow Cretaceous groundwater and the deep Cretaceous groundwater. Piper diagram and correlation analysis were used to reveal the hydrochemical characteristics of water resources. The dominant water type of the lake water was Cl–Na type, which was in accordance with hydrochemical characteristics of inland salt lakes; the predominant hydrochemical types for groundwater were HCO_3–Ca, HCO_3–Na and mixed HCO_3–Ca–Na–Mg types. The groundwater chemistry is mainly controlled by dissolution/precipitation of anhydrite, gypsum, halite and calcite. The dedolomitization and cation exchange are also important factors. Rock weathering is confirmed to play a leading role in the mechanisms responsible for the chemical composition of groundwater. The stable isotopic values of oxygen and hydrogen in groundwater are close to the local meteoric water line, indicating that groundwater is of modern local meteoric origin. Unlike significant differences in isotopic values between shallow groundwater and deep groundwater in the Habor Lake basin, shallow Cretaceous groundwater and deep Cretaceous groundwater have similar isotopic characteristics in the Subei Lake basin. Due to the evaporation effect and dry climatic conditions, heavy isotopes are more enriched in lake water than in groundwater. The low slope of the regression line of δ^{18}O and δD in lake water could be ascribed to a combination of mixing and evaporation under conditions of low humidity. Comparison of the regression line for δ^{18}O and δD showed that lake water in the Subei Lake basin contains more heavily isotopic composition than that in the Habor Lake basin, indicating that lake water in the discharge area has undergone stronger evaporation than lake water in the recharge area. Hydrochemical and isotopic information of utmost importance has been provided to decision makers by the present study so that a sustainable groundwater management strategy can be designed for the Ordos energy base.

1 Introduction

The Ordos Basin is located in Northwestern China, which covers an area of $28.2 \times 10^4 \, \text{km}^2$ in total and comprises the second largest coal reserves in China (Dai et al., 2006). It was authorized as a national energy base in 1998 by the former State Planning Commission (Hou et al., 2006). More than 400 lake basins with diverse sizes are distributed in the Ordos Basin. The Dongsheng–Shenfu coalfield, situated in the Inner Mongolia Autonomous Region, is an important component of the Ordos energy base. It is the largest explored coalfield with enormous potential for future development. The proven reserves of coal are 230 billion tons. The coal is extracted from Jurassic strata and subsurface mining is common. Local residents there mostly depend on groundwater on account of the serious shortage of surface water. Water resources support the exploitation of coal and development of related industries. In China, since 2011, all new construction projects must carry out an environment evaluation of groundwater consistent with the technical guidelines of the PRC Ministry of Environmental Protection (2011). It is of greatest significance in mining areas, because water resources are an essential component of the mining process (Agartan and Yazicigil, 2012). Over the past several decades, the quantity and quality of groundwater resources have been affected by the rapid development of coal mining. Haolebaoji well field of Subei Lake basin is a typical, large well field and acts as an important water source for this coalfield. However, large-scale and intensive groundwater exploitation could remarkably influence the hydrochemical field of groundwater systems in the study area. In recent years, with the fast development of Ordos energy base, more and more well fields have been built in some lake basins (including Haolebaoji well field newly built in the Subei Lake basin) in order to meet the increasing demand on water resources. However, due to a lack of adequate hydrogeological knowledge about these specific lake basins and reasonable groundwater management strategies, water resources in these specific lake basins are currently subject to increasing pressure from altered hydrology associated with water extraction for regional development and groundwater over-exploitation has taken place. If it continues, it may cause a series of negative impacts on the groundwater-dependent ecosystem around these lakes. Thus, studies about the lake basins are urgently needed so as to obtain comprehensive knowledge of the hydrochemical and isotopic characteristics, and geochemical evolution of groundwater under the background of intensive groundwater exploitation.

Research of groundwater and hydrogeology in the Ordos Basin has been conducted by numerous Chinese scholars and institutes because the Ordos Basin plays a vital role in natural resources exploitation and national economic development. Most importantly, China Geological Survey Bureau has conducted some regional-scale research on groundwater resources of Ordos Basin beginning in 1980s (Zhang et al., 1986; Hou et al., 2008). The previous research has clarified geology and hydrogeology and has provided a comprehensive overview of quantity and quality of groundwater in this region, laying a solid foundation for the present study. However, regional-scale groundwater investigations may not provide much accurate information on the groundwater flow characteristics in small basins (Toth, 1963). Hence, it is also significant to implement local groundwater resource investigations. As Winter (1999) concluded that lakes in different part of groundwater flow systems have different flow char-

acteristics. Data on hydrochemistry and stable isotopes of water were used to study the origin and geochemical evolution of groundwater in the Habor Lake basin (Yin et al., 2009), which is located in the recharge zone. But other lakes in the runoff and discharge area still have not been studied so far. Due to the particularity of the discharge area, a variety of hydrochemical effects such as evaporation, decarbonation, strong mixing action, etc., take place and result in extremely complicated hydrochemical and isotopic characteristics. In addition, intensive groundwater withdrawal has dramatically changed the local hydrologic cycle in these specific lake basins, groundwater flow field and hydrochemistry have been changed significantly, and a series of ecological environment problems have taken place. Therefore, given that these potential problems originate from human activity, it is essential to conduct hydrochemical and isotopic study of Subei Lake basin located in the discharge area.

Isotopic and geochemical indicators often serve as effective methods for solving multiple problems in hydrology and hydrogeology, especially in semi-arid and arid regions (Clark and Fritz, 1997; Cook and Herczeg, 1999). These techniques have been widely used to obtain groundwater information such as its source, recharge and the interaction between groundwater and surface water (De Vries and Simmers, 2002; Yuko et al., 2002; Yang et al., 2012a). The technique of stable isotopes as excellent tracers has been widely used by many scholars in the study of hydrological cycle (Chen et al., 2011; Cervi et al., 2012; Garvelmann et al., 2012; Yang et al., 2012a; Hamed and Dhahri, 2013; Kamdee et al., 2013). Greater knowledge on the origin and behavior of major ions in groundwater can enhance the understanding of the geochemical evolution of groundwater. Measurement of the relative concentration of major ions in groundwater from different aquifers can provide information on the geochemical reactions within the aquifer and the possible evolutionary pathways of groundwater (Cook and Herczeg, 1999).

The aim of the research is to recognize the origin and geochemical evolution of groundwater in the Subei Lake basin under the influence of human activities. The main objectives are to (1) ascertain the origin of groundwater and (2) determine the geochemical factors and mechanisms controlling the chemical composition of groundwater. In the context of a large number of well fields built in some lake basins in order to meet the increasing demand of water resources, the results of the present study will be valuable in obtaining a deeper insight into hydrogeochemical changes caused by human activity, and providing significant information on, for example, the water quality situation and geochemical evolution of groundwater to decision makers so that they can make sustainable groundwater management strategies for other similar small lake basins and even the Ordos energy base.

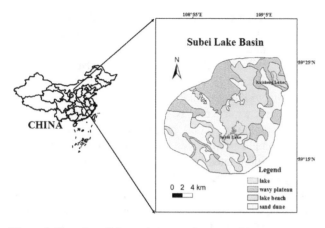

Figure 1. Location of the study area and geomorphic map.

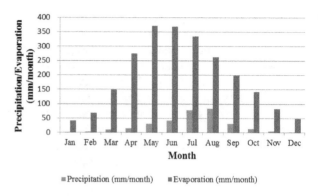

Figure 2. Average monthly precipitation and evaporation in the study area.

2 Study area

2.1 Physiography

The study area is situated in the northern part of the Ordos Basin, which is located at the junction of Uxin Banner, Hanggin Banner and Ejin Horo Banner in Ordos City and is mainly administratively governed by Uxin Banner of Ordos City. It covers an area of almost $400\,km^2$ within $39°13'30''$–$39°25'40''$ N and $108°51'24''$–$109°08'40''$ E. Its length is 23 km from east to west and its width is 22 km from north to south (Fig. 1).

The continental semi-arid to arid climate controls the whole study area, which is characterized by long, cold winters and short, hot summers (Li et al., 2010, 2011). According to the data of the Wushenzhao meteorological station, the average monthly temperature ranges from $-11.5\,°C$ in January to $21.9\,°C$ in July. The mean annual precipitation in the study area was $324.3\,mm\,yr^{-1}$ from 1985 to 2008. The total annual precipitation varied greatly from year to year with a minimum of 150.2 mm in 2000 and a maximum of 432.3 mm in 1985. The majority of the precipitation falls in the form of rain during the 3-month period from June to August, with more than 63.6 % of annual precipitation (Fig. 2). The mean annual evaporation is $2349.1\,mm\,yr^{-1}$ (from 1985 to 2008) at Wushenzhao station (Fig. 2), which far exceeds rainfall for the area. The average value of monthly evaporation is lowest in January ($42.4\,mm\,month^{-1}$) and highest from May to July, with maximum evaporation in May ($377.4\,mm\,month^{-1}$).

As a small-scale lake basin, the general geomorphic types of Subei Lake basin are wavy plateau, lake beach and sand dunes (Fig. 1). The terrain of Subei Lake's west, east and north sides is relatively higher with altitudes between 1370 and 1415 m; the terrain of its south side is slightly lower with elevations between 1290 and 1300 m. The topography of the center area of Subei Lake basin is flat and low-lying. There are no perennial or ephemeral rivers within the study area; the main surface water bodies are Subei and Kuisheng lakes, and they are situated in the same watershed considering actual hydrogeological conditions and groundwater flow field. In response to precipitation, diffuse overland flow and groundwater recharge the Subei and Kuisheng lakes (Hou et al., 2006; Wang et al., 2010). Subei Lake is located in the low-lying center of the study area (Fig. 1), which is an inland lake characterized by high alkalinity; Kuisheng Lake is also a perennial water body and it is located in northeastern corner of the study area, only covering $2\,km^2$ (Fig. 1).

2.2 Geologic and hydrogeologic setting

Subei Lake basin is a relatively closed hydrogeological unit given that a small quantity of lateral outflow occurs in a small part of southern boundary (Wang et al., 2010). The Quaternary sediments and Cretaceous formation can be observed in the study area. The Quaternary sediments are mainly distributed around the Subei Lake with relatively smaller thickness. Generally the thickness of Quaternary sediments is below 20 m. The Quaternary layer is chiefly composed of the interlaced layers of sand and mud. The Cretaceous formations mainly consist of sedimentary sandstones and generally outcrop in the regions with relatively higher elevation. The maximum thickness of Cretaceous rocks could be nearly 1000 m in the Ordos Plateau (Yin et al., 2009), so the Cretaceous formation composed of mainly sandstone is the major water-supplying aquifer of the investigated area. Calcite, dolomite, anhydrite, aragonite, gypsum, halite and feldspar are major minerals in the Quaternary and Cretaceous strata (Hou et al., 2006).

Groundwater resources are very abundant in the investigated area, and phreatic aquifer and confined aquifer can be observed in this region. According to Wang et al. (2010) and the data from Inner Mongolia Second Hydrogeology Engineering Geological Prospecting Institute, the phreatic aquifer is composed of Quaternary and Cretaceous sandstones, with its thickness ranging from 10.52 to 63.54 m. In terms of borehole data, the similar groundwater levels in the Quaternary and Cretaceous phreatic aquifers indicate a very close hy-

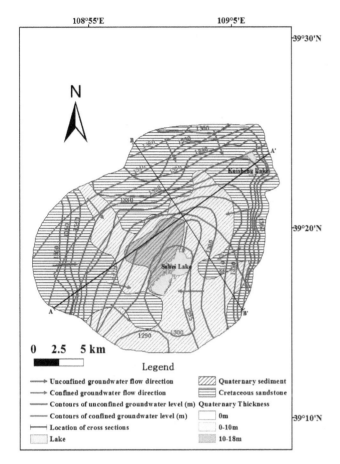

Figure 3. Hydrogeological map of the study area. Data were revised from original source (Inner Mongolia Second Hydrogeology Engineering Geological Prospecting Institute, 2010).

draulic connection between the Quaternary layer and Cretaceous phreatic aquifer, which could be viewed as an integrated unconfined aquifer in the area. The depth to water table in unconfined aquifer is influenced by the terrain change, of which the minimum value is below 1 m in the low-lying region, and the maximum value could be up to 13.24 m. The hydraulic conductivity of the aquifer changes between 0.16 and 17.86 m day^{-1}. The specific yield of unconfined aquifer varies from 0.058 to 0.155. The recharge source of groundwater in the unconfined aquifer is mainly the infiltration recharge of precipitation; it can be also recharged by lateral inflow from groundwater outside the study area. Besides the above recharge terms, leakage recharge from the underlying confined aquifer and infiltration recharge of irrigation water can also provide a small percentage of groundwater recharge. Evaporation is the main discharge way of the unconfined groundwater. In addition, lateral outflow, artificial exploitation and leakage discharge are also included in the main discharge patterns. Unconfined groundwater levels were contoured to illustrate the general flow field in the area (Fig. 3). Groundwater levels were monitored during September 2003.

As is shown in Fig. 3, lateral outflow occurs in a small part of southern boundary determined by analyzing the contours and flow direction of groundwater. The groundwater flows predominantly from surrounding uplands to low lands, which is under the control of topography. On the whole, groundwater in phreatic aquifer flows toward Subei Lake and recharges lake water (Fig. 3).

The unconfined and confined aquifers are separated by an uncontinuous aquitard. Generally speaking, permeable layers and aquitards intervein in the vertical profile of the aquifer system. Nevertheless, aquitards may pinch out in many places, so the aquifer system acts as a single hydrogeologic unit. In the present study, the covering aquitard is composed of the mudstone layer, which is mainly distributed in the second sand layer, and discontinued mudstone lens could also be observed in Cretaceous strata (Fig. 4). The phreatic aquifer is underlain by a confined aquifer composed of Cretaceous rocks. Due to huge thickness and high permeability of confined aquifer, it is regarded as the most promising water-supplying aquifer for domestic and industrial uses. The hydraulic conductivity of confined aquifer changes between 0.14 and 27.04 m day^{-1}. The hydraulic gradient varies from 0.0010 to 0.0045 and the storage coefficient changes between 2.17×10^{-5} and 1.98×10^{-3}. The confined aquifer primarily receives leakage recharge from the unconfined groundwater. The flow direction of confined groundwater is similar to that of unconfined groundwater (Fig. 3). Artificial exploitation is the major way in which confined groundwater is drained.

In the present study, the depth of sampling wells, in combination with hydrogeological map of the study area, is used to classify the groundwater as Quaternary groundwater, shallow Cretaceous groundwater and deep Cretaceous groundwater. As a research on an adjacent, specific, shallow groundwater system of Ordos Basin shows that the circulation depth is 120 m (Yin et al., 2009). It is difficult to determine the circulation depth of shallow groundwater in fact because the circulation depth of local flow systems changes depending on the topography and the permeability of local systems (Yin et al., 2009). In this study, Quaternary groundwater was defined on the basis of the distribution of Quaternary sediments thickness and depth of sampling wells. According to Hou et al. (2006), the maximum circulation depth of local groundwater flow system in the study area is also 120 m, determined by using a large amount of hydrochemical and isotopic data; 120 m is chosen as the maximum circulation depth of the local groundwater system and is used to divide the Cretaceous groundwater samples into two groups: samples taken in wells shallower than 120 m were classified as shallow Cretaceous groundwater, while samples taken in wells deeper than 120 m were deep Cretaceous groundwater.

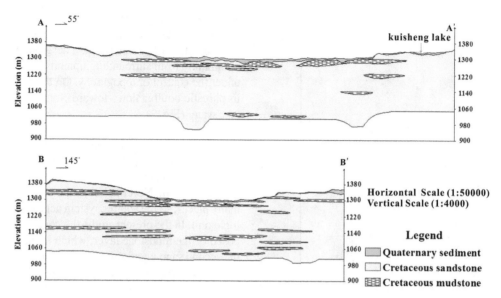

Figure 4. Geologic sections of the study area. Data were revised from original source (Inner Mongolia Second Hydrogeology Engineering Geological Prospecting Institute, 2010).

3   Methods

3.1   Water sampling

Two important sampling actions were conducted in the study area during August and December 2013, respectively. A total of 95 groundwater samples and seven lake water samples were collected. The first sampling action was during the rainy season and the other was during the dry season. The sampling locations are shown in Fig. 5. The water samples were taken from wells for domestic and agricultural purposes, ranging in depth from 2 to 300 m. The length of screen pipes in all sampling wells ranges from 1 to 10 m and every sampling well has only one screen pipe rather than multiple screens. The distance between the bottom of the screen pipe and the total well depth ranges from 0 to 3 m in the study area, and the bottom depth of screen pipe was assigned to the water samples. The samples from the wells were mostly taken using pumps installed in these wells and after removing several well volumes prior to sampling. The 100 and 50 mL polyethylene bottles were pre-rinsed with water sample three times before the final water sample was collected. Lake water samples were collected at Subei Lake, Kuisheng Lake and Shahaizi Lake. Cellulose membrane filters (0.45 μm) were used to filter samples for cation and anion analysis. All samples were sealed with adhesive tape so as to prevent evaporation. GPS was applied to locate the sampling locations.

3.2   Analytical techniques

Electrical conductivity (EC), pH value and water temperature of each sample were measured in situ using an EC/pH meter (WM-22EP, DKK-TOA, Japan), which was previously calibrated. Dissolved oxygen concentration and oxidation–reduction potential were also determined using a HACH HQ30d Single-Input Multi-Parameter Digital Meter. In situ hydrochemical parameters were monitored until these values reached a steady state.

The hydrochemical parameters were analyzed at the Center for Physical and Chemical Analysis of Institute of Geographic Sciences and Natural Resources Research, Chinese Academy of Sciences (IGSNRR, CAS). Major ion compositions were measured for each sample including K^+, Na^+, Ca^{2+}, Mg^{2+}, Cl^-, SO_4^{2-} and NO_3^-. An inductively coupled plasma optical emission spectrometer (ICP-OES) (Perkin-Elmer Optima 5300DV, USA) was applied to analyze major cations. Major anions were measured by ion chromatography (ICS-2100, Dionex, USA). HCO_3^- concentrations in all groundwater samples were determined by the titration method using 0.0048 M H_2SO_4 on the day of sampling; methyl orange endpoint titration was adopted with the final pH of 4.2–4.4. Due to the extremely high alkalinity of lake water samples, HCO_3^- concentrations in all lake water samples were analyzed by titration using 0.1667 M H_2SO_4. CO_3^{2-} concentrations were also analyzed by titration; phenolphthalein was used as an indicator of endpoint titration.

Hydrogen (δD) and oxygen ($\delta^{18}O$) composition in the water samples were analyzed using a liquid water isotope analyzer (LGR, USA) at the Institute of Geographic Sciences and Natural Resources Research, Chinese Academy of Sciences (IGSNRR, CAS). Results were expressed in the standard δ notation as per mil (‰) difference from Vienna standard mean ocean water (VSMOW, 0‰) with analytical precisions of ±1‰ (δD) and ±0.1‰ ($\delta^{18}O$).

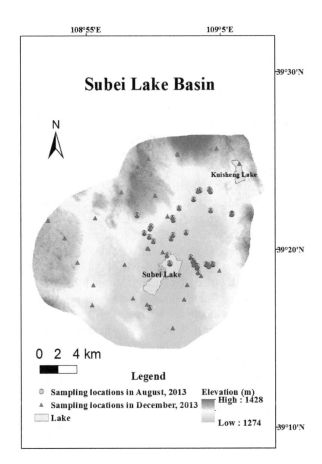

Figure 5. Sampling locations in August and December 2013.

Table 1. The chemical composition and isotopic data of lake water in August and December 2013.

ID	Date	EC (μS cm^{-1})	T (°C)	pH	DO (mg L^{-1})	ORP (mV)	K^+ (mg L^{-1})	Na^+ (mg L^{-1})	Ca^{2+} (mg L^{-1})	Mg^{2+} (mg L^{-1})	Cl^- (mg L^{-1})	SO_4^{2-} (mg L^{-1})	HCO_3^- (mg L^{-1})	CO_3^{2-} (mg L^{-1})	NO_3^- (mg L^{-1})	TDS (mg L^{-1})	δD (‰)	δ^{18}O (‰)
EEDS08	29 Aug 2013	130 400	22.5	10.11	11.06	−1.8	1956	42.020	2.28	3.01	37 440.28	22 066.83	6000.3	19 356.45	98.93	125 943.93	−1	19.4
EEDS09	29 Aug 2013	190 100	24.3	10.25	15.8	−14.8	6475	96.530	0.00	2.4	108 517.4	37 581.86	12 661.65	46 565.53	511.48	302 514.49	15	29.4
EEDS38	30 Aug 2013	1017	23.7	8.86	7.65	61.8	10.63	97.59	17.21	70.03	32.71	92.85	480.68	0	1.07	562.43	−45	−5.8
EEDS08	6 Dec 2013	120 400	1.8	8.9		17.6	1997.73	36 617.7	11.52	3.7	30 787.74	7513.4	5186.7	19 406.47	87.48	99 019.09	−18	5.8
EEDS09	6 Dec 2013	229 000	2.3	8.49		39.5	7567	77 840	36.34	11.39	113 003.44	11 593.8	13 754.59		207.99	223 494.4	−9	16.2
EEDS38	4 Dec 2013	4200	1.1	10.47	17.96	26.3	38.88	602	9.06	352.2	164.54	448.23	1423.8	900.3	9.53	3226.64	−28	−2.6
EEDS60	4 Dec 2013	14 080	2.7	9.04	10.58	23.6	56.154	3393.74	4.27	41.49	1418.76	386.04	1067.85	2600.87	11.77	8447.02	−28	−1.9

4 Results

4.1 Hydrochemical characteristics

In situ water quality parameters such as pH, electrical conductivity (EC), temperature, dissolved oxygen concentration (DO), oxidation–reduction potential (ORP) and total dissolved solids (TDS) as well as analytical data of the major ions composition in groundwater and lake water samples are shown in Table 1 and Table S1 in the Supplement. Based on the chemical data, hydrochemical characteristics of groundwater and lake water are discussed.

The chemical composition for lake water showed that Na^+ accounted for, on average, 93 % of total cations and Cl^- accounted for, on average, 58 % of total anions. Thus, Na^+ and Cl^- were the dominant elements (Fig. 6), which was in accordance with hydrochemical characteristics of inland salt lakes. This was also observed in lake water of Habor Lake basin located in the recharge area (Yin et al., 2009). The pH of lake water varied from 8.86 to 10.25 with an average of 9.74 in August and from 8.49 to 10.47 with an average of 9.23 in December; it can be seen that the pH was relatively stable and was always more than 8.4 without obvious seasonal variation, which indicated that the dissolved carbon-

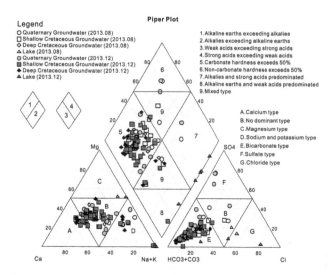

Figure 6. Piper diagram of groundwater and lake water in August and December 2013.

The hydrochemical data of groundwater were plotted on a Piper triangular diagram (Piper, 1953), which is perhaps the most commonly used method for identifying hydrochemical patterns of major ion composition (Fig. 6). With respect to cations, most of samples are scattered in zones A, B and D of the lower-left triangle, indicating that some are calcium-type, some are sodium-type water but most are of a mixed type; regarding anions, most groundwater samples are plotted in zone E of the lower-right triangle (Fig. 6), showing that bicarbonate-type water is predominant. The predominant hydrochemical types are HCO_3–Ca, HCO_3–Na and mixed HCO_3–Ca–Na–Mg types. Figure 6 also indicates that there are three groups of groundwater in the Subei Lake basin: the Quaternary groundwater, shallow Cretaceous groundwater and deep Cretaceous groundwater. The shallow Cretaceous groundwater refers to groundwater in the local groundwater system, and the deep Cretaceous groundwater refers to groundwater in the intermediate groundwater system of Ordos Basin. The hydrochemical characteristics of the three groups of groundwater indicate that they have undergone different degrees of mineralization.

With respect to the Quaternary groundwater, the pH varied from 7.64 to 9.04 with an average of 8.09 in August and changed from 7.49 to 9.26 with an average of 8.08 in December, indicating an alkaline nature. The TDS varied from 396 to $1202\,mg\,L^{-1}$, 314 to $1108\,mg\,L^{-1}$ with averages of 677 and $625\,mg\,L^{-1}$, respectively, in August and December. The major cations were Na^+, Ca^{2+} and Mg^{2+}, while the major anions were HCO_3^- and SO_4^{2-}.

As for the shallow Cretaceous groundwater ($< 120\,m$), the pH varied from 7.37 to 8.3 with an average of 7.77 in August and oscillated from 7.49 to 9.37 with an average of 8.14 in December. The TDS varied from 249 to $1383\,mg\,L^{-1}$ and from 217 to $1239\,mg\,L^{-1}$, with averages of 506 and $400\,mg\,L^{-1}$, respectively, in August and December.

For the deep Cretaceous groundwater ($> 120\,m$), the pH varied from 7.75 to 8.09 with an average of 7.85 in August and fluctuated from 7.99 to 8.82 with an average of 8.23 in December. The TDS varied from 266 to $727\,mg\,L^{-1}$, 215 to $464\,mg\,L^{-1}$ with averages of 377 and $296\,mg\,L^{-1}$, respectively, in August and December.

4.2 Stable isotopic composition in groundwater and surface water

In the present study, the results of the stable isotope analysis for groundwater and lake water are plotted in Fig. 7. In a previous study, the local meteoric water line (LMWL) in the northern Ordos Basin had been developed by Yin et al. (2010). The LMWL is $\delta D = 6.45\delta^{18}O - 6.51$ ($r^2 = 0.87$), which is similar to that developed by Hou et al. (2007) ($\delta D = 6.35\ \delta^{18}O - 4.69$). In addition, it is very clear in the plot that the LMWL is located below the global meteoric water line (GMWL) defined by Craig (1961) ($\delta D = 8\ \delta^{18}O + 10$), which suggests the occurrence of secondary

ates were in the HCO_3^- and CO_3^{2-} forms simultaneously. The temperatures of lake water ranged from 1.1 to 24.3 °C with large seasonal variations, implying that the surface water body was mainly influenced by hydrometeorological factors. The dissolved oxygen concentration of lake water showed an upward tendency from August (mean value: $11.50\,mg\,L^{-1}$) to December (mean value: $14.27\,mg\,L^{-1}$) because the relationship between water temperature and DO is inverse when oxygen content in the air stays relatively stable. With the decreasing water temperature, the dissolved oxygen value rises. The average value of ORP ranged from 15.1 mV in August to 26.8 mV in December, which was in accordance with the upward tendency of DO. It showed that lake water had stronger oxidation in December than that in August and there is a close relationship between DO and ORP. The average values of major ions concentrations showed a downward trend except for Ca^{2+}, Mg^{2+} from August to December. Specifically, the average values of Ca^{2+} and Mg^{2+} increased from 6.50 to $15.30\,mg\,L^{-1}$ and 25.15 to $102.20\,mg\,L^{-1}$, respectively; other ions concentrations were reduced to different degrees. The same variation trend of major ions from August to December could be found in the Habor Lake basin (Yin et al., 2009) as well. Before August, the strong evaporation capacity of lake water exceeded the finite recharge amount, which caused lake water to be enriched. After August, lake water was recharged and diluted by groundwater and a large amount of fresh overland flow from precipitation. The EC values varied between 1017 and $229\,000\,\mu S\,cm^{-1}$. This relatively large range of variation was closely related to the oscillation of the TDS values, which ranged from 0.56 to $302.5\,g\,L^{-1}$. The results showed that lake water chemistry was controlled by strong evaporation and recharge from overland flow and groundwater.

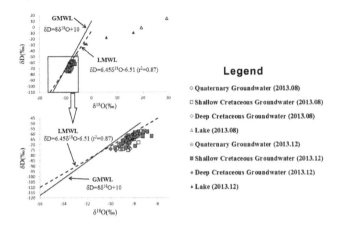

Figure 7. Relationship between hydrogen and oxygen isotopes in groundwater and lake water in August and December 2013.

evaporation during rainfall. The LMWL is controlled by local hydrometeorological factors, including the origin of the vapor mass, re-evaporation during rainfall and the seasonality of precipitation (Clark and Fritz, 1997).

The linear regression curve equation of $\delta^{18}O$ and δD in groundwater can be defined as $\delta D = 6.3\ \delta^{18}O - 13.0$ ($r^2 = 0.62$). Groundwater follows the LMWL in the study area, indicating that it is of modern local meteoric origin rather than the recharge from precipitation in paleoclimate conditions. In August, the stable isotope values in the Quaternary groundwater were found to range from -9.2 to -8.0% in ^{18}O with an average of -8.8% and from -74 to -62% in ^2H with an average of -71%; the shallow Cretaceous groundwater had $\delta^{18}O$ ranging from -9.3 to -7.5% and δD varying from -75 to -57%. The average values of $\delta^{18}O$ and δD of the shallow Cretaceous groundwater were -8.3 and -66%, respectively. $\delta^{18}O$ and δD of the deep Cretaceous groundwater ranged from -9.3 to -7.8% and from -74 to -61%, respectively. The average values of $\delta^{18}O$ and δD were -8.4 and -67%, respectively. In December, the stable isotope values in the Quaternary groundwater ranged from -8.9 to -7.2% in ^{18}O with an average of -8.2% and from -74 to -57% in ^2H with an average of -65%; $\delta^{18}O$ and δD of the shallow Cretaceous groundwater ranged from -9.7 to -6.5% and from -73 to -58%, respectively. The average values of $\delta^{18}O$ and δD were -8.2 and -64%, respectively. $\delta^{18}O$ of the deep Cretaceous groundwater varied from -10.0 to -7.5% and δD ranged from -75 to -60%. The average values of $\delta^{18}O$ and δD of the deep Cretaceous groundwater were -8.5 and -66%, respectively.

The regression curve equation of $\delta^{18}O$ and δD in lake water could be defined as $\delta D = 1.47\ \delta^{18}O - 29.09$ ($r^2 = 0.95$), where $\delta^{18}O$ ranged from -5.8 to 29.4% and δD ranged from -46 to 15% with averages of 14.3 and -10% in August, while in December, $\delta^{18}O$ and δD of lake water ranged from -2.6 to 16.2% and from -28 to -9%, respectively. The average values of $\delta^{18}O$ and δD were 4.4 and -21%, respec-

tively, in December. The low slope of the regression line of $\delta^{18}O$ and δD in lake water could be ascribed to a combination of mixing and evaporation under conditions of low humidity.

4.3 Linkage among geochemical parameters of groundwater

Correlations among groundwater-quality parameters are shown in Table 2 and Fig. 8. All of the major cations and anions are significantly correlated with TDS (Table 2), which shows that these ions have been dissolved into groundwater continuously and resulted in the rise of TDS.

As is shown in Table 2, the concentration of Mg^{2+} is correlated with HCO_3^- and SO_4^{2-}, with correlation coefficients of 0.582 and 0.819, respectively. The concentration of Ca^{2+} is well correlated with SO_4^{2-} with a correlation coefficient of 0.665. Cl^- has good correlation with Na^+ with a large correlation coefficient of 0.824.

The results of linear regression analysis of some pairs of ions are displayed in Fig. 8. There is good correlation between Cl^- and Na^+ in Quaternary groundwater and shallow Cretaceous groundwater; Ca^{2+} and SO_4^{2-} have a good positive correlation in Quaternary groundwater and shallow Cretaceous groundwater. Mg^{2+} is well correlated with HCO_3^- in shallow Cretaceous groundwater.

5 Discussion

Generally speaking, water–rock interactions are the most important factors influencing the observed geochemical composition of groundwater (Appelo and Willemsen, 1987); the geochemical and isotopic results of this work are no exception. In terms of dissolved minerals and the correlation of geochemical parameters, the dominant geochemical processes and formation mechanisms could be found (Su et al., 2009). The weathering and dissolution of minerals in the host rocks and ion exchange are generally the main source of ions in groundwater based on available research. The stable isotopes signatures in lake water can reveal the predominant mechanism controlling the chemical composition of lake water.

5.1 Geochemical processes of groundwater

As displayed in the correlation analysis of geochemical parameters, a good correlation between Mg^{2+} and HCO_3^- indicates that the weathering of dolomite releases ions to the groundwater, as expressed in Reaction (R1). The fact that Mg^{2+} is well correlated with HCO_3^- could be found in the Habor Lake basin of Ordos Plateau (Yin et al., 2009). Ca^{2+} has good correlation with SO_4^{2-}, implying that the dissolution of gypsum and anhydrite may be the key processes controlling the chemical composition of groundwater in the discharge area, which can be explained by Reaction (R2). Just as with the achievements obtained by Hou et al. (2006), gyp-

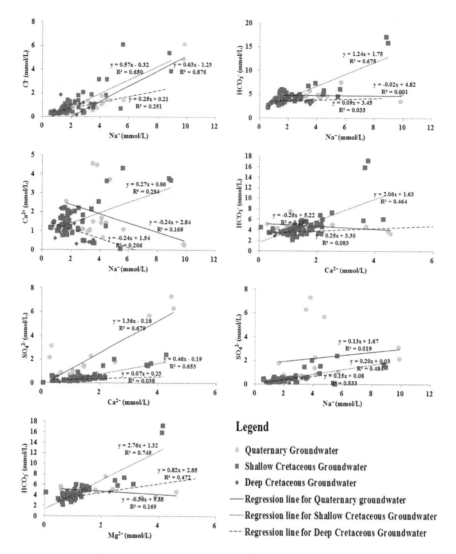

Figure 8. Relationship between some pairs of ions in groundwater.

sum and anhydrite are present in these strata, so it is reasonable to consider that gypsum and anhydrite are the source of the SO_4^{2-}. However, in Yin's study, there is poor correlation between Ca^{2+} and SO_4^{2-} in the Habor Lake basin (Yin et al., 2009). It can be explained by geochemical evolution of groundwater along flow path from the recharge area to the discharge area. There is poor correlation between Na^+ and SO_4^{2-}, suggesting that the weathering of Glauber's salt ($Na_2SO_4 \cdot 10H_2O$) may not be the major sources of such ions in groundwater. On the contrary, a good correlation between Na^+ and SO_4^{2-} can be found in the Habor Lake basin (Yin et al., 2009). It indicates that Glauber's salt may be more abundant in the recharge area (Habor Lake basin) than in the discharge area (Subei Lake basin). Although there is no obvious correlation between Ca^{2+} and HCO_3^-, it is reasonable to regard the dissolution of carbonate minerals as a source of Ca^{2+} and HCO_3^- due to the widespread occurrence of carbonate rocks in the study area, as conveyed in Reaction (R3).

The concentration of Mg^{2+} is well correlated with SO_4^{2-}, suggesting the possible dissolution of gypsum, followed by cation exchange. The pH is negatively correlated with Ca^{2+}; it is likely that the dissolution of carbonate minerals is constrained due to the reduction of the hydrogen ion concentration in water at higher pH. It can be judged from the above analysis that during groundwater flow, the following reactions are very likely to take place in the study area:

$$CaMg(CO_3)_2 + 2CO_2 + 2H_2O \Leftrightarrow Ca^{2+} + Mg^{2+} + 4HCO_3^-, \quad (R1)$$

$$CaSO_4 \Leftrightarrow Ca^{2+} + SO_4^{2-}, \quad (R2)$$

$$CaCO_3 + CO_2 + H_2O \Leftrightarrow Ca^{2+} + 2HCO_3^-. \quad (R3)$$

In order to explore the mechanism of salinity in semi-arid regions, the plot of Na^+ versus Cl^- is widely used (Magaritz et

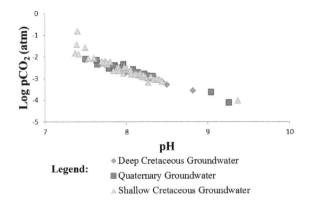

Figure 9. Geochemical relationship of pH vs. log (pCO_2) in groundwater.

al., 1981; Dixon and Chiswell, 1992; Sami, 1992). The concentration of Cl^- is well correlated with Na^+, suggesting that the dissolution of halite may be the major source of Na^+ and Cl^-. Theoretically, the dissolution of halite will release equal amounts of Na^+ and Cl^- into the solution. Nevertheless, the results deviate from the anticipated 1 : 1 relationship. Almost all samples have more Na^+ than Cl^-. The molar Na/Cl ratio varies from 0.68 to 16.00 with an average value of 3.48. A greater Na/Cl ratio may be ascribed to the feldspar weathering and the dissolution of other Na-containing minerals. The relatively high Na^+ concentration in the groundwater could also be illustrated by cation exchange between Ca^{2+} or Mg^{2+} and Na^+, as is discussed later.

The partial pressure of carbon dioxide (pCO_2) values were calculated by the geochemical computer code PHREEQC (Parkhurst and Appelo, 2004). The pCO_2 values of groundwater range from $10^{-0.82}$ to $10^{-4.1}$ atm. The vast majority of groundwater samples (about 96 %) have higher pCO_2 values than the atmospheric pCO_2, which is equal to $10^{-3.5}$ atm (Van der Weijden and Pacheco, 2003), indicating that groundwater has received CO_2 from root respiration and the decomposition of soil organic matter. Figure 9 indicates that the pCO_2 values are negatively correlated with pH values; the partial pressure values of CO_2 decrease as pH values increase (Rightmire, 1978; Adams et al., 2001). It likely has a connection with relatively longer aquifer residence time, more physical, chemical reactions with aquifer minerals and biological reactions of microorganism that produce CO_2 taking place. According to Hou et al. (2008), feldspars can be observed in the Cretaceous formations and it is possible that the following reaction occurs:

$$Na_2\,Al_2\,Si_6\,O_{16} + 2H_2\,O + CO_2 \rightarrow Na_2\,CO_3 \qquad (R4)$$
$$+ H_2\,Al_2\,Si_2\,O_8 + H_2\,O + 4\,SiO_2.$$

This reaction will consume CO_2 and give rise to the increase of the concentration of Na^+ and HCO_3^-. As a result, the partial pressure of CO_2 will decrease and the pH will increase. In terms of a statistical analysis, the average pH values of the

Quaternary groundwater and the shallow Cretaceous groundwater are 8.08 and 7.99, respectively, lower than that of the deep Cretaceous groundwater (8.11). However, the average pCO_2 values of the Quaternary groundwater and the shallow Cretaceous groundwater are $10^{-2.67}$ and $10^{-2.58}$ atm, respectively, higher than that of the deep Cretaceous groundwater (about $10^{-2.79}$ atm). The negative correlation characteristic between pCO_2 and pH shows that the dissolution of feldspar takes place along groundwater flow path. The phenomenon also occurs in the Habor Lake basin according to Yin et al. (2009).

Cation exchange is an important process of water–rock interactions that obviously influences the major ion composition of groundwater (Xiao et al., 2012). Although the cation exchange is widespread in the geochemical evolution of all groundwater, it is essential to know and identify the various changes undergone by water during their traveling processes in the groundwater system under the influence of anthropogenic activities. In the present study, the molar Na/Ca ratio changes between 0.5 and 106.09 with an average of 3.80, suggesting the presence of Na^+ and Ca^{2+} exchange. It can be conveyed in the following reaction:

$$Ca^{2+} + Na_2 - X = 2Na^+ + Ca - X, \qquad (R5)$$

where X is sites of cation exchange.

Schoeller proposed that chloro-alkaline indices could be used to study the cation exchange between the groundwater and its host environment during residence or travel (Schoeller, 1965; Marghade et al., 2012; Li et al., 2013). The Schoeller indices, such as CAI-I and CAI-II, are calculated by the following equations, where all ions are expressed in $mEq\,L^{-1}$:

$$CAI-I = \frac{Cl^- - (Na^+ + K^+)}{Cl^-}, \qquad (1)$$

$$CAI-II = \frac{Cl^- - (Na^+ + K^+)}{HCO_3^- + SO_4^{2-} + CO_3^{2-} + NO_3^-}. \qquad (2)$$

When the Schoeller indices are negative, an exchange of Ca^{2+} or Mg^{2+} in groundwater with Na^+ or K^+ in aquifer materials takes place, Ca^{2+} or Mg^{2+} will be removed from solution and Na^+ or K^+ will be released into the groundwater. At the same time, negative value indicates chloro-alkaline disequilibrium and the reaction is known as cation–anion exchange reaction. During this process, the host rocks are the primary sources of dissolved solids in the water. In another case, if the positive values are obtained, then the inverse reaction possibly occurs and it is known as base exchange reaction. In the present study, almost all groundwater samples had negative Schoeller index values (Table S1), which indicates cation–anion exchange (chloro-alkaline disequilibrium). The results indeed clearly show that Na^+ and K^+ are released by the Ca^{2+} and Mg^{2+} exchange, which is

Table 2. Correlation coefficient of major parameters in groundwater.

	K^+	Na^+	Ca^{2+}	Mg^{2+}	Cl^-	SO_4^{2-}	HCO_3^-	TDS	pH
K^+	1.000	0.538	0.309	0.560	0.553	0.300	0.572	0.534	−0.063
Na^+		1.000	0.217	0.651	0.824	0.485	0.602	0.728	−0.072
Ca^{2+}			1.000	0.754	0.375	0.665	0.478	0.796	−0.600
Mg^{2+}				1.000	0.655	0.819	0.582	0.939	−0.382
Cl^-					1.000	0.375	0.576	0.776	−0.144
SO_4^{2-}						1.000	0.160	0.770	−0.226
HCO_3^-							1.000	0.625	−0.398
TDS								1.000	−0.378
pH									1.000

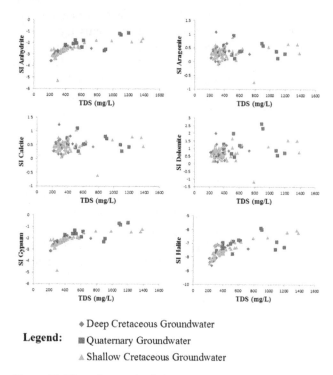

Legend:
- ◆ Deep Cretaceous Groundwater
- ■ Quaternary Groundwater
- ▲ Shallow Cretaceous Groundwater

Figure 10. Plots of saturation indices with respect to some minerals in groundwater.

a common form of cation exchange in the study area. This also further confirms that the cation exchange is one of the major contributors to higher concentrations of Na^+ in the groundwater, and it is still an important geochemical process of groundwater in the Subei Lake basin under the influence of human activities.

5.2 The formation mechanisms of groundwater and surface water

The saturation index is a vital geochemical parameter in the fields of hydrogeology and geochemistry, often useful for identifying the existence of some common minerals in the groundwater system (Deutsch, 1997). In this present study,

saturation indices (SIs) with respect to gypsum, anhydrite, calcite, dolomite, aragonite and halite were calculated in terms of the following equation (Lloyd and Heathcote, 1985):

$$SI = \log\left(\frac{IAP}{k_s(T)}\right) \tag{3}$$

where IAP is the relevant ion activity product, which can be calculated by multiplying the ion activity coefficient γ_i and the composition concentration m_i, and $k_s(T)$ is the equilibrium constant of the reaction considered at the sample temperature. The geochemical computer model PHREEQC (Parkhurst and Appelo, 2004) was used to calculate the saturation indices. When the groundwater is saturated with some minerals, SI equals zero; positive values of SI represent oversaturation, and negative values show undersaturation (Appelo and Postma, 1994; Drever, 1997).

Figure 10 indicates the plots of SI versus the total dissolved solids (TDS) for all the groundwater samples. The calculated values of SI for anhydrite, gypsum and halite oscillate between −5.27 and −1.11, between −4.8 and −0.65 and between −8.61 and −5.9, with averages of −2.62, −2.16 and −7.49, respectively. It shows that the groundwater in the study area was below the equilibrium with anhydrite, gypsum and halite, indicating that these minerals are anticipated to dissolve. However, the SIs of aragonite, calcite, and dolomite range from −0.74 to 1.09, −0.59 to 1.25 and −1.16 to 2.64, with averages of 0.32, 0.48 and 0.81, respectively. On the whole, the groundwater samples were dynamically saturated to oversaturated with aragonite, calcite and dolomite, implying that the three major carbonate minerals may have affected the chemical composition of groundwater in the Subei Lake basin. The results show that the groundwater may well produce the precipitation of aragonite, calcite and dolomite. Saturation of aragonite, calcite and dolomite could be attained quickly due to the existence of abundant carbonate minerals in the groundwater system.

The soluble ions in natural waters mainly derive from rock and soil weathering (Lasaga et al., 1994), anthropogenic input and partly from the precipitation input. In order to make an analysis of the formation mechanisms of hydrochemistry,

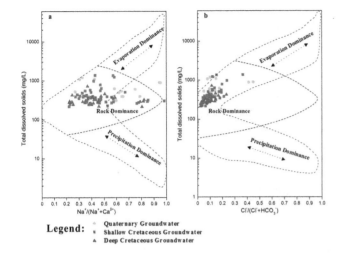

Legend:
· Quaternary Groundwater
■ Shallow Cretaceous Groundwater
▲ Deep Cretaceous Groundwater

Figure 11. Gibbs diagram of groundwater samples in the Subei Lake basin: **(a)** TDS vs. $Na^+ / (Na^+ + Ca^{2+})$, **(b)** TDS vs. $Cl^- / (Cl^- + HCO_3^-)$.

Gibbs diagrams have been widely used in hydrogeochemical studies (Feth and Gibbs, 1971; Naseem et al., 2010; Marghade et al., 2012; Yang et al., 2012b; Xing et al., 2013). Gibbs (1970) recommended two diagrams to assess the dominant effects of precipitation, rock weathering and evaporation on geochemical evolution of groundwater in semi-arid and arid regions. The diagrams show the weight ratios of $Na^+ / (Na^+ + Ca^{2+})$ and $Cl^- / (Cl^- + HCO_3^-)$ against TDS, and precipitation dominance, rock dominance, and evaporation dominance are included in the controlling mechanisms (Gibbs, 1970). The distributed characteristic of samples in Fig. 11 shows that rock weathering is the dominant mechanism in the geochemical evolution of the groundwater in the study area. The ratio of $Na^+ / (Na^+ + Ca^{2+})$ was mostly less than 0.5 in shallow and deep Cretaceous groundwater, with low TDS values (Fig. 11). It shows that rock weathering was the main mechanism controlling the chemical compositions of shallow and deep Cretaceous groundwater. In the Quaternary groundwater, about two-thirds of samples had a ratio of $Na^+ / (Na^+ + Ca^{2+})$ greater than 0.5 and higher TDS between 314 and 1202 mg/L, which indicated that the Quaternary groundwater was not only controlled by rock weathering, but also by the process of evaporation–crystallization. It is obvious that the weight ratio of $Na^+ / (Na^+ + Ca^{2+})$ spreads from low to high without a great variation of TDS, which indicated that cation exchange also played a role by increasing Na^+ and decreasing Ca^{2+} under the background of rock dominance. During the cation exchange process, the TDS values do not change obviously because $2\,mmol\,L^{-1}$ of Na^+ is released by $1\,mmol\,L^{-1}$ Ca^{2+} exchange, and the weight of $1\,mmol\,L^{-1}$ of Ca^{2+} $(40\,mg\,L^{-1})$ is nearly equal to that of $2\,mmol\,L^{-1}$ of Na^+ $(46\,mg\,L^{-1})$.

In August, the average isotopic values of deep Cretaceous groundwater ($\delta^{18}O$: $-8.4‰$, δD: $-67‰$) were enriched compared with the Quaternary groundwater ($\delta^{18}O$: $-8.8‰$, δD: $-71‰$), but in December, the average isotopic values of deep Cretaceous groundwater ($\delta^{18}O$: $-8.5‰$, δD: $-66‰$) were depleted compared with the Quaternary groundwater ($\delta^{18}O$: $-8.2‰$, δD: $-65‰$); the stable isotopic values of Quaternary groundwater had a wider range from August to December than those of deep Cretaceous groundwater. This may be explained by heavy isotope enrichment in the Quaternary groundwater caused by evaporation given that there was effectively no precipitation in the study area during the period from August to December; meanwhile, the deep Cretaceous groundwater may have been mainly recharged by lateral inflow from groundwater outside the study area, resulting in smaller seasonal fluctuations in the isotopic values.

Furthermore, the average values of $\delta^{18}O$ and δD of the shallow Cretaceous groundwater are -8.3 and $-66‰$ and -8.2 and $-64‰$, respectively, in August and December; meanwhile, the average values of $\delta^{18}O$ and δD of the deep Cretaceous groundwater are -8.4 and $-67‰$ and -8.5 and $-66‰$, respectively, in August and December. Thus, given the precision of the analysis, shallow Cretaceous groundwater and deep Cretaceous groundwater have similar isotopic characteristics in the Subei Lake basin, which indicates that they may be replenished by the similar water source due to the similar geological setting. This also validates the existence of leakage. The similar isotopic characteristic of groundwater from the Cretaceous aquifer may be ascribed to the increasingly close relationship between shallow Cretaceous groundwater and deep Cretaceous groundwater due to changes in the hydrodynamic field caused by intensive groundwater exploitation. Conversely, the phenomenon of deep groundwater depleted in heavy isotopes compared with shallow groundwater was found in the Habor Lake basin located in the recharge area (Yin et al., 2009).

The hydrogen and oxygen isotopes signatures in lake water show that it contains abnormally high levels of heavy isotopic composition. Compared with the stable isotopic values in groundwater, it is evident that lake water has undergone a greater degree of enrichment in heavy isotopes, which further illustrates that fractionation by strong evaporation is occurring predominantly in the lake water. This also proves to be in accordance with the unique hydrochemical characteristics of the lake water. In addition, the slope and intercept of the regression line for $\delta^{18}O$ and δD in lake water were 1.47 and -29.09, lower than the slope and intercept $(7.51, -7.12)$ observed for lake water in the Habor Lake basin (Yin et al., 2009). By comparison, it is clearly confirmed that lake water in the discharge area has undergone stronger evaporation than lake water in the recharge area. As a result, lake water in the Subei Lake basin contains more heavily isotopic composition than that in the Habor Lake basin.

6 Conclusions

The present study examines the hydrochemical and isotopic composition of the groundwater and surface water in the Subei Lake basin with various methods such as correlation analysis, saturation index, Piper diagram and Gibbs diagrams. The combination of major elements geochemistry and stable isotopes (δ^{18}O, δD) has provided a comprehensive understanding of the hydrodynamic functioning and the processes of mineralization that underpin the geochemical evolution of the whole water system. The hydrochemical data show that three groups of groundwater are present in the Subei Lake basin: the Quaternary groundwater, shallow Cretaceous groundwater and deep Cretaceous groundwater. The analysis of groundwater chemistry clarifies that the chemistry of lake water was controlled by strong evaporation and recharge from overland flow and groundwater; meanwhile the major geochemical processes responsible for the observed chemical composition in groundwater are the dissolution/precipitation of anhydrite, gypsum, halite and calcite and the weathering of feldspar and dolomite. Furthermore, the cation exchange has also played an extremely vital role in the groundwater evolution. The absolute predominance of rock weathering in the geochemical evolution of groundwater in the study area is confirmed by the analytical results of Gibbs diagrams. The stable isotopic data indicate that groundwater is of modern local meteoric origin rather than the recharge from precipitation in paleoclimate conditions. Unlike significant differences in isotopic values between shallow groundwater and deep groundwater in the Habor Lake basin, shallow Cretaceous groundwater and deep Cretaceous groundwater have similar isotopic characteristics in the Subei Lake basin. Due to the evaporation effect and dry climatic conditions, heavy isotopes are more enriched in lake water than groundwater. The low slope of the regression line of δ^{18}O and δD in lake water could be ascribed to a combination of mixing and evaporation under conditions of low humidity. A comparison of the regression line for δ^{18}O and δD shows that lake water in the Subei Lake basin contains more heavily isotopic composition than that in the Habor Lake basin, indicating that lake water in the discharge area has undergone stronger evaporation than lake water in the recharge area.

Much more accurate groundwater information has been obtained by conducting this study on Subei Lake basin, which will further enhance the knowledge of geochemical evolution of the groundwater system in the whole Ordos Basin and provide comprehensive understanding of Subei Lake basin, typical of lake basins in the discharge area where significant changes in the groundwater system have taken place under the influence of human activity. More importantly, it could provide valuable groundwater information for decision makers and researchers to formulate scientifically reasonable groundwater resource management strategies in these lake basins of Ordos Basin so as to minimize the negative impacts of anthropogenic activities on the water system. In addition, given that there have been a series of ecological and environmental problems, more ecohydrological studies in these lake basins are urgently needed from the perspective of the future sustainable development of natural resources.

Acknowledgements. This research was supported by the State Basic Research Development Program (973 Program) of China (grant no. 2010CB428805) and Greenpeace International. The authors are grateful for our colleagues for their assistance in sample collection and analysis. Special thanks go to the editor and the two anonymous reviewers for their critical reviews and valuable suggestions.

Edited by: S. Uhlenbrook

References

Adams, S., Titus, R., Pietersen, K., Tredoux, G., and Harris, C.: Hydrochemical characteristics of aquifers near Sutherland in the Western Karoo, South Africa, J. Hydrol., 241, 91–103, doi:10.1016/S0022-1694(00)00370-X, 2001.

Agartan, E. and Yazicigil, H.: Assessment of water supply impacts for a mine site in western Turkey, Mine Water Environ., 31, 112–128, doi:10.1007/s10230-011-0167-z, 2012.

Appelo, C. A. J. and Postma, D.: Geochemistry, groundwater and pollution, AA Balkema, Rotterdam, 1994.

Appelo, C. A. J. and Willemsen, A.: Geochemical calculations and observations on salt water intrusions, I. A combined geochemical/mixing cell model, J. Hydrol., 94, 313–330, 1987.

Cervi, F., Ronchetti, F., Martinelli, G., Bogaard, T. A., and Corsini, A.: Origin and assessment of deep groundwater inflow in the Ca' Lita landslide using hydrochemistry and in situ monitoring, Hydrol. Earth Syst. Sci., 16, 4205–4221, doi:10.5194/hess-16-4205-2012, 2012.

Chen, J., Liu, X., Wang, C., Rao, W., Tan, H., Dong, H., Sun, X., Wang, Y., and Su, Z.: Isotopic constraints on the origin of groundwater in the Ordos Basin of northern China, Environ. Earth Sci., 66, 505–517, doi:10.1007/s12665-011-1259-6, 2011.

Clark, I. D. and Fritz, P.: Environmental isotopes in hydrogeology, CRC press, Boca Raton, Florida, 1997.

Cook, P. G. and Herczeg, A. L.: Environmental tracers in subsurface hydrology, Kluwer, Dordrecht, 1999.

Craig, H.: Isotopic variations in meteoric waters, Science, 133, 1702–1703, doi:10.1126/science.133.3465.1702, 1961.

Dai, S. F., Ren, D. Y., Chou, C. L., Li, S. S., and Jiang, Y. F.: Mineralogy and geochemistry of the No. 6 coal (Pennsylvanian) in the Junger Coalfield, Ordos Basin, China, Int. J. Coal Geol., 66, 253–270, doi:10.1016/j.coal.2005.08.003, 2006.

Deutsch, W. J.: Groundwater geochemistry: fundamentals and applications to contamination, CRC press, Boca Raton, Florida, 1997.

De Vries, J. J. and Simmers, I.: Groundwater recharge: an overview of processes and challenges, Hydrogeol. J., 10, 5–17, doi:10.1007/s10040-001-0171-7, 2002.

Dixon, W. and Chiswell, B.: The use of hydrochemical sections to identify recharge areas and saline intrusions in alluvial aquifers, southeast Queensland, Australia, J. Hydrol., 135, 259–274, doi:10.1016/0022-1694(92)90091-9, 1992.

Drever, J. I.: The geochemistry of natural waters: surface and groundwater environments, Prentice Hall, New Jersey, 1997.

Feth, J. H. and Gibbs, R. J.: Mechanisms controlling world water chemistry: evaporation-crystallization process, Science, 172, 871–872, 1971.

Garvelmann, J., Külls, C., and Weiler, M.: A porewater-based stable isotope approach for the investigation of subsurface hydrological processes, Hydrol. Earth Syst. Sci., 16, 631–640, doi:10.5194/hess-16-631-2012, 2012.

Gibbs, R. J.: Mechanisms controlling world water chemistry, Science, 170, 1088–1090, doi:10.1126/science.170.3962.1088, 1970.

Hamed, Y. and Dhahri, F.: Hydro-geochemical and isotopic composition of groundwater, with emphasis on sources of salinity, in the aquifer system in Northwestern Tunisia, J. Afr. Earth Sci., 83, 10–24, doi:10.1016/j.jafrearsci.2013.02.004, 2013.

Hou, G., Zhao, M., and Wang, Y.: Groundwater investigation in the Ordos Basin, China Geological Survey, Beijing, 2006 (in Chinese).

Hou, G., Su, X., Lin, X., Liu, F., Yi, S., Dong, W., Yu, F., Yang, Y., and Wang, D.: Environmental isotopic composition of natural water in Ordos Cretaceous Groundwater Basin and its significance for hydrological cycle, J. Jilin Univ. (Earth Science Edition), 37, 255–260, 2007.

Hou, G., Liang, Y., Su, X., Zhao, Z., Tao, Z., Yin, L., Yang, Y., and Wang, X.: Groundwater systems and resources in the Ordos Basin, China, Acta Geol. Sin., 82, 1061–1069, 2008.

Kamdee, K., Srisuk, K., Lorphensri, O., Chitradon, R., Noipow, N., Laoharojanaphand, S., and Chantarachot, W.: Use of isotope hydrology for groundwater resources study in Upper Chi river basin, J. Radioanal. Nuclear Chem., 297, 405–418, doi:10.1007/s10967-012-2401-y, 2013.

Lasaga, A. C., Soler, J. M., Ganor, J., Burch, T. E., and Nagy, K. L.: Chemical-weathering rate laws and global geochemical cycles, Geochim. Cosmochim. Ac., 58, 2361–2386, doi:10.1016/0016-7037(94)90016-7, 1994.

Li, P., Wu, J., and Qian, H.: Groundwater quality assessment and the forming mechanism of the hydrochemistry in Dongsheng Coalfield of Inner Mongolia, J. Water Resour. Water Eng., 21, 38–41, 2010 (in Chinese).

Li, P., Qian, H., and Wu, J. H.: Application of set pair analysis method based on entropy weight in groundwater quality assessment – a case study in Dongsheng City, Northwest China, E-J. Chem., 8, 851–858, 2011.

Li, P., Qian, H., Wu, J., Zhang, Y., and Zhang, H.: Major ion chemistry of shallow groundwater in the Dongsheng Coalfield, Ordos Basin, China, Mine Water Environ., 32, 195–206, doi:10.1007/s10230-013-0234-8, 2013.

Lloyd, J. W. and Heathcote, J.: Natural inorganic hydrochemistry in relation to groundwater, Oxford University Press, New York, 1985.

Magaritz, M., Nadler, A., Koyumdjisky, H., and Dan, J.: The use of Na-Cl ratios to trace solute sources in a semi-arid zone, Water Resour. Res., 17, 602–608, doi:10.1029/Wr017i003p00602, 1981.

Marghade, D., Malpe, D. B., and Zade, A. B.: Major ion chemistry of shallow groundwater of a fast growing city of Central India, Environ. Monit. Assess., 184, 2405–2418, doi:10.1007/s10661-011-2126-3, 2012.

Naseem, S., Rafique, T., Bashir, E., Bhanger, M. I., Laghari, A., and Usmani, T. H.: Lithological influences on occurrence of high-fluoride groundwater in Nagar Parkar area, Thar Desert, Pakistan, Chemosphere, 78, 1313–1321, doi:10.1016/j.chemosphere.2010.01.010, 2010.

Parkhurst, D. L. and Appelo, C.: PHREEQC2 user's manual and program, Water-Resources Investigations Report, US Geological Survey, Denver, Colorado, 2004.

Piper, A. M.: A graphic procedure in the geochemical interpretation of water analysis, US Department of the Interior, Geological Survey, Water Resources Division, Ground Water Branch, Washington, 1953.

PRC Ministry of Environmental Protection: Technical guidelines for environment impact assessment-groundwater environment, China Environmental Science Press, Beijing, 2011.

Rightmire, C. T.: Seasonal-variation in pCO_2 and ^{13}C content of soil atmosphere, Water Resour. Res., 14, 691–692, doi:10.1029/Wr014i004p00691, 1978.

Sami, K.: Recharge mechanisms and geochemical processes in a semiarid sedimentary basin, Eastern Cape, South-Africa, J. Hydrol., 139, 27–48, doi:10.1016/0022-1694(92)90193-Y, 1992.

Schoeller, H.: Qualitative evaluation of groundwater resources, In: Methods and techniques of groundwater investigations and development, UNESCO, Paris, 54–83, 1965.

Su, Y., Zhu, G., Feng, Q., Li, Z., and Zhang, F.: Environmental isotopic and hydrochemical study of groundwater in the Ejina Basin, northwest China, Environ. Geol., 58, 601–614, doi:10.1007/s00254-008-1534-3, 2009.

Toth, J.: A theoretical analysis of groundwater flow in small drainage basins, J. Geophys. Res., 68, 4795–4812, doi:10.1029/Jz068i008p02354, 1963.

Van der Weijden, C. H., and Pacheco, F. A. L.: Hydrochemistry, weathering and weathering rates on Madeira island, J. Hydrol., 283, 122–145, doi:10.1016/S0022-1694(03)00245-2, 2003.

Wang, W., Yang, G., and Wang, G.: Groundwater numerical model of Haolebaoji well field and evaluation of the environmental problems caused by exploitation, South-to-North Water Trans. Water Sci. Technol., 8, 36–41, 2010 (in Chinese).

Winter, T. C.: Relation of streams, lakes, and wetlands to groundwater flow systems, Hydrogeol. J., 7, 28–45, doi:10.1007/s100400050178, 1999.

Xiao, J., Jin, Z., Zhang, F., and Wang, J.: Solute geochemistry and its sources of the groundwaters in the Qinghai Lake catchment, NW China, J. Asian Earth Sci., 52, 21–30, doi:10.1016/j.jseaes.2012.02.006, 2012.

Xing, L., Guo, H., and Zhan, Y.: Groundwater hydrochemical characteristics and processes along flow paths in the North China Plain, J. Asian Earth Sci., 70–71, 250–264, doi:10.1016/j.jseaes.2013.03.017, 2013.

Yang, L., Song, X., Zhang, Y., Han, D., Zhang, B., and Long, D.: Characterizing interactions between surface water and groundwater in the Jialu River basin using major ion chemistry and stable isotopes, Hydrol. Earth Syst. Sci., 16, 4265–4277, doi:10.5194/hess-16-4265-2012, 2012a.

Yang, L., Song, X., Zhang, Y., Yuan, R., Ma, Y., Han, D., and Bu, H.: A hydrochemical framework and water quality assessment of river water in the upper reaches of the Huai River Basin, China, Environ. Earth Sci., 67, 2141–2153, doi:10.1007/s12665-012-1654-7, 2012b.

Yin, L., Hou, G., Dou, Y., Tao, Z., and Li, Y.: Hydrogeochemical and isotopic study of groundwater in the Habor Lake Basin of the Ordos Plateau, NW China, Environ. Earth Sci., 64, 1575–1584, doi:10.1007/s12665-009-0383-z, 2009.

Yin, L., Hou, G., Su, X., Wang, D., Dong, J., Hao, Y., and Wang, X.: Isotopes (δD and δ^{18}O) in precipitation, groundwater and surface water in the Ordos Plateau, China: implications with respect to groundwater recharge and circulation, Hydrogeol. J., 19, 429–443, doi:10.1007/s10040-010-0671-4, 2010.

Yuko, A., Uchida, T., and Ohte, N.: Residence times and flow paths of water in steep unchannelled catchments, Tanakami, Japan, J. Hydrol., 261, 173–192, 2002.

Zhang, J., Fang, H., and Ran, G.: Groundwater resources assessment in the Ordos Cretaceous Artesian Basin, Inner Mongolia Bureau of Geology and Mineral Resources, Hohhot, 1986.

7

A lab in the field: high-frequency analysis of water quality and stable isotopes in stream water and precipitation

Jana von Freyberg[1,2], **Bjørn Studer**[1], and **James W. Kirchner**[1,2]

[1]Department of Environmental Systems Science, ETH Zurich, Zurich, Switzerland
[2]Swiss Federal Institute for Forest, Snow and Landscape Research (WSL), Birmensdorf, Switzerland

Correspondence to: Jana von Freyberg (jana.vonfreyberg@usys.ethz.ch)

Abstract. High-frequency measurements of solutes and isotopes (^{18}O and ^{2}H) in rainfall and streamflow can shed important light on catchment flow pathways and travel times, but the workload and sample storage artifacts involved in collecting, transporting, and analyzing thousands of bottled samples severely constrain catchment studies in which conventional sampling methods are employed. However, recent developments towards more compact and robust analyzers have now made it possible to measure chemistry and water isotopes in the field at sub-hourly frequencies over extended periods. Here, we present laboratory and field tests of a membrane-vaporization continuous water sampler coupled to a cavity ring-down spectrometer for real-time measurements of δ^{18}O and δ^2H combined with a dual-channel ion chromatograph (IC) for the synchronous analysis of major cations and anions. The precision of the isotope analyzer was typically better than 0.03‰ for δ^{18}O and 0.17‰ for δ^2H in 10 min average readings taken at intervals of 30 min. Carryover effects were less than 1.2 % between isotopically contrasting water samples for 30 min sampling intervals, and instrument drift could be corrected through periodic analysis of secondary reference standards. The precision of the ion chromatograph was typically $\sim 0.1\text{--}1$ ppm or better, with relative standard deviations of ~ 1 % or better for most major ions in stream water, which is sufficient to detect subtle biogeochemical signals in catchment runoff.

We installed the coupled isotope analyzer/IC system in an uninsulated hut next to a stream of a small catchment and analyzed stream water and precipitation samples every 30 min over 28 days. These high-frequency measurements facilitated a detailed comparison of event-water fractions via endmember mixing analysis with both chemical and isotope tracers.

For two events with relatively dry antecedent moisture conditions, the event-water fractions were < 21 % based on isotope tracers but were significantly overestimated (40 to 82 %) by the chemical tracers. These observations, coupled with the storm-to-storm patterns in precipitation isotope inputs and the associated stream water isotope response, led to a conceptual hypothesis for runoff generation in the catchment. Under this hypothesis, the pre-event water that is mobilized by precipitation events may, depending on antecedent moisture conditions, be significantly shallower, younger, and less mineralized than the deeper, older water that feeds baseflow and thus defines the "pre-event" endmember used in hydrograph separation. This proof-of-concept study illustrates the potential advantages of capturing isotopic and hydrochemical behavior at a high frequency over extended periods that span multiple hydrologic events.

1 Introduction

Environmental tracers are widely used in hydrology to investigate recharge processes, subsurface flow mechanisms, and streamflow components (Leibundgut and Seibert, 2011). The most common environmental tracers are the naturally occurring stable water isotopes ^{18}O and ^{2}H (Klaus and McDonnell, 2013). Solutes such as dissolved organic compounds, nutrients, and major ions are also widely used, together with stable isotopes, as indicators of flow paths and biogeochemical reactions (e.g., McGlynn and McDonnell, 2003; Vitvar and Balderer, 1997; Weiler et al., 1999). Environmental tracer studies typically involve manual or automated sample collection followed by transport, storage, and subsequent laboratory analysis. The time and effort involved in sample handling are often major constraints limiting the frequency and duration of sampling and thus the scope of tracer studies. While automated in situ analyzers for certain solutes and nutrients are becoming standard tools in environmental monitoring studies (e.g., Bende-Michl and Hairsine, 2010; Rode et al., 2016b), high-frequency analyses of isotopes and major ions over longer time periods remain challenging.

To date, isotope studies have maintained high sampling frequencies only during a few storm events (e.g., Berman et al., 2009; Lyon et al., 2008; Pangle et al., 2013) with the result that only limited ranges of catchment behavior have been explored. Long-term catchment studies capture a wider range of hydrologic events, but generally collect water samples at only weekly or monthly intervals for subsequent laboratory analysis (Buso et al., 2000; Darling and Bowes, 2016; Jasechko et al., 2016; Neal et al., 2011), making higher-frequency behavior unobservable. As pointed out by Kirchner et al. (2004), sampling at intervals much longer than the hydrological response times of a catchment may result in a significant loss of information. For instance, sub-daily sampling is required to capture diurnal fluctuations in stream water hydrochemistry, which reflect evapotranspiration effects or in-stream biological activity (e.g., Aubert and Breuer, 2016; Hayashi et al., 2012). In order to differentiate hydrological and biogeochemical catchment processes related to different water ages and flow pathways, long-term monitoring has to be complemented by additional high-frequency hydrochemical and isotope measurements. So far, only a few long-term studies have sampled stream water at daily or sub-daily intervals for on-site measurements or subsequent analysis in the laboratory. These include studies conducted at Plynlimon in Wales (Neal et al., 2012), at the Kervidy–Naizin catchment in western France (Aubert et al., 2013), and at the Selke River in Germany (Rode et al., 2016a). Such studies have yielded fundamental insights into catchment hydrological behavior, not only at a wide range of temporal scales, but also under varying hydroclimatic conditions (Benettin et al., 2015; Halliday et al., 2013; Harman, 2015; Kirchner and Neal, 2013; Riml and Worman, 2015).

The recent development of compact and robust isotope analyzers has fostered initial attempts to continuously measure δ^{18}O and δ^2H in stream water or precipitation directly in the field. The only previous field-based isotope monitoring over 4 contiguous weeks was carried out by Berman et al. (2009) with a customized liquid water isotope analyzer based on off-axis integrated cavity output spectroscopy (OA-ICOS; Los Gatos Research, San Jose, CA, USA), which measured δ^{18}O and δ^2H in 90 samples per day. As the system was based on repeated injections of samples into a vaporizer, daily maintenance (i.e., injection septa change, filter cleaning) was required to keep it running. An alternative approach uses a semi-permeable membrane to generate water vapor from a continuous sample throughflow, which is then transferred to a wavelength-scanned cavity ring-down spectrometer (CRDS)

(Herbstritt et al., 2012). Munksgaard et al. (2011) developed such a custom-made diffusion sampler and attached it to a CRDS (Picarro Inc., Santa Clara, CA, USA) that was used to measure δ^{18}O and δ^2H in precipitation at frequencies of up to 30 s over a 15-day period (Munksgaard et al., 2012) as well as to monitor the isotopic response at 1 min resolution in streamflow during a storm event (Tweed et al., 2016).

A similar diffusion sampling system has recently become commercially available (Continuous Water Sampler, or CWS; Picarro Inc., Santa Clara, CA, USA), which allows for quasi-continuous measurements of δ^{18}O and δ^2H in liquid water samples when coupled to a CRDS analyzer. Here, we present initial laboratory and field verification experiments with this device, which we have combined with a dual-channel ion chromatograph (IC; Metrohm AG, Herisau, Switzerland) for real-time analysis of major cations and anions. Laboratory experiments quantifying the precision and sample carryover memory effects of this system are presented in Sect. 3. Section 4 illustrates the performance of the system in the field using a 28-day deployment at a small catchment in Switzerland. Section 5 quantifies the fractions of event water that contributed to the flood hydrograph in eight storm events, illustrating one potential application of high-frequency measurements of isotopes and major ions.

2 Methodology

2.1 Isotope analysis and ion chromatography

For the analysis of the stable water isotopes ^{18}O and ^{2}H, the Continuous Water Sampler (CWS) was coupled to a wavelength-scanned cavity ring-down spectrometer (CRDS; model L2130-i; Picarro Inc., Santa Clara, CA, USA). In the CWS, the water sample flows at a rate of ~ 1 mL min^{-1} through an expanded polytetrafluoroethylene (ePTFE) membrane tube. This tube is mounted in a stainless steel chamber that is supplied with dry air to facilitate the steady diffusion of a small fraction of the through-flowing water as vapor through the membrane. Through the continuous flow of dry air over the outer surface of the membrane, the vapor is carried directly to the CRDS for isotope analysis. To minimize temperature-induced fractionation effects, the instrument keeps the temperatures of the membrane chamber and the inflowing water constant at (± 1 standard deviation) 45 ± 0.1 and 15 ± 0.1 °C, respectively. A solenoid diaphragm pump situated upstream of the membrane cartridge draws water samples from the sample container and pushes them through the membrane tube at a flow rate of approximately 1 mL min^{-1}. As we show in Sect. 3.1, preliminary tests showed that this pump is not sufficient for our purposes, so we substituted a programmable high-precision dosing unit (800 Dosino, hereafter simply "Dosino"; Metrohm AG, Herisau, Switzerland) in its place.

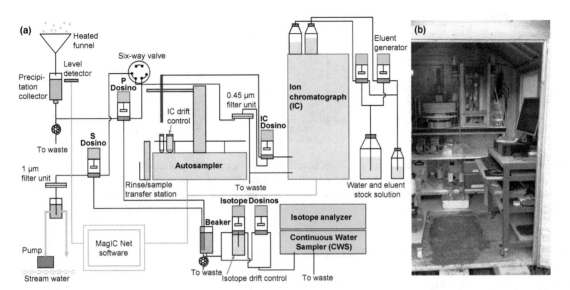

Figure 1. (a) Schematic overview of the coupled isotope analyzer/IC system for the collection and analysis of stream water and precipitation samples. The components of the sample distribution and the IC are shown in blue, while the isotope analyzer with the CWS is shown in green. Panel **(b)** shows a photo of the coupled isotope analyzer/IC system in the wooden hut during the field experiment.

Isotopic abundances are reported through the δ notation relative to the VSMOW-SLAP standards. For the laboratory experiments, we used the factory calibration of the isotope analysis system because only relative isotope values are needed for quantifying precision, drift, and carryover, and thus the absolute isotope values are unimportant. For the field experiment, however, we periodically measured two internal isotope standards (Fiji and Evian bottled water), which were calibrated by a Picarro L2130-i CRDS at the isotope laboratory of the University of Freiburg (Germany), to primary reference materials (IAEA standards SLAP, VSMOW, and GISP; instrument precision 0.16‰ for $\delta^{18}O$ and 0.6‰ for $\delta^{2}H$).

Major ions in liquid water samples, i.e., Na^+, K^+, NH_4^+, Ca^{2+}, Mg^{2+}, F^-, Cl^-, NO_3^-, SO_4^{2-}, and PO_4^{3-}, were analyzed with an ion chromatograph (940 Professional IC Vario, hereafter simply "IC"; Metrohm AG, Herisau, Switzerland) with a two-column configuration (anions, Metrosep A Supp 5-250/4.0; cations, Metrosep C6-250/4.0). Continuous operation of the instrument was possible due to fully automated eluent generation (941 Eluent Production Module; Metrohm AG, Herisau, Switzerland). To generate the full ion chromatograms of both anions and cations, approximately 28 min were required; thus, the sampling interval of the combined analysis system was fixed at 30 min.

2.2 Sample collection and distribution

The water samples were distributed between the analyzers with high-precision dosing units (Dosinos). Each Dosino contains a programmable piston that fills and empties a glass cylinder with up to 50 mL of sample at a resolution

of 10 000 increments (implying 5 µL increment^{-1}). The design of the dosing unit minimizes the dead volume and thus the potential for sample carryover. In the base of the glass cylinder sits a rotating valve disc that guides the liquid sample through one of four ports; thus, each Dosino functions as both a switching valve and a syringe pump.

Figure 1 depicts the schematic overview of the automatic sample collection and analysis system, showing how the different Dosinos distribute precipitation and stream water samples between the isotope analyzer, the IC, and an autosampler (which can be programmed to save individual samples for subsequent analysis in the laboratory). The sampling routine begins with a cleaning step during which either the P Dosino (which handles precipitation) or the S Dosino (which handles stream water) transports 10 mL of sample water for rinsing to a sample storage beaker. The Isotope Dosinos also eject any remaining sample into the beaker, after which the beaker is emptied. Then, 50 mL of fresh stream water or precipitation sample is transported (by either the S Dosino or the P Dosino for stream water or precipitation, respectively) into the rinsed beaker, from which one of the Isotope Dosinos draws 30 mL of water and injects it at a flow rate of 1 mL min^{-1} into the CWS for isotope analysis. The two Isotope Dosinos operate alternatingly to minimize the length of time that the sample flow into the CWS is interrupted. Meanwhile, either the P Dosino or the S Dosino takes up another 12 mL of water sample and pumps it through a 0.45 µm tangential filter into the IC Dosino, which discards the first 2 mL of the filtered sample. From the remaining filtered sample, 8 mL are filled into vials by the autosampler and 2 mL are delivered to the IC for direct ion analysis. During the ion analysis (ca. 28 min), the S Dosino, P Dosino, IC Dosino, the autosampler, and all

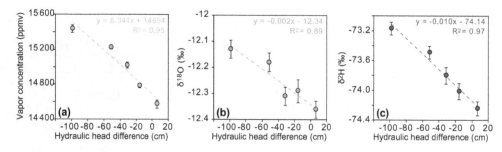

Figure 2. Laboratory experiment showing the isotope effects of sample injection into the Continuous Water Sampler (CWS). Panel (**a**) shows the measured vapor concentrations, and panels (**b**) and (**c**) show the raw, uncalibrated isotope values of a single water sample (nanopure water) as a function of the hydraulic head difference between the water level in the sample bottle and the waste outlet. Negative values of the hydraulic head difference indicate that the sample source was located below the waste outlet of the CWS.

tubing are rinsed with nanopure water to minimize carryover effects. The entire sampling routine is programmed with the IC control software MagIC Net (Metrohm, Herisau, Switzerland), which facilitates detailed data logging and documentation of the sample handling.

3   Laboratory experiments

3.1   Optimization of sample injection into the Continuous Water Sampler (CWS)

In the original design of the CWS, water samples are transported by a small solenoid diaphragm pump between the inlet port and the membrane cartridge at a flow rate of approximately $1\,\text{mL}\,\text{min}^{-1}$. During preliminary tests, however, we observed that raising or lowering the sample container detectably altered the reported isotope ratios. In order to quantify the sensitivity of the instrument to hydraulic head differences (i.e., the height of the water table in the sample bottle relative to the waste outlet of the CWS), we changed the elevation of the sample container relative to the instrument while continuously analyzing a single water sample (nanopure water). We measured the vapor concentration, $\delta^{18}\text{O}$ and $\delta^{2}\text{H}$ for the same water sample at five different elevations, ranging from 7 cm above to 98 cm below the waste outlet. The end of the waste outlet tube was always freely draining. Each configuration was measured for 1 h, and the average values and standard deviations of the uncalibrated 6 s measurements of vapor concentration, $\delta^{18}\text{O}$, and δ^{2} were calculated from the last 10 min of each 1 h configuration.

The results of this experiment are summarized in Fig. 2, which shows clear linear relationships between the hydraulic head differences and both the vapor concentrations and the isotope measurements. Lowering the sample source relative to the outflow results in systematically heavier isotopic values in the vapor measured by the instrument. The vapor concentrations show a similar trend; i.e., more vapor was generated at lower positions of the sample source. These observations suggest that the hydraulic head difference directly affected the flow rate of the liquid sample through the CWS membrane tube. Because the water is much colder than the surrounding air as it enters the membrane chamber, it is continuously warming as it travels through the membrane tube. At greater head gradients (and thus smaller flow rates), the sample will travel more slowly through the membrane chamber and will warm up more. At higher water temperatures, water should diffuse more rapidly through the membrane, and the resulting vapor will be less fractionated relative to the liquid phase (Kendall and McDonnell, 1998), as observed in Fig. 2.

It is unknown whether the empirical linear relationships shown in Fig. 2 are generally applicable or are specific to each individual membrane or to the properties of the sample. Nevertheless, for this membrane and this sample, the results indicate that changing the hydraulic head by 50 cm changes the reported isotope values by approximately 0.12‰ for $\delta^{18}\text{O}$ and 0.52‰ for $\delta^{2}\text{H}$. This flow-rate artifact might become particularly important for applications in which isotope standards and samples are drawn from sample containers at different elevations relative to the waste outlet of the CWS (e.g., shipboard sampling). In such cases, a vapor concentration correction relative to a reference height would have to be carried out. Alternatively, a different injection system could be used to deliver a specified flow rate independent of the position of the source relative to the CWS. We used the 800 Dosino for this purpose, since it functions as a high-precision syringe pump with a delivery rate specified by the pulse rate of the stepper motor, independent of the hydraulic head gradient.

Because of the limited volume of each Dosino's glass cylinder (50 mL), a sample could be injected at a flow rate of $1\,\text{mL}\,\text{min}^{-1}$ for a maximum of 50 min. For longer injections, or to switch samples, a second Dosino had to take over the sample delivery. The handoff between the Dosinos interrupted the sample flow to the CWS for around 2 s. This interruption was reflected in a sharp but brief increase in vapor concentrations and isotope values, which returned back to stable values approximately 10 min after the injection started

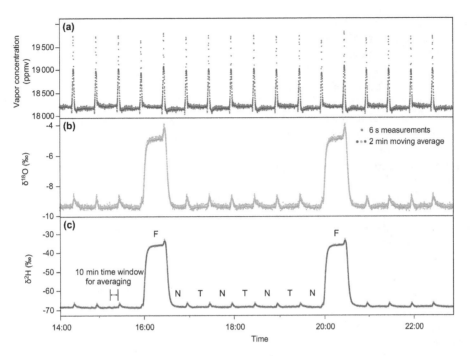

Figure 3. A 9 h excerpt showing the raw, uncalibrated data of vapor concentrations (panel **a**) and isotope measurements (panels **b** and **c**) in tap water (T), nanopure water (N), and Fiji bottled water (F) during the 48 h laboratory experiment. The samples were injected alternately with two Dosinos for 30 min each at a flow rate of 1 mL min^{-1}.

(see Fig. 3 for an example). For our application, i.e., synchronous IC measurements, we programmed a 30 min injection period for the isotope analysis. To obtain the final isotope values of a liquid sample, we averaged the individual 6 s measurements reported by the CRDS during the last 10 min of each 30 min injection period, using the first 20 min to minimize any memory effects from the previous sample or from Dosino changeover. The advantage of the Dosino-based sample handling system is the very steady, pressure-independent sample injection.

3.2 Performance of the isotope analyzer with the Continuous Water Sampler (CWS)

We quantified the precision, drift coefficients, and carryover effects of the isotope analyzer with the CWS and Dosino-based sample injection system using a continuous 48 h laboratory experiment that alternated between three water samples (i.e., to mimic stream water, precipitation, and a reference standard). The sample handling system was as shown in Fig. 1, except that the precipitation collector was replaced with a 10 L bottle of nanopure water and the stream water sampler was replaced by a 10 L bottle of tap water. The sampling system alternated between these two sources, and for each eighth injection it introduced an isotopically heavier secondary standard (Fiji bottled water) (Fig. 3). The isotopic differences between Fiji bottled water and tap water were about ($\pm$standard error; SE) 4.54 ± 0.02 and 32.67 ± 0.08 ‰ for $\delta^{18}O$ and δ^2H, respectively. The isotopic differences be-

tween tap water and nanopure water were much smaller (0.05 ± 0.01 ‰ for $\delta^{18}O$ and 0.12 ± 0.03 ‰ for δ^2H) because the nanopure water was generated from the same tap water by reverse osmosis.

The precision of the isotope values, as quantified by the standard deviations of the individual 6 s measurements during the last 10 min of each injection period, was better than 0.08 ‰ for $\delta^{18}O$ and 0.18 ‰ for δ^2H. These standard deviations imply that the standard errors of the 10 min averages should be better than 0.008 and 0.018 ‰ for $\delta^{18}O$ and δ^2H, respectively. These standard errors overestimate the repeatability of successive measurements, however. As a measure of sample-to-sample repeatability, the standard deviations of the 10 min averages for the entire 48 h experiment were 0.03 ‰ ($\delta^{18}O$) and 0.17 ‰ (δ^2H) or better for each of the three water samples (excluding two outliers associated with an interruption in the sampling routine), much larger than the calculated standard errors. Thus, the major uncertainties in the 10 min averages do not arise from the counting statistics of the instrument itself, but rather, we suspect, from sample-to-sample variability in the performance of the vaporizer. We use these larger estimates of uncertainty (0.03 ‰ for $\delta^{18}O$ and 0.17 ‰ for δ^2H) in the error propagation calculations presented in Sect. 5.1.

Instrument drift was analyzed through linear regression of the 10 min averages from the ends of each 30 min injection period. The instrument drift for $\delta^{18}O$ was statistically indistinguishable from zero for two of the three waters, averaging ($\pm$SE) -0.009 ± 0.008, -0.009 ± 0.006, and

$-0.015 \pm 0.007 \permil$ day^{-1} for Fiji, nanopure, and tap water, respectively. The instrument drift for δ^2H was slow but statistically significant for two of the three waters, averaging 0.133 ± 0.040, 0.084 ± 0.016, and $-0.021 \pm 0.021 \permil$ day^{-1} for Fiji, nanopure, and tap water, respectively. Thus, the accumulated drift over 1 day was typically smaller than the measurement precision for individual 10 min averages for either isotope. As explained in Sect. 4.2, substantially faster drift occurred during the field experiment that could, however, easily be measured and corrected using regularly injected reference standards. This faster drift can be explained by biofilm growth on the membrane, which could be observed on the inside of the membrane tube during preliminary tests with stream water samples at the field site.

Between-sample memory mainly arises from small remnants of previously injected samples that remain in the sample handling system (e.g., tubes, membrane, valves, and pumps) or the analyzer itself, and are carried over to the following analysis. We quantified the between-sample memory effect of the isotope analyzer using two isotopically contrasting samples, Fiji water and nanopure water. The true isotopic difference was obtained from the seventh (and last) injection of nanopure water, which was measured around 3 h after the reference standard (Fiji) and was thus assumed to be free of any memory effects. We calculated the memory coefficient (X) as a measure of carryover effects using Gupta et al. (2009):

$$X = \frac{C_i - C_{i-1}}{C_{\text{true}} - C_{i-1}}, \tag{1}$$

where C denotes the isotope ratio (or the solute concentration), the indices (i) and (i-1) denote the current and the previous injection, and (true) denotes the true value taken from the last value of multiple injections. Based on the 10 min averages from the end of each 30 min injection period, the average carryover from the Fiji bottled water to the next sample was $100\% \cdot (1 - X) \approx 0.9\%$ for δ^{18}O and 1.2 % for δ^2H (Table 1). The carryover during the first and second 10 min of each 30 min injection period was, however, much larger (up to 53 and 6 %, respectively), implying that our 30 min sampling cycle is indeed necessary to prevent unacceptably large carryover effects.

3.3   Performance of the ion chromatograph (IC)

With the IC, a 48 h laboratory experiment was carried out as well. However, the sampling sequence differed slightly from that of the isotope analyzer described previously: each measurement of tap water or Fiji water was followed by two to six samples of nanopure water, which mimics precipitation samples with generally very low solute concentrations. Due to the low solute concentrations in the nanopure water, the carryover effects can be quantified efficiently.

The average concentrations of the major anions and cations during the 48 h experiment are reported in Table 1

along with their absolute and relative standard deviations. For tap water and Fiji water, relative standard deviations were < 5 % for all constituents with concentrations above the limit of quantification (LOQ) and $\sim 1\%$ or less for most major ions, indicating that the IC measurements were stable over the 48 h period and that they were sufficiently precise to detect even subtle biogeochemical signals in stream water. The drift effects in the instrument were not statistically significant ($p > 0.05$) for most constituents in Fiji water and tap water. For Cl$^-$, NO$_3^-$, and SO$_4^{2-}$ in the Fiji water, the linear drift was statistically significant but also very slow: the accumulated drift over 24 h was never much larger than the LOQ (Table 1). The average percent of carryover ($100\% \cdot (1 - X)$; Eq. 1) in the nanopure water sample, following immediately after a sample of tap water or Fiji water, was $\leq 3.8\%$.

4   Application in the field

4.1   Setup

For the field experiment, the system was installed in a hut (area $1.7\,\text{m} \times 1.7\,\text{m}$) next to a small perennial stream flowing behind the Swiss Federal Institute for Forest, Snow and Landscape Research (WSL) near Zurich, Switzerland. The creek drains an area mainly covered with open grassland, grain fields, and suburban residential neighborhoods (Fig. 4). The dominant soil type is colluvial, partly gleyic brown soil (GIS-ZH, 2016).

The hut was connected to the electricity grid to allow for the continuous operation of all instruments. Stream stage, temperature and electrical conductivity were recorded in the stream every 10 min using a data logging sonde (model DL/N70; STS Sensor Technik Sirnach AG, Sirnach, Switzerland). The volumetric discharge was not gauged, but we assume that the times of the highest stream stage coincided with peak flow, and thus we use both terms synonymously. Once a day at 07:30, daily precipitation was measured with a heated collector and snow depth was recorded. For higher temporal resolution, we used the hourly CombiPrecip data set (MeteoSwiss), a grid–data product that combines radar estimates and rain gauge measurements to compute precipitation rates at $1\,\text{km}^2$ spatial resolution. Good agreement ($R^2 = 0.86$) between the measured daily precipitation at our field site and the daily sums of hourly CombiPrecip data indicate that the CombiPrecip data set is a reasonable proxy for precipitation variability at the field site. To distinguish rain and snowfall events, air temperature was recorded near the instrument hut every 10 min (Haeni, 2016; Schaub et al., 2011). The uninsulated hut was not temperature controlled; however, the instruments produced heat so that inside air temperatures were on average 12 °C higher than outside. Outside air temperature variations were reflected inside the hut, where air temperatures ranged from 7 to 23 °C.

Table 1. Average isotope values and solute concentrations as well as standard deviations (and relative standard deviations; RSDs) of three water samples analyzed during two different 48 h laboratory experiments with the isotope analyzer and the IC, respectively. In Fiji bottled water, tap water, and nanopure water, concentrations of F^-, Li^+, K^+, NH_4^+, and PO_4^{3-} were mostly below the limit of quantification (LOQ) and thus were not included in the table. The calculation of the average memory coefficient is described in the text (Eq. 1). The uncertainties in the IC measurements were obtained by simple linear regression analysis of the average value and the standard deviation of the respective constituent.

	Isotope analyzer 48 h laboratory experiment		IC 48 h laboratory experiment					
	$\delta^{18}O$ (‰)	δ^2H (‰)	Na^+ (mg L^{-1})	Mg^{2+} (mg L^{-1})	Ca^{2+} (mg L^{-1})	Cl^- (mg L^{-1})	NO_3^- (mg L^{-1})	SO_4^{2-} (mg L^{-1})
Limit of quantification (LOQ)	–	–	0.1	0.1	0.1	0.05	0.05	0.05
Measurement uncertainty	0.03	0.17	$0.053 + 0.005 \cdot C$	$0.008 + 0.006 \cdot C$	$0.087 + 0.009 \cdot C$	$0.027 + 0.003 \cdot C$	$0.028 + 0.002 \cdot C$	$0.037 + 0.006 \cdot C$
Water sample	Fiji bottled water		Fiji bottled water					
Number of measurements	12	12	10	10	10	10	10	10
Average value	−4.86	−35.89	21.6	15.7	24.3	9.69	1.05	1.56
Standard deviation	0.06	0.26	0.1	0.1	0.3	0.06	0.05	0.03
RSD	–	–	0.5 %	0.4 %	1.1 %	0.60 %	4.3 %	1.80 %
Linear drift per 24 h	−0.009 ±	0.133 ±	0.129 ±	0.058 ±	0.093 ±	0.088 ±	−0.078 ±	0.045 ±
(mean ± standard error)	0.008	0.040	0.056[a]	0.036[b]	0.160[c]	0.019	0.008	0.007
Water sample	Tap water		Tap water					
Number of measurements	34	34	18	18	18	18	18	18
Average value	−9.40	−68.55	10.9	34.4	133.2	12.41	4.96	17.29
Standard deviation	0.03	0.12	0.2	0.2	1.3	0.057	0.03	0.14
RSD	–	–	1.6 %	0.6 %	1.0 %	0.5 %	0.7 %	0.8 %
Water sample	Nanopure water		Nanopure water (last sample)					
Number of measurements	43	43	27	27	27	27	27	27
Average value	−9.44	−68.67	<LOQ	0.1	0.6	<LOQ	<LOQ	0.09
Standard deviation	0.02	0.18	0.02	0.003	0.1	0.03	0.02	0.05
Carryover	0.9 %	1.2 %	2.8 %	3.3 %	3.8 %	2.1 %	1.9 %	2.3 %

[a] $p > 0.05$. [b] $p > 0.15$. [c] $p > 0.50$.

A submersible pump (EHEIM GmbH & Co KG, Deizisau, Germany) continuously pumped stream water at a rate of 6 L min^{-1} into a throughflow bucket inside the hut. The volume of the bucket was 10 L; thus, every several minutes the contents of the bucket were effectively exchanged. Every 30 min, water was drawn from the bucket by the S Dosino through a 1 µm cellulose filter to supply the isotope analyzer, IC, and autosampler (Fig. 1). Precipitation was collected with a heated 45 cm diameter funnel installed 2.5 m above the ground. Precipitation flowed into a Teflon®-coated collector with a level detector. The status of the level detector was queried before the end of each measurement routine, and a precipitation sample was taken only if the threshold volume of 72 mL (equaling roughly 0.5 mm of precipitation) was exceeded. For the initial filtration of the precipitation sample, a ceramic frit filter was attached to the suction tube of the P Dosino that drew the sample from the precipitation collector. After precipitation was sampled, a peristaltic pump emptied

the precipitation collector to avoid mixing fresh and old precipitation samples. The sampling routine was programmed to always alternate between stream water and precipitation samples in order to obtain enough stream water samples during storm periods. To reduce biofilm growth on the membrane in the CWS, copper wool was placed in the beaker from which the Isotope Dosinos drew the samples. Sampling was interrupted approximately once a week for basic maintenance (i.e., replacing the filter membranes, cleaning the Dosinos, and refilling the reference standards and eluent stock solutions).

To correct for instrument drift, internal reference standards were analyzed every 3 h. For the five samples between two bracketing measurements of the same reference standard, the following equation was applied:

$$C_{corr} = C_{raw} + \left(C_{true} - \frac{C_{std,i} + C_{std,j}}{2}\right), \quad (2)$$

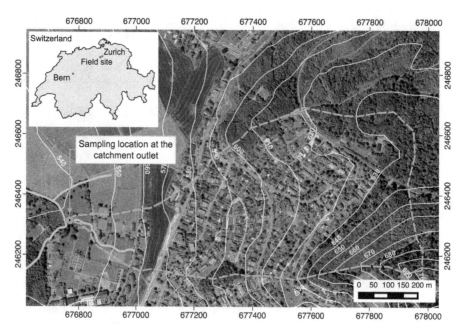

Figure 4. Location of the field site at a small creek on the property of the Swiss Federal Institute for Forest, Snow and Landscape Research (WSL) near Zurich, Switzerland. The catchment boundaries are approximate.

with C denoting the solute concentration or the isotope ratio, respectively. The indices represent the corrected value (corr), the current raw measurement (raw), the true value of the reference standard (true), and the previous and successive measurements of the same reference standard (std) measured at time i and 3 h later at time j. For the isotope analyzer, Fiji bottled water was used as an internal reference standard, which was injected directly from a container by one of the Isotope Dosinos (Fig. 1). The measurements of the IC were drift-corrected with another reference standard (Evian bottled water) that was transferred directly to the IC by the IC Dosino. Evian bottled water was used, as its mineral composition resembles that of stream water more closely than Fiji bottled water does.

4.2 Temporal high-resolution measurements of stable isotopes and major ions in precipitation and stream water

The measurement system was deployed at the field site from 13 February 2016 to 11 March 2016, and more than 1000 stream water and precipitation samples were analyzed for stable water isotopes and major ions, capturing a wide range of hydrological and hydrochemical conditions. Table 2 provides an overview of the eight storm events during that period. Air temperature measurements at the site and daily observations of the snow height showed that precipitation during Events 1–7 was mostly rainfall. Snowfall occurred occasionally after 1 March, while during Event 8 most precipitation fell as snow.

We calculated the response time of streamflow as the time difference between the first detection of precipitation and the first significant increase in stream water level relative to the initial conditions. The response times were between 0 and 2.5 h (Table 2), suggesting fast runoff from the residential area in the eastern part of the catchment. The most delayed streamflow response (2.5 h) was observed after the snowfall Event 8, reflecting delayed snowmelt. As illustrated by Fig. 5, a 30 min sampling interval was sufficient to resolve the temporal patterns of stable isotopes and solutes in streamflow during the rising limb of the hydrograph, even during low-intensity precipitation periods such as Event 5.

Compared to the laboratory experiment with the isotope analyzer, during the field experiment we observed carryover effects in the isotope measurements of up to $100\% \cdot (1 - X) = 3\%$, which can be explained by the copper wool in the beaker from which the Isotope Dosinos drew the water samples. Despite the rinsing routine of the beaker, the wool retained small volumes of sample from previous injections that affected the isotopic composition in the fresh sample. Consequently, the wool was removed and the prior isotope measurements were adjusted with $X = 97\%$ and Eq. (1). Further, the instrument drift was substantially faster at the beginning of the field experiment due to biofilm growth in the membrane tube. For instance, during the first week, the instrument drift for raw $\delta^{18}O$ and δ^2H measurements in the Fiji bottled water was statistically significant, averaging ($\pm$SE) -0.185 ± 0.006 and $-0.288 \pm 0.015\%_o$ day^{-1}, respectively. The variations in air temperature outside and inside the hut were not reflected in the isotope measurements because the

Table 2. Characteristics of precipitation events and antecedent moisture conditions during the field experiment. The initial stream stage is used here as a proxy for the initial discharge.

Event	Start of event	Total precipitation (mm)	Total precipitation until peak flow (mm)	Response time (h:min)	48 h antecedent precipitation (mm)	24 h antecedent precipitation (mm)	Initial stream stage (m)
1	14 February 2016 11:00	5.8	2.2	01:10	8.3	2.7	0.44
2	20 February 2016 10:00	11.5	8.8	00:30	1.9	0.5	0.36
3	23 February 2016 08:00	5.8	3.5	00:00	0.8	0.8	0.37
4	24 February 2016 15:00	14.3	8.1	01:00	6.6	5.0	0.41
5	29 February 2016 13:00	10.5	2.0	00:00	0.0	0.0	0.38
6	2 March 2016 13:00	8.7	6.8	01:10	12.3	1.9	0.46
7	5 March 2016 04:00	11.5	9.4	02:10	4.6	0.9	0.45
8	7 March 2016 23:00	8.4	8.4	02:30	0.6	0.0	0.45

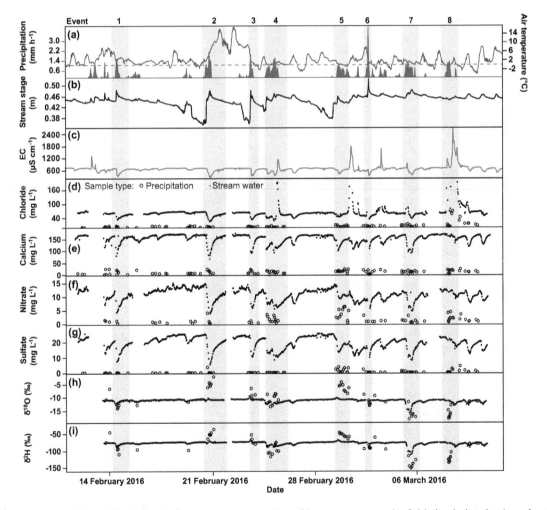

Figure 5. Time series of (a) precipitation and air temperature as well as (b) stream stage at the field site during the 4-week study period. Panel (c) shows stream water EC, whereas panels (d–g) show the chloride, calcium, nitrate, and sulfate concentrations, respectively. Panels (h) and (i) show the isotopic compositions of precipitation and stream water samples. The stream water samples are shown by the blue dots, and the precipitation samples are shown by the open circles. The vertical grey bars indicate the periods of the eight precipitation events used for hydrograph separation.

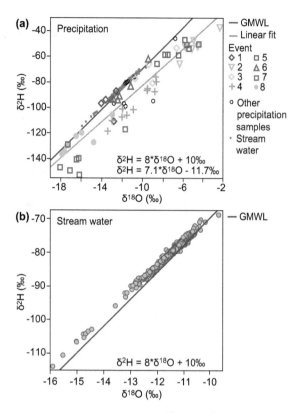

Figure 6. Dual-isotope plot of all $\delta^{18}O$ and δ^2H values measured in **(a)** precipitation and **(b)** stream water during the field experiment. The stream water samples are also plotted in grey in the upper panel for comparison (note the difference in scales). The global meteoric water line (GMWL; Craig, 1961) and the linear fit to the precipitation data are shown in blue and in grey, respectively.

CWS regulates inlet air and water temperatures using Peltier thermoelectric controllers.

Figure 6a illustrates that the isotopic composition of precipitation varied over a range of 15.72‰ in $\delta^{18}O$ and 115.63‰ in δ^2H. By capturing many precipitation events over weeks to months, our isotope analysis system provides a more detailed insight into the variability of precipitation isotopes compared to previous studies that only monitored individual storms at high frequency (e.g., Moerman et al., 2013; Pangle et al., 2013; Tweed et al., 2016). At our site, a correlation between air temperature and the isotopic composition of precipitation is evident for most storm events. Figure 5 shows that, for instance, precipitation samples became isotopically heavier during Events 2 and 8 when air temperature increased, while the precipitation samples became isotopically lighter during Events 1, 3, and 5, when air temperature decreased. During Events 4, 6, and 7, however, the correlation with temperature was not as distinct as during the other five events.

The isotopic composition of stream water varied by less than half as much as that of precipitation, i.e., by 6.24‰ for $\delta^{18}O$ and by 45.11‰ for δ^2H (Fig. 6b). For all eight

events, the isotopic signature of pre-event stream water was relatively constant, averaging -11.04 ± 0.21‰ for $\delta^{18}O$ and -76.97 ± 1.46‰ for δ^2H ($\pm$standard deviation; $n = 8$). During the events, $\delta^{18}O$ and δ^2H in stream water changed by up to 4.80 and 36.38‰, respectively (Event 7).

For the IC, memory effects were negligible during the field experiment (because the sample did not make contact with the copper wool), so the measurements were corrected only for drift effects. The solute concentrations in precipitation and stream water varied widely, as shown in Fig. 5. For Li^+, NH_4^+, K^+, F^-, and PO_4^{3-} in stream water as well as Mg^{2+} in precipitation, measured concentrations were generally below the LOQ. Ca^{2+}, NO_3^-, and SO_4^{2-} in stream water exhibited clear dilution patterns during all precipitation events (Fig. 5e–g). The concentrations of Ca^{2+}, NO_3^-, and SO_4^{2-} in precipitation during the eight events were on average ($\pm$standard deviation) 12.1 ± 2.9, 1.5 ± 1.1, and 0.5 ± 0.8 mg L^{-1}, respectively. The solute concentrations in pre-event stream water were on the order of ($\pm$ standard deviation) 160.8 ± 9.7 mg L^{-1} for Ca^{2+}, 11.7 ± 1.8 mg L^{-1} for NO_3^-, and 21.5 ± 3.3 mg L^{-1} for SO_4^{2-}, whereas the concentrations during storm events dropped to values as low as 64.6 mg L^{-1} (Ca^{2+}), 3.73 mg L^{-1} (NO_3^-), and 5.12 mg L^{-1} (SO_4^{2-}). In contrast, EC and the concentrations of Cl^- (and Na^+, not shown) in stream water showed dilution patterns until Event 3 and then showed distinct enrichment patterns thereafter (Fig. 5c–d), likely associated with road salt wash-off. Due to possible road-salt effects on Na^+ and Cl^-, we will focus on Ca^{2+}, NO_3^-, and SO_4^{2-} in the analysis below.

5   Comparison of event-water fractions estimated from isotopic and chemical tracers

5.1   Hydrograph separation methodology and uncertainty analysis

To illustrate a potential application of high-frequency isotope and chemical measurements, here we quantify the event-water fractions of the major events captured during the 1-month observation period. We used a two-component end-member mixing analysis by applying the conventional mass balance equation (Pinder and Jones, 1969):

$$F_E = \frac{Q_E}{Q_S} = \frac{C_S - C_P}{C_E - C_P}. \tag{3}$$

The fraction of event water relative to total streamflow ($F_E = Q_E/Q_S$) was calculated from the isotope values or solute concentrations in total streamflow (C_S), event precipitation (C_E), and pre-event streamflow (C_P). Here, C_P was obtained for each event from the average of the five stream water samples immediately before the onset of precipitation. The value of C_E was the incremental, volume-weighted mean (McDonnell et al., 1990) of all precipitation samples that were collected before the respective streamflow sample:

$$C_{E,j} = \frac{\sum_{i=k}^{j} P_i C_i}{\sum_{i=k}^{j} P_i}, \tag{4}$$

with P_i being the precipitation depth associated with the isotope value (or solute concentration) C_i collected at time i since the start time k of the precipitation event.

Uncertainty in the hydrograph separation was quantified with Gaussian error propagation (Genereux, 1998), using calculated standard errors (SEs) arising from the analytical uncertainties and the temporal variability of the isotope values (or solute concentrations). Because C_E is a volume-weighted mean, the standard error SE_{C_E} is calculated with

$$SE_{C_{E,j}} = \left[\frac{\sum_{i=k}^{j} P_i (C_i - C_{E,j})^2}{(j-k) \sum_{i=k}^{j} P_i} \right]^{\frac{1}{2}}, \tag{5}$$

where $C_{E,j}$ denotes the volume-weighted mean, C_i denotes the ith concentration that comprises that mean, and (j) is the number of samples included in the volume-weighted mean. The standard error of C_S, which is given here as SE_{C_S}, arises from the measurement uncertainties given in Table 1. For SE_{C_P}, the same measurement uncertainties are applied, as well as the temporal variability of the five measurements comprising C_P. The standard error of the event-water fraction (SE_{F_E}) can then be obtained by Gaussian error propagation:

$$SE_{F_E} = \left\{ \left[\frac{-1}{C_P - C_E} SE_{C_S} \right]^2 + \left[\frac{C_S - C_E}{(C_P - C_E)^2} SE_{C_P} \right]^2 \right.$$
$$\left. + \left[\frac{C_P - C_S}{(C_P - C_E)^2} SE_{C_E} \right]^2 \right\}^{1/2} . \tag{6}$$

Isotope hydrograph separation (IHS) was performed using both δ^{18}O and δ^2H, whereas chemical hydrograph separation (CHS) was carried out with the three constituents Ca^{2+}, NO_3^-, and SO_4^{2-} (Cl^- and Na^+ were not used for CHS due to the influence of road salt at the site) as well as stream water EC. EC was used here since several studies have applied EC in lieu of chemical concentrations for hydrograph separation (e.g., Dzikowski and Jobard, 2012; Matsubayashi et al., 1993; Muñoz-Villers and McDonnell, 2012; Pellerin et al., 2008). As we did not measure EC in precipitation directly, we had to estimate it empirically. For this, we used a standard conversion equation, i.e., the pseudo-linear approach following Sposito (2008), to calculate EC in precipitation from the ionic strength of the major cations and anions in the precipitation samples. We assume that the ion concentrations measured by the IC account for the great majority

of the ionic strength. In order to estimate the uncertainty of this method, we also calculated the EC values in stream water and compared them with the actual measurements of the EC probe in the stream. The (absolute value) difference between the calculated and measured stream water EC values averaged $20\,\mu S\,cm^{-1}$.

For the uncertainty analysis of the calculated event-water fractions, the analytical uncertainties in the isotope measurements were assumed to be 0.03 and 0.17‰ for δ^{18}O and δ^2H, respectively (Sect. 3.2; Table 1). The relative uncertainties in the IC measurements were $0.006 \cdot C + 0.087\,mg\,L^{-1}$ for Ca^{2+}, $0.028 \cdot C + 0.002\,mg\,L^{-1}$ for NO_3^-, and $0.037 \cdot C + 0.006\,mg\,L^{-1}$ for SO_4^{2-} (where C is the concentration in $mg\,L^{-1}$; Table 1). For the EC values, a measurement uncertainty of 2 % was assumed for the EC probe based on the specifications given by the EC probe's manufacturer. The assumed uncertainty in the EC values in precipitation was $20\,\mu S\,cm^{-1}$, as calculated above.

5.2 Event-water fractions for eight storm events

A mixing analysis for two endmembers, event water and pre-event water, was carried out for eight storm events between 20 February and 8 March 2016 based on isotopic and chemical tracers. Event 8, where precipitation fell partly as snow, was included in the analysis because river discharge and stream water EC responded within 4 h after the onset of precipitation (Table 2). Hence, the temporal change in the snowmelt isotopic signal due to fractionation was assumed to be negligible. Two storm events are analyzed in more detail, followed by a general discussion of the hydrograph separation results based on all eight events.

5.2.1 Two storm events

Figures 7 and 8 show the hydrologic, isotopic, and chemical responses in stream water and precipitation during Events 1 and 2. During Event 1, total rainfall was 6.8 mm within 6 h, while 11.5 mm of rain fell within 13 h during Event 2. Antecedent moisture conditions, as inferred from the total rainfall within 48 and 24 h before the event as well as the initial stream water level, were relatively wet for Event 1 and relatively dry for Event 2 (Table 2).

For Event 1, δ^{18}O and δ^2H in stream water followed the observed patterns in precipitation; i.e., stream water became isotopically lighter over time. Isotope hydrograph separation (IHS) for this event yielded maximum event-water fractions ($F_{E,max}$) of 80 ± 11 and 59 ± 14 % for δ^{18}O and δ^2H, respectively. This is similar to the results obtained from the chemical tracers Ca^{2+}, NO_3^-, and SO_4^{2-} (57 ± 1, 65 ± 2, and 65 ± 3 %) and EC (56 ± 3 %; Fig. 7d and e). The larger uncertainties in the IHS compared to CHS can be explained by the large temporal variability of the isotope values in precipitation, which substantially exceeds the analytical uncertainty. During Event 1, the fraction of event water increased

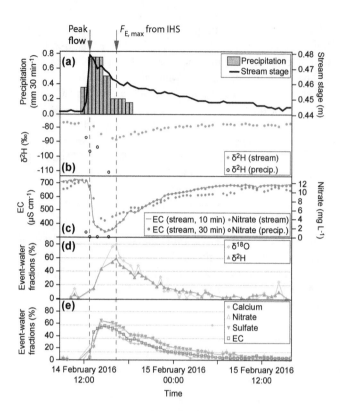

Figure 7. Precipitation Event 1 together with the **(a)** hydrologic, **(b)** isotopic, and **(c)** chemical responses in stream water. Panels **(d)** and **(e)** show the fractions of event water based on isotopic and chemical hydrograph separation, respectively, which are similar for both types of tracers. However, the timing of the maximum event-water fraction ($F_{E,max}$) differs, with the isotopes indicating the largest contribution of event water around 3 h after the peak flow was reached. In panel **(e)**, the gaps in the F_E time series based on calcium concentrations are due to measurement outliers.

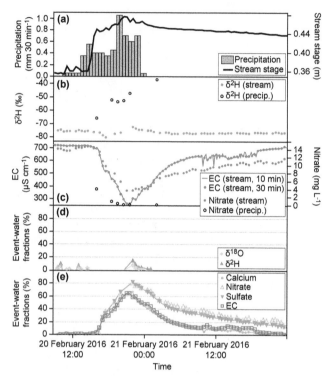

Figure 8. Precipitation Event 2 and the **(a)** hydrologic, **(b)** isotopic, and **(c)** chemical responses in stream water. Panels **(d)** and **(e)** show the fractions of event water (F_E) based on isotopic and chemical hydrograph separation. Chemical tracers greatly exaggerate the event-water fraction.

rapidly after the start of rainfall and declined continuously as the stream stage receded. A difference in timing of $F_{E,max}$ was evident for both tracer types (Fig. 7d–e): $F_{E,max}$ based on the chemical tracers occurred 1 h after the peak flow, whereas $F_{E,max}$ based on the isotope tracers was delayed by roughly 3 h, possibly because the isotopic signature in precipitation became lighter as the event progressed. Consequently, if C_S values at the time of peak flow were used to perform hydrograph separation (Eq. 3), the isotope-based F_E values would be substantially smaller (i.e., 13 ± 4 and $15 \pm 3\%$ for $\delta^{18}O$ and δ^2H, respectively) than the $F_{E,max}$ values reported above.

During Event 2, the solutes in stream water showed a clear dilution signal (Fig. 8c) similar to Event 1. The isotopic composition in stream water, by contrast, showed only a very weak and inconsistent response to precipitation. For instance, δ^2H in precipitation increased continuously through the event, whereas δ^2H in stream water first decreased and then began to increase again several hours after the onset of precipitation. Consequently, IHS and CHS yielded substantially different interpretations of Event 2. The maximum

event-water fractions based on CHS ranged from $67 \pm 1\%$ (Ca^{2+}) to $82 \pm 3\%$ (SO_4^{2-}), similar to Event 1. In contrast, $F_{E,max}$ values based on IHS ranged from 8 ± 1 to $15 \pm 3\%$, indicating that pre-event water was the dominant source of stream water during peak flow.

How can such a large discrepancy between the event-water fractions calculated from different environmental tracers be explained? From Fig. 5 it can be seen that precipitation was isotopically lighter than stream water in the 6 days leading up to Event 2. Thus, the initial decrease in the $\delta^{18}O$ and δ^2H values in stream water during Event 2 suggests the release of isotopically lighter soil water and groundwater that were recharged during previous events. An activation of this pre-event water storage might have been triggered by enhanced infiltration after relatively dry antecedent moisture conditions (AMC) compared to the previous event, whereas wet AMC would be more consistent with surface runoff generation. This hypothesis is further supported by the isotopic responses in stream water during Event 5, another isotopically heavy event with dry AMC, following earlier inputs of isotopically lighter precipitation. In Event 5, small event-water fractions (12 ± 1 and $21 \pm 1\%$ for $\delta^{18}O$ and δ^2H, respectively; Fig. S1 in the Supplement) were again obtained, indicating that pre-event water dominated streamflow, similar to Event 2. In Event 5, just as in Event 2, the chemical trac-

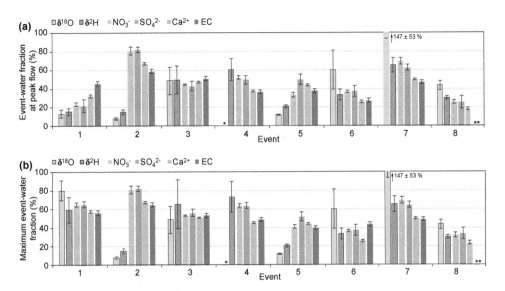

Figure 9. Event-water fractions (F_E) based on isotopic and chemical hydrograph separation for eight storm events. Panel **(a)** shows F_E during peak flow, and panel **(b)** shows the maximum event-water fractions ($F_{E,max}$) of each event. Unrealistic F_E and $F_{E,max}$ values based on $\delta^{18}O$ were obtained for Event 4 because the isotopic signatures in precipitation and pre-event stream water were too similar (*). For Event 8, the wash-off of road salt resulted in unrealistic F_E and $F_{E,max}$ values based on EC: -96 ± 6 and -95 ± 6% (**), respectively. The larger uncertainties in the IHS results compared to CHS can be explained by the large temporal variability of the isotope values in precipitation, which substantially exceeds the analytical uncertainty during most events.

ers showed strong dilution, leading to an overestimate of the maximum event-water fraction ($> 40 \pm 2$%). In both Event 2 and Event 5, the chemical and isotopic data indicate a large contribution from recent soil water or groundwater that had not yet become highly mineralized, rather than from either event precipitation or from older groundwater that presumably accounted for most of the pre-event baseflow.

5.2.2 General discussion of hydrograph separation results

Figure 9 summarizes the estimated event-water fractions for all eight events, based on IHS and CHS, for two points in time during each event: the time with the largest isotopic or chemical response (i.e., $F_{E,max}$) and the time of peak flow. Maximum event-water fractions varied greatly across the eight events (for example, from 15 ± 3 to 73 ± 17% based on δ^2H; Fig. 9; Tables S1 and S2 in the Supplement). Also, within individual events, hydrograph separations based on different isotopic and chemical tracers differed, often by much more than their uncertainties. The inconsistencies between the estimated event-water fractions can be explained by the fact that different tracers are shaped by different hydrochemical processes and flow pathways and thus may describe different endmembers (Richey et al., 1998; Wels et al., 1991). While stable water isotopes are considered to be ideal conservative tracers, chemical tracers are altered by biogeochemical processes on their way through hydrological systems. These biogeochemical processes also vary over time, as they depend on antecedent conditions and precipitation

characteristics. The high-frequency analysis of environmental tracers can document this temporal variability, which, in turn, helps to constrain conceptual catchment models. As illustrated by Events 2 and 5, comparing chemical and isotopic tracers can be useful in identifying the temporally variable contributions of different water storages in the subsurface.

For Event 7, IHS based on $\delta^{18}O$ resulted in event-water fractions of > 100%, which can be explained by the fact that the first precipitation sample from this event was isotopically very similar to the pre-event water signature ($C_E = -11.69$‰, $C_P = -11.09$‰). The incremental, volume-weighted mean of the event-water endmember was thus isotopically heavier than the stream water endmember, resulting in a smaller difference from the pre-event water endmember signature (Eq. 3). The precipitation samples after this first, less $\delta^{18}O$-depleted sample had an average $\delta^{18}O$ value of -16.86 ± 0.73‰ ($\pm$standard deviation; $n = 6$). For δ^2H, such a strong effect did not occur, and we could obtain reasonable isotope-based hydrograph separation results similar to the chemical hydrograph separation.

Figure 9 illustrates further that for three events (2, 5, and 8), the estimated event-water fractions for the two isotopes $\delta^{18}O$ and δ^2H differed significantly (i.e., by more than twice their pooled uncertainties). These differences did not follow any particular pattern; for instance, $F_E(\delta^{18}O) > F_E(\delta^2H)$ for Event 8, while $F_E(\delta^{18}O) < F_E(\delta^2H)$ for Events 2 and 5. Such discrepancies might be caused by temporally variable $\delta^{18}O$–δ^2H relations (d excess) of contributing water sources (groundwater, soil water, and overland flow), resulting in different event-water fractions based on both isotopes. An al-

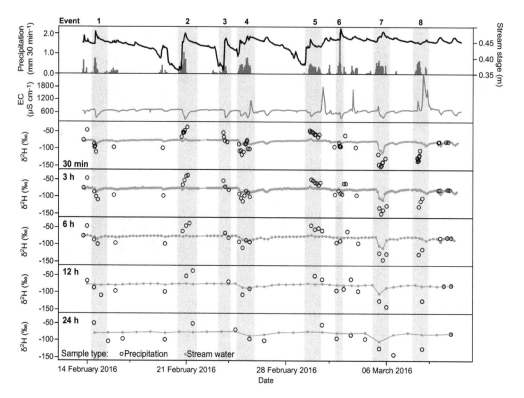

Figure 10. Time series of precipitation, stream stage, and stream water EC, as well as δ^2H values in stream water and precipitation at sampling intervals of 30 min and 3, 6, 12, and 24 h. The stream water isotope values at 3–24 h temporal resolution were obtained by subsampling from the 30 min time series. To mimic the effects of integrated bulk precipitation samples, the isotope values in precipitation were calculated by volume-weighted averaging the 30 min data over the corresponding time intervals. The vertical grey bars indicate the periods of the eight precipitation events used for hydrograph separation.

ternative explanation is that the pre-event streamflow signature (C_P) may not reflect the isotopic signature of the entire pre-event water storage, but only of the components that feed baseflow (Klaus and McDonnell, 2013). Another way of viewing this problem is that the precipitation event may have mobilized a third pre-event water storage with unknown isotopic composition (Tetzlaff et al., 2014). This conjecture is strongly supported by the initial shift toward isotopically lighter streamflow early in Event 2, even though the event precipitation was isotopically heavier than the pre-event baseflow. Event 5 also showed divergent event-water fractions between the two isotopes, and like Event 2, it also had strongly contrasting pre-event precipitation inputs. Thus, the history of both events suggests that pre-event storage in this catchment was isotopically heterogeneous. This observation is unsurprising given the pervasive heterogeneity of typical catchments, but a more detailed explanation is not possible with our spatially limited data set. Spatially distributed measurements, such as from groundwater and soil water storages, would help to constrain the individual endmembers that contribute to streamflow (e.g., Hangen et al., 2001). Additional high-frequency time series of the groundwater table and soil moisture profiles would allow for documenting of the effects of antecedent moisture conditions on the response

times and on the activation of different storages at the site. Finally, a spatially distributed precipitation sampling network might help to fully quantify the uncertainty inherent in the event-water signature (Fischer et al., 2017; Lyon et al., 2009).

5.3 The role of the sampling frequency in capturing hydrological and hydrochemical catchment processes

A sampling frequency can be considered optimal when the gain of information from additional measurements is marginal (Kirchner et al., 2004; Neal et al., 2012). With our high-resolution data set, we can thus investigate the potential of different sampling frequencies for capturing hydrological and hydrochemical catchment processes by subsampling the 30 min time series at smaller sampling frequencies, i.e., at 3, 6, and 12 h and daily intervals. To mimic the effects of integrated bulk precipitation samples, we calculated the volume-weighted averages of concentrations and isotope values in precipitation over the corresponding time intervals.

Figure 10 shows that 3 h sampling frequencies would still be sufficient to capture the isotopic variations in stream water, including during low-intensity precipitation events. However, the short-term variability within single storm periods,

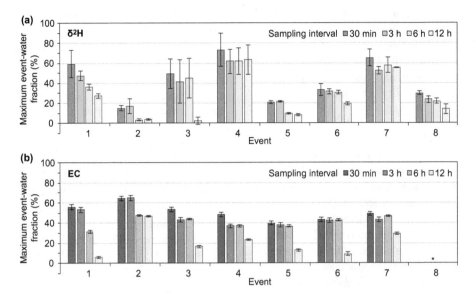

Figure 11. Maximum event-water fractions at sampling intervals of 30 min and 3 h, 6 h, and 12 h based on **(a)** δ^2H and **(b)** EC. With lower sampling frequencies, the event-water fractions are often underestimated or become unrealistic, as the likelihood increases that the point of the largest δ^2H or EC variations in streamflow will be missed (stream water δ^2H and EC time series were subsampled at 3 h, 6 h, and 12 h and daily intervals; concentrations of integrated bulk precipitation samples were calculated from the volume-weighted averages over the respective time interval). For Event 8, the wash-off of road salt resulted in unrealistic $F_{E,max}$ values based on EC ($\ast$).

as well as the rapid changes in precipitation isotope values, cannot be resolved at this lower sampling frequency. Thus, even sampling intervals of 3 h can result in a significant loss of information relative to 30 min sampling, and at sampling intervals of 12 h or longer, diurnal fluctuations and some isotopic and chemical responses to low-intensity precipitation events would also be lost. Likewise, the 6 h or 12 h bulk precipitation samples shown in Fig. 10 fail to reflect the large isotopic variability revealed by the 30 min samples.

To further illustrate the effect of lower sampling frequencies, we performed hydrograph separation with the subsampled data sets, for which illustrative results of the maximum event-water fractions are shown for δ^2H and EC in Fig. 11. With a sampling frequency of 3 h, maximum event-water fractions similar to those for the 30 min sampling can still be obtained, except for Events 3 (EC) and 4 (EC), where $F_{E,max}$ is underestimated. Longer sampling intervals (6 h, 12 h) result in much smaller event-water fractions for most events. Because the hydrologic response times in this catchment were only between 0 h and 2.5 h, the durations of the maximum hydrochemical variations were similarly short. Consequently, sampling at longer time intervals increases the risk of missing this critical peak response; if the sample is taken before or after the maximum hydrochemical response, the event-water signal in stream water (C_S) may be too weak, which will inevitably underestimate the event-water fractions or even lead to unrealistic negative values. Furthermore, the rapid changes observed in precipitation isotopic composition (Fig. 6) suggest that high-frequency measurements are crucial to adequately represent the signature of the event-water

endmember. Capturing the short-term responses of environmental tracers also helps to better quantify transit time distributions (Birkel et al., 2012; Stockinger et al., 2016; Timbe et al., 2015) and to constrain concentration–discharge models (Stelzer and Likens, 2006; Jones et al., 2012).

Our data also show that peak flow is not always a reliable predictor of the time at which F_E becomes largest. As can be seen, for example, during Event 1 (Fig. 7), $F_{E,max}$ based on IHS occurred up to 3.0 ± 1.0 h after the peak flow. The timing of the peak flow and the $F_{E,max}$ values for chemical and isotope tracers coincided for only four events (Events 2, 6, 7, and 8). During the remaining events, the tracer signal showed the strongest responses up to 2.5 ± 1.0 h after the peak flow, indicating that the time window for sample collection at our site must extend more than 3 h before and after the peak flow in order to capture the whole range of event-water dynamics. In the case of snowmelt Event 8, when the maximum EC response occurred 5 h before the peak flow, an even longer sampling period would be required in order to capture unusual events, such as the inflow of water contaminated by road salt.

6 Concluding remarks

This paper presents the first field hydrology application of Picarro's Continuous Water Sampler (CWS), which was coupled to a L2130-i wavelength-scanned cavity ring-down spectrometer to measure δ^{18}O and δ^2H in stream water and precipitation at a temporal resolution of 30 min. We combined this real-time isotope analysis system with a dual-channel ion chromatograph for the synchronous analysis of major cations and anions. Good instrument performance and high measurement precision could be achieved during continuous 48 h laboratory experiments and a 28-day deployment in the field at a small, partly urbanized catchment in central Switzerland.

Problematic issues, such as sample degradation during storage and transportation that arise in conventional sampling for catchment tracer studies, become irrelevant with the system presented here. At the same time, potential registration errors arising during the collection and handling of a large number of water samples are avoided. Conversely, two major limitations of the coupled isotope analyzer/IC system are its high cost and the need for line power, which constrains its use in remote locations. However, the laboratory analysis of conventionally collected grab samples is also cost-intensive, and the autosamplers used in conventional sampling schemes also require a reliable energy supply (though at much lower power levels).

The results of the high-frequency analysis system are presented here to provide a proof-of-concept and an illustration of its functionality in the field, rather than to fully document the hydrological and biogeochemical processes at this field site. A more detailed interpretation would require additional measurements of soil water and groundwater isotopes and chemistry in order to better constrain the endmembers in the mixing analysis. Nevertheless, our 1-month field experiment demonstrates the marked short-term variability of several natural tracers in a small, highly dynamic watershed. The hydrograph separation exercise clearly showed that long-term, high-frequency isotopic and chemical analyses are essential for capturing the "unusual but informative" events that shed light on catchment storage and flow processes. We further showed that the right timing for capturing peak event-water contributions can easily be missed with conventional grab sampling strategies at time intervals longer than 3 h, resulting in an underestimation of the event-water fraction. In addition, the relative timing of the isotopic and chemical responses was highly variable, demonstrating the challenge of capturing the right moments with episodic snapshot campaigns or long-term monitoring with daily, weekly, or even monthly sampling intervals.

As was shown here and elsewhere (e.g., Kirchner, 2003), the short-term responses of streamflow and environmental tracers may follow distinctly different patterns, which helps to constrain streamflow generation mechanisms and quantifying short transit times. Thus, high-frequency isotopic and chemical measurements also have great potential for catchment model validation. Potential future applications of the

system could include sites with rapid hydrologic responses, such as urban streams (e.g., Jarden et al., 2016; Jefferson et al., 2015; Soulsby et al., 2014), wastewater and drinking water systems (e.g., Houhou et al., 2010; Kracht et al., 2007), or agricultural catchments with artificial drainage networks (e.g., Doppler et al., 2012; Heinz et al., 2014). By eliminating the errors associated with the handling, transportation, and storage of individual bottles, our analysis system may also achieve better precision than conventional field sampling followed by laboratory analyses. As a result, our system may be able to detect subtle isotopic and biogeochemical signals (associated with, e.g., evaporation effects or in-stream biological processes) that would be missed by conventional approaches to sampling and analysis. Thus, this system can potentially shed new light on the linkages between hydrological, biological, and geochemical processes.

Competing interests. The authors declare that they have no conflict of interest.

Acknowledgements. We thank Anton Burkhardt and the facility staff of the Swiss Federal Institute for Forest, Snow and Landscape Research (WSL) for logistical support, and Matthias Haeni from the Long-term Forest Ecosystem Research Programme (LWF) at WSL for providing air temperature data. We also thank Barbara Herbstritt of the isotope laboratory at the University Freiburg (Germany) for the analysis of the isotope reference standards, as well as Kate Dennis and David Kim-Hak of Picarro Inc. (Santa Clara, CA, USA) for technical advice.

Edited by: M. Weiler

References

Aubert, A. H., Gascuel-Odoux, C., Gruau, G., Akkal, N., Faucheux, M., Fauvel, Y., Grimaldi, C., Hamon, Y., Jaffrézic, A., Lecoz-Boutnik, M., Molénat, J., Petitjean, P., Ruiz, L., and Merot, P.: Solute transport dynamics in small, shallow groundwater-dominated agricultural catchments: insights from a high-frequency, multisolute 10 yr-long monitoring study, Hydrol. Earth Syst. Sci., 17, 1379–1391, doi:10.5194/hess-17-1379-2013, 2013.

Aubert, A. H. and Breuer, L.: New seasonal shift in in-stream diurnal nitrate cycles identified by mining high-frequency data, PLoS ONE, 11, e0153138, doi:10.1371/journal.pone.0153138, 2016.

Bende-Michl, U. and Hairsine, P. B.: A systematic approach to choosing an automated nutrient analyser for river monitoring, J. Environ. Monitor., 12, 127–134, 2010.

Benettin, P., Kirchner, J. W., Rinaldo, A., and Botter, G.: Modeling chloride transport using travel time distributions at Plynlimon, Wales, Water Resour. Res., 51, 3259–3276, doi:10.1002/2014WR016600, 2015.

Berman, E. S. F., Gupta, M., Gabrielli, C., Garland, T., and McDonnell, J. J.: High-frequency field-deployable isotope analyzer for hydrological applications, Water Resour. Res., 45, W10201, doi:10.1029/2009wr008265, 2009.

Birkel, C., Soulsby, C., Tetzlaff, D., Dunn, S., and Spezia, L.: High-frequency storm event isotope sampling reveals time-variant transit time distributions and influence of diurnal cycles, Hydrol. Process., 26, 308–316, doi:10.1002/hyp.8210, 2012.

Buso, D. C., Likens, G. E., and Eaton, J. S.: Chemistry of precipitation, streamwater, and lakewater from the Hubbard Brook Ecosystem Study: a record of sampling protocols and analytical procedures, USDA Forest Service, Northeastern Research Station, USDA Forest Service, Newtown Square, PA., Gen. Tech. Rep. NE-275, 52 pp., 2000.

Craig, H.: Isotopic variations in meteoric waters, Science, 133, 1702–1703, 1961.

Darling, W. G. and Bowes, M. J.: A long-term study of stable isotopes as tracers of processes governing water flow and quality in a lowland river basin: the upper Thames, UK, Hydrol. Process., 30, 2178–2195, doi:10.1002/hyp.10779, 2016.

Doppler, T., Camenzuli, L., Hirzel, G., Krauss, M., Lück, A., and Stamm, C.: Spatial variability of herbicide mobilisation and transport at catchment scale: insights from a field experiment, Hydrol. Earth Syst. Sci., 16, 1947–1967, doi:10.5194/hess-16-1947-2012, 2012.

Dzikowski, M. and Jobard, S.: Mixing law versus discharge and electrical conductivity relationships: application to an alpine proglacial stream, Hydrol. Process., 26, 2724–2732, doi:10.1002/Hyp.8366, 2012.

Fischer, B. M. C., van Meerveld, I., and Seibert, J.: Spatial variability in the isotopic composition of rainfall in a small headwater catchment and its effect on hydrograph separation, J. Hydrol., doi:10.1016/j.jhydrol.2017.01.045, 2017.

Genereux, D.: Quantifying uncertainty in tracer-based hydrograph separations, Water Resour. Res., 34, 915–919, 1998.

GIS-ZH: Geographisches Informationssystem des Kantons Zürich (GIS-ZH), Amt für Raumentwicklung, Abteilung Geoinformation, GIS-Produkte GIS-Browser, Map, available at: http://maps.zh.ch/, last access: 4 October 2016.

Gupta, P., Noone, D., Galewsky, J., Sweeney, C., and Vaughn, B. H.: Demonstration of high-precision continuous measurements of water vapor isotopologues in laboratory and remote field deployments using wavelength-scanned cavity ring-down spectroscopy (WS-CRDS) technology, Rapid Commun. Mass Sp., 23, 2534–2542, doi:10.1002/rcm.4100, 2009.

Haeni, M.: Air Temperature and Precipitation data (1 February to 14 March 2016) from the WSL Uitikon forest meteorological station, in 10 and 30 minute time resolution, dataset, available at: https://doi.org/10.13140/RG.2.2.11510.60480, 2016.

Halliday, S. J., Skeffington, R. A., Wade, A. J., Neal, C., Reynolds, B., Norris, D., and Kirchner, J. W.: Upland streamwater nitrate dynamics across decadal to sub-daily timescales: a case study of Plynlimon, Wales, Biogeosciences, 10, 8013–8038, doi:10.5194/bg-10-8013-2013, 2013.

Hangen, E., Lindenlaub, M., Leibundgut, C., and von Wilpert, K.: Investigating mechanisms of stormflow generation by natural tracers and hydrometric data: a small catchment study in the Black Forest, Germany, Hydrol. Process., 15, 183–199, 2001.

Harman, C. J.: Time-variable transit time distributions and transport: Theory and application to storage-dependent transport of chloride in a watershed, Water Resour. Res., 51, 1–30, doi:10.1002/2014WR015707, 2015.

Hayashi, M., Vogt, T., Mächler, L., and Schirmer, M.: Diurnal fluctuations of electrical conductivity in a pre-alpine river: Effects of photosynthesis and groundwater exchange, J. Hydrol., 450, 93–104, doi:10.1016/J.Jhydrol.2012.05.020, 2012.

Heinz, E., Kraft, P., Buchen, C., Frede, H. G., Aquino, E., and Breuer, L.: Set Up of an Automatic Water Quality Sampling System in Irrigation Agriculture, Sensors-Basel, 14, 212–228, doi:10.3390/S140100212, 2014.

Herbstritt, B., Gralher, B., and Weiler, M.: Continuous in situ measurements of stable isotopes in liquid water, Water Resour. Res., 48, W03601, doi:10.1029/2011wr011369, 2012.

Houhou, J., Lartiges, B. S., France-Lanord, C., Guilmette, C., Poix, S., and Mustin, C.: Isotopic tracing of clear water sources in an urban sewer: A combined water and dissolved sulfate stable isotope approach, Water Res., 44, 256–266, doi:10.1016/j.watres.2009.09.024, 2010.

Jarden, K. M., Jefferson, A. J., and Grieser, J. M.: Assessing the effects of catchment-scale urban green infrastructure retrofits on hydrograph characteristics, Hydrol. Process., 30, 1536–1550, doi:10.1002/hyp.10736, 2016.

Jasechko, S., Kirchner, J. W., Welker, J. M., and McDonnell, J. J.: Substantial proportion of global streamflow less than three months old, Nat. Geosci., 9, 126–129, doi:10.1038/Ngeo2636, 2016.

Jefferson, A. J., Bell, C. D., Clinton, S. M., and McMillan, S. K.: Application of isotope hydrograph separation to understand contributions of stormwater control measures to urban headwater streams, Hydrol. Process., 29, 5290–5306, doi:10.1002/hyp.10680, 2015.

Jones, A. S., Horsburgh, J. S., Mesner, N. O., Ryel, R. J., and Stevens, D. K.: Influence of Sampling Frequency on Estimation of Annual Total Phosphorus and Total Suspended Solids Loads, J. Am. Water Resour. Assoc., 48, 1258–1275, doi:10.1111/j.1752-1688.2012.00684.x, 2012.

Kendall, C. and McDonnell, J. J.: Isotope tracers in catchment hydrology, Elsevier, Amsterdam, New York, xxix, 839 pp., 1998.

Kirchner, J. W.: A double paradox in catchment hydrology and geochemistry, Hydrol. Process., 17, 871–874, doi:10.1002/Hyp.5108, 2003.

Kirchner, J. W., Feng, X. H., Neal, C., and Robson, A. J.: The fine structure of water-quality dynamics: the (high-frequency) wave of the future, Hydrol. Process., 18, 1353–1359, doi:10.1002/Hyp.5537, 2004.

Kirchner, J. W. and Neal, C.: Universal fractal scaling in stream chemistry and its implications for solute transport and water quality trend detection, P. Natl. Acad. Sci. USA, 110, 12213–12218, doi:10.1073/Pnas.1304328110, 2013.

Klaus, J. and McDonnell, J. J.: Hydrograph separation using stable isotopes: Review and evaluation, J. Hydrol., 505, 47–64, doi:10.1016/j.jhydrol.2013.09.006, 2013.

Kracht, O., Gresch, M., and Gujer, W.: A Stable Isotope Approach for the Quantification of Sewer Infiltration, Environ. Sci. Technol., 41, 5839–5845, doi:10.1021/es062960c, 2007.

Leibundgut, C. and Seibert, J.: Tracer Hydrology, in: The Science of Hydrology, edited by: Uhlenbrook, S., Treatise on Water Science, Elsevier, Amsterdam, 215–236, 2011.

Lyon, S. W., Desilets, S. L. E., and Troch, P. A.: Characterizing the response of a catchment to an extreme rainfall event using hydrometric and isotopic data, Water Resour. Res., 44, W06413, doi:10.1029/2007wr006259, 2008.

Lyon, S. W., Desilets, S. L. E., and Troch, P. A.: A tale of two isotopes: differences in hydrograph separation for a runoff event when using delta D versus delta O-18, Hydrol. Process., 23, 2095–2101, doi:10.1002/hyp.7326, 2009.

Matsubayashi, U., Velasquez, G. T., and Takagi, F.: Hydrograph separation and flow analysis by specific electrical conductance of water, J. Hydrol., 152, 179–199, doi:10.1016/0022-1694(93)90145-Y, 1993.

McDonnell, J. J., Bonell, M., Stewart, M. K., and Pearce, A. J.: Deuterium Variations in Storm Rainfall – Implications for Stream Hydrograph Separation, Water Resour. Res., 26, 455–458, doi:10.1029/WR026i003p00455, 1990.

McGlynn, B. L. and McDonnell, J. J.: Quantifying the relative contributions of riparian and hillslope zones to catchment runoff, Water Resour. Res., 39, 1310, doi:10.1029/2003wr002091, 2003.

Moerman, J. W., Cobb, K. M., Adkins, J. F., Sodemann, H., Clark, B., and Tuen, A. A.: Diurnal to interannual rainfall $\delta 18O$ variations in northern Borneo driven by regional hydrology, Earth Planet. Sc. Lett., 369–370, 108–119, doi:10.1016/j.epsl.2013.03.014, 2013.

Munksgaard, N. C., Wurster, C. M., and Bird, M. I.: Continuous analysis of delta O-18 and delta D values of water by diffusion sampling cavity ring-down spectrometry: a novel sampling device for unattended field monitoring of precipitation, ground and surface waters, Rapid Commun. Mass Sp., 25, 3706–3712, doi:10.1002/rcm.5282, 2011.

Munksgaard, N. C., Wurster, C. M., Bass, A., and Bird, M. I.: Extreme short-term stable isotope variability revealed by continuous rainwater analysis, Hydrol. Process., 26, 3630–3634, doi:10.1002/hyp.9505, 2012.

Muñoz-Villers, L. E. and McDonnell, J. J.: Runoff generation in a steep, tropical montane cloud forest catchment on permeable volcanic substrate, Water Resour. Res., 48, W09528, doi:10.1029/2011WR011316, 2012.

Neal, C., Reynolds, B., Norris, D., Kirchner, J. W., Neal, M., Rowland, P., Wickham, H., Harman, S., Armstrong, L., Sleep, D., Lawlor, A., Woods, C., Williams, B., Fry, M., Newton, G., and Wright, D.: Three decades of water quality measurements from the Upper Severn experimental catchments at Plynlimon, Wales: an openly accessible data resource for research, modelling, environmental management and education, Hydrol. Process., 25, 3818–3830, doi:10.1002/hyp.8191, 2011.

Neal, C., Reynolds, B., Rowland, P., Norris, D., Kirchner, J. W., Neal, M., Sleep, D., Lawlor, A., Woods, C., Thacker, S., Guyatt, H., Vincent, C., Hockenhull, K., Wickham, H., Harman, S., and Armstrong, L.: High-frequency water quality time series in precipitation and streamflow: From fragmentary signals to scientific challenge, Sci. Total Environ., 434, 3–12, doi:10.1016/j.scitotenv.2011.10.072, 2012.

Pangle, L. A., Klaus, J., Berman, E. S. F., Gupta, M., and McDonnell, J. J.: A new multisource and high-frequency approach to measuring $\delta 2H$ and $\delta 18O$ in hydrological field studies, Water Resour. Res., 49, 7797–7803, doi:10.1002/2013WR013743, 2013.

Pellerin, B. A., Wollheim, W. M., Feng, X., and Vörörsmarty, C. J.: The application of electrical conductivity as a tracer for hydrograph separation in urban catchments, Hydrol. Process., 22, 1810–1818, doi:10.1002/hyp.6786, 2008.

Pinder, G. F. and Jones, J. F.: Determination of the groundwater component of peak discharge from the chemistry of total runoff, Water Resour. Res., 5, 438–445, doi:10.1029/WR005i002p00438, 1969.

Richey, D. G., McDonnell, J. J., Erbe, M. W., and Hurd, T. M.: Hydrograph separations based on chemical and isotopic concentrations: A critical appraisal of published studies from New Zealand, North America and Europe, J. Hydrol., 37, 95–111, 1998.

Riml, J. and Worman, A.: Spatiotemporal decomposition of solute dispersion in watersheds, Water Resour. Res., 51, 2377–2392, doi:10.1002/2014WR016385, 2015.

Rode, M., Angelstein, S. H. N., Anis, M. R., Borchardt, D., and Weitere, M.: Continuous In-Stream Assimilatory Nitrate Uptake from High Frequency Sensor Measurements, Environ. Sci. Technol., 50, 5685–5694, 2016a.

Rode, M., Wade, A. J., Cohen, M. J., Hensley, R. T., Bowes, M. J., Kirchner, J. W., Arhonditsis, G. B., Jordan, P., Kronvang, B., Halliday, S. J., Skeffington, R. A., Rozemeijer, J. C., Aubert, A. H., Rinke, K., and Jomaa, S.: Sensors in the Stream: The High-Frequency Wave of the Present, Environ. Sci. Technol., 50, 10297–10307, doi:10.1021/acs.est.6b02155, 2016b.

Schaub, M., Dobbertin, M., Krauchi, N., and Dobbertin, M. K.: Preface-long-term ecosystem research: understanding the present to shape the future, Environ. Monit. Assess., 174, 1–2, 2011.

Soulsby, C., Birkel, C., and Tetzlaff, D.: Assessing urbanization impacts on catchment transit times, Geophys. Res. Lett., 41, 442–448, 2014.

Sposito, G.: The chemistry of soils, 2nd ed., Oxford University Press, Oxford; New York, xii, 329 pp., 2008.

Stelzer, R. S. and Likens, G. E.: Effects of sampling frequency on estimates of dissolved silica export by streams: The role of hydrological variability and concentration-discharge relationships, Water Resour. Res., 42, W07415, doi:10.1029/2005WR004615, 2006.

Stockinger, M. P., Bogena, H. R., Lücke, A., Diekkrüger, B., Cornelissen, T., and Vereecken, H.: Tracer sampling frequency influences estimates of young water fraction and streamwater transit time distribution, J. Hydrol., 541, Part B, 952–964, doi:10.1016/j.jhydrol.2016.08.007, 2016.

Tetzlaff, D., Birkel, C., Dick, J., Geris, J., and Soulsby, C.: Storage dynamics in hydropedological units control hillslope connectivity, runoff generation, and the evolution of catchment transit time distributions, Water Resour. Res., 50, 969–985, 2014.

Timbe, E., Windhorst, D., Celleri, R., Timbe, L., Crespo, P., Frede, H.-G., Feyen, J., and Breuer, L.: Sampling frequency tradeoffs in the assessment of mean transit times of tropical montane catchment waters under semi-steady-state conditions, Hydrol. Earth Syst. Sci., 19, 1153–1168, doi:10.5194/hess-19-1153-2015, 2015.

Tweed, S., Munksgaard, N., Marc, V., Rockett, N., Bass, A., Forsythe, A. J., Bird, M. I., and Leblanc, M.: Continuous monitoring of stream delta O-18 and delta H-2 and stormflow hydrograph separation using laser spectrometry in an agricultural catchment, Hydrol. Process., 30, 648–660, 10.1002/hyp.10689, 2016.

Vitvar, T. and Balderer, W.: Estimation of mean water residence times and runoff generation by O-18 measurements in a pre-Alpine catchment (Rietholzbach, eastern Switzerland), Appl. Geochem., 12, 787–796, 1997.

Weiler, M., Scherrer, S., Naef, F., and Burlando, P.: Hydrograph separation of runoff components based on measuring hydraulic state variables, tracer experiments, and weighting methods, Integrated Methods in Catchment Hydrology: Tracer, Remote Sensing and New Hydrometric Techniques, 249–255, 1999.

Wels, C., Cornett, R. J., and Lazerte, B. D.: Hydrograph Separation – a Comparison of Geochemical and Isotopic Tracers, J. Hydrol., 122, 253–274, doi:10.1016/0022-1694(91)90181-G, 1991.

8

The importance of snowmelt spatiotemporal variability for isotope-based hydrograph separation in a high-elevation catchment

Jan Schmieder[1], Florian Hanzer[1], Thomas Marke[1], Jakob Garvelmann[2], Michael Warscher[2], Harald Kunstmann[2], and Ulrich Strasser[1]

[1]Institute of Geography, University of Innsbruck, 6020 Innsbruck, Austria
[2]Institute of Meteorology and Climate Research – Atmospheric Environmental Research, Karlsruhe Institute of Technology, 82467 Garmisch-Partenkirchen, Germany

Correspondence to: Jan Schmieder (jan.schmieder@uibk.ac.at)

Abstract. Seasonal snow cover is an important temporary water storage in high-elevation regions. Especially in remote areas, the available data are often insufficient to accurately quantify snowmelt contributions to streamflow. The limited knowledge about the spatiotemporal variability of the snowmelt isotopic composition, as well as pronounced spatial variation in snowmelt rates, leads to high uncertainties in applying the isotope-based hydrograph separation method. The stable isotopic signatures of snowmelt water samples collected during two spring 2014 snowmelt events at a north- and a south-facing slope were volume weighted with snowmelt rates derived from a distributed physics-based snow model in order to transfer the measured plot-scale isotopic composition of snowmelt to the catchment scale. The observed δ^{18}O values and modeled snowmelt rates showed distinct inter- and intra-event variations, as well as marked differences between north- and south-facing slopes. Accounting for these differences, two-component isotopic hydrograph separation revealed snowmelt contributions to streamflow of 35 ± 3 and $75 \pm 14\,\%$ for the early and peak melt season, respectively. These values differed from those determined by formerly used weighting methods (e.g., using observed plot-scale melt rates) or considering either the north- or south-facing slope by up to 5 and $15\,\%$, respectively.

1 Introduction

In many headwater catchments, seasonal water availability is strongly dependent on cryospheric processes and understanding these processes becomes even more relevant in a changing climate (APCC, 2014; IPCC, 2013; Weingartner and Aschwanden, 1992). The seasonal snow cover is an important temporary water storage in alpine regions. The timing and amount of water released from this storage is important to know for water resources management, especially in downstream regions where the water is needed (drinking water, snow making, hydropower, irrigation water) or where it represents a potential risk (flood, drought). Environmental tracers are a common tool to investigate the hydrological processes, but scientific studies are still rare for high-elevation regions because of the restricted access and high risk for field measurements in these challenging conditions.

Two-component isotope-based hydrograph separation (IHS) is a technique to separate streamflow into different time source components (event water, pre-event water) (Sklash et al., 1976). The event component depicts water that enters the catchment during an event (e.g., snowmelt) and is characterized by a distinct isotopic signature, whereas pre-event water is stored in the catchment prior to the onset of the event (i.e., groundwater and soil water, which form baseflow) and is characterized by a different isotopic signature (Sklash and Farvolden, 1979; Sklash et al., 1976). The technique dates back to the late 1960s (Pinder and Jones, 1969) and was initially used for separating storm hydrographs in humid catchments. The first snowmelt-based studies were conducted in the 1970s by Dinçer et al. (1970) and Martinec et al. (1974). These studies showed a large pre-event water fraction (> 50 %) of streamflow that changed the understanding of the processes in catchment hydrology fundamentally (Klaus and McDonnell, 2013; Sklash and Farvolden, 1979) and forced a paradigm shift, especially for humid temperate catchments. However, other snowmelt-based studies in permafrost or high-elevation catchments (Huth et al., 2004; Liu et al., 2004; Williams et al., 2009) revealed a large contribution of event water (> 70 %), depending on the system state (e.g., frost layer thickness and snow depth), catchment characteristics, and runoff generation mechanisms.

Klaus and McDonnell (2013) highlighted the need to quantify and account for the spatial variability of the isotope signal of event water, which is still a vast uncertainty in snowmelt-based IHS. In the literature inconclusive results prevail with respect to the variation of the isotopic signal of snowmelt. Spatial variability of snowmelt isotopic composition was statistically significant in relation to elevation (Beaulieu et al., 2012) in a catchment in British Columbia, Canada, with 500 m relief. Moore (1989) and Laudon et al. (2007) found no statistical significant variation in their snowmelt $\delta^{18}O$ data, due to the low gradient and small elevation range (approximately 30 and 290 m) in their catchments, which favors an isotopically more homogenous snow cover. The effect of the aspect of the hillslopes on isotopic variability and IHS results in topographically complex terrain has been rarely investigated. Dahlke and Lyon (2013) and Dietermann and Weiler (2013) surveyed the snowpack isotopic composition and showed a notable spatial variability in their data, particularly between north- and south-facing slopes. They conclude that the spatial variability of snowmelt could be high and that the timing of meltwater varies with the morphology of the catchment. Dietermann and Weiler (2013) also concluded that an elevation effect (decrease of snowpack isotopic signature with elevation), if observed, is disturbed by fractionation due to melt/refreeze processes during the ablation period. Aspect and slope are therefore important factors that affect the isotopic evolution of the snow cover and its melt (Cooper, 2006). In contrast, there have been various studies that have investigated the temporal variability of the snowmelt isotopic signal, e.g., with the use of snow lysimeters (Hooper and Shoemaker, 1986; Laudon et al., 2002; Liu et al., 2004; Maulé and Stein, 1990; Moore, 1989; Williams et al., 2009). During the ablation season the isotopic composition of the snowpack changes due to percolating rain and meltwater, and fractionation caused by melting, refreezing and sublimation (Dietermann and Weiler, 2013; Lee et al., 2010; Unnikrishna et al., 2002; Zhou et al., 2008), which leads to a homogenization of the isotopic profile of the snowpack (Árnason et al., 1973; Dinçer et al., 1970; Stichler, 1987) and an increase in heavy isotopes of meltwater throughout the freshet period (Laudon et al., 2007; Taylor et al., 2001, 2002; Unnikrishna et al., 2002). Therefore, the characterization and the use of the evolving isotopic signal

of snowmelt water instead of single snow cores is crucial for applying IHS (Taylor et al., 2001, 2002).

There have been various approaches to cope with the temporal variability of the input signal. If one uses more than one $\delta^{18}O$ snowmelt sample for applying the IHS method, it is important to weight the values with appropriate melt rates, e.g., measured from the outflow of a snow lysimeter. Common weighting methods are the volume-weighted average approach (VWA), as used by Mast et al. (1995), and the current meltwater approach (CMW), applied by Hooper and Shoemaker (1986). Laudon et al. (2002) developed the runoff-corrected event water approach (runCE), which accounts for both, the temporal isotopic evolution and temporary storage of meltwater in the catchment and overcomes the shortcoming of the exclusion of residence times by VWA and CMW. This method was also deployed in several other snowmelt-based IHS (Beaulieu et al., 2012; Carey and Quinton, 2004; Laudon et al., 2004, 2007).

Tracers have successfully been used in modeling studies to provide empirical insights into runoff generation processes and catchment functioning (Birkel and Soulsby, 2015; Birkel et al., 2011; Capell et al., 2012; Uhlenbrook and Leibundgut, 2002), but the combined use of distributed modeling and isotope tracers in snow-dominated environments is rare. Ahluwalia et al. (2013) used an isotope and modeling approach to derive snowmelt contributions to streamflow and determined differences between the two techniques of 2 %. Distributed modeling can provide areal melt rates that can be used for weighting the measured isotopic composition of meltwater. Pomeroy et al. (2003) described the differences of insolation between north- and south-facing slopes in complex terrain that lead to spatial varying melt rates of the snowpack throughout the freshet period. The use of the areal snowmelt data from models will likely reduce the uncertainty that arises from the representativeness of measured melt rates at the plot-scale.

The overall goal of our study was to quantify the contribution of snowmelt to streamflow and hence to improve the knowledge of hydroclimatological processes in high-elevation catchments. This study aims to enhance the reliability of isotope-based hydrograph separation by considering the distinct spatiotemporal variability of snowmelt and its isotopic signature in a high-elevation study region. This study has the following three objectives: (1) the estimation of the spatiotemporal variability of snowmelt and its isotopic composition, (2) the quantification of the impact of the spatial variability in snowmelt rates and its isotopic composition on IHS, and (3) to assess the combined use of a physically based snowmelt model and traditional IHS to determine snowmelt contributions to streamflow. Distributed melt rates provided by a surface energy balance model were used to weight the measured isotopic composition of snowmelt in order to characterize the event water isotopic composition. Traditional weighting methods (e.g., using plot-scale observed melt rates) were compared with the model approach.

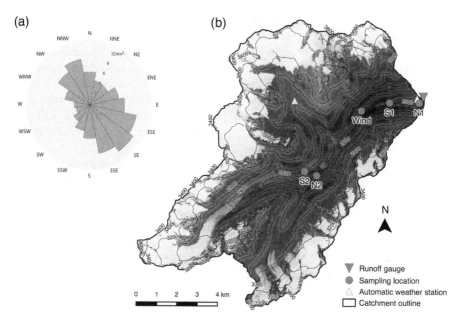

Figure 1. (a) Distribution of slope aspects in the study area; **(b)** study area (Rofen valley) with underlying orthophoto, sampling, and measurement locations.

2   Study area

The 98 km^2 high-elevation catchment of the Rofenache stream is located in the central eastern Alps (Oetztal Alps, Austria), close to the main Alpine ridge. The basin ranges in elevation from approximately 1900 to 3770 m.a.s.l. The average slope is 25° and the average elevation is 2930 m.a.s.l. (calculated from a 50 m digital elevation model). A narrow riparian zone (< 100 m width) is located in the valley floor. The predominantly south- (southeast) and north-facing (north-northwest) slopes form the main valley (Fig. 1a), which trends roughly from southwest to northeast (Fig. 1b). The study area has a dry inner-alpine climate. Mean annual precipitation is 800 mm yr^{-1}, of which 44 % falls as snow. The mean annual temperature at the gauging station in Vent (1890 m.a.s.l., reference period: 1982–2003) is 2 °C. Seasonal snow cover typically lasts from October to the end of June at the highest regions of the valley.

The bedrock consists of mainly paragneiss and mica schist and is overlain by a mantle of glacial deposits and thin soils (< 1 m). The bedrock outcrops and unconsolidated bare rocks cover the largest part (42 %) of the catchment (CLC, 2006). Glaciers cover approximately a third of the Rofenache catchment (35 %), while pastures and coniferous forests are located in the lowest parts of the catchment and cover less than 0.5 % (CLC, 2006). Sparsely vegetated areas and natural grassland cover 15 and 7.5 %, respectively (CLC, 2006). Besides seasonally frozen ground at slopes of various expositions, permafrost is likely to occur at an elevation over 2600 m.a.s.l. at the north-facing slopes (Haeberli, 1975). The annual hydrograph reveals a highly seasonal flow regime.

The mean annual discharge is 4.5 m^3 s^{-1} (reference period: 1971–2009) and is dominated by snow and glacier melt during the ablation season, which typically lasts from May to September. The onset of the early snowmelt season in the lower part of the basin is typically in April.

3   Methods

3.1   Field sampling, measurements, and laboratory analysis

The field work was conducted during the 2014 snowmelt season between the beginning of April and the end of June. Two short-term melt events (3 days) were investigated to illustrate the difference between early spring season melt and peak melt. The events were defined as warm and precipitation-free spells, with clear skies and dry antecedent conditions (i.e., no precipitation was observed 48 h prior to the event). Low discharge and air temperatures with a small diurnal variation and low melt rates, as well as a snow-covered area (SCA) of about 90 % in the basin (Fig. 2a), characterize the conditions of the early melt event at the end of April (Fig. 3b). In contrast, the peak melt period at the end of June is characterized by high discharge and melt rates, a flashy hydrograph, high air temperatures with remarkable diurnal variations (Fig. 3c), and a strongly retreated snow line (SCA: 66 %; Fig. 2c). Discharge data are available at an hourly resolution for the gauging station in Vent and meteorological data are obtained by two automatic weather stations (hourly resolution) located in and around the basin (Fig. 1).

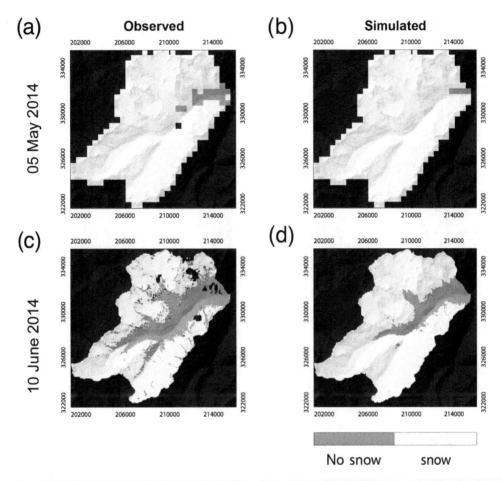

Figure 2. Comparison of observed and simulated snow distributions for **(a, b)** 5 May (MODIS scene) and **(c, d)** 10 June 2014 (Landsat scene).

The stream water sampling for stable isotope analysis consisted of pre-freshet baseflow samples at the beginning of March, sub-daily samples (temporal resolution ranges between 1 and 4 h) during the two studied events and a post-event sample in July as indicated in Fig. 3a (gray-shaded area). Samples of snowmelt, snowpack, and surface overland flow (if observed) were collected at the south- (S1, S2) and north-facing slope (N1, N2), as well as on a wind-exposed ridge (Fig. 1b) using a snowmelt collector. At each test site, a snow pit was dug to install a 0.1 m² polyethylene snowmelt collector at the ground–snowpack interface. The snowmelt collector consists of a pipe that drains the percolating meltwater into a fixed plastic bag. Tests yield a preclusion of evaporation for this sampling method. Composite daily snowmelt water samples (bulk sample) were collected in these bags and transferred to polyethylene bottles in the field before the onset of the diurnal melt cycle. Furthermore, sub-daily grab melt samples were collected at S1 (on 23 April) and at N2 (on 7 June) to define the diurnal variability of the respective melt event. Unfortunately further sub-daily snowmelt sampling was not feasible. The pit face was covered with white styrofoam to protect it from direct

sunlight. Stream, surface overland flow, and grab snowmelt water samples were collected in 20 mL polyethylene bottles. Snow samples from snow pit layers were filled in airtight plastic bags and melted below room temperature before being transferred into bottles. Overall, 144 samples were taken during the study period. Snow water equivalent (SWE), snow height, snow density, and various snowpack observations (wetness and hand hardness index) were observed before the onset of the diurnal melt cycle at the study plots (Fig. 1). Mean SWE was determined by averaging five snow-tube measurements within an area of 20 m² at each site. Daily melt rates were calculated by subtracting succeeding SWE values. Sublimation was neglected, as it contributes only a small percentage ($\sim 10\%$) to the seasonal water balance in high-altitude catchments in the Alps (Strasser et al., 2008).

All samples were treated by the guidelines proposed by Clark and Fritz (1997) and were stored in the dark and kept cold until analysis. The isotopic composition of the samples ($\delta^{18}O$, δD) was measured with cavity ring-down spectroscopy (Picarro L1102-i). Results are expressed in the delta notation as parts per thousand relative to the Vienna Standard Mean Ocean Water (VSMOW2). The mean laboratory pre-

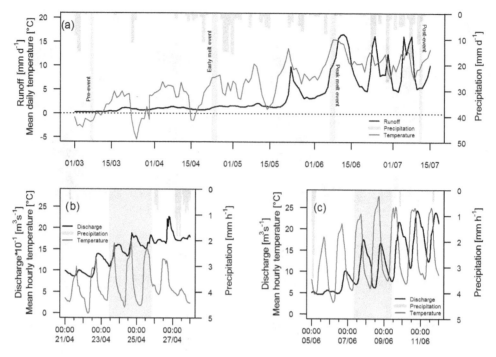

Figure 3. (a) Daily precipitation, air temperature, and discharge at the outlet of the catchment during the complete study period; hourly hydroclimatologic data of a 7-day period around the **(b)** early melt and **(c)** peak melt event. Gray-shaded areas indicate the investigated events.

cision (replication of eight measurements) for all measured samples was 0.06‰ for $\delta^{18}O$. Due to the covariance of δ^2H (δD) and $\delta^{18}O$ (Fig. 5), all analyses were done with oxygen-18 values.

3.2 Model description

For the simulation of the daily melt rates, the non-calibrated, distributed, and physically based hydroclimatological model AMUNDSEN (Strasser, 2008) was applied. Model features include interpolation of meteorological fields from point measurements (Marke, 2008; Strasser, 2008); simulation of shortwave and longwave radiation, including topographic and cloud effects (Corripio, 2003; Greuell et al., 1997); parameterization of snow albedo depending on snow age and temperature (Rohrer, 1991); modeling of forest snow and meteorological processes (Liston and Elder, 2006; Strasser et al., 2011); lateral redistribution of snow due to gravitational- (Gruber, 2007) and wind-induced (Helfricht, 2014; Warscher et al., 2013) processes; and determination of snowmelt using an energy balance approach (Strasser, 2008). Besides having been applied for various other Alpine sites in the past (Hanzer et al., 2014; Marke et al., 2015; Pellicciotti et al., 2005; Strasser, 2008; Strasser et al., 2004, 2008), AMUNDSEN has recently been set up and extensively validated for the Oetztal Alps region (Hanzer et al., 2016). This setup was also used to run the model in this study for the period 2013–2014 using a temporal resolution of 1 h and a spatial

resolution of 50 m. In order to determine the model performance during the study period, catchment-scale snow distribution by satellite-derived binary snow-cover maps and plot-scale observed SWE data were used for the validation (cf. Sect. 4.2). Therefore, the spatial snow distribution as simulated by AMUNDSEN was compared with a set of MODIS (500 m spatial resolution) and Landsat (30 m resolution, subsequently resampled to the 50 m model resolution) snow maps with less than 10 % cloud coverage over the study area using the methodology described in Hanzer et al. (2016). Model results were evaluated using the performance measures BIAS, accuracy (ACC) and critical success index (CSI) (Zappa, 2008). ACC represents the fraction of correctly classified pixels (either snow covered or snow free both in the observation and the simulation). CSI describes the number of correctly predicted snow-covered pixels divided by the number of times where snow is predicted in the model and/or observed, and BIAS corresponds to the number of snow-covered pixels in the simulation divided by the respective number in the observation. ACC and CSI values range from 0 to 1 (where 1 is a perfect match), while BIAS values below 1 indicate underestimations of the simulated snow cover, and values above 1 indicate overestimations. At the plot-scale, observed SWE values were compared with AMUNDSEN SWE values represented by the underlying pixels at the location of the snow course. Catchment-scale melt rates are calculated by subtracting two consecutive daily SWE grids, neglecting sublimation losses, which is also done to achieve observed melt rates at the plot scale. Subsequently, the digital elevation model was used to calculate an aspect grid and further to divide the catchment into two parts: grid cells with aspects ranging from ≥ 270 to $\leq 90°$ were classified as "north facing", while the remaining cells were attributed to the class "south facing". Finally, these two grids were combined to derive melt rates for the south-facing (melt$_s$) and for the north-facing slope (melt$_n$).

3.3 Isotopic hydrograph separation, weighting approaches, and uncertainty analysis

IHS is a steady-state tracer mass balance approach, and several assumptions underlie this simple principle, which are described and reviewed in Buttle (1994) and Klaus and McDonnell (2013):

1. The isotopic compositions of event and pre-event water are significantly different.

2. The event water isotopic signature has no spatiotemporal variability, or variations can be accounted for.

3. The pre-event water isotopic signature has no spatiotemporal variability, or variations can be accounted for.

4. Contributions from the vadose zone must be negligible or soil water should be isotopically similar to groundwater.

5. There is no or minimal discharge contribution from surface storage.

The focus of this study is on one of the assumptions: the spatiotemporal variability of the event water isotopic signature is absent or can be accounted for. The fraction of event water (f_e) contributing to streamflow was calculated from Eq. (1).

$$f_e = \frac{(C_p - C_s)}{(C_p - C_e)} \tag{1}$$

The tracer concentration of the pre-event component (C_p) is the $\delta^{18}O$ composition of baseflow prior to the onset of the freshet period, constituted mainly by groundwater and potentially by soil water, which was assumed to have the same isotopic signal as groundwater. Tracer concentration C_s is the isotopic composition of stream water for each sampling time. The isotopic compositions of snowmelt samples were weighted differently to obtain the event water tracer concentration (C_e) using the following five weighting approaches:

1. volume weighted with observed plot-scale melt rates (VWO);

2. equally weighted, assuming an equal melt rate on north- and south-facing slopes (VWE);

3. no weighting, only south-facing slopes considered (SOUTH);

4. no weighting, only north-facing slopes considered (NORTH);

5. volume weighted with simulated catchment-scale melt rates (VWS).

Equation (2) is the VWS approach with simulated melt rates for north- and south-facing slope as described in Sect. 3.2, where M is the simulated melt rate (in $mm\,d^{-1}$), $\delta^{18}O$ is the isotopic composition of sampled snowmelt, and subscripts s and n indicate north and south, respectively. For obtaining the value of C_e a daily time step (t) is used, considering daily melt rates and the isotopic composition of the daily bulk snowmelt samples.

$$C_e(t) = \frac{M_s(t)\delta^{18}O_s(t) + M_n(t)\delta^{18}O_n(t)}{M_s(t) + M_n(t)} \tag{2}$$

An uncertainty analysis (Eq. 3) was performed according to the Gaussian standard error method proposed by Genereux (1998):

$$W_{f_e} = \left\{ \left[\frac{C_p - C_s}{(C_p - C_e)^2} W_{C_e} \right]^2 + \left[\frac{C_s - C_e}{(C_p - C_e)^2} W_{C_p} \right]^2 + \left[\frac{-1}{(C_p - C_e)^2} W_{C_s} \right]^2 \right\}^{1/2}, \tag{3}$$

where W is the uncertainty, C is the isotopic composition, f is the fraction, and the subscripts p, s, and e refer to the pre-event, stream, and event component, respectively. This assumes negligible errors in the discharge measurements and the melt rates (modeled and observed). The uncertainty of streamflow (W_{C_s}) is assumed to be equal to the laboratory precision (0.06‰). For the uncertainty of the event component (W_{C_e}), the diurnal temporal variability (standard deviation) of the snowmelt isotopic signal (from one site and 1 day) was multiplied by the appropriate value of the two-tailed t-table (dependent on sample number) and used for the event, as proposed by Genereux (1998). This resulted in different uncertainty values for the early melt event ($W_{C_e} = 0.2$‰) and the peak melt event ($W_{C_e} = 0.5$‰). An error of 0.04‰ was assumed for the pre-event component (W_{C_p}), which reflects the standard deviation of two baseflow samples. A 95 % confidence level was used. Spatial variation in snowmelt and its isotopic composition were not considered in this error calculation method as they represent the hydrologic signal of interest.

4 Results

4.1 Spatiotemporal variability of streamflow and stable isotopic signature of sampled of water sources

Two major snowmelt pulses (mid-May and beginning of June) and four less pronounced pulses between mid-March

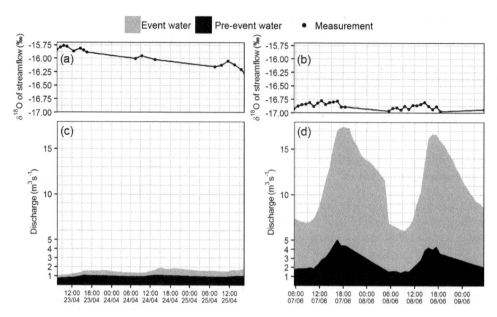

Figure 4. Linearly interpolated stream isotopic content of Rofenache for **(a)** the early melt and **(b)** the peak melt event. Dots indicate measurements. Event and pre-event water contributions during **(c)** the early melt and **(d)** the peak melt event calculated with the VWS approach.

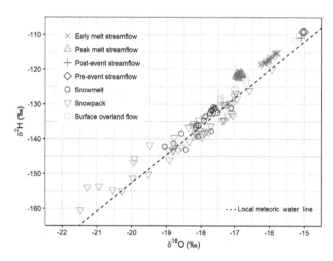

Figure 5. Relationship between δ^2H and δ^{18}O of water sources sampled during the snowmelt season of 2014 in the Rofen valley, Austrian Alps.

to early May occurred during the snowmelt season (Fig. 3a). Peak melt occurred at the beginning of June with maximum daily temperatures and runoff of 15 °C and 18 mm d^{-1}, respectively. The following high flows were affected by rain (Fig. 3a) and glacier melt due to the strongly retreated snow line and snow-free ablation area of the glaciers in July. Diurnal variations in discharge were strongly correlated with diurnal variations in air temperature (Fig. 3b and c) with a time lag of 3–5 h for the early melt event and 2–3 h for the peak melt event. An inverse relationship between stream-

flow δ^{18}O and discharge was found for the early melt event (Fig. 4a and c). Small diurnal responses of streamflow δ^{18}O were identified for both events, but were masked due to missing data during the recession of the hydrograph.

The quality control of the isotopic data was performed by the δ^2H–δ^{18}O plot (Fig. 5), which indicated no shift in the linear regression line and thus no secondary fractionation effects (evaporation) during storage and transport of the samples. The slope of the linear regression (slope = 8.5, $n = 144$, $R^2 = 0.93$) of the measurement data slightly deviates from that of the global meteoric (slope = 8) and local meteoric water line (slope = 8.1) based on monthly data from the Austrian Network of Isotopes in Precipitation sampling site in Obergurgl, which is located in an adjacent valley (reference period: 1991–2014). The small deviation (visible in Fig. 5) of the sampled water (i.e., snowpack and snowmelt) could indicate fractionation effects induced by phase transition (i.e., melt/refreeze and sublimation). The significant differences between the isotopic signatures of pre-event streamflow and snowmelt water enabled the IHS.

Overall, the δ^{18}O values ranged from −21.5 to −15.0‰, while snowpack samples were characterized by the most negative and pre-event baseflow samples by the least negative values. Snowpack samples showed a wide isotopic range, while streamflow samples revealed the narrowest spread, reflecting a composite isotopic signal mixing of the water components. Figure 6 shows the δ^{18}O data of the water samples grouped into different categories and split into early and peak melt data. It shows the different δ^{18}O ranges and medians of the sampled water sources (Fig. 6a), as well as marked spatiotemporal variations in the isotopic signal (Fig. 6b and c).

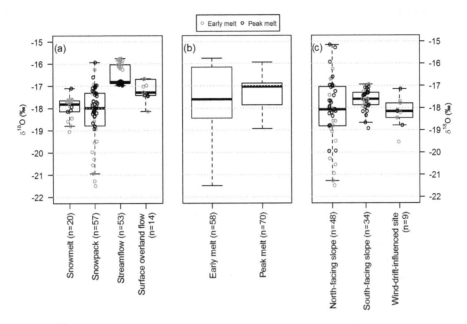

Figure 6. Jittered dot plots for $\delta^{18}O$ of collected water samples split into **(a)** water sources, **(b)** stage of snowmelt and **(c)** spatial origin. Gray circles indicate early melt samples and black circles peak melt samples. The gray and black line represents the median of early and peak melt data, respectively. n_e is the number of early melt samples and n_p is the number of peak melt samples.

It is apparent that the snowpack $\delta^{18}O$ values have a larger variation compared to the snowmelt data due to homogenization effects (Fig. 6a), as was also shown by Árnason et al. (1973), Dinçer et al. (1970) and Stichler (1987). The median of the $\delta^{18}O$ of snowmelt was higher than that of the snowpack, which indicates fractionation. The median $\delta^{18}O$ of surface overland flow was higher than that of snowmelt (Fig. 6a) for the early and peak melt period. Overall, the peak melt $\delta^{18}O$ values (Fig. 6b) were less variable and had a higher median than the early melt values, because fractionation effects (due to melt/refreeze and sublimation) most likely altered the isotopic composition of the snowpack over time (cf. Taylor et al., 2001, 2002). One major finding was that the $\delta^{18}O$ values on the north-facing slope had a larger range and a lower median compared to the opposing slope (Fig. 6c). Samples from the wind-drift-influenced site (also south exposed) were more depleted in heavy isotopes compared to the south-facing slope samples (Fig. 6c).

In general, the average snowmelt and snowpack isotopic composition was more depleted for the early melt period (Table 1) and changed over time because fractionation likely altered the snowpack and its melt. It is obvious that the isotopic evolution (gradually enrichment) on the south-facing slope took place earlier in the annual melting cycle of the snow, and indicates a premature snowpack concerning the enrichment of isotopes and earlier ripening compared to the north-facing slope.

Table 1 shows that meltwater sampling throughout the entire snowmelt period is required to account for the temporal variation in the isotopic composition of the snowpack

(cf. Taylor et al., 2001, 2002). In detail, the snowpack and snowmelt $\delta^{18}O$ data highlighted a marked spatial inhomogeneity between north- and south-facing slopes throughout the study period. The snowpack isotopic composition from both sampled slopes was statistically different for the early melt, but not for the peak melt (with Kruskal–Wallis test at 0.05 significance level), whereas the snowmelt $\delta^{18}O$ showed a significant difference throughout the study period (Fig. 7).

Sub-daily snowmelt samples ($n = 5$) at S1 (23 April 2014) had a range of 0.1‰ in $\delta^{18}O$, and the bulk sample (integrating the entire diurnal melt cycle) was within the scatter of those values (Fig. 8). The intra-daily variability of snowmelt ($n = 3$) at N2 (7 June 2014) was relatively higher with values ranging from -17.9 to $-18.1‰$. The bulk sample ($-17.9‰$) was at the upper end of those values (Fig. 8).

Stream water isotopic composition was more enriched in heavy isotopes during the early melt period and successively became more depleted throughout the freshet period, resulting in more negative values during peak melt (Table 2). The standard deviation and range of stream water $\delta^{18}O$ during early melt was higher and could be related to an increasing snowmelt contribution throughout the event and larger diurnal amplitudes of snowmelt contribution compared to peak melt (Table 2).

4.2 Snow model validation and snowmelt variability

Figure 9 shows the values for the selected performance measures based on the available MODIS and Landsat scenes during the period March–July 2014. The results indicate a

Table 1. Average isotopic composition of snowpack and snowmelt with standard deviation for north- and south-facing slopes during the early and the peak melt event. Values are averages of 3 consecutive days.

	North-facing slope		South-facing slope	
	Snowpack $\delta^{18}O$ (‰)	Snowmelt $\delta^{18}O$ (‰)	Snowpack $\delta^{18}O$ (‰)	Snowmelt $\delta^{18}O$ (‰)
Early melt event	-19.7 ± 0.6 ($n = 12$)	-18.8 ± 0.2 ($n = 3$)	-17.3 ± 0.3 ($n = 4$)	-17.4 ± 0.2 ($n = 8$)
Peak melt event	-17.6 ± 0.4 ($n = 18$)	-17.9 ± 0.1 ($n = 3$)	-17.9 ± 0.1 ($n = 15$)	-17.1 ± 0.0 ($n = 2$)

Table 2. Descriptive statistics of streamflow isotopic composition at the outlet of the Rofenache during events of the snowmelt season 2014.

	Pre-event	Early melt	Peak melt	Post-event
Date	7 Mar	23–25 Apr	7–9 Jun	11 Jul
Average ($\delta^{18}O$‰)	-15.02	-15.97	-16.87	-15.09
Standard deviation ($\delta^{18}O$‰)	0.04	0.16	0.05	–
Range ($\delta^{18}O$‰)	0.05	0.50	0.20	–
Number of samples	2	17	30	1

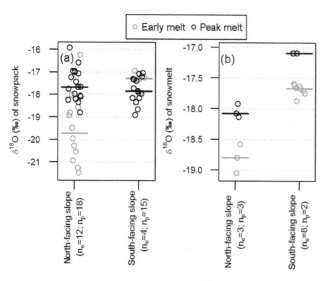

Figure 7. Jittered dot plots for $\delta^{18}O$ of **(a)** snowpack and **(b)** snowmelt of north- and south-facing slopes. Gray circles indicate early melt samples and black circles are for peak melt samples. The gray and black lines indicates the median of the early and peak melt data, respectively. n_e is the number of early melt samples and n_p is the number of peak melt samples.

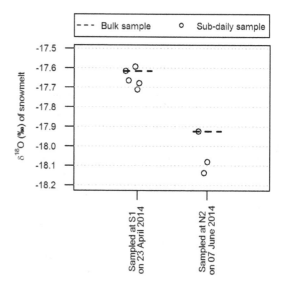

Figure 8. Comparison of snowmelt $\delta^{18}O$ between the bulk sample (dashed line) and sub-daily samples (circles) for the two sites (S1, N2).

reasonable model performance with a tendency to slightly overestimate the snow cover during the peak melt season (BIAS > 1). In general the CSI does not drop below 0.7, and 80 % of the pixels are correctly classified (ACC) throughout the study period. Figure 2 shows the observed and simulated spatial snow distribution around the time of the two events. Despite a higher SCA during the early melt season (Fig. 2a and b) compared to the peak melt season (Fig. 2c and d) one can see the overestimation of the simulated SCA compared to the observed (MODIS/Landsat) SCA. Table 3

shows the observed and simulated SWE values at the plot scale. The model slightly underestimated SWE during peak melt, but generally appears to be in quite good agreement, suggesting well-simulated snowpack processes. Throughout the study period the model deviates by 13 % from the observed SWE values, but the representativeness (small-scale effects) of SWE values for the respective 50 m pixels should be considered.

Snowmelt (observed and simulated daily losses of SWE) showed a distinct spatial variation between the north-facing and the south-facing slope for the early melt (23/24 April) period, but less marked variations for the peak melt (7/8 June) period (Fig. 10). Relative day-to-day differences are more

Table 3. Comparison of observed and simulated (represented by the underlying pixel) SWE values.

Site	Date	Stage of snowmelt season	SWE [mm] Observed	SWE [mm] Simulated	Difference between observed and simulated SWE [%]
S1	23 Apr 2014	Early melt	141	151	7
N1	23 Apr 2014	Early melt	351	356	1
Wind	24 Apr 2014	Early melt	201	229	14
S1	25 Apr 2014	Early melt	113	78	−31
N1	25 Apr 2014	Early melt	270	293	9
N2	7 Jun 2014	Peak melt	594	477	−20
N2	8 Jun 2014	Peak melt	568	435	−23
N2	9 Jun 2014	Peak melt	537	390	−27
Mean deviation between observed and simulated SWE					13

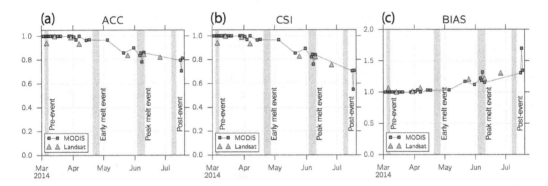

Figure 9. Performance measures of **(a)** accuracy (ACC), **(b)** critical success index (CSI), and **(c)** BIAS as calculated by comparing AMUND-SEN simulation results with satellite-derived (MODIS/Landsat) snow maps.

pronounced for the early melt season. Both simulated and observed melt rates are higher for the peak melt event on the south-facing slope, but not for the north-facing slope. Simulated melt intensity on the south-facing slope at the end of April was twice the rate on the north-facing slope, while simulated melt rates were approximately the same for the opposing slopes during peak melt. Simulated (catchment scale) snowmelt rates were markedly lower during the early melt (23 and 24 April) on the north-facing slope compared to the observed (plot scale) melt rates (Fig. 10a), but differences between them were small during peak melt for both slopes (7 and 8 June; Fig. 10).

4.3 Weighting techniques and isotope-based hydrograph separation

Differences between the applied snowmelt weighting techniques, induced by the high spatial variability of snowmelt (Sect. 4.2), led to different event water isotopic compositions (C_e) for the IHS analyses (Table 4). The event water component was depleted in $\delta^{18}O$ by roughly 0.3‰ for the second day (24 April) of the early melt event compared to the preceding day, but inter-daily variation during the peak melt is

Table 4. Isotopic composition of the event water component for the applied weighting techniques.

	Event water isotopic composition ($\delta^{18}O$‰)			
	23 Apr	24 Apr	7 Jun	8 Jun
VWS	−17.9	−18.2	−17.5	−17.5
VWO	−18.3	−18.6	−17.4	−17.5
VWE	−18.1	−18.3	−17.5	−17.5
NORTH	−18.6	−18.8	−17.9	−17.9
SOUTH	−17.6	−17.9	−17.1	−17.1

almost absent. Especially during early melt (23 to 24 April), strong deviations between observed plot-scale melt rates and distributed (areal) melt rates obtained by AMUNDSEN occurred (Fig. 11), and led to more different event water isotopic compositions between the VWS and the VWO approach (Table 4).

The hydrograph and the results of the IHS applied with the VWS method for the early and peak melt event are presented in Fig. 4 and highlight the lower flow rates and higher pre-

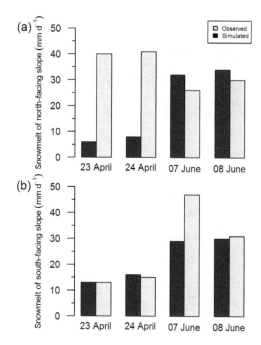

Figure 10. Observed (plot scale) and simulated (catchment scale) daily snowmelt on **(a)** the north-facing and **(b)** the south-facing slope for the early melt (23/24 April) and peak melt (7/8 June).

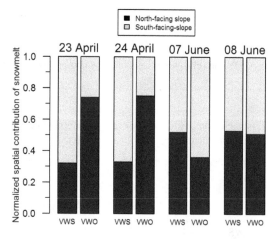

Figure 11. Relative contribution of the north- and south-facing slope $\delta^{18}O$ values to the catchment average. VWS: volume weighted with simulated (areal) melt rates. VWO: volume weighted with observed (plot-scale) melt rates.

Table 5. Discharge characteristics of the Rofenache for the early and peak melt event.

	Event	
	Early melt	Peak melt
Date	23–25 Apr	7–9 Jun
Mean discharge	$1.5\,\mathrm{m^3\,s^{-1}}$	$11.5\,\mathrm{m^3\,s^{-1}}$
Peak discharge	$1.9\,\mathrm{m^3\,s^{-1}}$	$17.4\,\mathrm{m^3\,s^{-1}}$
Volume runoff	3.3 mm	20.7 mm
Mean-event water fraction	$35 \pm 3\,\%$	$75 \pm 14\,\%$
Peak-event water fraction	$44 \pm 4\,\%$	$78 \pm 15\,\%$

event than for the early melt event because the difference between isotopic composition of pre-event water and event water was smaller than for the early melt event (uncertainty: 3 %) (cf. Tables 2 and 4).

Throughout the early melt event, the snowmelt fraction increased from 25 to 44 % (Fig. 4c; Table 5). This trend mirrors the stream isotopic composition, which became more depleted (Fig. 4a). Event water contributions during peak melt were generally higher but had a smaller range (70 to 78 %; Fig. 4d). Diurnal isotopic variations of stream water were small for both events (Fig. 4a and b), and could not clearly be obtained due to missing data on the falling limb of the hydrographs.

The use of the different weighting approaches led to strongly varying estimated snowmelt fractions of streamflow (Fig. 12). Especially the differences between the SOUTH and the NORTH approach during both investigated events (up to 24 %), and the differences between the VWS and the VWO approach (5 %) during early melt (Fig. 12a) are notable. Event water contributions estimated by the different weighting methods ranged from 21–28 % at the beginning of the early melt event up to 31–55 % at the end of the event (Fig. 12a, Table 6). Minimum event water contributions during the peak melt were estimated at 60–84 % and maxima ranged between 67 and 94 % for the different weighting methods (Table 6, Fig. 12b). Beside these intra-event variations in snowmelt contribution, the volumetric variations at the event-scale were smaller and ranged between 28–40 and 66–90 %, for the early and peak melt event, respectively (Table 6).

Considering only spatial variation of snowmelt isotopic signatures (i.e., comparing the NORTH/SOUTH approach with the VWE approach) for IHS led to differences in estimated event water fractions up to 7 and 14 % for the early and peak melt period, respectively (Table 6). However, considering only spatial variation in snowmelt rates (i.e., comparing the VWS/VWO approach with the VWE approach) led to differences in event water fraction up to 3 and 2 % for the early and peak melt period, respectively (Table 6).

Surface overland flow was not considered in the IHS analyses, but if applied, it would most likely increase the cal-

event fractions during early melt (Fig. 4c) and vice versa for the peak melt period (Fig. 4d). The total runoff volume during the peak melt period was approximately 6 times higher than in the early melt period. The fractions of snowmelt (volume) estimated with the VWS approach were 35 and 75 % with calculated uncertainties (95 % confidence level) of ± 3 and ± 14 % for the early and peak melt event, respectively. The uncertainty calculated from Eq. (3) of the IHS applied with the VWS method was higher (14 %) for the peak melt

Table 6. Event water contribution to streamflow based on the different weighting techniques. The error indicates the variability (standard deviation) and the values in parentheses depict the range.

	Event water contribution (%)				
	VWS	VWO	VWE	NORTH	SOUTH
Early melt event	35 ± 6	30 ± 4	33 ± 5	28 ± 3	40 ± 9
	(25–44)	(22–35)	(24–39)	(21–31)	(28–55)
Peak melt event	75 ± 2	78 ± 3	76 ± 2	66 ± 2	90 ± 3
	(70–78)	(71–82)	(70–78)	(60–67)	(84–94)

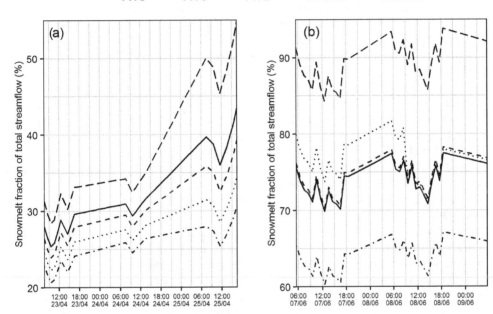

Figure 12. Comparison of the IHS results for the different weighting techniques used for **(a)** early melt and **(b)** peak melt. Scale of y axis in **(b)** differs from that in **(a)**.

culated snowmelt fraction slightly. Furthermore, snowmelt samples from the wind-exposed site were not used in the IHS analyses because this site was only sampled on the south-facing slope during early melt and is not representative for the catchment due to its limited coverage. However, incorporation of this data would decrease the calculated snowmelt fraction by approximately 2 %.

5 Discussion

5.1 Temporal variation in streamflow during the melting season

Snowmelt is a major contributor to streamflow during the spring freshet period in alpine regions and large amounts of snowmelt water infiltrate into the soil and recharge ground-water (Penna et al., 2014). The hydrological response of the stream followed the variations of air temperature, as already

observed by Braithwaite and Olesen (1989) (Fig. 3a). The observed time lags (Fig. 3b and c) between maximum daily air temperature and daily peak flow are common in mountain catchments (Engel et al., 2016; Schuler, 2002). During peak melt, the flashy hydrograph revealed less variation in the timing of peak discharge of 7-day data (Fig. 3c) compared to the early melt, as reported by Lundquist and Cayan (2002). The increase in discharge coincides with decreasing streamflow $\delta^{18}O$ during the early melt event (Fig. 4a and c) and confirms the earlier findings of Engel et al. (2016), who identified inverse relationships between streamflow $\delta^{18}O$ and discharge during several 24 h events in an adjacent valley on the southern side of the main Alpine ridge, although their findings rely on streamflow contributions from snow and glacier melt. The lower stream water isotopic composition during peak melt suggests a remarkable contribution of more depleted snowmelt to streamflow and therefore confirms the results of the IHS.

5.2 Spatiotemporal variability of snowmelt and its isotopic signature

The rate of snowmelt varies spatially in catchments with complex topography (Carey and Quinton, 2004; Dahlke and Lyon, 2013; Pomeroy et al., 2003). This was also demonstrated for the Rofen valley in this study (Fig. 10, Table 3). Snowmelt results from a series of processes (e.g., energy exchange between snow–atmosphere) that are spatially variable – especially in complex terrain. This also becomes obvious when comparing the snowmelt rates on 23 April 2014 in Fig. 10a. Differences of observed and simulated snowmelt rates might result from the non-representativeness of point measurements for catchment averages and refer to the scale issue of data collection. The peak melt period was characterized by less spatial and day-to-day variation in observed melt rates (Fig. 10). The modeled daily snowmelt during this period was similar for north- and south-facing slopes, likely because of higher melt rates but also a smaller snow-covered area of the south-facing slope in contrast to the north-facing slope during peak melt (Fig. 11). The model performance was good for SWE (Table 3) and snow-cover extent (Figs. 2 and 9). The spatial variation of snowpack isotopic composition are significant, as can be seen in the differences for north- and south-facing slopes, and also shown by Carey and Quinton (2004), Dahlke and Lyon (2013), and Dietermann and Weiler (2013) in their studied high-gradient catchments, whereas there are unclear differences for the spatial variation of snowmelt isotopic signals in the literature. It is not clear to which extent altitude is important, as Dietermann and Weiler (2013) stated that a potential elevation effect (decrease in snowmelt $\delta^{18}O$ with elevation) is likely to be disturbed by melting processes (isotopic enrichment) depending on catchment morphology (aspect, slope) during the ablation period. Beaulieu et al. (2012) detected elevation as a predictor, which explained most of the variance they observed in snowmelt $\delta^{18}O$ from four distributed snow lysimeters. Moore (1989) and Laudon et al. (2007) found no significant difference of $\delta^{18}O$ in their lysimeter outflows, which was likely due to the small elevation gradient of their catchments that favor an isotopically homogenous snowpack, whereas Unnikrishna et al. (2002) found remarkable small-scale spatial variability. An altitudinal gradient was not considered in this study, but possible effects on IHS are discussed in Sect. 5.6. The difference of snowmelt (not snowpack) isotopic signature between north- and south-facing slopes was clearly shown in this study. The dataset is small, but reveals clear differences induced by varying magnitudes and timing of melt due to differences in solar radiation on the opposing slopes (Fig. 7). Temporal variability in snowmelt isotopic composition is greater for the north-facing slope compared to the south-facing slope (Fig. 7), which was also pointed out by Carey and Quinton (2004) in their subarctic catchment. Earlier homogenization in the isotopic profile of the snowpack and earlier melt out are responsible for this phenomenon (cf.

Dinçer et al., 1970; Unnikrishna et al., 2002). Fractionation processes likely controlled this homogenization of the snowpack between the two investigated melt events. The isotopic homogenization of the snowpack on the south-facing slope started earlier in the melting period and caused a smaller spatial and temporal variation compared to the north-facing snowpack, as was also reported by Unnikrishna et al. (2002) and Dinçer et al. (1970). The differences between these investigated snowpacks were larger in the early melt season than in the peak melt season. This affects the IHS results, especially because the snowmelt contributions from the south- and north-facing slope – with marked isotopic differences – were distinct. Due to melt, fractionation processes proceeded and the snowpack likely became more homogenous throughout the snowmelt season. However, inter-daily variations of snowpack isotopic composition, especially for the north-facing slope, were still observable during the peak melt period. The gradual isotopic enrichment of the snowpack was also observed for snowmelt, as described by many others (Feng et al., 2002; Shanley et al., 2002; Taylor et al., 2001, 2002; Unnikrishna et al., 2002).

Intra-daily variations of snowmelt $\delta^{18}O$ could be quantified for two sites (Fig. 8). At S1 on the south-facing slope during the early melt event, the 0.1 ‰ range in $\delta^{18}O$ ($n = 5$) was smaller than the range at N2 on the north-facing slope during the peak melt event ($n = 3$, range $= 0.2$ ‰). This sub-daily variability is markedly smaller than the differences between the investigated slopes (cf. Table 1), which ranged from 0.8 ‰ (peak melt) to 1.4 ‰ (early melt). Unnikrishna et al. (2002) described significant temporal variations of snowmelt $\delta^{18}O$ during large snowmelt events (peak melt). However, these findings could not be confirmed within in this study, probably due to the temporally limited data and should be tested with a larger dataset. The bulk sample at S1 (23 April 2014) was isotopically closer to the sub-daily values compared to the bulk sample at N2 (7 June 2014) that was at the upper range of the sub-daily samples (Fig. 8). Therefore, one could argue that for the south-facing slope there is a negligible uncertainty if one uses a single snowmelt value (at one time) for IHS instead of using a bulk sample, but this is not the case for the north-facing slope (Fig. 8, site N2). Unfortunately the sample numbers are small, because more frequent and more distributed sampling (at different sites) was not feasible due to logistical issues. Hence, these results should be used with caution and should be investigated in further studies. If the focus and the scale of the study is not on the sub-daily variability, the authors recommend the use of bulk samples, because these integrate (automatically weighed with snowmelt rate) the diurnal variations.

5.3 Validity of isotopic hydrograph separation

The validity of IHS relies on several assumptions (cf. Sect. 3.3; Buttle, 1994; Klaus and McDonnell, 2013).

The assumption that the isotopic composition of event and pre-event water differ significantly (assumption 1) was successfully proven, because the snowmelt isotopic values were markedly lower than pre-event baseflow values (cf. Tables 2 and 4, Fig. 5). Spatiotemporal variations of event water isotopic composition (assumption 2) were accounted for by collecting daily and sub-daily samples during both events throughout the freshet period and meltwater sampling at a north- and south-facing slope, respectively. The spatially variable input of event water was considered by dividing the catchment into two parts – a north- and a south-facing slope. This study supports the findings of Dahlke and Lyon (2013) and Carey and Quinton (2004), emphasizing the highly variable snowpack/snowmelt isotopic composition in complex topography catchments due to enrichment. The temporal variability of event water isotopic composition was considered by using bulk daily samples, which integrate snowmelt from the entire diurnal melting cycle, but smooth out a sub-daily signal. Because the focus of this study was more on the inter-event than the intra-daily scale, this approach seemed reasonably reliable. The spatiotemporal variability of the isotopic composition of pre-event water (assumption 3) is a major limitation and could not be clearly identified due to a lack of data and was therefore assumed to be constant. Small differences between the pre-event samples (-15.00 and $-15.05\,\%_{o}$ for $\delta^{18}O$) and post-event stream water isotopic composition support this assumption (Table 2). The assumption of soil water having the same isotopic composition as groundwater in time and space (assumption 4) is critical. Some studies reveal no significant differences (e.g., Laudon et al., 2007), whereas others do (e.g., Sklash and Farvolden, 1979). Isotopic differences between groundwater and soil water were not considered due to a lack of data. Furthermore, it is not known to which degree the vadose zone contributes to baseflow in the study area. Winter baseflow used in the analyses is assumed to integrate mainly groundwater and partly soil water. Soil water could be hypothesized to have a negligible contribution to baseflow during winter due to the recession of the soil water flow in autumn and frozen soils in winter. The assumption that no or minimal surface storage occurs (assumption 5) is plausible because water bodies like lakes or wetlands do not exist in the study catchment and due to the steep topography detention storage is likely limited. The transit time of snowmelt was assumed to be less than 24 h. This short travel time is characteristic for headwater catchments (Lundquist et al., 2005) with high in-channel flow velocities, steep hillslopes, a high drainage density with snow-fed tributaries, thin soils, most snowmelt originating from the edge of the snow line (small average travel distances), partly frozen soil, and observed surface overland flow. The state-of-the-art method (runCE) to include residence times of snowmelt in the event water reservoir proposed by Laudon et al. (2002) was applied in several IHS studies (Beaulieu et al., 2012; Carey and Quin-

ton, 2004; Petrone et al., 2007), but was not feasible due to the short-term character and temporally limited data.

5.4 Hydrograph separation results and inferred runoff generation processes

Large contributions from snowmelt to streamflow are common in high-elevation catchments. Daily contributions between 35 and 75 % in the Rofen valley are comparable to the results of studies conducted in other mountainous regions, mostly outside the European Alps. Beaulieu et al. (2012) estimated snowmelt contributions ranging from 7 to 66 % at the seasonal scale for their 2.4 km^2 catchment and reported contributions of 34 and 62 %, for the early melt and peak melt, respectively. The hydrograph was dominated by pre-event water during early melt in April (Fig. 4c), which is in accordance with the results obtained by other IHS studies (Beaulieu et al., 2012; Laudon et al., 2004, 2007; Moore, 1989). The snowmelt contribution increased as the freshet period progressed and peaked with high contributions at the beginning of June. Beaulieu et al. (2012) and Sueker et al. (2000) reported comparable results for their physically similar catchments during peak melt with 62 and up to 76 % event water contributions to streamflow, respectively. At the event-scale comparable studies are rare. Engel et al. (2016) report a maximum daily snowmelt contribution estimated with a three-component hydrograph separation of 33 % for an 11 km^2 southwest of the Rofen valley with similar physiographic characteristics, but on the southern side of the main Alpine ridge. It should be mentioned that in their study, runoff was fed by three components (snowmelt, glacier melt, and groundwater) and lower snowmelt contributions were prevalent because most of the catchment area (69 %) was snow free.

Initial snowmelt events flush the pre-event water reservoir as snowmelt infiltrates into the soil and causes the pre-event water to exfiltrate and contribute to the streamflow. As the soil and groundwater reservoir becomes gradually filled with new water (snowmelt), the event water fraction in the stream increases. The system is also wetter during peak melt. The dominance of event water in the hydrograph is interpreted as an outflow of pre-event water stored in the subsurface and the gradual replenishment of the soil and groundwater reservoirs by event water. The higher water table – compared to the early melt period – could cause a transmissivity feedback mechanism (Bishop, 1991). This is a common mechanism in catchments with glacial till (Bishop et al., 2011) characterized by higher transmissivities and hence increased lateral flow velocities towards to the surface. Runoff generation is spatially very variable in the study area. There are areas (meadow patches between rock fields) where saturation excess overland flow is dominant (observed mainly at plots S1, S2, and Wind) and areas (with larger rocks and debris) where rapid shallow subsurface flow can be assumed (plot N2). Catchment morphology controls various hydrologic processes and hence the shape of the hydrograph. Upslope residence times of snowmelt are usually smaller due to thin soils (observed during the field work), steeper slopes (Sueker et al., 2000), and higher contributing areas of glaciers with impermeable ice (Behrens et al., 1978), and would be indicators for the more flashy hydrograph during the peak melt season.

5.5 Impact of spatial varying snowmelt and its isotopic composition on isotope-based hydrograph separation and assessment of weighting approaches

Klaus and McDonnell (2013) stress in their review paper the need to investigate the effects of the spatially varying snowmelt and its isotopic composition on IHS. This study quantified the impact of the spatially varying isotopic composition of snowmelt between north- and south-facing slopes on IHS results for the first time. The IHS results were more sensitive to the spatial variability of snowmelt $\delta^{18}O$ than to spatial variations of snowmelt rates (Table 6). This is even more pronounced for the peak melt period, because snowmelt rates were similar for the north- and south-facing slope, probably due to a ripe snow cover throughout the catchment. The difference in volumetric snowmelt contribution to streamflow at the event-scale determined using the five different weighting methods for IHS is maximum 24 % (NORTH approach vs. SOUTH approach). The data show that the variations between the weighting approaches (VWS, VWO, and VWE) are higher throughout the early melt season (Table 6), because small-scale variability of snowmelt and its isotopic composition are more pronounced in the early melt season. Thus, the influence of spatial variability of snowmelt and its isotopic composition on the event water fraction calculated with IHS is larger during this time. Melt rates strongly differ between the south- and the north-facing slope (Fig. 11), which were deceptively gathered by manually measured SWE, likely due to micro-topographic effects. As the contributions from both slopes are used in Eq. (3), they strongly influence the average isotopic composition of event water. The weighting method SOUTH (or NORTH) represents the hypothetical and most extreme scenario in which only one sampling site is used for the IHS analysis. Because snowmelt is more enriched in $\delta^{18}O$ and closer to pre-event water isotopic composition on the south-facing slope during peak melt, this scenario has the greatest effect on IHS and leads to the strongest deviation in estimated snowmelt fractions (up to 15 % overestimation compared to the VWS approach). These scenarios (NORTH/SOUTH) are theoretical and it is obvious that it is not recommended to conduct a IHS analysis by using only samples from either north- or south-facing slopes in catchments with complex terrain. Similar to the VWE method, snowmelt isotopic data were not volume weighted in other studies (e.g., Engel et al., 2016) where snowmelt data were not available. This has a more distinct effect on IHS during the early melt season be-

cause of the higher spatiotemporal variability in snowmelt (and its isotopic composition) compared to the peak melt season and led to a deviation in the snowmelt fraction in streamflow of 2 and 3 % compared to the VWS and VWO approaches, respectively. These differences are small, because the differing snowmelt and isotopic values offset each other in this particular case (Table 6). Nevertheless, the results of VWS are more correct for the right reason, because single observed plot-scale melt rates do not represent distributed snowmelt contribution at the catchment scale. Therefore, one can hypothesize that distributed simulated melt rates enhance the reliability of IHS, whereas plot-scale weighting introduces a large error caused by the difficulty in finding locations that represent the average melt rate in complex terrain.

5.6 Limitations of the study

Collecting water samples in high-elevation terrain is challenging due to limited access and high risk (e.g., avalanches), limiting high-frequency sampling. Hence, some limitations are inherent in this study. Potential elevation effects on snowmelt isotopic composition were not tested. The opposing sampling sites (S1–N1 and S2–N2) were at the same elevation (Fig. 1). It was assumed that the differences in north- and south-facing slopes were much greater than a possible altitudinal gradient in snowmelt isotopic composition. This hypothesis was not tested, but based on the results of other studies (Dietermann and Weiler, 2013). However, accounting for a potential altitudinal gradient (decrease in snowmelt $\delta^{18}O$ with elevation) would lead to more depleted isotopic signatures of event water and hence to lower event water fractions.

Another disadvantage is that no snow survey was conducted prior to the onset of snowmelt (peak accumulation) to estimate spatial variability in bulk snow $\delta^{18}O$. Because snowmelt is used for applying IHS, it is not clear to which degree the spatial variability of the snowpack isotopic composition is important. Two-component isotope-based hydrograph separation was successfully applied using the snowmelt and baseflow endmembers, but potential contributions of glacier melt were neglected (here defined as ice/firn melt). Because glaciers in the catchment were still covered by snow during the peak melt season, a significant contribution from ice/firn melt was assumed to be unlikely. Nevertheless, negligible amounts of basal (ice) meltwater could originate from temperate glaciers. No samples could be collected during the recession of the hydrograph (at night). Even though the spatial variability of the event water signal was the focus of the study, only temporal variability was considered in the Genereux-based uncertainty analyses. Although the temporal variability of winter baseflow isotopic composition seems to be insignificant, the sample number ($n = 2$) could be too small to characterize the pre-event component and should be clearly investigated in future work. Penna et al. (2016) used two approaches to determine the isotopic composition of pre-event water and described differences in the estimated event water contributions during snowmelt events. They advise to take pre-event samples prior to the onset of the melt season because pre-event samples taken prior to the onset of the diurnal melt cycle could be affected by snowmelt water from the previous melt pulses and therefore could lead to underestimated snowmelt fractions and high uncertainties. Furthermore, model results and observed discharges were assumed to be free of error in the analyses. As pointed out, instrumentation and accessibility are major problems for high-elevation studies. For this study it turned out that composite snowmelt samples were easier to collect, representing the day-integrated melt signal. A denser network of melt collectors would be desirable, as well as a snow lysimeter to gain high-frequency data automatically. Representative samples of the elevation zones and different vegetation belts could be important too, especially in partly forested catchments with a distinct relief (cf. Unnikrishna et al., 2002).

6   Conclusions

This study provides new insights into the variability of the isotopic composition in snowmelt and highlights its impact on IHS results in a high-elevation environment. The spatial variability in snowmelt isotopic signature was considered by experimental investigations on south- and north-facing slopes to define the isotopic composition of the snowmelt endmember with greater accuracy. This study clearly shows that distributed snowmelt rates obtained from a model based on meteorological data from local automatic weather stations affect the weighting of the event water isotopic signal, and hence the estimation of the snowmelt fraction in the stream by IHS. The study provides a variety of relevant findings that are important for hydrologic research in high-alpine environments. There was a distinct spatial variability in snowmelt between north- and south-facing slopes, especially during the early melt season. The isotopic composition of snowmelt water was significantly different between north-facing and south-facing slopes, which resulted in a pronounced effect on the estimated snowmelt contributions to streamflow with IHS. The IHS results were more sensitive to the spatial variability of snowmelt $\delta^{18}O$ than to spatial variation of snowmelt rates. The differences in the estimated snowmelt fraction due to the weighting methods used for IHS were as large as 24 %. This study also shows that it is hardly possible to characterize the event water signature of larger slopes based on plot-scale snowmelt measurements. Applying a distributed model reduced the uncertainty of the spatial snowmelt variability inherent to point-scale observations. Hence, applying the VWS method provided more reasonable results than the VWO method. This study highlighted that the selection of sampling sites has a major effect on IHS results. Sampling at least north-facing and south-facing slopes in complex terrain and using distributed melt rates to weight the snowmelt isotopic composition of the differing exposures is therefore highly recommended for applying snowmelt-based IHS.

Acknowledgements. The authors wish to thank the Institute of Atmospheric and Cryospheric Sciences of the University of Innsbruck, the Zentralanstalt für Meteorologie and Geodynamik, the Hydrographic Service of Tyrol and the TIWAG-Tiroler Wasserkraft AG for providing hydrological and meteorological data, the Amt der Tiroler Landesregierung for providing the digital elevation model, the Center of Stable Isotopes (CSI) for laboratory support, as well as many other individuals, who have helped to collect data in the field. We also thank the reviewers for their valuable suggestions that have much improved the manuscript, and the editor for the careful handling of the manuscript.

Edited by: I. van Meerveld

References

Ahluwalia, R. S., Rai, S. P., Jain, S. K., Kumar, B., and Dobhal, D. P.: Assessment of snowmelt runoff modelling and isotope analysis: a case study from the western Himalaya, India, Ann. Glaciol., 54, 299–304, doi:10.3189/2013AoG62A133, 2013.

APCC: Austrian Assessment Report (AAR14). Summary for Policymakers (SPM), Austrian Panel on Climate Change, Vienna, Austria, 2014.

Árnason, B., Buason, T., Martinec, J., and Theodorson, P.: Movement of water through snowpack traced by deuterium and tritium, in: The role of snow and ice in hydrology, Proc. Banff Symp., edited by: UNESCO-WMO-IAHS, IAHS Publ. No. 107, 1973.

Beaulieu, M., Schreier, H., and Jost, G.: A shifting hydrological regime: a field investigation of snowmelt runoff processes and their connection to summer base flow, Sunshine Coast, British Columbia, Hydrol. Process., 26, 2672–2682, doi:10.1002/hyp.9404, 2012.

Behrens, H., Moser, H., Oerter, H., Rauert, W., Stichler, W., and Ambach, W.: Models for the runoff from a glaciated catchment area using measurements of environmental isotope contents, Isotope Hydrology Vol. ll, W-05, Proceedings of a Symposium, Neuherberg, 19–23 June 1978, IAEA, Vienna, IAEA-SM-228/41, 2, 829–846, 1978.

Birkel, C. and Soulsby, C.: Advancing tracer-aided rainfall-runoff modelling: a review of progress, problems and unrealised potential, Hydrol. Process., 29, 5227–5240, doi:10.1002/hyp.10594, 2015.

Birkel, C., Tetzlaff, D., Dunn, S. M., and Soulsby, C.: Using time domain and geographic source tracers to conceptualize streamflow generation processes in lumped rainfall-runoff models,

Water Resour. Res., 47, W02515, doi:10.1029/2010WR009547, 2011.

Bishop, K.: Episodic increase in stream acidity, catchment flow pathways and hydrograph separation, PhD thesis, University of Cambridge, 246 pp., 1991.

Bishop, K., Seibert, J., Nyberg, L., and Rodhe, A.: Water storage in a till catchment. II: Implications of transmissivity feedback for flow paths and turnover times, Hydrol. Process., 25, 3950–3959, doi:10.1002/hyp.8355, 2011.

Braithwaite, R. J. and Olesen, O. B.: Calculation of glacier ablation from air temperature, West Greenland, in: Glacier Fluctuations and Climatic Change, Glaciology and Quaternary Geology, edited by: Oerlemans, J., Kluwer Academic Publisher, Dordrecht, 1989.

Buttle, J. M.: Isotope hydrograph separations and rapid delivery of pre-event water from drainage basins, Prog. Phys. Geog., 18, 16–41, doi:10.1177/030913339401800102, 1994.

Capell, R., Tetzlaff, D., and Soulsby, C.: Can time domain and source area tracers reduce uncertainty in rainfall-runoff models in larger heterogeneous catchments?, Water Resour. Res., 48, W09544, doi:10.1029/2011WR011543, 2012.

Carey, S. K. and Quinton, W. L.: Evaluating snowmelt rnoff generation in a discontinuous permafrost catchment using stable isotope, hydrochemical and hydrometric data, Nord. Hydrol., 35, 309–324, 2004.

Clark, I. D. and Fritz, P.: Environmental Isotopes in Hydrogeology, Lewis Publishers, Ney York, 342 pp., 1997.

CLC: Corine Land Cover 2006 raster data, European Environment Agency. The European Topic Centre on Land Use and Spatial Information, available at: http://www.eea.europa.eu/data-and-maps/data/clc-2006-raster (last access: 10 December 2015), 2006.

Cooper, L. W.: Isotopic fractionation in snow cover, in: Isotope tracers in catchment hydrology, edited by: Kendall, C. and McDonnell, J. J., Elsevier Science, Amsterdam, Netherlands, 119–136, 2006.

Corripio, J. G.: Vectorial algebra algorithms for calculating terrain parameters from DEMs and the position of the sun for solar radiation modelling in mountainous terrain, Int. J. Geogr. Inf. Sci., 17, 1–23, 2003.

Dahlke, H. E. and Lyon, S. W.: Early melt season snowpack isotopic evolution in the Tarfala valley, northern Sweden, Ann. Glaciol., 54, 149–156, doi:10.3189/2013AoG62A232, 2013.

Dietermann, N. and Weiler, M.: Spatial distribution of stable water isotopes in alpine snow cover, Hydrol. Earth Syst. Sci., 17, 2657–2668, doi:10.5194/hess-17-2657-2013, 2013.

Dinçer, T., Payne, B. R., Florkowski, T., Martinec, J., and Tongiorgi, E.: Snowmelt runoff from measurements of tritium and oxygen-18, Water Resour. Res., 6, 110–124, doi:10.1029/WR006i001p00110, 1970.

Engel, M., Penna, D., Bertoldi, G., Dell'Agnese, A., Soulsby, C., and Comiti, F.: Identifying run-off contributions during melt-induced run-off events in a glacierized alpine catchment, Hydrol. Process., 30, 343–364, doi:10.1002/hyp.10577, 2016.

Feng, X., Taylor, S., Renshaw, C. E., and Kirchner, J. W.: Isotopic evolution of snowmelt 1. A physically based one-dimensional model, Water Resour. Res., 38, 35-31–35-38, doi:10.1029/2001WR000814, 2002.

Genereux, D.: Quantifying uncertainty in tracer-based hydrograph separations, Water Resour. Res., 34, 915–919, doi:10.1029/98WR00010, 1998.

Greuell, W., Knap, W. H., and Smeets, P. C.: Elevational changes in meteorological variables along a midlatitude glacier during summer, J. Geophys. Res.-Atmos., 102, 25941–25954, doi:10.1029/97JD02083, 1997.

Gruber, S.: A mass-conserving fast algorithm to parameterize gravitational transport and deposition using digital elevation models, Water Resour. Res., 43, W06412, doi:10.1029/2006WR004868, 2007.

Haeberli, W.: Untersuchungen zur Verbreitung von Permafrost zwischen Flüelapass und Piz Grialetsch (Graubünden), Mitteilungen der Versuchsanstalt für Wasserbau, Hydrologie und Glaziologie der ETH Zürich, 1975.

Hanzer, F., Marke, T., and Strasser, U.: Distributed, explicit modeling of technical snow production for a ski area in the Schladming region (Austrian Alps), Cold Reg. Sci. Technol., 108, 113–124, doi:10.1016/j.coldregions.2014.08.003, 2014.

Hanzer, F., Helfricht, K., Marke, T., and Strasser, U.: Multilevel spatiotemporal validation of snow/ice mass balance and runoff modeling in glacierized catchments, The Cryosphere, 10, 1859–1881, doi:10.5194/tc-10-1859-2016, 2016.

Helfricht, K.: Analysis of the spatial and temporal variation of seasonal snow accumulation in Alpine catchments using airborne laser scanning. Basic research for the adaptation of spatially distributed hydrological models to mountain regions, PhD, University of Innsbruck, Innsbruck, 134 pp., 2014.

Hock, R.: Temperature index melt modelling in mountain areas, J. Hydrol., 282, 104–115, doi:10.1016/S0022-1694(03)00257-9, 2003.

Hooper, R. P. and Shoemaker, C. A.: A Comparison of Chemical and Isotopic Hydrograph Separation, Water Resour. Res., 22, 1444–1454, doi:10.1029/WR022i010p01444, 1986.

Huth, A. K., Leydecker, A., Sickman, J. O., and Bales, R. C.: A two-component hydrograph separation for three high-elevation catchments in the Sierra Nevada, California, Hydrol. Process., 18, 1721–1733, doi:10.1002/hyp.1414, 2004.

IPCC: Summary for Policymakers. Climate Change 2013: The Physical Science Basis. Contribution of Working Group I to the Fifth Assessment Report of the Intergovernmental Panel on Climate Change, Cambridge, United Kindom and New York, NY, USA, 2013.

Klaus, J. and McDonnell, J. J.: Hydrograph separation using stable isotopes: Review and evaluation, J. Hydrol., 505, 47–64, doi:10.1016/j.jhydrol.2013.09.006, 2013.

Laudon, H., Hemond, H. F., Krouse, R., and Bishop, K. H.: Oxygen 18 fractionation during snowmelt: Implications for spring flood hydrograph separation, Water Resour. Res., 38, 40-41–40-10, doi:10.1029/2002WR001510, 2002.

Laudon, H., Seibert, J., Köhler, S., and Bishop, K.: Hydrological flow paths during snowmelt: Congruence between hydrometric measurements and oxygen 18 in meltwater, soil water, and runoff, Water Resour. Res., 40, W03102, doi:10.1029/2003WR002455, 2004.

Laudon, H., Sjöblom, V., Buffam, I., Seibert, J., and Mörth, M.: The role of catchment scale and landscape characteristics for runoff generation of boreal streams, J. Hydrol., 344, 198–209, doi:10.1016/j.jhydrol.2007.07.010, 2007.

Lee, J., Feng, X., Faiia, A. M., Posmentier, E. S., Kirchner, J. W., Osterhuber, R., and Taylor, S.: Isotopic evolution of a seasonal snowcover and its melt by isotopic exchange between liquid water and ice, Chem. Geol., 270, 126–134, doi:10.1016/j.chemgeo.2009.11.011, 2010.

Liston, G. E. and Elder, K.: A Distributed Snow-Evolution Modeling System (SnowModel), J. Hydrometeorol., 7, 1259–1276, doi:10.1175/JHM548.1, 2006.

Liu, F., Williams, M. W., and Caine, N.: Source waters and flow paths in an alpine catchment, Colorado Front Range, United States, Water Resour. Res., 40, 1–16, doi:10.1029/2004WR003076, 2004.

Lundquist, J. D. and Cayan, D. R.: Seasonal and Spatial Patterns in Diurnal Cycles in Streamflow in the Western United States, J. Hydrometeorol., 3, 591–603, doi:10.1175/1525-7541(2002)003<0591:SASPID>2.0.CO;2, 2002.

Lundquist, J. D., Dettinger, M. D., and Cayan, D. R.: Snow-fed streamflow timing at different basin scales: Case study of the Tuolumne River above Hetch Hetchy, Yosemite, California, Water Resour. Res., 41, W07005, doi:10.1029/2004WR003933, 2005.

Marke, T.: Development and Application of a Model Interface to couple Regional Climate Models with Land Surface Models for Climate Change Risk Assessment in the Upper Danube Watershed, Dissertation der Fakultät für Geowissenschaften, Digitale Hochschulschriften der LMU München, München, 2008.

Marke, T., Strasser, U., Hanzer, F., Stötter, J., Wilcke, R. A. I., and Gobiet, A.: Scenarios of Future Snow Conditions in Styria (Austrian Alps), J. Hydrometeorol., 16, 261–277, doi:10.1175/JHM-D-14-0035.1, 2015.

Martinec, J., Siegenthaler, U., Oeschger, H., and Tongiorgi, E.: New insights into the run-off mechanism by environmental isotopes, in: Isotope techniques in groundwater hydrology, Proceedings of an International Symposium, IAEA, Vienna, Austria, 1974.

Mast, A. M., Kendall, K., Campbell, D. H., Clow, D. W., and Back, J.: Determination of hydrologic pathways in an alpine-subalpine basin using isotopic and chemical tracers, Loch Vale Watershed, Colorado, USA, in: Biogeochemistry of Seasonally Snow-Covered Catchments, edited by: Tonnessen, K., William, M., and Tranter, M., Int. Assoc. of Hydrol. Sci. Proc., Boulder, Colorado, 1995.

Maulé, C. P. and Stein, J.: Hydrologic Flow Path Definition and Partitioning of Spring Meltwater, Water Resour. Res., 26, 2959–2970, doi:10.1029/WR026i012p02959, 1990.

Moore, R. D.: Tracing runoff sources with deuterium and oxygen-88 during spring melt in a headwater catchment, southern Laurentians, Quebec, J. Hydrol., 112, 135–148, doi:10.1016/0022-1694(89)90185-6, 1989.

Pellicciotti, F., Brock, B., Strasser, U., Burlando, P., Funk, M., and Corripio, J.: An enhanced temperature-index glacier melt model including the shortwave radiation balance: development and testing for Haut Glacier d'Arolla, Switzerland, J. Glaciol., 51, 573–587, 2005.

Penna, D., Engel, M., Mao, L., Dell'Agnese, A., Bertoldi, G., and Comiti, F.: Tracer-based analysis of spatial and temporal variations of water sources in a glacierized catchment, Hydrol. Earth Syst. Sci., 18, 5271–5288, doi:10.5194/hess-18-5271-2014, 2014.

Penna, D., van Meerveld, H. J., Zuecco, G., Dalla Fontana, G., and Borga, M.: Hydrological response of an Alpine catchment to rainfall and snowmelt events, J. Hydrol., 537, 382–397, doi:10.1016/j.jhydrol.2016.03.040, 2016.

Petrone, K., Buffam, I., and Laudon, H.: Hydrologic and biotic control of nitrogen export during snowmelt: A combined conservative and reactive tracer approach, Water Resour. Res., 43, 1–13, doi:10.1029/2006WR005286, 2007.

Pinder, G. F. and Jones, J. F.: Determination of the groundwater component of peak discharge from the chemistry of total runoff, Water Resour. Res., 5, 438–445, doi:10.1029/WR005i002p00438, 1969.

Pomeroy, J. W., Toth, B., Granger, R. J., Hedstrom, N. R., and Essery, R. L. H.: Variation in Surface Energetics during Snowmelt in a Subarctic Mountain Catchment, J. Hydrometeorol., 4, 702–719, doi:10.1175/1525-7541(2003)004<0702:VISEDS>2.0.CO;2, 2003.

Rohrer, M. B.: Die Schneedecke im Schweizer Alpenraum und ihre Modellierung, PhD thesis, Swiss Federal Institute of Technology in Zurich, Switzerland, 178 pp., 1992.

Schuler, T.: Investigation of water drainage through an alpine glacier by tracer experiments and numerical modeling, PhD thesis, Swiss Federal Institute of Technology in Zurich, Switzerland, 140 pp., 2002.

Shanley, J. B., Kendall, C., Smith, T. E., Wolock, D. M., and McDonnell, J. J.: Controls on old and new water contributions to stream flow at some nested catchments in Vermont, USA, Hydrol. Process., 16, 589–609, doi:10.1002/hyp.312, 2002.

Sklash, M. G. and Farvolden, R. N.: The role of groundwater in storm runoff, J. Hydrol., 43, 45–65, doi:10.1016/0022-1694(79)90164-1, 1979.

Sklash, M. G., Farvolden, R. N., and Fritz, P.: A conceptual model of watershed response to rainfall, developed through the use of oxygen-18 as a natural tracer, Can. J. Earth Sci., 13, 271–283, doi:10.1139/e76-029, 1976.

Stichler, W.: Snowcover and Snowmelt Processes Studied by Means of Environmental Isotopes, in: Seasonal Snowcovers: Physics, Chemistry, Hydrology, edited by: Jones, H. G. and Orville-Thomas, W. J., D. Reidel Publishing Company, Dordrecht, Holland, 673–726, 1987.

Strasser, U.: Modelling of the mountain snow cover in the Berchtesgaden National Park, Research Rep. 55, Berchtesgaden, 2008.

Strasser, U., Corripio, J., Pellicciotti, F., Burlando, P., Brock, B., and Funk, M.: Spatial and temporal variability of meteorological variables at Haut Glacier d'Arolla (Switzerland) during the ablation season 2001: Measurements and simulations, J. Geophys. Res.-Atmos., 109, D03103, doi:10.1029/2003JD003973, 2004.

Strasser, U., Bernhardt, M., Weber, M., Liston, G. E., and Mauser, W.: Is snow sublimation important in the alpine water balance?, The Cryosphere, 2, 53–66, doi:10.5194/tc-2-53-2008, 2008.

Strasser, U., Warscher, M., and Liston, G. E.: Modeling Snow–Canopy Processes on an Idealized Mountain, J. Hydrometeorol., 12, 663–677, doi:10.1175/2011JHM1344.1, 2011.

Sueker, J. K., Ryan, J. N., Kendall, C., and Jarrett, R. D.: Determination of hydrologic pathways during snowmelt for alpine/subalpine basins, Rocky Mountain National Park, Colorado, Water Resour. Res., 36, 63–75, doi:10.1029/1999WR900296, 2000.

Taylor, S., Feng, X., Kirchner, J. W., Osterhuber, R., Klaue, B., and Renshaw, C. E.: Isotopic evolution of a seasonal snowpack and its melt, Water Resour. Res., 37, 759–769, doi:10.1029/2000WR900341, 2001.

Taylor, S., Feng, X., Williams, M., and McNamara, J.: How isotopic fractionation of snowmelt affects hydrograph separation, Hydrol. Process., 16, 3683–3690, doi:10.1002/hyp.1232, 2002.

Uhlenbrook, S. and Leibundgut, C.: Process-oriented catchment modelling and multiple-response validation, Hydrol. Process., 16, 423–440, doi:10.1002/hyp.330, 2002.

Unnikrishna, P. V., McDonnell, J. J., and Kendall, C.: Isotope variations ni a Sierra Nevada snowpack and their relation to meltwater, J. Hydrol., 260, 38–57, 2002.

Warscher, M., Strasser, U., Kraller, G., Marke, T., Franz, H., and Kunstmann, H.: Performance of complex snow cover descriptions in a distributed hydrological model system: A case study for the high Alpine terrain of the Berchtesgaden Alps, Water Resour. Res., 49, 2619–2637, doi:10.1002/wrcr.20219, 2013.

Weingartner, R. and Aschwanden, H.: Discharge regime – the basis for the estimation of average flows, in: Hydrological Atlas of Switzerland, Plate 5.2, Bern, 1992.

Williams, M. W., Seibold, C., and Chowanski, K.: Storage and release of solutes from a subalpine seasonal snowpack: soil and stream water response, Niwot Ridge, Colorado, Biogeochemistry, 95, 77–94, doi:10.1007/s10533-009-9288-x, 2009.

Zappa, M.: Objective quantitative spatial verification of distributed snow cover simulations – an experiment for the whole of Switzerland, Hydrolog. Sci. J., 53, 179–191, doi:10.1623/hysj.53.1.179, 2008.

Zhou, S., Nakawo, M., Hashimoto, S., and Sakai, A.: The effect of refreezing on the isotopic composition of melting snowpack, Hydrol. Process., 22, 873–882, doi:10.1002/hyp.6662, 2008.

9

Soil water migration in the unsaturated zone of semiarid region in China from isotope evidence

Yonggang Yang[1,2] **and Bojie Fu**[2]

[1]Institute of Loess Plateau, Shanxi University, Taiyuan, Shanxi, 030006, China
[2]Research Center for Eco-Environmental Sciences, Chinese Academy of Sciences, Beijing, 100085, China

Correspondence to: Bojie Fu (bfu@rcees.ac.cn)

Abstract. Soil water is an important driving force of the ecosystems, especially in the semiarid hill and gully region of the northwestern Loess Plateau in China. The mechanism of soil water migration in the reconstruction and restoration of Loess Plateau is a key scientific problem that must be solved. Isotopic tracers can provide valuable information associated with complex hydrological problems, difficult to obtain using other methods. In this study, the oxygen and hydrogen isotopes are used as tracers to investigate the migration processes of soil water in the unsaturated zone in an arid region of China's Loess Plateau. Samples of precipitation, soil water, plant xylems and plant roots are collected and analysed. The conservative elements deuterium (D) and oxygen (^{18}O) are used as tracers to identify variable source and mixing processes. The mixing model is used to quantify the contribution of each end member and calculate mixing amounts. The results show that the isotopic composition of precipitation in the Anjiagou River basin is affected by isotopic fractionation due to evaporation. The isotopic compositions of soil waters are plotted between or near the local meteoric water lines, indicating that soil waters are recharged by precipitation. The soil water migration is dominated by piston-type flow in the study area and rarely preferential flow. Water migration exhibited a transformation pathway from precipitation to soil water to plant water. δ^{18}O and δD are enriched in the shallow (< 20 cm depth) soil water in most soil profiles due to evaporation. The isotopic composition of xylem water is close to that of soil water at the depth of 40–60 cm. These values reflect soil water signatures associated with *Caragana korshinskii* Kom. uptake at the depth of 40–60 cm. Soil water from the surface soil layer (20–40 cm) comprised 6–12 % of plant xylem water, while soil water at the depth of 40–60 cm is the largest component of plant xylem water (ranging from 60 to 66 %), soil water below 60 cm depth comprised 8–14 % of plant xylem water and only 5–8 % is derived directly from precipitation. This study investigates the migration process of soil water, identifies the source of plant water and finally provides a scientific basis for identification of model structures and parameters. It can provide a scientific basis for ecological water demand, ecological restoration, and management of water resources.

1 Introduction

Water in the soil environment plays a crucial role as a carrier of dissolved and solid species and as a reservoir in the hydrological cycle. Soil water represents a small proportion (only 0.05 %) of the hydrological cycle, but it is vital for ecosystems and affects spatial and temporal processes at different scales (Koeniger et al., 2016; Busari et al., 2013). Understanding soil water migration in the unsaturated zone is essential to describe the movement of salt, carbon, nitrogen and other nutrients. Soil water migration plays an important role in the processes of infiltration, evaporation, transpiration and percolation, hydraulic conductivity and water uptake capacity of soils in the unsaturated zone. The traditional methods have been carried out to study the movement of soil water such as hydrologic experiments, intensive observations, modeling and remote sensing (Luo et al., 2013; Yang et al., 2013; Carucci et al., 2012). However, soil water migration is a complex nonlinear and inhomogeneous flow process. It is difficult to model on the basis of Darcy's law exclusively or using other techniques. Climatic conditions, soil texture and structure, antecedent moisture and vegetation cover exert influence on the movement of soil water (Bose et al., 2016; Kidron and Gutschick, 2013).

Isotopic tracers can provide valuable information on complex hydrological problems, such as the runoff processes, residence time, runoff pathway and the origin and contribution of each runoff component (Arny et al., 2013; Ohlanders et al., 2013; Yang et al., 2012a; McInerney et al., 2011; Maurya et al., 2011; Kevin et al., 2010). The isotopes of D and ^{18}O are conservative and do not react with clay minerals and other soil materials. D and ^{18}O are widely used to investigate ecological and hydrological processes. Stable isotopes can provide information about mixing, transport processes and residence time of water within a soil profile in the unsaturated zone.

Different flow mechanisms result in different isotopic profiles. Stable isotope compositions of soil water, plant xylems and precipitation can be obtained to identify soil water migration processes such as infiltration, evaporation, transpiration and percolation (Caley and Roche, 2013; Yang et al., 2012b; Catherne et al., 2010; Stumpp et al., 2009). Zimmerman et al. (1967) first applied stable isotopes to study the soil water profile, showing that evaporation at the surface of a saturated soil column causes deuterium enrichment near the surface that decreases exponentially with depth. Robertson and Gazis (2006) studied the seasonal trends of oxygen isotope composition in soil water fluxes at two sites along the climate gradient. The hydrogen and oxygen isotopes are used to study the transforming of precipitation, soil water and groundwater of typical vegetation in area of the Taihang Mountains (Song et al., 2009). Gazis and Feng (2004) compared the oxygen isotope compositions of precipitation and soil water from profiles at six sites with different soil textures. δD and δ^{18}O in soil water have been observed in many studies, including column experiments using sand or soil columns and field observations of unsaturated soils (Singh, 2013; Návar, 2011; Brodersen et al., 2015; Gehrels et al., 1998).

Stable isotopes provide evidence for assessing plant water sources according to their variations caused by equilibrium and kinetic isotopic fractionation mechanisms (Orlowski et al., 2016a; Haverd et al., 2011; Zhao et al., 2013; Bhatia et al., 2011). Several studies (Liu et al., 2013; Mathieu and Bariac, 1996; Allison and Barnes, 1985) have shown that transpiration do not cause isotopic fractionation of soil water. The isotopes δD and δ^{18}O can be a powerful tool to determine the water sources utilized by plants in the field, since δD and δ^{18}O values of xylem water reflect those of the water sources utilized (Gierke et al., 2016; Yang et al., 2011; Lu et al., 2011). The naturally occurring vertical gradients of δD and δ^{18}O in soil water provide similar information about plant water uptake depth from soils. A number of studies have used isotopes to characterize soil water movement in one location and thus one climatic regime (Orlowski et al.,

2016b; Manzoni et al., 2013; Liu et al., 2011; Stumpp et al., 2009).

Soil moisture is an important driving force of ecosystems, especially in the northwestern Loess Plateau. Characterized by dry climate, less precipitation, more evaporation and thicker soil layers, groundwater in this region is difficult to use due to the depth of water table. Thus, soil water is almost the only water resource in the study area, and has become the only factor controlling agricultural production and ecological restoration. It is necessary to investigate the mechanism of soil water migration in the Loess Plateau. Stable isotopes can provide valuable information on the mechanism of soil water migration. At present, much research focuses on the soil water recharge by precipitation and its isotope variation characteristics. There is some research on the interaction of soil water, plant water and precipitation. These studies are mostly on the individual scale in one specific region (Lin and Horita, 2016; Heathman et al., 2012; William and Eric, 2010; Ferretti et al., 2003). However, the research on the migration process of soil water, plant water, precipitation and groundwater in the unsaturated zone based on the isotopic technique is still rare. At present, the mechanism of soil water migration in the Loess Plateau is a key scientific problem to be solved.

Therefore, this study investigates soil water migration processes in the unsaturated zone in the hill and gully region of Loess Plateau using isotopes, integrated with sampling in the field, experimental observations and laboratory analysis. Samples include the soil water, plant xylem and root, etc. The objectives of this research are (1) to probe the migration process and variation of plant water, soil water and precipitation and (2) to identify each potential water source uptake by plant, and evaluate their contributions of each potential water source in the unsaturated zone. It can provide a scientific basis for ecological water demand, ecological construction, and management of water resources.

2   Materials and methods

2.1   The experimental site

The study area is located in the Anjiagou River basin in area of the city of Dingxi, China, at 34°26′–35°35′ N and 103°52′–104°39′ E. The climate is semiarid, with an average annual temperature of 6.3°. The annual mean precipitation is 420 mm. The mean annual evaporation is 1510 mm. The aridity index is 1.15. Precipitation is low and unevenly distributed temporally and spatially. This area is a part of the typical semiarid hill and gully region of the Loess Plateau, with altitudes ranging from 1900 to 2250 m.

The watershed area is 8.91 km^2, which belongs to the hill and gully region of Loess Plateau. Gully density is 3.14 km km^{-2}, and the ditch depth ranges from 30 to 50 m. The soil are of the yellow loessial and saline soil types, and

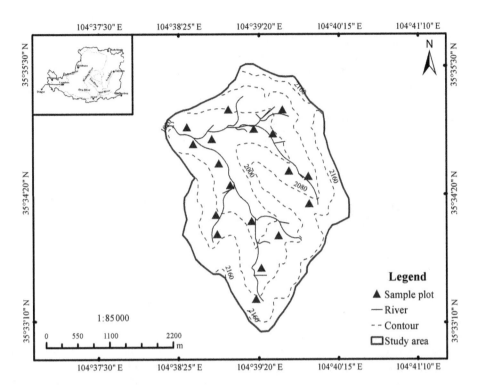

Figure 1. Location of the sampling sites in the study area.

the average thickness ranges from 40 to 60 m. The density of the soil layer ranges from 1.1 to 1.4 g cm^{-3}, average soil porosity is 55 %. Soil structure has a vertical joint, and the nature of soil is loose. The most dominant ecosystems in the Anjiagou River basin are grass and shrubland.

The study area has broken terrain and serious soil erosion, with an overall gully- and valley-filled landscape. Geological structure is the uplift zone between the eastern part of Qilian fold system and the western Qinling fold system, at an altitude of 1700–2580 m, with the gully density of 3–5 km km^{-2}, ditch slope of 5–10 %, and mountain slope of generally 20 to 50 %. The soil parent material is quaternary eolian loess, and the zonal soil mainly is yellow spongy soils, sierozem, which belongs to the typical semiarid loess hill and gully region. It has soft soil, homogeneous structure, thicker soil layers, good water performance and the widest distribution. The average thickness is 40–60 m. Clay soil is between 33.12–42.17 %, organic matter content is between 0.37–1.34 %, soil bulk density is 1.17 g cm^{-3}, wilting moisture content is 7.3 % and the saturated moisture content is 21.9 % at 0–20 cm. The soil bulk density is 1.09–1.36 g cm^{-3}, and the porosity is 50–55 % at the 2 m soil layer. The soil has vertical joint and strong collapsibility, so the soil erosion is easily happened, and the soil erosion modulus is 5000 t km^{-2} a^{-1}.

The vegetation type belongs to arid grassland vegetation type, with infrequent natural tree coverage. The grassland and shrubland ecosystems were the most extensive dominant ecosystems. Woodland area is minimal, most is open forest

land. Areas with crown density of greater than 0.2 have only *Caragana intermedia* and *Pinus tabulaeformis*. The vegetation is sparse, and species diversity is relatively poor because of long-term influence by human activities. The artificial forest vegetation mainly consists of *Caragana intermedia*, *Hippophae rhamnoides*, *Pinus tabulaeformis*, *Platycladus orientalis*, *Stipa bungeana*, etc.

2.2 Sample collection and field experiment

The research was carried out in the Anjiagou River basin in the hill and gully region of Loess Plateau from May 2013 to October 2015. Samples of precipitation, soil, xylem and root were collected in the study area. The locations of the sampling sites are shown in Fig. 1. Precipitation was collected after each rainfall event. Precipitation was filtered and transferred to sealed glass vials prior to analysis. Plant xylem and root of *Caragana korshinskii* Kom. were collected at each sampling site, respectively. For sap water analysis, a plant twig (0.5 to 1 cm in diameter and about 5 cm in length) was cut from a main branch or trunk, cortical and phloem were peeled off. All samples were kept in 8 mL glass bottles and sealed with packaging tape to reduce evaporation. Five glass bottles were collected each plant at each sampling site. All samples were collected at least monthly. Altogether 27 samples sites and 396 samples were collected, and sealed with paraffin. The samples were taken back to the laboratory and preserved at 4 °C, and samples of soil, plant xylem, plant stem and leaf were refrigerated at −20 °C until analysis.

Soil was sampled at 10 cm intervals for the first 40 cm, 20 cm intervals from 40 to 100 cm and 30 cm intervals from 100 to 130 cm. Maximum depths of sampling ranged up to 130 cm (plant root is rarely found below 100 cm in the study area). At each sampling site, soil moisture (volumetric soil water content) was obtained with time domain reflectometry (TDR) in the field manually at 0–10, 10–20, 20–30, 30–40, 40–60, 60–80, 80–100 and 100–130 cm. Soil moisture content was determined by oven drying simultaneously.

2.3 Laboratory analysis

Water was extracted from soil, xylem, root, stem and leaf by cryogenic vacuum distillation method, and the extracted water (2 to 10 mL) was trapped at liquid nitrogen temperature. Vacuum distillation was considered a reliable and acceptable method. The moisture in the soil or plants under the condition of vacuum (vacuum below 60 mTorr), was heated to 105° after evaporation. Water vapor of evaporation in −50° (liquid nitrogen) was collected with frozen water collecting pipe (top-down frozen, in order to increase the collecting rate), and the extraction precision was within δD and δ^{18} are ±3 and ±0.3‰, respectively. Water samples were filtered through 0.2 μm Millipore membrane for trace elements analyses. Isotopes $\delta^{18}O$ and δD were measured in the Key Laboratory of Ecohydrology and River Basin Science of Cold and Arid Region Environmental and Engineering Institute, Chinese Academy of Sciences. $\delta^{18}O$ of samples was analyzed by a Euro-PyrOH elemental analyzer at a temperature of 1300 °C, and δD was analyzed at a temperature of 1030 °C. At both temperatures the reaction products were analyzed on a GV Isoprime continuous-flow isotope ratio mass spectrometer. In order to eliminate the memory effect of online and continues flow, every sample was analyzed six times. The analytical precision of δD calculations was ±1‰, and that of $\delta^{18}O$ ±0.2‰. Isotopic concentration was expressed as δ per thousand (‰) relative to the Vienna Standard Mean Ocean Water, according to the follow equation:

$$\delta^{18}O(\text{‰}) = \frac{\left(^{18}O/^{16}O\right)_{\text{sample}} - \left(^{18}O/^{16}O\right)_{\text{SMOW}}}{\left(^{18}O/^{16}O\right)_{\text{SMOW}}} \times 1000,$$

$$\delta D(\text{‰}) = \frac{(D/H)_{\text{sample}} - (D/H)_{\text{SMOW}}}{(D/H)_{\text{SMOW}}} \times 1000.$$

2.4 Statistical analysis

When n isotope is used to determine the proportional contributions of $i + 1$ sources to a mixture, standard linear mixing models can be used to mathematically solve for the unique combination of source proportions that conserves mass balance for all n isotopes (Phillips et al., 2013). In this study, the conservative elements D and ^{18}O are utilized to calculate mixing amounts of potential water source. The mixing model is used to calculate the contributions of each respective end member. The mixing model is described on the basis of the mass balance equation:

$$\delta D_p = \sum_{i=1}^{n} f_i \delta D_i,$$

$$\delta^{18}O_p = \sum_{i=1}^{n} f_i \delta^{18}O_i,$$

$$I = \sum_{i=1}^{n} f_i,$$

where δD and $\delta^{18}O$ are concentrations of the tracers; I refers to potential water source; δD_p and $\delta^{18}O_p$ refer to plant waters; f_i is the fraction of each component contributing to plant water.

3 Results

3.1 Isotopic composition of precipitation

Yurtsever and Gat (1981) modified Craig's global water line, and made a more accurate global precipitation linear relationship between δD and $\delta^{18}O$ of $\delta D = 8.17\delta^{18}O + 10.56$, which is the global meteoric water line (GMWL). Figure 2 shows the relationship between mean $\delta^{18}O$ and mean δD in precipitation, which is defined as the local meteoric water line (LMWL) with the equation $\delta D = 7.41\delta^{18}O + 7.22$, with $R^2 = 0.95$. The $\delta^{18}O$ and δD compositions of the precipitation in the Anjiagou River catchment are extremely variable, ranging from −118.36 to −22.32‰ and from −16.35 to −2.19‰, respectively. Thirty-four precipitation samples were collected from May 2013 to October 2015. There are significant differences in the composition of D and ^{18}O under different precipitation events. The difference reflects the extreme nature of the climate and the complexity of moisture sources in the arid region. Compared with GMWL, the slope and intercept of LMWL are lower than that of GMWL (Fig. 2), indicating that they are affected by the local climate and environment with less precipitation and lower humidity. When the slope is low, the implication is that precipitation is subject to evaporation. The Anjiagou River basin is located in the northwestern inland region, and it is difficult for the sea water vapor to reach directly. This indicates that the significant imbalance of isotope dynamic fractionation exists, which are strongly affected by strong solar radiation and evaporation in the local environment, leading to the enrichment in $\delta^{18}O$ and δD. Therefore, the slope of LMWL is low, and the intercept of LMWL decreases with the slope.

3.2 The variations of δD and $\delta^{18}O$ in soil water vertical profiles

Stable isotope compositions of soil water are presented in Table 1. The measured δD and $\delta^{18}O$ of soil water range from −72.42 to −37.05‰ and −11.74 to −3.57‰, respectively. By comparing the isotopic composition of soil water with

Table 1. Statistic characteristics of isotopic compositions for precipitation and soil water in the study area.

Type	Depth (cm)	$\delta^{18}O$ (‰)					δD (‰)				
		n	Min	Max	Mean	SD	n	Min	Max	Mean	SD
Precipitation		34	−16.35	−2.19	−8.26	1.9	34	−118.36	−22.32	−55.15	5.7
Soil water	10	27	−10.29	−7.36	−9.73	0.7	27	−67.27	−49.70	−53.24	3.9
	20	27	−8.80	−3.57	−7.20	1.2	27	−72.71	−44.32	−48.32	4.1
	40	27	−11.35	−7.09	−10.36	0.6	27	−74.21	−61.53	−71.31	1.7
	60	27	−12.79	−9.75	−11.24	0.9	27	−86.91	−74.8	−76.80	2.4
	80	27	−11.53	−8.92	−10.35	0.5	27	−78.82	−57.22	−62.42	1.8
	100	27	−10.75	−8.34	−9.31	0.6	27	−78.11	−51.89	−64.89	1.5

that of precipitation, the isotope compositions of soil water in the unsaturated zone are relatively enriched. Therefore, it could be beneficial to identify the influence of various water bodies on different precipitation events. As Fig. 2 shows, isotope compositions of most soil waters are plotted in the bottom of the LMWL and are close to the LMWL. The variable trends of δD and $\delta^{18}O$ values of soil water are similar to that of precipitation. There is a strong linear relationship ($R^2 = 0.96$) for δD and $\delta^{18}O$ between precipitation and soil water, which indicates that soil water is originally recharged by precipitation. The soil water is increasingly enriched in D and ^{18}O, which are influenced by evaporation (Table 1). It has been reported that soil water is normally more enriched because precipitation enters the soil and mixes with antecedent soil water that has been modified by evaporation.

Vertical profiles of soil water δD and ^{18}O in the Anjiagou River basin are shown in Fig. 3. The surface layer (0–20 cm) showed larger variations and higher δD and ^{18}O values than the deeper layer. δD and ^{18}O values of soil water are greatest at the surface layer (Fig. 3, Table 1). δD and ^{18}O values of shallow soil water at 0–10 cm are −10.29 to −7.36‰ and −72.71 to −35.45‰, respectively. The δD and ^{18}O values of shallower soil water at 10–20 cm are also high (ranging from −8.08 to −3.57‰ and from −67.27 to −49.70‰, respectively). The net effect of evaporation is the enrichment of heavy isotopes near the soil surface. Soil profiles (Fig. 3) show the increase in the D–^{18}O with soil depths from 0 to 20 cm, suggesting the occurrence of evaporation. The large variation in isotope compositions of shallow depths (0–20 cm depth) may be caused by the precipitation and evapotranspiration effect. The correlation of the isotopic composition between soil water and precipitation is weak. These have been reported by many other studies (Gazis and Feng, 2004; Robertson and Gazis, 2006).

The $\delta^{18}O$ and δD compositions of soil water at 40 cm depths range from −11.35 to −7.09‰ and −74.21 to −54.06‰, respectively. Soil water from 20 to 40 cm depths is easily influenced by precipitation infiltration and evapotranspiration, which leads to the extremely variable δD and

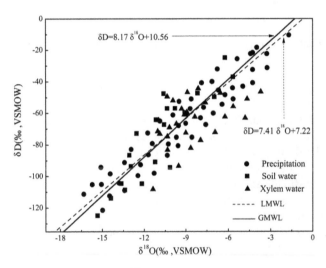

Figure 2. Plot of δD vs. $\delta^{18}O$ for the various samples in the study area.

$\delta^{18}O$ values of soil water (Fig. 3). The impact of precipitation on soil water becomes weaker with increasing depths. The isotopic composition of soil water is strongly enriched in δD and $\delta^{18}O$ due to less precipitation and evaporation in the study area. The δD and $\delta^{18}O$ decrease exponentially with the increasing depths, and the variation trends are larger. Previous studies have also shown that evaporation decreases with the increasing soil depth, and its influence is generally within tens of centimeters. δD and ^{18}O values declined with depth to 60 cm but remained constant at deeper depths. When it reaches 60 cm depth, the variation trend tends to be stationary. The $\delta^{18}O$ and δD compositions of soil water at 60 cm depths ranged from −12.79 to −9.75‰ and from −96.91 to −57.27‰, respectively. There is an abrupt peak value in the plant root zone. The isotopic fluctuations quickly disappear due to the dispersive effects of the deep root system.

However, the values of δD and $\delta^{18}O$ gradually decrease from 60 to 80 cm because the recharge of precipitation decreases, and the influence of antecedent isotope value of soil water increases. As Fig. 3 shows, D and ^{18}O values of soil water at depths of 80 cm range from −11.53 to −8.92‰ and

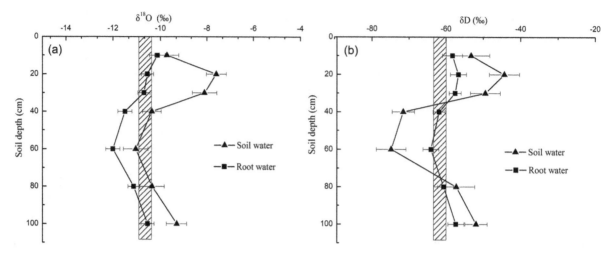

Figure 3. Isotope profiles of $\delta^{18}O$ and δD (piston-type flow) in soil water and xylem water. Values are mean $\pm$ 1 standard deviation ($n = 27$).

−78.82 to −62.80‰, respectively. Compared with the large variation in the δD and $\delta^{18}O$ of surface soil water, soil water deeper than 80 cm is almost constant. The low value zone is at 100 cm in the isotopic profiles of soil water, which is the deepest layer that is recharged by precipitation. Variations of $\delta^{18}O$ and δD of soil water in the deep layer are non-significant. Evaporation is larger than precipitation in the study area. Precipitation is difficult to infiltrate to 80 cm depth in the shrubland ecosystems, except in the high flow years. Therefore, it is difficult to cause the isotope fractionation of soil water blow 80 cm.

3.3 Variability in soil moisture content

Values of soil moisture obtained by TDR in the Anjiagou River basin are shown in Fig. 5. The variation of soil moisture content in the shallow soil layer (0–40 cm) is extremely large, ranging from 5.94 to 17.51 %, which belongs to the highly variable layer. The variation of soil moisture content at the 40–60 cm soil depth is relatively small, which belongs to the less active layer. The variation of soil moisture content is relatively stable at 60–100 cm soil depth, ranging from 17.78 to 29.47 %, which belongs to the relatively stable layer. Soil water content also exhibits depth variation. Variability of water content is larger in the surface horizons more than 40 cm depth in the soil profile. The shallow soil layers are impacted by evaporation and precipitation recharge more than the deeper soil layer. Therefore, the high evapotranspiration rate and precipitation recharge are the main factors controlling soil moisture, especially at the surface horizon.

As the soil moisture profile shows (Fig. 4), soil moisture content is low in the shallow layer (0–20 cm) and relatively high at the active layer (40–60 cm), and then tends to be stable. Figure 4 shows that soil moisture content is different in different sample sites, but soil moisture content is low under 60 cm soil depths in all the sample sites. Soil moisture con-

tent generally increases with the increase of soil depth in the variable greatly layer. Soil moisture content decreases with the increase of soil depth in the active layer, and then tends to be stable.

As Fig. 4 shows, water content increases and soil water $\delta^{18}O$–D values increase in the shallow layer (0–40 cm), which are impacted by evapotranspiration. This is the dominant process for most of the sites. Mean values of the soil moisture content increase with depth in the soil profile (40–60 cm depth), while soil water $\delta^{18}O$–D values decrease with depth, where it mixes with antecedent moisture. Low water content and more positive ^{18}O–D values (Fig. 4) occur at the deep soil layer (below 60 cm depth), which is impacted by recharge. Single precipitation events cannot easily infiltrate deeper than 80 cm, when drought lasts for a long time, especially when soil water content is lower than 20 %. Extremely high values of soil water content at the depth of 60 cm are observed in the Anjiagou River basin. The higher water content and more negative $\delta^{18}O$ values are observed in the active layer (40–60 cm; Fig. 4). The variability of water content and the ^{18}O–D values are relatively larger for the surface horizons than deeper in the soil profile. The profile of soil moisture content is highly responsive to evaporation and precipitation. The average water content and soil water $\delta^{18}O$–D values are representative of the trends found in most of the sample sites.

4 Discussion

4.1 The migration process of soil water in the unsaturated zone

Precipitation infiltration and evapotranspiration in the vertical direction are the main processes of the water cycle, playing an important role in the transformation process of the water cycle in the unsaturated zone. Different migration mecha-

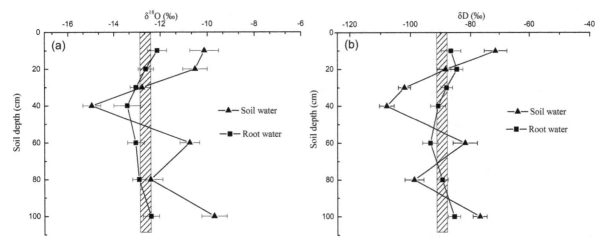

Figure 4. Isotope profiles of $\delta^{18}O$ and δD (preferential flow) in soil water and xylem water. Values are mean ± 1 standard deviation ($n = 27$).

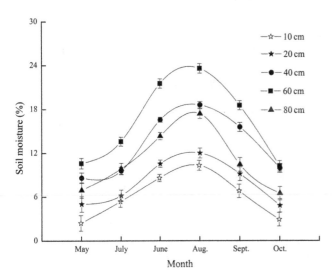

Figure 5. The variation of soil moisture content at different soil depths.

nisms of soil water result in different isotopic profiles. Therefore, isotope δD and $\delta^{18}O$ of soil water and precipitation can be obtained to identify soil water migration processes such as infiltration, evaporation, transpiration and percolation through the soil waters at different depths. These different isotopic profiles may occur as a result of several processes, including evaporation, change in isotopic composition of precipitation and mixing of new and old water. While evaporation is widely recognized, we believe that mixing is also an important factor for controlling the isotopic composition of soil water at these sites. This is possibly caused by the infiltration of summer rainwater through the soil layer. Generally, there are two major infiltration mechanisms (piston-type flow and preferential flow) that can be identified by comparing the isotope compositions of precipitation and soil water.

4.1.1   Evidence for piston-type flow

The δD and $\delta^{18}O$ content of soil water at different depths are shown in Fig. 4. By comparing the isotopic compositions of precipitation and soil water, the mechanisms of soil water movement can be identified. AT most of the sampling sites in the Anjiagou River basin (except at sites on 20 July 2014 and 24 September 2014), the isotopic composition of soil water in shallow layer are enriched in D and ^{18}O due to evaporation. The isotopic composition of soil water changes with depth in the soil profile, especially after continuous precipitation between 18 and 20 August 2014. Precipitation infiltrated and completely mixed with free water. The vertical trend in the δD and $\delta^{18}O$ profiles of soil water is simple. There is an abrupt peak value in the isotopic profiles, which suggests that the older water is pushed downward by the new water and infiltrates to the deep soil. That is, while some soil water remains stationary, the mobile soil water successively displaces pre-existing mobile soil water, pushing it downward. Piston-type flow after an isotopic distinct rainfall event results in an abrupt "isotopic front" within the soil (Gehrels et al., 1998; Song et al., 2009). The evidence for mixing is that abrupt changes in the soil water δD and $\delta^{18}O$ with depth in the soil profile indicates that piston flow has occurred. A relatively flat pattern is observed, slightly enriched in accordance with precipitation. δD and $\delta^{18}O$ declined with depth to 60 cm but remained constant at deeper depths. The older soil water is pushed downward by newer rainwater. Precipitation infiltration exhibits piston infiltration characteristics (Zimmerman et al., 1967).

4.1.2   Evidence for preferential flow

At the sample sites on 20 July 2014 and 24 September 2014, the vertical trend of δD and $\delta^{18}O$ in the soil profile indicates a uniform infiltration process (Fig. 5). There are two abrupt peak values in the isotopic vertical profiles of soil water. The soil water at depths between 20 and 40 cm decreases quickly, caused by the strong transpiration effect of plants roots. Abrupt change occurred at 20–40 cm depth of the isotopic profile where the soil water moisture increased quickly (ranging from −14.97 to −12.78 ‰and −107.93 to −101.21 ‰, respectively). There are two continuous rainfall events on 20 July 2014 and 24 September 2014 (the rainfall event on 20 July 2014, with precipitation of 25 mm, lasted for 3 days. The rainfall event on 24 September 2014, with the precipitation of 16 mm, lasted for 2 days), which infiltrated uniformly into deep depths, as shown by the reducing δD and δ^{18}O values with time after the precipitation (Fig. 4). The data do not show the obvious correlation with depth at the 20 July 2014 and 24 September 2014 samples sites in the Anjiagou River basin (Fig. 4). This is mainly caused by the mixture effect occurred during the process of infiltration. That is to say, the newer water, which did not replace the old water totally during the process of infiltration, mixed with the old water stored in the soil.

The second peak of δ^{18}O and δD values of soil water is at the depth of 80 cm (ranging from −12.41 to −10.75 ‰and −98.87 to −82.77 ‰, respectively), caused most likely by macropore flow. The non-existence of a significant difference between the isotopic content of precipitation and soil water suggests that the infiltrating summer precipitation passes the root zone quickly and contributes to the soil water recharge. The deep soil water is recharged by precipitation in the form of preferential flow, which passes the soil porosity and quickly reaches the deep layer, and does not mix with older water. Variations of δ^{18}O and δD of deep soil water are only due to mixing, rarely from evaporation. A number of studies documenting tracer movement in the soil column have pointed out the existence of preferential fluid flow, which can be caused by cracks (in dried soil), macropores (Beven and Germann, 1982), plant roots, earthworm burrows, etc. (Vincent et al., 2001; Bronswijk, 1991). However, the preferred flow is rarely found in the Loess Plateau, except from macropores, caused by plant root system or animal invasion, etc. The deep soil water is quickly recharged by precipitation in the form of preferential flow through observing the excavated soil profile. The macropores can be found in the 100–200 cm soil layer in the study area, which are created by plant roots. The cracks in the Loess Plateau can also provide an important path for preferential flow. This is supported by the variations of soil water content. The soil water content at 80 cm depth increases quickly. The soil water content ranges from 34.5 to 36.8 %. At corresponding depths of the isotope profile, the isotope values are relatively depleted. Significant soil water content increase in the deep soil is similar to variability of isotopic composition of soil water. The isotope profile can provide some evidence for preferential flow.

4.2 Origin and contribution of each potential water source to plants

4.2.1 Plant water sources

Zimmermann et al. (1967) studied the effect of transpiration with an experiment in which the root water uptake of various plant species is monitored, and no fractionation is found. Allison and Barnes (1985) tested a much greater fraction of root water uptake and found no conclusive proof of fractionation by root water uptake. Isotope fractionation is not caused when soil water is taken up by plant roots. By comparing the δD and δ^{18}O of plant water and each water sources, the water origin and contribution of each component for xylem could be identified.

The isotopic composition of the soil layer is similar to the xylem water, confirmed by comparing δD and δ^{18}O of xylem water and soil water, which reflect the signatures of soil at the depth of soil water uptake by plants. All the soil and xylem samples are taken simultaneously in the Anjiagou River basin. The isotopes of δD and δ^{18}O are conservative and do not react with clay minerals and other soil materials. During the process of transpiration, the soil water content changes, but the isotopic composition of soil water remains constant. The isotopic composition of plant xylem water is a mixture of soil water at different soil depths. The isotopic composition of xylem water, soil water and precipitation in the Anjiagou River basin is analyzed. The isotopic composition of plant xylem water is a mixture of soil water at different soil depths. Comparing the signatures δD and δ^{18}O of both soil water and xylem water with a simple graphical interference approach indicates a distinct difference between both isotopes. Graphical "best match" approach is suitable to illustrate water uptake depth as plants often withdraw water from more than one distinct soil depth. The δD and δ^{18}O in xylem water ranged from −11.64 to −7.95 ‰and from −81.62 to −55.21 ‰, respectively (Fig. 6, Table 2). Plant xylem water δD and δ^{18}O for both species is similar to that of soil water at 40–60 cm depth (Fig. 6). It indicates that soil water at the 40–60 cm depth is mainly used by *Caragana korshinskii* Kom.

4.2.2 The contribution of each potential water source to plants water

Hooper (2003) and Christophersen and Hooper (1992) introduced the end-member mixing analysis (EMMA), which is an often used method for analyzing possible source area contributions to flow. Multiple-source mixing models (Parnell et al., 2010; Phillips et al., 2013) account for water uptake from more than one discrete soil layer and weigh the importance of certain layers for water uptake by incorporating soil water potentials into the calculation. In this study, the mixing model is used to identify potential source areas and mixing processes, and to quantify the contribution of each end mem-

Table 2. Statistic characteristics of isotopic compositions for root water and xylem water in the study area.

Type	Depth (cm)	$\delta^{18}O$ (‰)					δD (‰)				
		n	Min	Max	Mean	SD	n	Min	Max	Mean	SD
Xylem water		27	−11.64	−7.95	−10.61	0.66	27	−81.62	−55.21	−69.94	4.9
Root water	10	27	−14.21	−5.12	−8.34	0.64	27	−62.39	−52.99	−62.37	5.4
	20	27	−10.27	−5.97	−9.21	1.5	27	−56.21	−44.18	−59.38	3.3
	40	27	−11.79	−6.91	−10.1	0.89	27	−73.20	−61.56	−67.05	4.2
	60	27	−12.47	−8.08	−10.6	0.77	27	−83.49	−65.95	−73.14	2.9
	80	27	−11.99	−8.92	−9.7	0.81	27	−81.18	−57.72	−66.75	1.7
	100	27	−9.80	−7.3	−8.9	0.64	27	−70.57	−59.27	−60.62	2.5

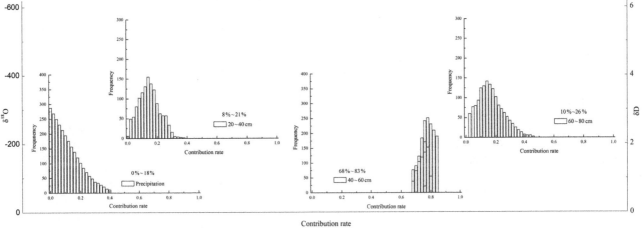

Figure 6. The contributions of each potential water source to plants.

ber using mixture fractions. The conservative elements δD and $\delta^{18}O$ are utilized as tracers to calculate mixing amounts.

It is more likely that plant xylem water is a mixture of soil water from several soil depths. Multiple-source mass-balance mixing models of proportional contributions (%) of plant xylem water (Parnell et al., 2010) is used to evaluate the contributions of each potential water source (Fig. 6). The results show that soil water from the surface horizons (20–40 cm) comprised 8–21 % of plant xylem water, while soil water at 40–60 cm soil depth comprised the largest portion of plant xylem water (ranging from 68 to 83 %). Soil water below 60 cm depth comprised 10–26 % of plant xylem water and only 0–18 % comes directly from precipitation. Water source is dominated by soil water at a depth of 40–60 cm.

5   Conclusions

The following conclusions can be drawn from the present study:

1. $\delta^{18}O$ and δD enrichment in the shallow (< 20 cm depth) soil water, observed in most soil profiles, is due to evaporation. The isotopic composition of xylem water

is close to that of soil water at a depth of 40–60 cm. These values reflect soil water signatures associated with shrubland uptake at a depth of 40–60 cm.

2. A sharp isotopic front at approximately 40 cm depth observed shortly after an isotopic distinct rainfall event suggests that infiltration into soil occurred as piston-type flow with newer water pushing older water downward in the soil profile. Soil waters are recharged from precipitation. The soil water migration is dominated by piston-type flow in the study area and rarely preferential flow, except where there are macrospores in the Loess Plateau, caused by plant root or animal invasion, etc. Water migration exhibited a transformation pathway from precipitation to soil water to plant water.

3. Soil water from the surface horizons (20–40 cm) comprised 8–21 % of plant xylem water, while soil water at 40–60 cm soil depth was the is the largest component of plant xylem water (ranging from 68 to 83 %). Soil water below 60 cm depth comprised 10–26 % of plant xylem water and only 0–18 % comes directly from precipitation. Water source is dominated by soil water at depth of 40–60 cm.

Competing interests. The authors declare that they have no conflict of interest.

Acknowledgements. This research is supported by National Natural Science Foundation of China (41390464;41201043), and China Postdoctoral Science Foundation Funded Project (2014M550095). The authors are grateful to the experimental research station staff and all participants in the field for their contributions to the progress of this study. We also express our appreciation to the anonymous reviewers of the manuscript.

Edited by: B. Hu

References

Allison, G. B. and Barnes, C. J.: Estimation of evaporation from the normally "dry" Lake Frome in South Australia, J. Hydrol., 78, 229–242, 1985.

Arny, E. S., Hans, C. S., Thorsteinn, J., and Sigfus, J. J.: Monitoring the water vapor isotopic composition in the temperate North Atlantic, J. Geophys. Res.-Abstr., 15, 2013–5376, 2013.

Beven, K. and Germann, P.: Macropores and water flow in soils, Water Resour. Res., 18, 1311–1325, 1982.

Bhatia, M. P., Das, S. B., and Kujawinski, E. B.: Seasonal evolution of water contributions to discharge from a Greenland outlet glacier: insight from a new isotope-mixing model, J. Glaciol., 57, 929–941, 2011.

Bose, T., Sengupta, S., Chakraborty, S., and Borgaonkar, H.: Reconstruction of soil water oxygen isotope values from tree ring cellulose and its implications for paleoclimate studies, Quatern. Int., 425, 387–398, 2016.

Brodersen, C., Pohl, S., Lindenlaub, M., Leibundgut, C., and Wilpert, K. V.: Influence of vegetation structure on isotope content of throughfall and soil water, Hydrol. Process., 14, 1439–1448, 2015.

Bronswijk, J. J. B.: Magnitude, modelling and significance of swelling and shrinkage processes in clay soils, PhD Thesis, 1991.

Busari, M. A., Salako, F. K., Tuniz, C., and Zuppi, G. M.: Estimation of soil water evaporative loss after tillage operation using the stable isotope technique, Int. Agrophys., 27, 257–264, 2013.

Caley, T. and Roche, D. M.: δ^{18}O water isotope in the iLOVE-CLIM model (version 1.0) – Part 3: A palaeo-perspective based on present-day data–model comparison for oxygen stable isotopes in carbonates, Geosci. Model Dev., 6, 1505–1516, doi:10.5194/gmd-6-1505-2013, 2013.

Carucci, V., Petitta, M., and Aravena, R.: Interaction between shallow and deep aquifers in the Tivoli Plain enhanced by groundwater extraction: A multi-isotope approach and geochemical modeling, Appl. Geochem., 27, 266–280, 2012.

Catherne, C. G., Alan, F. M., and Katharine, J. M.: Hydrological processes and chemical characteristics of low-alpine patterned wetlands, south-central New Zealand, J. Hydrol., 385, 105–119, 2010.

Christophersen, N. and Hooper, R. P.: Multivariate analysis of stream water chemical data: the use of components analysis for the end-member mixing problem, Water Resour. Res., 28, 99–107, 1992.

Ferretti, D. F., Pendall, E., Morgan, J. A., Nelson, J. A., LeCain, D., and Mosier, A. R.: Partitioning evapotranspiration fluxes from Colorado grassland using stable isotopes: Seasonal variations and ecosystem implications of elevated atmospheric CO_2, Plant Soil, 254, 291–303, 2003.

Gazis, C. and Feng, X.: A stable isotope study of soil water: evidence for mixing and preferential flow paths, Geoderma, 119, 97–111, 2004.

Gehrels, J. C., Peeters, E. M., De Vries, J. J., and Dekkers, M.: The mechanism of soil water movement as inferred from ^{18}O stable isotope studies, Hydrolog. Sci. J., 43, 579–594, 1998.

Gierke, C., Newton, B. T., and Phillips, F. M.: Soil-water dynamics and tree water uptake in the Sacramento Mountains of New Mexico: a stable isotope study, Hydrogeol. J., 24, 1–14, 2016.

Haverd, V., Cuntz, M., and Griffith, D.: Measured deuterium in water vapour concentration does not improve the constraint on the partitioning of evapotranspiration, using a soil vegetation atmosphere transfer model, Agr. Forest Meteorol., 151, 645–654, 2011.

Heathman, G. C., Cosh, M. H., Merwade, V., and Han, E.: Multiscale temporal stability analysis of surface and subsurface soil moisture within the Upper Cedar Creek Watershed, Indiana, Catena, 95, 91–103, 2012.

Hooper, R. P.: Diagnostic tools for mixing models of stream water chemistry, Water Resour. Res., 39, 249–256, 2003.

Kevin, W. T., Brent, B. W., and Thomas, W. D. E.: Characterizing the role of hydrological processes on lake water balances in the old Crow Flats, Yukon Territory, using water isotope tracers, J. Hydrol., 386, 103–117, 2010.

Kidron, G. J. and Gutschick, V. P.: Soil moisture correlates with shrub-grass association in the Chihuahuan Desert, Catena, 107, 71–79, 2013.

Koeniger, P., Gaj, M., Beyer, M., and Himmelsbach, T.: Review on soil water isotope-based groundwater recharge estimations, Hydrol. Process., 30, 2817–2834, 2016.

Lin, Y. and Horita, J.: An experimental study on isotope fractionation in a mesoporous silica-water system with implications for vadose-zone hydrology, Geochim. Cosmochim. Ac., 184, 257–271, 2016.

Liu, H., Tian, F., Hu, H. C., Hu, H. P., and Sivapalan, M.: Soil moisture controls on patterns of grass green-up in Inner Mongolia: an index based approach, Hydrol. Earth Syst. Sci., 17, 805–815, doi:10.5194/hess-17-805-2013, 2013.

Liu, Y., Xu, Z., Duffy, R., Chen, W., An, S., Liu, S., and Liu, F.: Analyzing relationships among water uptake patterns, rootlet biomass distribution and soil water content profile in a subalpine shrubland using water isotopes, Eur. J. Soil Biol., 47, 380–386, 2011.

Lu, N., Chen, S. P., Wilske, B., Sun, G., and Chen, J.: Evapotranspiration and soil water relationships in a range of disturbed and undisturbed ecosystems in the semi-arid Inner Mongolia, China, J. Plant Ecol., 4, 49–60, 2011.

Luo, G. J., Kiese, R., Wolf, B., and Butterbach-Bahl, K.: Effects of soil temperature and moisture on methane uptake and nitrous oxide emissions across three different ecosystem types, Biogeosciences, 10, 3205–3219, doi:10.5194/bg-10-3205-2013, 2013.

Manzoni, S., Giulia, V., Gabriel, K., Sari, P., and Robert, B. J.: Hydraulic limits on maximum plant transpiration and the emergence of the safety–efficiency trade-off, New Phytol., 198, 169–178, 2013.

Mathieu, R. and Bariac, T.: An isotopic study (^{2}H and ^{18}O) of water movements in clayey soils under a semiarid climate, Water Resour. Res., 32, 779–789, 1996.

Maurya, A. S., Shah, M., and Deshpande, R. D.: Hydrograph separation and precipitation source identification using stable water isotopes and conductivity: River Ganga at Himalayan foothills, Hydrol. Process., 25, 1521–1530, 2011.

McInerney, F. A., Helliker, B. R., and Freeman, K. H.: Hydrogen isotope ratios of leaf wax nalkanes in grasses are insensitive to transpiration, Geochim. Cosmochim. Ac., 75, 541–554, 2011.

Návar, J.: Stemflow variation in Mexico's northeastern forest communities: Its contribution to soil moisture content and aquifer recharge, J. Hydrol., 408, 35–42, 2011.

Ohlanders, N., Rodriguez, M., and McPhee, J.: Stable water isotope variation in a Central Andean watershed dominated by glacier and snowmelt, Hydrol. Earth Syst. Sci., 17, 1035–1050, doi:10.5194/hess-17-1035-2013, 2013.

Orlowski, N., Pratt, D. L., and Mcdonnell, J. J.: Intercomparison of soil pore water extraction methods for stable isotope analysis, Hydrol. Process., 30, 3434–3449, 2016a.

Orlowski, N., Breuer, L., and McDonnell, J. J.: Critical issues with cryogenic extraction of soil water for stable isotope analysis, Ecohydrology, 9, 3–10, 2016b.

Parnell, A. C., Inger, R., Bearhop, S., and Jackson, A. L.: Source Partitioning Using Stable Isotopes: Coping with Too Much Variation, Plos One, 5, e9672, doi:10.1371/journal.pone.0009672, 2010.

Phillips, S. J., Anderson, R. P., and Schapire, R. E.: Maximum entropy modeling of species geographic distribution, Ecol. Model., 19, 231–259, 2013.

Robertson, J. A. and Gazis, C. A.: An oxygen isotope study of seasonal trends in soil water fluxes at two sites along a climate gradient in Washington state (USA), J. Hydrol., 328, 375–387, 2006.

Singh, B. P.: Isotopic composition of water in precipitation in a region or place, J. Appl. Radiat. Isot., 75, 22–25, 2013.

Song, X. F., Wang, S. Q., Xiao, G. Q., Wang, Z. M., Liu, X., and Wang, P.: A study of soil water movement combining soil water potential with stable isotopes at two sites of shallow groundwater areas in the North China Plain, Hydrol. Process., 23, 1376–1388, 2009.

Stumpp, C., Maloszewski, P., and Stichler, W.: Environmental isotope (δ^{18}O) and hydrological data to assess water flow in unsaturated soils planted with different crops, J. Hydrol., 36–, 198-208, 2009.

Vincent, M., Didon-Lescot, J. F., and Couren, M.: Investigation of the hydrological processes using chemical and isotopic tracers in a small Mediterranean forested catchment during autumn recharge, J. Hydrol., 247, 215–229, 2001.

William, T. P. and Eric, E. S.: The influence of spatial patterns of soil moisture on the grass and shrub responses to a summer rainstorm in a Desert ecotone, Ecosystems, 13, 511–525, 2010.

Yang, Y. G., Xiao, H. L., Wei, Y. P., and Zou, S. B.: Hydrologic processes in the different landscape zones in the alpine cold region during the melting period, J. Hydrol., 409, 149–156, 2011.

Yang, Y. G., Xiao, H. L., Wei, Y. P., and Zou, S. B.: Hydrochemical and hydrological processes in the different landscape zones of alpine cold region in China, Environ. Earth Sci., 65, 609–620, 2012a.

Yang, Y. G., Xiao, H. L., Wei, Y. P.. and Zou, S. B.: Hydrological processes in the different landscape zones of alpine cold regions in the wet season, combining isotopic and hydrochemical tracers, Hydrol. Process., 25, 1457–1466, 2012b.

Yang, Y. G., Xiao, H. L., and Zou, S. B.: Hydrogen and oxygen isotopic records in monthly scales variations of hydrological characteristics in the different landscape zones, J. Hydrol., 499, 124–131, 2013.

Yurtsever, Y. and Gat, J. R.: Atmospheric waters, in: Stable Isotope Hydrology: Deuterium and Oxygen-18 in the Water Cycle, Technical Report Series, IAEA, Vienna, 103–142, 1981.

Zhao, L. J., Wang, L. X., and Xiao, H. L.: The effects of short-term rainfall variability on leaf isotopic traits of desert plants in sand-binding ecosystems, Ecol. Eng., 60, 116–125, 2013.

Zimmermann, U., Ehhalt, D., and Muennich, K. O.: Soil–water movement and evapotranspiration: Changes in the isotopic composition of the water, International Atomic Energy Agency, Vienna, 567–585, 1967.

10

Understanding runoff processes in a semi-arid environment through isotope and hydrochemical hydrograph separations

V. V. Camacho Suarez[1], **A. M. L. Saraiva Okello**[1,2], **J. W. Wenninger**[1,2], **and S. Uhlenbrook**[1,2]

[1]Department of Water Science and Engineering, UNESCO-IHE Institute for Water Education, P.O. Box 3015, 2601 DA Delft, the Netherlands
[2]Section of Water Resources, Delft University of Technology, P.O. Box 5048, 2600 GA Delft, the Netherlands

Correspondence to: V. V. Camacho Suarez (viviancamacho@gmail.com)

Abstract. The understanding of runoff generation mechanisms is crucial for the sustainable management of river basins such as the allocation of water resources or the prediction of floods and droughts. However, identifying the mechanisms of runoff generation has been a challenging task, even more so in arid and semi-arid areas where high rainfall and streamflow variability, high evaporation rates, and deep groundwater reservoirs may increase the complexity of hydrological process dynamics. Isotope and hydrochemical tracers have proven to be useful in identifying runoff components and their characteristics. Moreover, although widely used in humid temperate regions, isotope hydrograph separations have not been studied in detail in arid and semi-arid areas. Thus the purpose of this study is to determine whether isotope hydrograph separations are suitable for the quantification and characterization of runoff components in a semi-arid catchment considering the hydrological complexities of these regions. Through a hydrochemical characterization of the surface water and groundwater sources of the catchment and two- and three-component hydrograph separations, runoff components of the Kaap catchment in South Africa were quantified using both isotope and hydrochemical tracers. No major disadvantages while using isotope tracers over hydrochemical tracers were found. Hydrograph separation results showed that runoff in the Kaap catchment is mainly generated by groundwater sources. Two-component hydrograph separations revealed groundwater contributions of between 64 and 98 % of total runoff. By means of three-component hydrograph separations, runoff components were further separated into direct runoff, shallow and deep groundwater components. Direct runoff, defined as the direct precipitation on the stream channel and overland flow, contributed up to 41 % of total runoff during wet catchment conditions. Shallow groundwater defined as the soil water and near-surface water component (and potentially surface runoff) contributed up to 45 % of total runoff, and deep groundwater contributed up to 84 % of total runoff. A strong correlation for the four studied events was found between the antecedent precipitation conditions and direct runoff. These findings suggest that direct runoff is enhanced by wetter conditions in the catchment that trigger saturation excess overland flow as observed in the hydrograph separations.

1   Introduction

Understanding runoff processes facilitates the evaluation of surface water and groundwater risks with respect to quality and quantity (Uhlenbrook et al., 2002). It assists in quantifying water resources for water allocation, hydropower production, design of hydraulic structures, environmental flows, drought and flood management, and water quality purposes (Blöschl et al., 2013). The need for understanding runoff processes has led to the development of tools such as hydrograph separation techniques that identify runoff components in streamwater, flowpaths, residence times and contributions to total runoff (Klaus and McDonnell, 2013; Hrachowitz et al., 2009; Weiler et al., 2003). Several hydrograph separation studies using environmental isotopes and geochemical tracers have been carried out in forested, semi-humid environments, which have led to new insights into runoff processes in these areas (e.g., Pearce et al., 1986; Bazemore et al., 1994;

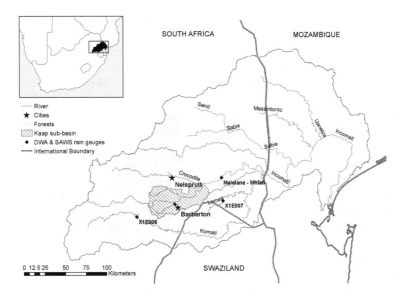

Figure 1. Location of the Kaap catchment in the Incomati Basin displaying nearby cities, and DWA and SAWS rain gauges.

Tetzlaff and Soulsby, 2008; Uhlenbrook et al., 2002; Burns et al., 2001). But there is still a need for understanding runoff generation mechanisms in tropical, arid and semi-arid areas, as they were much less investigated (Burns, 2002).

Studying runoff processes in arid and semi-arid regions may be a challenging task due to the high temporal and spatial variability of rainfall, high evaporation rates, deep groundwater resources, poorly developed soils, and in some cases the lack of surface runoff (Blöschl et al., 2013; Hrachowitz et al., 2011; Wheater et al., 2008). Although these challenges may not be applicable in various instances (e.g., a reduced vegetation cover results in less importance of the interception process), arid and semi-arid regions may still face extra difficulties due to their remoteness and financial constraints.

Arid and semi-arid regions are characterized by their sporadic, high-energy, and low-frequency precipitation occurrence (Camarasa-Belmonte and Soriano, 2014; Wheater et al., 2008). Dry spells can last for years, and rain events may vary from a few millimeters to hundreds of millimeters per year. High-intensity storms may generate most if not all the season's runoff (Love et al., 2010; Van Wyk et al., 2012). These events can also increase erosion, reduce soil infiltration capacity and enhance surface runoff (Camarasa-Belmonte and Soriano, 2014). On the contrary, the lack of precipitation may result in reduced to non-existent groundwater recharge. Compared to humid regions, where evaporation is generally limited by the amount of energy available, evaporation in arid and semi-arid areas is usually limited by the water availability in the catchment (Wang et al., 2013). Evaporation becomes the dominant factor in driving the hydrology of arid and semi-arid areas. Understanding the impact of evaporation on stream runoff processes becomes more complex due to the spatial variability of vegetation. An

increase in vegetation cover due to a wetter rainfall season may result in higher evaporation rates, reduced streamflow and increased soil infiltration capacity (Hughes et al., 2007; Mostert et al., 1993). Transmission losses through the stream channel bed may also reduce the total runoff and increase the volume of recharged groundwater. This occurrence is evident in the entire Incomati Basin, where downstream areas (e.g., Mozambique) benefit from transmission losses and return flows of upstream areas (Nkomo and van der Zaag, 2004; Sengo et al., 2005).

This paper explores the runoff processes, including surface–groundwater interactions in the Kaap catchment, South Africa, by describing the spatial hydrochemical characterization of the catchment, separating the runoff components through isotope and geochemical tracer analysis, and determining the suitability of isotopic tracers for the characterization of runoff components in the catchment.

2 Study area

The Kaap catchment is located in the northeast of South Africa in the Mpumalanga province and has an area of approximately 1640 km^2 (Fig. 1). Nelspruit, the provincial capital, and Barberton are the closest urban areas to the Kaap, with populations of approximately 125 000 and 35 000 inhabitants, respectively (GRIP, 2012). The study area is predominantly located in the low-elevation sub-tropical region of South Africa, Swaziland and Mozambique known as the Lowveld region, with elevations ranging from 300 to 1800 m a.s.l. (above sea level). The average slope is 18 %.

The geology dates back to Archean times. Biotite granite is the predominant formation in the valley (Fig. 2c). Headwater streams originate on the weathered granite, whose felsic properties indicate high concentrations of dissolved silica. Surrounding granite, lava formations or Onverwacht formations contain basaltic and peridotitic komatiite, which are low in silicates and high in magnesium. The Onverwacht formation is one of the oldest formations in the area. Formed in an ocean, it is rich in quartz, volcanic rocks and chert horizons (de Wit et al., 2011). Sandstones and shales are found in closer proximity to the Kaap River and at the southern section of the catchment. In addition to the gneiss formation observed near the outlet, other formations include ultramafic (high in iron and low in silicates) rocks, quartzite and dolomite (Sharpe et al., 1986). Borehole logs near the upper Suidkaap and Noordkaap tributaries displayed a top layer of weathered granite (approximately 25 to 37 m in depth) followed by a thinner, less fractured granite layer and hard rock granite. Borehole logs analyzed in closer proximity to the catchment outlet presented more diverse formations including layers of clay, sand, greywacke and weathered shale.

Bushveld and grasslands are the predominant land cover types in the Kaap Valley, covering up to 68 % of the catchment (Fig. 2b). In the upstream region (western part of the catchment), approximately 25 % of the total catchment consists of pine and eucalyptus plantations used for paper and timber production. Sugar cane, citrus trees, and other cash crops are found in the downstream region where many diversion channels for irrigation are present. Irrigation demand in the Kaap catchment is approximately 56 mm a^{-1} (Mallory and Beater, 2009).

The climate is semi-arid with cool dry winters and hot wet summers. Precipitation ranges from 583 to 1243 mm a^{-1} (WRC, 2005) with an average annual precipitation of 742 mm a^{-1}, and a mean annual runoff coefficient of 0.14. Between 2001 and 2012, recorded minimum and maximum daily air temperatures at the Barberton meteorological station ranged from 3 to 42 °C, with a long-term average of 20 °C (SASRI, 2013). The wet season lasts typically from October to March, as shown in Fig. 3, where the climate diagram shows the monthly averages (from 2001 to 2012) of precipitation, pan A evaporation from four weather stations (X1E006, X1E007, Barberton, and Malelane) and maximum and minimum temperatures at the Barberton station. A dry season is observed from May to September. Class-A-pan evaporation rates largely exceed precipitation during most parts of the year. The range of long-term potential evapotranspiration (PET) shown in Fig. 2f for the catchment (1950–2000) is between 1500 and 1900 mm a^{-1} (Schulze, 1997). The PET data show that most of the catchment is semiarid, according to the UNEP definition (UNEP, 1997), as illustrated in Fig. 2e (aridity index = mean annual precipitation/mean annual potential evaporation). However, according to the Köppen–Geiger classification, the catchment is subtropical.

The Kaap catchment contains three main tributaries: the Queens, the upper Suidkaap, and the Noordkaap. The highest monthly average flow during the year occurs in February,

with an average of 9.2 m^3 s^{-1}. The lowest monthly flow during the year occurs at the end of the dry season in September, reaching an average of 0.8 m^3 s^{-1}. Minimum and maximum daily average flows recorded between 1961 and 2012 at the Kaap outlet range from 0 to 483 m^3 s^{-1}. The long-term mean flow at the outlet is 3.7 m^3 s^{-1}, which is equivalent to 55 mm a^{-1}.

Although analytical methods for hydrograph separation have been carried out in the Kaap River, no accurate estimations of runoff components were retrieved in the area. Thus, this paper also provides a baseline for understanding surface and groundwater dynamics in the Incomati trans-boundary river system. The Kaap River is a major contributor of flow to the Crocodile River, which flows into the Incomati transboundary river. The Incomati waters are shared by South Africa, Swaziland and Mozambique, where tensions related to the management of water resources have led to the development of water-sharing agreements such as the Tripartite Interim Agreement on Water Sharing of the Maputo and Incomati Rivers (Van der Zaag and Carmo Vaz, 2003). The need for reliable data and understanding of the hydrological functioning of the system has been highlighted in these agreements (Slinger et al., 2010). In addition, the Kaap River and the neighboring catchments experienced devastating floods in February 2000 and March 2014 with return periods exceeding 200 years (Smithers et al., 2001).

3 Data and methods

3.1 Long-term data sets

Hydrological data in the catchment, including precipitation, evaporation, streamflow and groundwater records, were collected from the Department of Water Affairs (DWA), the South African Weather Service (SAWS), the South African Sugarcane Research Institute (SASRI), and the In-Situ Groundwater Consulting firm (http://www.insituconsulting.co.za). Geological, topographical and land use GIS (geographic information systems) data were obtained from the 2005 Water Research Commission study (WRC, 2005).

The average catchment precipitation was obtained by studying seven weather stations with daily rainfall data from 2001 to 2012. Only four stations were selected based on data availability and proximity to the catchment. These stations were X1E006, X1E007, Barberton and Malelane (Fig. 1). Missing rainfall values for Barberton (2 %) and X1E007 (33 %) were estimated by regression analysis. Malelane and X1E006 did not contain missing data. Using a Thiessen polygon distribution, the average rainfall was calculated for the catchment.

Average actual evaporation was calculated from daily Class-A-pan evaporation values from the Barberton and Malelane stations and daily Class-S-pan evaporation from X1E006 and X1E007 stations from 2003 to 2012. Daily pan

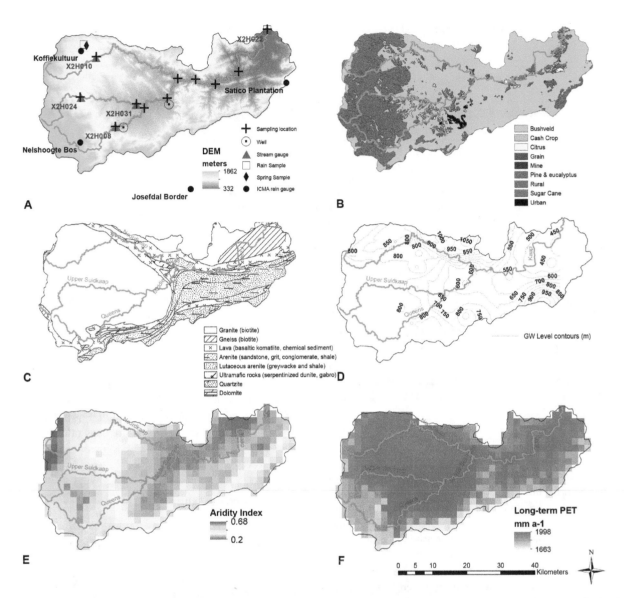

Figure 2. (**a**) Digital elevation model (DEM) of the Kaap catchment with sampling locations and stream and rain gauge locations, (**b**) land use map, (**c**) geological map, (**d**) contour map of static groundwater levels, (**e**) aridity index (< 0.03 hyper arid, 0.03–0.2 arid, 0.2–0.5 semi-arid, 0.5–0.65 dry sub-humid, > 0.65 humid) and (**f**) long-term mean potential evapotranspiration (PET). GIS layers are courtesy of the Water Research Commission (2005), South Africa.

evaporation values were aggregated to monthly pan evaporation values. Class-S-pan evaporation was converted to Class-A-pan evaporation following the WR90 Water Resources of South Africa study (Midgley et al., 1994). Class-A evaporation was converted to reference evaporation using the guidelines for crop water requirements (Allen et al., 1998) and reference evaporation was corrected for the specific land uses using data from the land satellite imagery collected from the Incomati Water Availability Assessment Study (Mallory and Beater, 2009). Using a long-term water balance from 2003 to 2012, actual mean evaporation rates were found.

To analyze the streamflow response at the outlet and tributaries, daily discharges at the X2H022 (outlet), X2H008

(Queens), X2H031 and X2H024 (Suidkaap) and X2H010 (Noordkaap) stream gauges were obtained from the DWA. The locations of the stations are shown in Fig. 2a.

3.2 Field and laboratory methods

3.2.1 General

A field campaign from 20 November 2013 to 4 February 2014 was carried out to obtain an overview of the hydrochemistry of the catchment prior to the rainy season and to collect data for hydrograph separation studies.

Table 1. UNESCO-IHE laboratory equipment used in chemical analysis of Kaap catchment samples.

	Parameter(s) analyzed	Equipment	Number of samples	Preservation method	Analytical uncertainty (σ)
Environmental isotopes	$^{18}O, ^2H$	LRG DLT-100 isotope analyzer	116	None	$\pm 0.2, \pm 1.5$ (‰)
Cations	$Ca^{2+}, Mg^{2+}, Na^+, K^+$	Thermo Fisher Scientific XSeries 2 ICP-MS	116	Nitric acid (HNO_3)	± 0.2 (mg L^{-1})
Anions	$Cl^-, NO_3^- - N, SO_4^{2-}$ PO_4^{3-}	Dionex ICS-1000	116	Refrigerated at $< 4\,^\circ C$	± 0.2 (mg L^{-1})

Stream discharge collected from DWA data loggers (water levels converted to stream discharge using the DWA rating curve) were retrieved at the outlet with a frequency of 12 min (0.2 h) from 30 October 2013 to 17 February 2014. Hourly precipitation rates were obtained from the Incomati Catchment Management Agency (ICMA) rain gauges at Koffiekultuur, Nelshoogte Bos, Satico, and the Josefdal border from 1 October 2013 to 28 February 2014 (see locations in Fig. 2a).

3.2.2 Water samples

Water samples were collected from the tributaries, the main river, one spring, and two drinking water wells as shown in Fig. 2a. Each location was sampled twice during dry weather conditions. Each sample of approximately 250 mL was collected in polyethylene bottles, rinsed three times before the final sample was taken to avoid contamination, and refrigerated for sample preservation. Electrical conductivity (EC), pH and temperature were measured in situ using a Wissenschaftlich-Technische-Werkstätten (WTW) conductivity meter.

3.2.3 Rain sampling

To obtain the isotopic and hydrochemical reference of rainfall, bulk rain samples were collected in the upstream and downstream parts of the catchment. The rain samplers were constructed according to standards of the International Atomic Energy Agency (IAEA) to avoid re-evaporation (Gröning et al., 2012). Thus, an average of upstream and downstream samples per rain event was used for the rainfall end-member concentrations for each hydrograph separation.

Rainfall characteristics, including duration, total rain amount, and maximum and average intensity, and the Antecedent Precipitation Index (API) were estimated for each rain event. A rainfall event was defined as a rainfall occurrence with rainfall intensity greater than 1 mm h^{-1} and in-

termittence less than 4 h, as observed in a similar study in a semi-arid area by Wenninger et al. (2008). The API for n days prior to the event was calculated using Eq. (1):

$$API_{-n} = \sum_{i=1}^{7} P_{(-n-1+i)}(0.1i), \quad (1)$$

where P in mm h^{-1} stands for precipitation and i is the number corresponding to the day of rainfall. For this study, APIs were calculated for the 7, 14, and 30 days prior to the event. Peak flow, runoff depth, and time to peak were determined for each event.

3.2.4 Automatic sampler

During the 2013–2014 rainy season, four events that occurred on 12–13 December (Event 1), 28–30 December (Event 2), 13 January (Event 3), and 30–31 January (Event 4) were sampled using an automatic sampler manufactured by the University of KwaZulu-Natal (UKZN). The first two events were sampled on a volume basis obtaining 22 samples for Event 1 and 5 samples for Event 2 (a smaller number of samples were obtained for Event 2 due to photo sensor failure in the automatic sampler). Events 3 and 4 were sampled using a time-based strategy obtaining 13 samples for Event 3, and 36 samples for Event 4. A total volume of approximately 100 mL was obtained for each sample.

3.2.5 Chemical analysis of water samples

All samples were refrigerated, filtered, and analyzed for HCO_3 and Cl using a Hach Digital Titrator, and SiO_2 using a Hach DR890 Portable Colorimeter, within 48 h. Then, samples were transported to the UNESCO-IHE laboratory in the Netherlands for further chemical analysis. The samples were analyzed for major anions, cations, and stable isotopes as listed in Table 1.

3.3 Data analysis

3.3.1 Groundwater analysis

Groundwater chemical data for 240 boreholes and 18 borehole logs were obtained from In-Situ Groundwater Consultants covering the different geological formations (granite, lava, arenite, and gneiss). For 27 out of the 240 boreholes, pH, $CaCO_3$, Mg, Ca, Na, K, Cl, NO_3–N, F, SO_4, SiO_2, Al, Fe, and Mn data were available. The remaining boreholes only had information on EC, static water table depth, and physical characteristics of the borehole.

Borehole chemical data were classified according to the geological formations. The classified data distribution was observed using GIS, and basic statistical analysis was carried out to determine the control of geology over the hydrochemistry of groundwater.

To gain better insights with regard to groundwater flow, groundwater contour lines were created using an inversed distance-weighted (IDW) interpolation of the static water tables from the boreholes.

3.3.2 End-member mixing analysis (EMMA)

Suitable parameters for hydrograph separation were identified by creating mixing diagrams of EC (μS cm^{-1}), SiO_2, $CaCO_3$, Cl, SO_4, Na, Mg, K, Ca (in mg L^{-1}) and δ^2H and δ^{18}O (‰ VSMOW). Parameters were plotted against discharge to observe dilution and hysteresis effects. A principal component analysis was carried out based on the method described by Christophersen and Hooper (1992). Only non-statistically correlated parameters were used. From these, the possibility of three end-members was explored. The three runoff components identified were direct runoff, deep groundwater and shallow groundwater. Direct runoff was defined according to the conceptual model by Uhlenbrook and Leibundgut (2000) where direct runoff (or the quick runoff component) was generated from direct precipitation on the stream channel, and overland flow from sealed and saturated areas and from highly fractured outcrops. The deep groundwater component was considered to be the portion of runoff generated from deeper highly weathered granite aquifers, and the shallow groundwater component was considered to be the intermediate component from perched groundwater tables.

The mixing plot for δ^2H and K is presented in Fig. 11. The direct runoff end-member was characterized by the upstream and downstream rain samples. Potassium was used as an indicator of the shallow groundwater component due to the main sources of potassium, which are the weathering of minerals from silicate rocks, application of fertilizers, and the decomposing of organic material. The mobilization of potassium is linked to the flushing of the soil and shallow subsurface layers of vegetated areas. This was also observed by Winston and Criss (2002). The direct runoff samples had a low K average (0.5 mg L^{-1}) and depleted δ^{18}O and

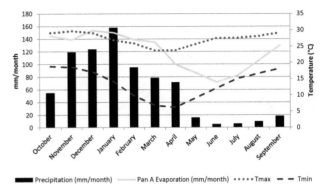

Figure 3. Average monthly precipitation, pan-A evaporation (from stations X1E006, X1E007, Barberton and Malelane), and maximum and minimum temperatures at Barberton station from 2001 to 2012.

δ^2H values (-4.8‰ for δ^{18}O; -27.5‰ for δ^2H). A spring sample was used to characterize the deep groundwater component that contained more enriched δ^{18}O and δ^2H values (-0.9‰ for δ^{18}O; -2.2‰ for δ^2H) and low K concentration (0.7 mg L^{-1}). The shallow groundwater end-member was estimated considering the high K concentrations (4 mg L^{-1}) and slightly less depleted δ^{18}O and δ^2H (-3.5‰ for δ^{18}O; -7.0‰ for δ^2H) observed in the stream samples. The error interval for the direct runoff in Fig. 11 is $\pm$ the standard deviation of the rain samples. For the groundwater end-members, the error intervals were estimated as ± 10 % of the measured values. While these errors are arbitrary, they were chosen as they are more conservative than the alternative analytical errors of ± 0.2 mg L^{-1} for K and ± 1.5‰ for δ^2H and because there were no additional samples from which to derive the standard deviation.

3.3.3 Hydrograph separation

Isotope and hydrochemical data were combined with discharge data to perform a multi-component hydrograph separation based on steady-state mass balance equations as described, for instance, in Uhlenbrook et al. (2002). The number of tracers ($n - 1$) was dependent on the number of runoff components (n). Equations (3) and (4) were applied in dividing the total runoff, Q_T, into two and three runoff components.

$$Q_T = Q_1 + Q_2 ... + Q_n, \tag{2}$$

$$c_T Q_T = c_1 Q_1 + c_2 Q_2 ... + c_n Q_n, \tag{3}$$

where Q_1, Q_2 and Q_n are the runoff components in m^3 s^{-1} and c_T, c_1, c_2 and c_n are the concentrations of total runoff, and runoff components.

3.3.4 Uncertainty estimation

Uhlenbrook and Hoeg (2003) showed that during the quantification of runoff components, uncertainties due to tracer

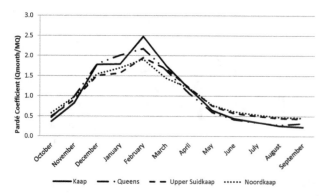

Figure 4. Annual flow regimes at X2H022 (outlet), X2H008 (Queens), X2H031 and X2H024 (Suidkaap) and X2H010 (Noordkaap) based on long-term flow data.

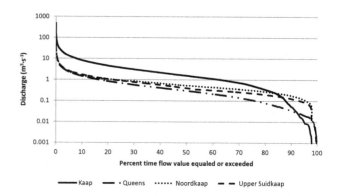

Figure 5. Flow duration curves for X2H022 (outlet), X2H008 (Queens), X2H031 and X2H024 (Suidkaap) and X2H010 (Noordkaap) based on long-term flow data.

and analytical measurements, intra-storm variability, elevation and temperature, solution of minerals, and the spatial heterogeneity of the parameter concentrations occur. For the Kaap River hydrograph separations, these uncertainties were accounted for by the spatial hydrochemical characterization of the catchment and by sampling rainfall during each event and at different locations. Moreover, tracer end-members and analytical uncertainties were estimated using a Gaussian error propagation technique and a confidence interval of 70 % as described by Genereux (1998) and Liu et al. (2004).

$$W = \left\{ \left[\frac{\partial y}{\partial x_1} W_{x1} \right]^2 + \left[\frac{\partial y}{\partial x_2} W_{x2} \right]^2 + \ldots + \left[\frac{\partial y}{\partial x_n} W_{xs} \right]^2 \right\}^{\frac{1}{2}} \quad (4)$$

W is the estimated uncertainty of each runoff component (e.g., direct runoff, shallow and deep groundwater components). W_{x1} and W_{x2} are the standard deviations of the end-members. W_{xs} is the analytical uncertainty and the partial derivatives $\frac{\partial y}{\partial x_1}$, $\frac{\partial y}{\partial x_2}$ and $\frac{\partial y}{\partial x_n}$ are the uncertainties of the runoff component contributions with respect to the tracer concentrations.

4　Results

4.1　Hydrology, hydrogeochemistry and groundwater flow

One of the characteristics of semi-arid areas is the high variability of flows. This large variability is observed at the Kaap outlet and tributaries (Table 2), where the highest and lowest flows recorded at the Kaap outlet are 483 and $0\,\mathrm{m}^3\,\mathrm{s}^{-1}$, respectively. Pardé coefficients (Fig. 4) reflect the seasonal flow behavior showing the dominance of one rainy season per hydrological year with the largest flows occurring in February. Moreover, the flat slopes observed at the upper end of the flow duration curves (Fig. 5) are evidence of groundwater storage areas located in the upstream part of the catchment.

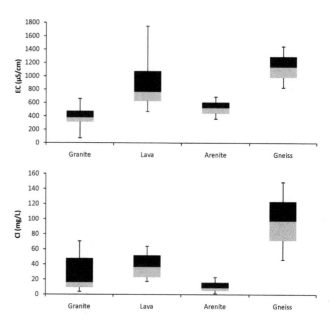

Figure 6. Boxplots of borehole water quality parameters at different geological locations in the Kaap catchment.

The variability of the catchment's groundwater quality parameters was studied from borehole data. Electrical conductivities in the granite region had the lowest electrical conductivity (EC) values (average $383\,\mu\mathrm{S\,cm}^{-1}$), while the gneiss formation, near the outlet, had the largest EC average of $1140\,\mu\mathrm{S\,cm}^{-1}$. Lava and arenite formations had mean EC values of 938 and $525\,\mu\mathrm{S\,cm}^{-1}$, respectively. The gneiss and lava formations had higher concentration averages of chloride and calcium carbonate than the granite and arenite formations. These can be seen in the boxplots in Fig. 6.

Groundwater contour lines followed the topographical relief. The highest water tables were observed at the northern boundary of the catchment, with water tables up to $1150\,\mathrm{m}$ (Fig. 2d). From the groundwater contour map, it was observed that groundwater moves toward the stream, indicating a gaining river system. Time series data from boreholes did

Table 2. Physical and hydrological characteristics of the Kaap tributaries and outlet.

Tributary name	Kaap outlet	Queens	Upper Suidkaap	Noordkaap
Station ID	X2H022	X2H008	X2H031	X2H010
Reach length (km)	45.7	41.2	42.5	57.5
Sub-basin area (km^2)	1640	291	256	315
Data analyzed	1961–2012	1949–2012	1967–2012	1970–2012
Period (years)	51	63	45	42
% data missing	5%	0%	3%	6%
Highest flow measured HHQ ($m^3 s^{-1}$)	483	96	123	28
Lowest flow measured NNQ ($m^3 s^{-1}$)	0	0	0	0
Mean of yearly highest flows MHQ ($m^3 s^{-1}$)	65	13	19	6
Mean of yearly lowest flows MNQ ($m^3 s^{-1}$)	0.4	0.1	0.3	0.2
Mean flow MQ ($m^3 s^{-1}$)	3.6	0.6	1.1	0.6
Variability ratio	180	186	65	31
Specific discharge ($L s^{-1} Km^{-1}$)	3.0	2.2	4.2	1.9
Maximum and average days of no flow per year	139; 8	12; 1	23; 1	17; 1

not show a significant change in water tables due to seasonal or long-term changes.

4.2 Spatial hydrochemical characterization

The upstream rain sample average had a more depleted isotopic signature (−5.1‰ for $\delta^{18}O$; −30.2‰ for δ^2H) than the lower-elevation rain sample average (−4.4‰ for $\delta^{18}O$; −24.7‰ for δ^2H). Upstream and downstream delta deuterium values ranged from a minimum of −30.2‰ to a maximum of −21.8‰ and delta oxygen-18 ranged from −5.14 to −3.72‰. Baseflow at the catchment outlet (X2H022) was characterized by analyzing DWA long-term water quality data and by field sampling prior to the 2013–2014 rainy season. Results from the field sampling are shown in Table 3.

The upper section of the catchment, mainly dominated by granite, is characterized by low to moderate electrical conductivities. Long-term mean electrical conductivities (sampled monthly by the DWA from 1984 to 2012) for the upper Suidkaap and Noordkaap tributaries were 75 and $104 \mu S cm^{-1}$, respectively. On the contrary, the catchment outlet had a higher long-term average EC of $572 \mu S cm^{-1}$ (DWA long-term monthly average from 1977 to 2012).

4.3 Rainfall–runoff observations

Table 4 summarizes the rainfall–runoff observations for the four studied events. The events had distinctive characteristics showing large variability in peak flows, API, rainfall duration, rain depth and maximum and average intensities. Event 1 had the highest peak flow at $124 m^3 s^{-1}$, while Event 3 had the smallest peak flow at $6.5 m^3 s^{-1}$. APIs, especially API_{-7}, differed from very wet conditions during Event 1 (39 mm) to very dry conditions (1 mm) during Event 2. Event 1 was a relatively short event (7 h) with high

Table 3. List of mean values of hydrochemical parameters obtained during the 2013–2014 field campaign.

Parameter	Location			
	Suidkaap	Queens	Noordkaap	Outlet
EC ($\mu S cm^{-1}$)	84.0	128.7	92.9	443.0
SiO_2 ($mg L^{-1}$)	22.4	17.0	20.9	24.1
$CaCO_3$ ($mg L^{-1}$)	38.5	59.5	41.3	154.0
Cl ($mg L^{-1}$)	3.8	3.6	2.8	15.5
SO_4 ($mg L^{-1}$)	1.8	4.1	1.6	47.2
Na ($mg L^{-1}$)	7.5	7.1	7.3	29.3
Mg ($mg L^{-1}$)	2.8	7.4	3.7	25.3
Ca ($mg L^{-1}$)	7.9	9.1	6.8	27.6
δ^2H (‰ VSMOW)	−12.1	−12.4	−12.7	−8.9
$\delta^{18}O$ (‰ VSMOW)	−3.2	−3.1	−3.5	−2.7

antecedent precipitation conditions and high rain intensities generating the largest amount of runoff at the outlet. In contrast, Event 3 was a short event with average rain intensity that generated the lowest peak flow.

4.4 Response of isotopes and hydrochemical parameters

During the storm events, most hydrochemical parameters (EC, Ca, Mg, Na, SiO_2 and Cl) and water isotopes (δ^2H and $\delta^{18}O$) showed dilution responses, except for potassium (Fig. 7). The first flood was the largest event sampled, reaching a peak flow of $124 m^3 s^{-1}$ where a larger contribution of direct runoff was observed. In this event, a larger degree of dilution of the sampled hydrochemical parameters is also observed. The following events had smaller peak flows of 27.6, 6.5, and $7.1 m^3 s^{-1}$ for Events 2, 3, and 4, respectively. Thus, smaller dilution effects were observed for Events 2, 3, and 4.

Table 4. Rainfall–runoff relationships observed during the 2013–2014 wet season for the Kaap catchment at the outlet (X2H022 stream gauge) and average precipitation from Koffiekultuur, Nelshoogte Bos, Satico, and the Josefdal border rain stations.

		Event 1	Event 2	Event 3	Event 4
Runoff	Peak flow time and date	13 Dec 2013, 18:24	30 Dec 2013, 06:12	16 Jan 2014, 03:48	31 Jan 2014, 17:00
	Maximum river depth (m)	2.0	1.0	0.5	0.5
	Peak flow ($m^3\,s^{-1}$)	124.0	27.6	6.5	7.1
	Runoff volume (mm)	3.2	2.6	0.1	0.4
	Time to peak after rainfall started (h)	24.4	31.2	60.8	22.0
Rainfall	Rain start date and time	12 Dec 2013, 18:00	28 Dec 2013, 23:00	13 Jan 2014, 15:00	30 Jan 2014, 19:00
	Rain duration (h)	7	39	7	26
	Rain depth (mm)	24	78	17	20
	Average rain intensity ($mm\,h^{-1}$)	3.4	2.0	2.5	0.8
	Maximum rain intensity ($mm\,h^{-1}$)	9.8	12	5	10
	Antecedent Precipitation Index API_{-7} (mm)	38.7	1.3	7.8	24.9
	API_{-14} (mm)	118.1	12.8	20.0	67.9
	API_{-30} (mm)	390.2	220.8	192.4	223.8

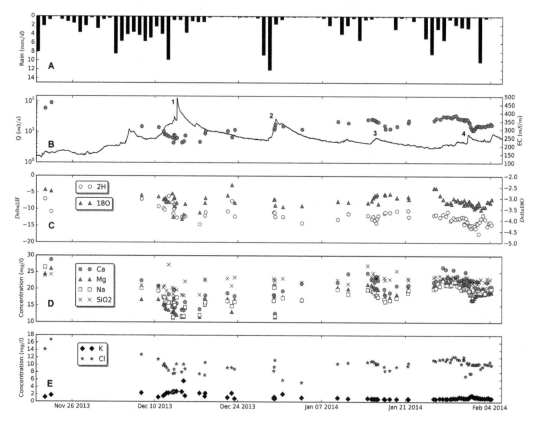

Figure 7. Kaap catchment: (**a**) average precipitation in mm day^{-1}, (**b**) discharge at the outlet in $m^3\,s^{-1}$ and electrical conductivity $\mu S\,cm^{-1}$, (**c**) delta deuterium and delta oxygen-18 in ‰ VSMOW, (**d**) calcium, magnesium, sodium and silica concentrations at the outlet in mg L^{-1}, and (**e**) chloride and potassium concentrations at the outlet in mg L^{-1}.

The smaller peak flows and lower direct runoff contributions for the latter events may explain the temporal variability observed in the increased concentrations of the hydrochemical parameters over time. During Event 1, EC's initial value of $317\,\mu S\,cm^{-1}$ decreased to $247\ \mu S\,cm^{-1}$ during peak flow. Similarly, CaCO$_3$ and SiO$_2$ decreased from

115 to 82 mg L^{-1} and from 21.1 to 19.6 mg L^{-1}, respectively. $\delta^{18}O$ ($-2.9‰$) and δ^2H ($-7.0‰$) decreased to -3.2 and $-12.6‰$, respectively. Potassium concentrations increased from 1.3 to 2.8 mg L^{-1}. For Event 2, a smaller number of samples were collected due to malfunctions of the automatic sampler. However, dilution of SiO$_2$ and Cl, and an increase

Table 5. Percentages of direct runoff (DR) and groundwater (GW) contributions and 70 % uncertainty percentages (W) from two-component hydrograph separations for the 2013–2014 wet season, Kaap catchment, South Africa.

Tracer	Event 1			Event 2			Event 3			Event 4		
	DR	GW	W	DR	GW	W	DR	GW	W	DR	GW	W
EC	22	78	6.8	5	95	7.9	6	94	7.0	27	73	4.2
SiO_2	21	79	2.6	6	94	2.5	12	88	2.2	21	79	2.6
$CaCO_3$	29	71	6.3	9	91	6.9	6	94	6.8	24	76	4.6
Mg	22	78	5.6	13	87	6.0	8	92	5.3	24	76	4.0
^{18}O	23	77	8.6	8	92	3.3	10	90	3.1	36	64	12.4
^{2}H	19	81	5.6	5	95	15.0	2	98	19.4	21	79	24.9

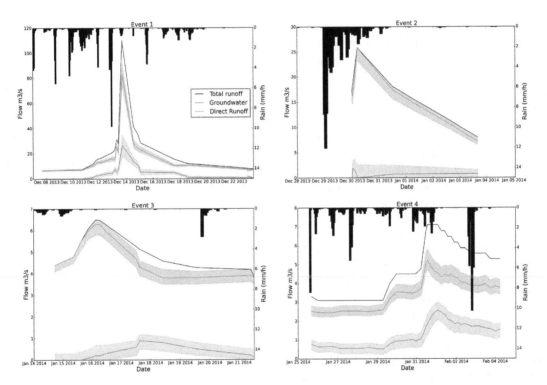

Figure 8. Two-component hydrograph separations using electrical conductivity as a tracer. Events 1 and 4 had a larger direct runoff contribution coinciding with the total runoff peak. Events 2 and 3 had a smaller direct runoff contribution.

in potassium concentrations, were observed. Events 3 and 4 were smaller events, but a smaller sampling interval showed the same dilution behavior of the sampled parameters and the increase in potassium concentrations.

4.5 Two-component hydrograph separation

Event and pre-event components were separated using $\delta^{18}O$ and $\delta^{2}H$, and direct runoff and groundwater were separated using EC, SiO^2, $CaCO_3$, and Mg. For simplicity, the two-component hydrograph separation components in this study are referred to as direct runoff and groundwater components. Direct runoff (quick-flow component), defined in the methods section as the portion of direct precipitation and infiltration excess overland flow, was characterized using the

rain samples collected upstream and downstream inside the catchment. Groundwater end-members were obtained from the initial streamwater samples before the rainfall started. Events 1 and 4 had the largest contributions of direct runoff among the four events, accounting for 29 % in the case of Event 1 and up to 36 % for Event 4 (Table 5). Events 2 and 3 had lower direct runoff contributions ranging from 5 to 13 % for Event 2 and 2 to 12 % for Event 3. Figure 8 shows the two-component hydrograph separations for the four events.

4.6 Isotope hydrograph separation versus hydrochemical hydrograph separation

Hydrochemical tracers usually separate runoff from source areas, while isotopes generally separate old water from new

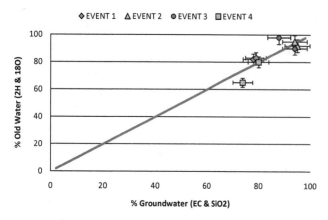

Figure 9. Percentages of groundwater and old water contributions using environmental isotopes (δ^2H and δ^{18}O) and hydrochemical (EC and SiO$_2$) tracers.

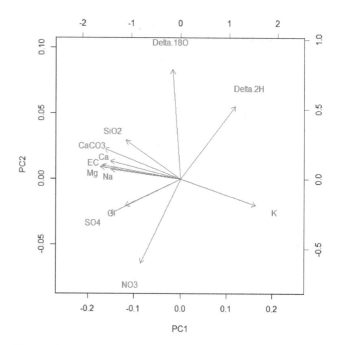

Figure 10. Biplot of principal components generated during PCA of streamwater samples using EC, SiO$_2$, CaCO$_3$, Cl, NO$_3$–N, SO$_4$, Na, Mg, K, Ca, δ^2H, and δ^{18}O.

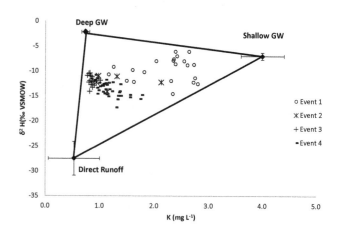

Figure 11. Mixing diagram of δ^2H and K showing streamwater samples at the outlet for four rain events during the 2013–2014 wet season.

water. The definition of Klaus and McDonnell (2013) was used for this study, stating that pre-event water (or old water as referred to in this section of the study) is the water stored in the catchment before the rainfall event. This component may not be representative of deep groundwater sources, but it may be water stored from the same rainfall season but from previous rainfall events. Thus, a comparison between "old-water" and "groundwater" components obtained during the four events was carried out to investigate to what extent these components are similar. This allowed us to determine the suitability of isotopic hydrograph separations versus hydrochemical separations for semi-arid environments. Figure 9 presents the percentages of groundwater and old-water contributions using environmental isotopes (δ^2H and δ^{18}O) and hydrochemical (EC, SiO$_2$, CaCO$_3$, and Mg) tracers for the four investigated events. It is noted that Events 1 and 4 have smaller contributions of groundwater than Events 3 and 4. During Event 4 and Event 2, old water resembles groundwater. The data points above the line present instances where old water is not necessarily groundwater, but water stored before the event. No major differences are observed from using hydrochemical or isotope tracers for the hydrograph separation.

4.7 End-member mixing analysis (EMMA)

To further differentiate the runoff components, a principal component analysis (PCA) was carried out on 12 solutes (EC, SiO$_2$, CaCO$_3$, Cl, NO$_3$–N, SO$_4$, Na, Mg, K, Ca, δ^{18}O, and δ^2H) using the R statistical software (R Development Core Team, 2014). The correlation matrix was used for the PCA. Results indicated that 90 % of the variability is explained by two principal components (m). Thus, the number of end-members (n) can be chosen as $n = m + 1$, leading to a three-component hydrograph separation (Christophersen and Hooper, 1992). Figure 10 shows the biplot of principal components where the orthogonal vectors indicate

no dependency between parameters. This is observed for δ^{18}O, δ^2H, K, and NO$_3$. The clustering of the hydrochemical parameters reveals the strong correlation between these parameters (SiO$_2$, CaCO$_3$, Ca, EC, Mg, Na, Cl, and SO$_4$). Potassium shows a negative strong correlation with the clustered parameters but not with the water isotopes and NO$_3$. Thus, for the three-component hydrograph separations, orthogonal vectors with weak Pearson correlations were selected. These are K and δ^{18}O ($r = -0.28$) and K and δ^2H ($r = 0.45$). The latter shown in Fig. 11. Nitrate was not selected due to its non-conservative properties. Potassium was identified as a useful tracer due to its increasing concentra-

Table 6. Direct runoff (DR), shallow groundwater (GW$_S$), and deep groundwater (GW$_D$) contributions in % and 70 % uncertainty of three-component hydrograph separations in %.

Tracers	Event 1			Event 2			Event 3			Event 4		
	DR	GW$_S$	GW$_D$	DR	GW$_S$	GW$_D$	DR	GW$_S$	GW$_D$	DR	GW$_S$	GW$_D$
K and ^{18}O	28	45	26	7	19	74	16	6	78	41	21	37
70 % uncertainty (%)	7.2	5.3	5.8	7.4	3.2	5.1	5.3	3.0	3.9	7.9	6.2	5.8
K and ^{2}H	22	45	33	14	19	67	11	5	84	37	20	42
70 % uncertainty (%)	4.8	6.6	6.4	3.8	3.9	5.5	3.0	2.8	4.0	6.3	6.2	7.6

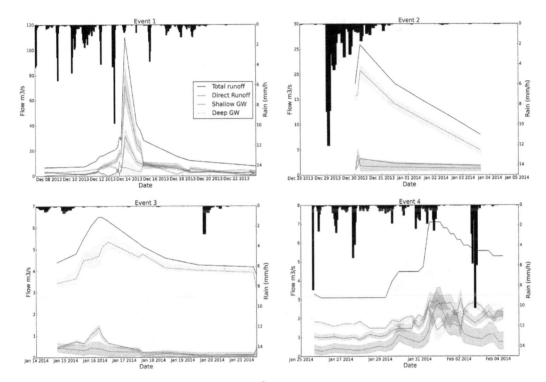

Figure 12. Three-component hydrograph separations using K and ^{2}H.

tions during runoff peaks. This high potassium concentration suggested the presence of soil water influenced by mobilization of fertilizer and/or organic material. To account for additional near-surface water, this component is referred to as the "shallow groundwater component" during this study. It is important to note that the shallow groundwater component could be a mix of surface runoff and near-surface water since potassium was used as an indicator of shallow groundwater, and this element can also be found in surface runoff.

4.8 Three-component hydrograph separation

Direct runoff contributions obtained during the three-component hydrograph separations (Table 6 and Fig. 12) concur with the two-component hydrograph separations. Events 1 and 4 were characterized by higher contributions of direct runoff than Events 2 and 3. Moreover, Event 1 also

had a higher contribution of shallow groundwater that peaked during the total runoff peak. Events 2, 3, and 4 had higher deep groundwater contributions. Uncertainties for the three-component hydrograph separations can be seen in Table 6.

5 Discussion

5.1 Runoff processes in the Kaap catchment

From the mixing diagrams, groundwater analysis and spatial hydrochemical characterization of the catchment, the runoff components were identified and characterized. The groundwater analysis suggested two sources of groundwater of different ionic content at the upper and lower sections of the catchment. In the upstream area, granite is the dominant formation explaining the lower ionic content in groundwater, in contrast to the downstream areas, where geologically diverse

Table 7. Runoff studies with the number of events studied.

Study name	Reference	Number of events
Hydrograph separation using stable isotopes, silica and electrical conductivity: an alpine example	Laudon and Slaymaker (1997)	5
The role of soil water in stormflow generation in a forested headwater catchment: synthesis of natural tracer and hydrometric evidence	Bazemore et al. (1994)	2
Quantifying contributions to storm runoff through end-member mixing analysis and hydrologic measurements at the Panola Mountain Research Watershed (Georgia, USA)	Burns et al. (2001)	2
On the value of combined event runoff and tracer analysis to improve understanding of catchment functioning in a data-scarce semi-arid area	Hrachowitz et al. (2011)	28
Quantifying uncertainties in tracer-based hydrograph separations: a case study for two-, three- and five-component hydrograph separations in a mountainous catchment	Uhlenbrook and Hoeg (2003)	4
Hydrograph separations in a mesoscale mountainous basin at event and seasonal timescales	Uhlenbrook et al. (2002)	2
Identification of runoff generation processes using combined hydrometric, tracer and geophysical methods in a headwater catchment in South Africa	Wenninger et al. (2008)	3
Runoff generation in a steep, tropical montane cloud forest catchment on permeable volcanic substrate	Muñoz-Villers and McDonnell (2012)	13
Quantifying the relative contributions of riparian and hillslope zones to catchment runoff	McGlynn and McDonnell (2003)	2
Dynamics of nitrate and chloride during storm events in agricultural catchments with different subsurface drainage intensity (Indiana, USA)	Kennedy et al. (2012)	2
Investigation of hydrological processes using chemical and isotopic tracers in a mesoscale Mediterranean forested catchment during autumn recharge	Marc et al. (2001)	3

formations and land use increase the ionic content of groundwater. The weathered granite layer allows rain to infiltrate to the deeper groundwater reservoir through preferential flowpaths with less contact time for weathering processes to occur. This explains the hydrochemical signature of the deep groundwater component, which is characterized by its moderate electrical conductivities, moderate to high dissolved silica, lower ionic content, and low potassium concentrations. The chemical signature of the shallow groundwater component is characterized by the high electrical conductivities, alkalinity, sulfates, potassium, and nitrates that are washed from top geological layers with large ionic content and land uses such as agriculture and mining that are more predominant in the downstream region of the catchment.

The three-component hydrograph separations suggest that the shallow groundwater component (potentially including surface runoff) is quickly activated during rainfall events, and its contribution increases as the antecedent precipitation increases as observed during Events 1 and 4, where the shallow groundwater contributions were 45 and 20–21 %, respectively. Moreover, a connection between surface and groundwater is evident from the groundwater contour map (Fig. 2d), which shows a gaining river system, and from the flow duration curves, which indicate exfiltrating groundwater storages to the streams. Further literature (Hughes, 2010) suggests that most of South Africa's groundwater is stored in secondary aquifers and that surface flow may be nourished by

lateral flow from semi-saturated fracture systems after storm events.

Other studies (Petersen, 2012) in the nearby Kruger National Park (KNP) have shown that groundwater recharge occurs mostly during the wet season and groundwater flow travels in accordance with the topographical relief. Petersen (2012) studied a granite-dominated area and a basaltic-rock-dominated area, approximately 30 km east of the Kaap outlet. The study found that the granite region was mainly characterized by the steep topography, which favors overland flow that infiltrates through depressions, cracks and fractures by preferential pathways, while the southern basaltic section with a flatter topography showed piston flow processes to be more predominant. The Petersen (2012) findings, covering studies of approximately 1011 boreholes in the KNP, support the findings in the Kaap catchment where high fracturing in the granite section allows recharge of deeper groundwater reservoirs through preferential flowpaths.

It is important to note that the inferences drawn from this study are based on four events sampled during the 2013–2014 wet season but supported by historical meteorological, hydrological and water quality data, groundwater analysis and a spatial hydrochemical study of the catchment. In addition, Table 7 shows runoff studies with a similar number of events studied.

5.2 The catchment's response dependency on antecedent precipitation

Hydrograph separation results suggested that there is a direct runoff contribution (2–36 %) to total runoff during storm events for the Kaap River. Similar results have been obtained for other catchments in semi-arid areas. For instance, Hrachowitz et al. (2011) in their study in four nested catchments in Tanzania found event runoff coefficients of 0.09. Similarly, Munyaneza et al. (2012) found groundwater contributions up to 80 % of total runoff in the Mingina catchment in Rwanda during the two- and three-hydrograph separations in a 258 km^2 catchment. The importance of subsurface flow in semi-arid catchments is also illustrated in Wenninger et al. (2008) in the Weatherley catchment in the Eastern Cape in South Africa.

From the several variables considered such as geology, topography and rainfall characteristics studied for the four events, the direct runoff component was most sensitive to the API. This is observed during Events 1 and 4, where API$_{-7}$ values are the largest among the four events and direct runoff contributions are also the largest for these events. The relationship between API$_{-7}$ and direct runoff generation is supported by a strong Pearson correlation (0.76–0.94). This suggests that direct runoff is enhanced by wetter conditions in the catchment due to saturation in the subsurface triggering saturation overland flow.

5.3 Complexities of runoff process understanding in semi-arid areas

The combination of climatic and hydrological processes influenced by topography, geology, soils and land use makes catchments complex systems. Although the opposite may be true for particular situations, in general, catchments become more non-linear as aridity increases and runoff processes become more spatially and temporally heterogeneous than in humid regions (Blöschl et al., 2013; Farmer et al., 2003). Thus, understanding hydrological processes in arid catchments becomes more difficult due to high variability of rainfall and streamflow, high evaporation losses, long infiltration pathways, permeable stream channel beds and often deep groundwater reservoirs (Hughes, 2007; Trambauer et al., 2013).

The high variability of rainfall enhances the difficulties of runoff prediction by triggering different runoff responses. For instance, high-intensity storms tend to generate overland flow in the form of infiltration excess overland flow (Smith and Goodrich, 2006), while high antecedent precipitation conditions enhance saturation excess overland flow. This effect is visible in this study during Event 1, where the high API suggested saturation of the subsurface, thus reducing the infiltration capacity and enhancing saturation excess overland flow. The opposite is observed for Events 2 and 3, where the low soil moisture conditions allow more rainfall to infiltrate, activating other runoff processes such as preferential vertical flow.

Although not included in this study, inter-annual variability, evaporation, hydraulic connectivity, permeable stream beds and interception have been shown to change the behavior of runoff processes in arid and semi-arid areas. For instance, inter-annual rainfall variability is closely related to high evaporation losses. The Mostert et al. (1993) study in a Namibian basin found that during wetter seasons, vegetation cover and total evaporation increased, thus reducing the amount of runoff reaching the outlet. Similarly, hydraulic connectivity in arid environments is limited by the reduced soil moisture conditions in these areas, leading to reduced groundwater recharge. Other fluxes such as interception and flow through permeable stream beds pose a greater challenge to the understanding of runoff processes in semi-arid areas. Interception can further decrease the hydrologic connectivity, breaking the link between meteoric water and groundwater as observed in the Zhulube catchment in Zimbabwe, where interception accounted for up to 56 % of rainfall during the dry season (Love et al., 2010). Similarly, transmission losses due to the high degree of fracturing of stream beds can significantly reduce streamflow but increase recharge of groundwater systems.

Thus, this study illustrated the effects of temporal rainfall variability during the wet season, suggesting the influence of antecedent precipitation conditions on direct runoff generation. However, studying the effects of spatial and inter-annual rainfall variability, high evaporation and transpiration (from unsaturated zones, alluvial aquifers, and riparian zones) fluxes, the spatial variability of vegetation, and deep groundwater resources on streamflow generation is still required for better understanding of runoff processes in semi-arid areas. More monitoring of groundwater levels and aquifers would assist in bridging this gap of knowledge, such as in Van Wyk et al. (2012). Emphasis is placed on studying the region during dry weather for further understanding of evaporation and transpiration from deeper layers of soil moisture that in some cases can reach even into groundwater systems (e.g., eucalyptus trees).

6 Conclusions

The Kaap catchment has suffered devastating floods that affect greatly the trans-boundary Incomati Basin, in particular downstream areas in South Africa, Swaziland and Mozambique, where recent floods have caused significant economic and social losses. Runoff processes were poorly understood in the Kaap catchment, limiting rainfall–runoff models to lead to better informed water management decisions. Through hydrometric measurements, tracers and groundwater observations, runoff components and main runoff generation processes were identified and quantified in the Kaap catchment for the 2013–2014 wet season. The suitability of isotope hydrograph separation was tested by comparing it to hydrochemical hydrograph separations showing no major differences between these tracers. Hydrograph separations showed that groundwater was the dominant runoff component for the 2013–2014 wet season. Three component hydrograph separations suggested a third component that we addressed as the shallow groundwater component. However, further research is still necessary to make a clear distinction between surface runoff and shallow groundwater. A strong correlation between direct runoff generation and antecedent precipitation conditions was found for the studied events. Direct runoff was enhanced by high antecedent precipitation activating saturation excess overland flow. Similar groundwater contributions have been observed in other studies in semi-arid areas (Hrachowitz et al., 2011; Munyaneza et al., 2012; Wenninger et al., 2008). The understanding of runoff generation mechanisms in the Kaap catchment contributes to the limited number of hydrological process studies and in particular hydrograph separation studies in semi-arid regions for the proper management of water resources. Moreover, this study was carried out during the wet season, and in order to gather a better understanding of the hydrological system, further studies focusing on the dry season are still needed, particularly on the dependency of runoff generation on soil moisture and vegetation.

Acknowledgements. This study was carried out under the umbrella of the RISK-based Operational water MANagement for the Incomati River basin (RISKOMAN) project. The authors would like to thank the RISKOMAN partners, UNESCO-IHE, Universidade Eduardo Mondlane (Mozambique), University of KwaZulu-Natal (UKZN, South Africa), the Komati River Basin Authority (Swaziland) and Incomati-Usuthu Catchment Management Agency (IUCMA, South Africa), for their financial and technical cooperation. This research has also been supported by the International Foundation for Science (IFS) Stockholm, Sweden, through a grant (W/5340-1) to Aline M. L. Saraiva Okello. Gratitude is also expressed to Ilyas Masih (UNESCO-IHE), Thomas Gyedu Ababio (IUCMA), Graham Jewitt (UKZN), Cobus Pretorius (UKZN), Eddie Riddell (SANParks), Gareth Bird (Independent), the ICMA staff, In-Situ Groundwater Consulting, IAEA, UNESCO-IHE laboratory staff, and the advanced class program at UNESCO-IHE funded by UNEP-DHI.

Edited by: A. Bronstert

References

Allen, R. G., Pereira, L. S., Raes, D., and Smith, M.: Crop evapotranspiration – Guidelines for computing crop water requirements, FAO Irrigation and drainage paper 56, Food and Agriculture Organization of the United Nations, Rome, 1998.

Bazemore, D. E., Eshleman, K. N. and Hollenbeck, K. J.: The Role of Soil-Water in Stormflow Generation in a Forested Headwater Catchment – Synthesis of Natural Tracer and Hydro-

metric Evidence, J. Hydrol., 162, 47–75, doi:10.1016/0022-1694(94)90004-3, 1994.

Blöschl, G., Sivapalan, M., Wagener, T., Viglione, A., and Savenije, H.: Runoff prediction in ungauged basins, Cambridge University Press, 2013.

Burns, D. A., McDonnell, J. J., Hooper, R. P., Peters, N. E., Freer, J. E., Kendall, C., and Beven, K.: Quantifying contributions to storm runoff through end-member mixing analysis and hydrologic measurements at the Panola Mountain Research Watershed (Georgia, USA), Hydrol. Process., 15, 1903–1924, doi:10.1002/Hyp.246, 2001.

Burns, D. A.: Stormflow-hydrograph separation based on isotopes: the thrill is gone – what's next?, Hydrol. Process., 16, 1515–1517, doi:10.1002/Hyp.5008, 2002.

Camarasa-Belmonte, A. M. and Soriano, J.: Empirical study of extreme rainfall intensity in a semi-arid environment at different time scales, J. Arid Environ., 100–101, 63–71, 2014.

Christophersen, N. and Hooper, R. P.: Multivariate-Analysis of Stream Water Chemical-Data – the Use of Principal Components-Analysis for the End-Member Mixing Problem, Water Resour. Res., 28, 99–107, doi:10.1029/91wr02518, 1992.

de Wit, M. J., Furnes, H., and Robins, B.: Geology and tectonostratigraphy of the Onverwacht Suite, Barberton Greenstone Belt, South Africa, Precambrian Res., 168, 1–27, 2011.

Farmer, D., Sivapalan, M., and Jothityangkoon, C.: Climate, soil, and vegetation controls upon the variability of water balance in temperate and semiarid landscapes: Downward approach to water balance analysis, Water Resour. Res., 39, 1035, doi:10.1029/2001WR000328, 2003.

Genereux, D.: Quantifying uncertainty in tracer-based hydrograph separations, Water Resour. Res., 34, 915–919, doi:10.1029/98wr00010, 1998.

GRIP: Groundwater Resources Information Programme, Polokwane, South Africa, 2012.

Gröning, M., Lutz, H. O., Roller-Lutz, Z., Kralik, M., Gourcy, L., and Pöltenstein, L.: A simple rain collector preventing water re-evaporation dedicated for ^{18}O and Deuterium analysis of cumulative precipitation samples, J. Hydrol., 448–449, 195–200, 2012.

Hrachowitz, M., Soulsby, C., Tetzlaff, D., Dawson, J. J. C., Dunn, S. M., and Malcolm, I. A.: Using long-term data sets to understand transit times in contrasting headwater catchments, J. Hydrol., 367, 237–248, doi:10.1016/j.jhydrol.2009.01.001, 2009.

Hrachowitz, M., Bohte, R., Mul, M. L., Bogaard, T. A., Savenije, H. H. G., and Uhlenbrook, S.: On the value of combined event runoff and tracer analysis to improve understanding of catchment functioning in a data-scarce semi-arid area, Hydrol. Earth Syst. Sci., 15, 2007–2024, doi:10.5194/hess-15-2007-2011, 2011.

Hughes, D. A.: Modelling semi-arid and arid hydrology and water resources – the southern African experience, in: Hydrological Modelling in Arid and Semi-Arid Areas, Cambridge University Press, Cambridge, 2007.

Hughes, D. A.: Unsaturated zone fracture flow contributions to stream flow: evidence for the process in South Africa and its importance, Hydrol. Process., 24, 767–774, doi:10.1002/hyp.7521, 2010.

Hughes, J. D., Khan, S., Crosbie, R. S., Helliwell, S., and Michalk, D. L.: Runoff and solute mobilization processes in a semi-arid headwater catchment, Water Resour. Res., 43, W09402, doi:10.1029/2006wr005465, 2007.

Kennedy, C. D., Bataille, C., Liu, Z. F., Ale, S., VanDe-Velde, J., Roswell, C. R., Bowling, L. C., and Bowen, G. J.: Dynamics of nitrate and chloride during storm events in agricultural catchments with different subsurface drainage intensity (Indiana, USA), J. Hydrol., 466–467, 1–10, doi:10.1016/j.jhydrol.2012.05.002, 2012.

Klaus, J. and McDonnell, J. J.: Hydrograph separation using stable isotopes: review and evaluation, J. Hydrol., 505, 47–64, doi:10.1016/j.jhydrol.2013.09.006, 2013.

Laudon, H. and Slaymaker, O.: Hydrograph separation using stable isotopes, silica and electrical conductivity: an alpine example, J. Hydrol., 201, 82–101, doi:10.1016/S0022-1694(97)00030-9, 1997.

Liu, F. J., Williams, M. W., and Caine, N.: Source waters and flow paths in an alpine catchment, Colorado Front Range, United States, Water Resour. Res., 40, W09401, doi:10.1029/2004wr003076, 2004.

Love, D., Uhlenbrook, S., Corzo-Perez, G., Twomlow, S., van der Zaag, P.: Rainfall–interception–evaporation–runoff relationships in a semi-arid catchment, northern Limpopo basin, Zimbabwe, Hydrolog. Sci. J., 55, 687–703, 2010.

Mallory, S. and Beater, A.: Inkomati Water Availability Assessment Study (IWAAS), Department of Water Affairs and Forestry (DWAF), Pretoria, 2009.

Marc, V., Didon-Lescot, J. F., and Michael, C.: Investigation of the hydrological processes using chemical and isotopic tracers in a small Mediterranean forested catchment during autumn recharge, J. Hydrol., 247, 215–229, doi:10.1016/S0022-1694(01)00386-9, 2001.

McGlynn, B. L. and McDonnell, J. J.: Quantifying the relative contributions of riparian and hillslope zones to catchment runoff, Water Resour. Res., 39, SWC21–SWC220, 2003.

Midgley, D. C., Pittman, W. V., and Middleton, B. J.: Surface Water Resources of South Africa 1990, Water Research Commission Pretoria Report No. 298/1/94, Water Research Commission, Pretoria, 1994.

Mostert, A., McKenzie, R., and Crerar, S.: A rainfall/runoff model for ephemeral rivers in an arid or semi-arid environment, 6th South African National Hydrology Symposium, Pietermaritzburg, 219–224, 1993.

Muñoz-Villers, L. E. and McDonnell, J. J.: Runoff generation in a steep, tropical montane cloud forest catchment on permeable volcanic substrate, Water Resour. Res., 48, W09528, doi:10.1029/2011WR011316, 2012.

Munyaneza, O., Wenninger, J., and Uhlenbrook, S.: Identification of runoff generation processes using hydrometric and tracer methods in a meso-scale catchment in Rwanda, Hydrol. Earth Syst. Sci., 16, 1991–2004, doi:10.5194/hess-16-1991-2012, 2012.

Nkomo, S. and van der Zaag, P.: Equitable water allocation in a heavily committed international catchment area: the case of the Komati Catchment, Phys. Chem. Earth A/B/C, 29, 1309–1317, doi:10.1016/j.pce.2004.09.022, 2004.

Pearce, A. J., Stewart, M. K., and Sklash, M. G.: Storm Runoff Generation in Humid Headwater Catchments, 1. Where Does the Water Come From?, Water Resour. Res., 22, 1263–1272, 1986.

Petersen, R.: A conceptual understanding of groundwater recharge processes and surface-groundwater interactions in the Kruger National Park Master in the Faculty of Natural Sciences, Department of Earth Sciences, University of the Western Cape, Bellville, 2012.

R Development Core Team: R: A Language and Environment for Statistical Computing: http://www.R-project.org, last access: 1 August 2014.

Sengo, D. J., Kachapila, A., van der Zaag, P., Mul, M., and Nkomo, S.: Valuing environmental water pulses into the Incomati estuary: Key to achieving equitable and sustainable utilisation of transboundary waters, Phys. Chem. Earth A/B/C, 30, 648–657, doi:10.1016/j.pce.2005.08.004, 2005.

Schulze, R. E.: South African atlas of agrohydrology and -climatology, Water Research Commission, Pretoria, 1997.

Sharpe, M. R., Sohnge, A. P., Zyl, J. S. V., Joubert, D. K., Mulder, M. P., ClubleyArmstrong, A. R., Plessis, C. P. D., Eeden, O. R. V., Rossouw, P. J., Visser, D. J. L., Viljoen, M. J., and Taljaard, J. J.: 2530 Barberton, Department of Minerals and Energy Affairs, Government Printer, South Africa, 1986.

Slinger, J. H., Hilders, M., and Dinis, J.: The practice of transboundary decision making on the Incomati River, Ecol. Soc., 15, 1, 2010.

Smith, R. E. and Goodrich, D. C.: Rainfall Excess Overland Flow, in: Encyclopedia of Hydrological Sciences, edited by: Anderson, M. G., John Wiley & Sons, Ltd, doi:10.1002/0470848944.hsa117, 2006.

Smithers, J., Schulze, R., Pike, A., and Jewitt, G.: A hydrological perspective of the February 2000 floods: A case study in the Sabie River Catchment, Water South Africa, Water SA, 325–332, doi:10.4314/wsa.v27i3.4975, 2001.

SASRI – South African Sugarcane Research Institute: http://portal.sasa.org.za/weatherweb/weatherweb.ww_menus.menuframe?menuid=1, last access: 1 October 2013.

Tetzlaff, D. and Soulsby, C.: Sources of baseflow in larger catchments – Using tracers to develop a holistic understanding of runoff generation, J. Hydrol., 359, 287–302, doi:10.1016/j.jhydrol.2008.07.008, 2008.

Trambauer, P., Maskey, S., Winsemius, H., Werner, M., and Uhlenbrook, S.: A review of continental scale hydrological models and their suitability for drought forecasting in (sub-Saharan) Africa, Phys. Chem. Earth, 66, 16–26, 2013.

Uhlenbrook, S., and Hoeg, S.: Quantifying uncertainties in tracer-based hydrograph separations: a case study for two-, three- and five-component hydrograph separations in a mountainous catchment, Hydrol. Process., 17, 431–453, doi:10.1002/Hyp.1134, 2003.

Uhlenbrook, S. and Leibundgut, C.: Development and validation of a process oriented catchment model based on dominating runoff generation processes, Phys.Chem. Earth B, 25, 653–657, doi:10.1016/S1464-1909(00)00080-0, 2000.

Uhlenbrook, S., Frey, M., Leibundgut, C., and Maloszewski, P.: Hydrograph separations in a mesoscale mountainous basin at event and seasonal timescales, Water Resour. Res., 38, 311–314, doi:10.1029/2001WR000938, 2002.

UNEP: World atlas of desertification 2ED, United Nations Environment Programme (UNEP), London, UK, 1997.

Van der Zaag, P. and Carmo Vaz, A.: Sharing the Incomati water:cooperation and competition in the balance, Water Policy, 5, 346–368, 2003.

Van Wyk, E., van Tonder, G. J., and Vermeulen, D.: Characteristics of local groundwater recharge cycles in South African semi-arid hard rock terrains: Rainfall–groundwater interaction, Water SA, 38, 747–754, 2012.

Wang, R., Kumar, M., and Marks, D.: Anomalous trend in soil evaporation in a semi-arid, snow-dominated watershed, Adv. Water Resour., 57, 32–40, 2013.

Weiler, M., McGlynn, B. L., McGuire, K. J., and McDonnell, J. J.: How does rainfall become runoff? A combined tracer and runoff transfer function approach, Water Resour. Res., 39, 1315, doi:10.1029/2003WR002331, 2003.

Wenninger, J., Uhlenbrook, S., Lorentz, S., and Leibundgut, C.: Idenfication of runoff generation processes using combined hydrometric, tracer and geophysical methods in a headwater catchment in South Africa, Hydrolog. Sci. J., 53, 65–80, doi:10.1623/hysj.53.1.65, 2008.

Wheater, H., Sorooshian, S., and Sharma, K. D.: Hydrological modelling in arid and semi-arid areas, University Press, Cambridge, UK, 195 pp., 2008.

Winston, W. E. and Criss, R. E.: Geochemical variations during flash flooding, Meramec River basin, May 2000, J. Hydrol., 265, 149–163, doi:10.1016/S0022-1694(02)00105-1, 2002.

WRC: Water resources of South Africa, 2005 study (WR2005), PretoriaTT381, Pretoria, 2005.

11

Stable water isotope tracing through hydrological models for disentangling runoff generation processes at the hillslope scale

D. Windhorst[1], **P. Kraft**[1], **E. Timbe**[1,2], **H.-G. Frede**[1], and **L. Breuer**[1]

[1]Institute for Landscape Ecology and Resources Management (ILR), Research Centre for BioSystems, Land Use and Nutrition (IFZ), Justus-Liebig-Universität Gießen, Gießen, Germany
[2]Departamento de Recursos Hídricos y Ciencias Ambientales, Universidad de Cuenca, Cuenca, Ecuador

Correspondence to: D. Windhorst (david.windhorst@umwelt.uni-giessen.de)

Abstract. Hillslopes are the dominant landscape components where incoming precipitation becomes groundwater, streamflow or atmospheric water vapor. However, directly observing flux partitioning in the soil is almost impossible. Hydrological hillslope models are therefore being used to investigate the processes involved. Here we report on a modeling experiment using the Catchment Modeling Framework (CMF) where measured stable water isotopes in vertical soil profiles along a tropical mountainous grassland hillslope transect are traced through the model to resolve potential mixing processes. CMF simulates advective transport of stable water isotopes ^{18}O and ^{2}H based on the Richards equation within a fully distributed 2-D representation of the hillslope. The model successfully replicates the observed temporal pattern of soil water isotope profiles (R^2 0.84 and Nash–Sutcliffe efficiency (NSE) 0.42). Predicted flows are in good agreement with previous studies. We highlight the importance of groundwater recharge and shallow lateral subsurface flow, accounting for 50 and 16 % of the total flow leaving the system, respectively. Surface runoff is negligible despite the steep slopes in the Ecuadorian study region.

1 Introduction

Delineating flow path in a hillslope is still a challenging task (Bronstert, 1999; McDonnell et al., 2007; Tetzlaff et al., 2008; Beven and Germann, 2013). However, a more complete understanding of the partitioning of incoming water to surface runoff, lateral subsurface flow components or percolation allows better understanding, for example, the impact of climate and land use change on hydrological processes. Models are often used to test different rainfall–runoff generation processes and the mixing of water in the soil (e.g., Kirkby, 1988; Weiler and McDonnell, 2004). Due to the prevailing measurement techniques and therefore the available data sets, it has become common practice to base the validation of modeled hillslope flow processes on quantitative data on storage change. In the simplest case, system-wide storage changes are monitored by discharge and groundwater level measurements or, on more intensively instrumented hillslopes, the storage change of individual soil compartments is monitored by soil moisture sensors. In the typical 2-D flow regime of a slope, such models bear the necessity to account not only for the vertical but also for the lateral movements of water within the soil (Bronstert, 1999). Quantitative data on storage change in this regard are only suitable to account for the actual change in soil water volume, but not to assess the source or flow direction. Knowing tracer compositions of relevant hydrological components along a hillslope allows one to account for mixing processes and thereby to delineate the actual source of the incoming water. Over the years a number of artificial, e.g., fluorescence tracers like uranine, and natural tracers, e.g., chloride or stable water isotopes, have emerged. While the application of the artificial tracers is rather limited in space and time (Leibundgut et al., 2011), the latter ones can be used over a wide range of scales (Barthold et al., 2011; Genereux and Hooper, 1999; Leibundgut et al., 2011; Muñoz-Villers and McDonnell, 2012; Soulsby et al., 2003). Stable water isotopes such as oxygen-18 (^{18}O) and hydrogen-2 (^{2}H) are integral parts of water molecules and consequently ideal tracers of water. Over the last decades isotope tracer studies have proven to provide reliable results on varying scales (chamber, plot, hillslope to catchment scale) and surface types (open water, bare soils, vegetated areas) to delineate or describe flow processes under field experimental or laboratory conditions (Garvelmann et al., 2012; Hsieh et al., 1998; Sklash et al., 1976; Vogel et al., 2010; Zimmermann et al., 1968).

Although the first 1-D process-orientated models to describe the dynamics of stable water isotope profiles for open water bodies (Craig and Gordon, 1965) were developed as early as in the mid-1960s, and a bit later for soils (Zimmermann et al., 1968), fully distributed 2-D to 3-D hydrological tracer models benefitting from the additional information to be gained by stable water isotopes are still in their early development stages (Davies et al., 2013) or use strong simplifications of the flow processes (e.g., TAC^D using a kinematic wave approach; Uhlenbrook et al., 2004). This can be attributed to the high number of interwoven processes affecting the soil water isotope fluxes not only in the soil's liquid phase but also in its vapor phase. The more process-based 1-D models (Braud et al., 2005; Haverd and Cuntz, 2010) therefore simultaneously solve the heat balance and the mass balance simultaneously for the liquid and the vapor phase and are thereby describing the

- convection and molecular diffusion in the liquid and vapor phase,

- equilibrium fractionation between liquid and vapor phase,

- fractionation due to evaporation,

- non-fractionated flux due to percolation and transpiration.

To obtain and compute the data required to apply these kind of models beyond the plot scale is still challenging. However, due to emerging measuring techniques the availability of sufficient data is currently becoming more realistic. Increasing computational power and especially the cavity ring-down spectroscopy (CRDS) – a precise and cost-effective method to analyze the signature of stable water isotopes (Wheeler et al., 1998) – promise progress.

Hence, it is tempting to investigate the suitability of isotope tracers to delineate hydrological flow paths using a more physical modeling approach. Recent research in this direction includes the work of McMillan et al. (2012) and Hrachowitz et al. (2013) using chloride as a tracer to study the fate of water in catchments in the Scottish Highlands. Even though some processes affecting the soil water isotope transport are still represented in a simplified manner or, due to their limited effect/importance of the respective process within the given study site, could be omitted, this approach allows us to determine the potential of soil water isotope modeling in catchment hydrology and highlights the future need for research.

This study is conducted in a 75 km^2 montane rain forest catchment in south Ecuador, the upper part of the Rio San Francisco, which has been under investigation since 2007 (Bogner et al., 2014; Boy et al., 2008; Bücker et al., 2011; Crespo et al., 2012; Fleischbein et al., 2006; Goller et al., 2005; Timbe et al., 2014; Windhorst et al., 2013b). The findings of those studies (briefly synthesized in Sect. 2.3) will (a) ease the setup of chosen model, (b) let us define suitable boundary conditions for the chosen modeling approach and (c) serve as a reference for the delineated flow bath. The additional information from previous studies conducted in the study area will therefore highlight the potential of this new model approach to delineate hydrological flow paths under natural conditions and support our preliminary hydrological process understanding retrieved from more classical methods conducted in the past.

Within this catchment we selected a hillslope with a distinct drainage area and nearly homogenous land use and established an experimental sampling scheme to monitor the isotopic signatures of the soil water of three soil profiles using passive capillary fiberglass wick samplers (PCaps). Based on the proposed modeling approach a 2-D virtual hillslope representation of this hillslope was then implemented using the Catchment Modeling Framework (CMF; Kraft et al., 2011). Due to the necessity to mix the flows in accordance with the observed soil water isotope signatures, we are confident that the degree of certainty for the modeled flow path will be higher than for conventional modeling approaches relying solely on quantitative information to evaluate the modeled data. Replacing the calibration target bears now the necessity to mix the right amount and signature of any given flow component, whereas the quantitative change only relies on the actual amount of water leaving or entering any given compartment. We will quantify the following flow components to disentangle the runoff generation processes: surface runoff, lateral subsurface flow in the vadose zone and percolation to groundwater. The lateral subsurface flow will be further subdivided into near-surface lateral flow and deep lateral flow.

To validate the chosen modeling approach and assess our process understanding, we tested the following hypotheses:

1. Under the given environmental conditions – high precipitation and humidity – (Bendix et al., 2008) and full vegetation cover (Dohnal et al., 2012; Vogel et al., 2010) only non-fractionating and advective water transport of isotopes is relevant.

2. Gaseous advection and diffusive process in the gaseous as well as the liquid phase and the enrichment due to evaporation are negligible; hence the stable water isotopes behave like a conservative tracer.

3. Large shares of the soil water percolate to deeper horizons, thereby creating long mean transit times (MTT) (Crespo et al., 2012; Timbe et al., 2014).

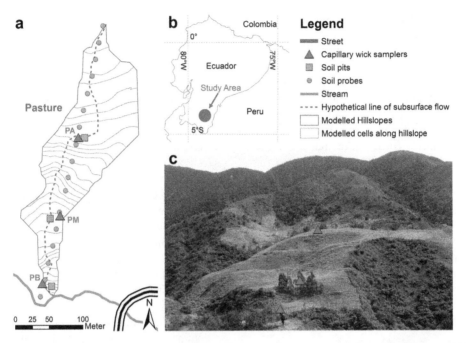

Figure 1. (a) Outline of the modeled hillslope and its virtual discretization into cells. **(b)** Location of the study area within Ecuador. **(c)** Photograph showing the location of the wick samplers (P represents pasture, B represents bajo/lower level, M represents medio/middle level, A represents alto/top level sampler).

4. Due to the high saturated conductivities of the top soil layers, the occurrence of Hortonian overland flow is unlikely to have an important contribution to the observed flows (Crespo et al., 2012).

5. Fast near-surface lateral flow contributes essentially to downhill water flows and plays a relevant role to understand the overall hydrological system (Bücker et al., 2010).

2 Materials and methods

2.1 Study area

The hillslope under investigation is located within the catchment of the Rio San Francisco in south Ecuador (3°58′30″ S, 79°4′25″ W) at the eastern outskirts of the Andes and encompasses an area of 75 km². Close to the continental divide the landscape generally follows a continuous eastward decline towards the lowlands of the Amazon basin (Fig. 1b). Due to the high altitudes (1720–3155 m a.s.l.), the deeply incised valleys (slopes are on average 25–40° over the entire watershed), the low population density and the partly protected areas of Podocarpus National Park, the human impact within the catchment is relatively low. The southern flanks of the Rio San Francisco are covered by an almost pristine tropical mountain cloud forest and lie mostly within Podocarpus National Park. At lower elevations the northern flanks have mostly been cleared by natural or slash-and-burn fires

during the last decades and are now partially used for extensive pasture (*Setaria sphacelata* Schumach.); reforestation sites (*Pinus patula*) are covered by shrubs or invasive weeds (especially tropical bracken fern; *Pteridium aquilinum* L.). The climate exhibits a strong altitudinal gradient, creating relatively low temperatures and high rainfall amounts (15.3 °C and 2000 mm a⁻¹ at 1960 m a.s.l. to 9.5 °C and > 6000 mm a⁻¹ at 3180 m a.s.l.) with the main rainy season in the austral winter (Bendix et al., 2008). A comprehensive description of the soils, climate, geology and land use has been presented by Beck et al. (2008), Bendix et al. (2008) and Huwe et al. (2008).

2.2 Experimental hillslope

To test our understanding of hydrological processes within the study area, we chose a hillslope with a nearly homogenous land use (Fig. 1). It is located on an extensive pasture site with low-intensity grazing by cows and dominated by *Setaria sphacelata*. *Setaria sphacelata* is an introduced tropical C4 grass species that forms a dense tussock grassland with a thick surface root mat (Rhoades et al., 2000). This grass is accustomed to high annual rainfall intensities (> 750 mm a⁻¹), has a low drought resistance and tolerates water logging to a greater extent than other tropical grass types (Colman and Wilson, 1960; Hacker and Jones, 1969). The hillslope has a drainage area of 0.025 km², a hypothetical length of the subsurface flow of 451 m and an elevation gradient of 157 m with an average slope of 19.2°. The soil catena of the slope was recorded by Pürckhauer sampling and soil pits. To investigate the passage of water through the hillslope, a series of three wick samplers has been installed along the line of subsurface flow.

Climate forcing data with an hourly resolution of precipitation, air temperature, irradiation, wind speed and relative humidity were collected by the nearby (400 m) climate station "ECSF" at similar elevation. Isotopic forcing data were collected manually for every rainfall event from October 2010 until December 2012 using a Ø25 cm funnel located in close proximity to the chosen hillslope at 1900 m a.s.l. (Timbe et al., 2014). To prevent any isotopic fractionation after the end of a single rainfall event (defined as a period of 30 min without further rainfall), all samples were directly sealed with a lid and stored within a week in 2 mL amber glass bottles for subsequent analysis of the isotopic signature as described in Sect. 2.4.1 (all samples < 2 mL were discarded).

2.3 Current process understanding at the catchment scale

The catchment of the Rio San Francisco has been under investigation since 2007 (Bücker et al., 2011; Crespo et al., 2012; Timbe et al., 2014; Windhorst et al., 2013b) and was complemented by a number of studies on forested micro-catchments ($\approx 0.1\,\text{km}^2$) within this catchment (Bogner et al., 2014; Boy et al., 2008; Fleischbein et al., 2006; Goller et al., 2005). Studies on both scales identify the similar hydrological processes as being active within the study area.

Studies on the micro-scale (Boy et al., 2008; Goller et al., 2005), supported by solute data and end member mixing analysis at the meso-scale (Bücker et al., 2011; Crespo et al., 2012), showed that fast "organic horizon flow" in forested catchments dominates during discharge events if the mineral soils are water saturated prior to the rainfall. Due to an abrupt change in saturated hydraulic conductivity (K_{sat}) between the organic ($38.9\,\text{m d}^{-1}$) and the near-surface mineral layer ($0.15\,\text{m d}^{-1}$), this organic horizon flow can contribute up to 78 % of the total discharge during storm events (Fleischbein et al., 2006; Goller et al., 2005). However, the overall importance of this organic horizon flow is still disputable because the rainfall intensity rarely gets close to such a high saturated hydraulic conductivity. In 95 % of the measured rainfall events between June 2010 and October 2012 the intensity was below $0.1\,\text{m d}^{-1}$ ($\approx 4.1\,\text{mm h}^{-1}$) and was therefore 15 times lower than the saturated hydraulic conductivity of the mineral soil layer below the organic layer under forest vegetation and around 30 times lower than the saturated hydraulic conductivity of the top soil under pasture vegetation (Zimmermann and Elsenbeer, 2008; Crespo et al., 2012). The same conclusion holds true for the occurrence of surface runoff due to infiltration access on pasture (lacking a significant organic layer). Solely based on rainfall intensities, surface runoff is therefore relatively unlikely to

contribute to a larger extent in rainfall–runoff generation. The reported K_{sat} values are based on measurements of $250\,\text{cm}^3$ undisturbed soil core samples vertically extracted from the center of each respective layer. Due to the chosen sampling method and the limited size of the soil cores, the effective saturated hydraulic conductivity will be even higher and can vary for the horizontal flow component. When and to which extent a subsurface saturated prior to the rainfall event would still trigger surface runoff on pastures therefore remains to be investigated.

Bücker et al. (2010) and Timbe et al. (2014) were able to show that base flow, on the other hand, has a rather large influence on the annual discharge volume across different land use types, accounting for > 70 and > 85 %, respectively. These findings are also supported by the long MTT of the base flow for different sub-catchments of the Rio San Francisco in comparison to the fast runoff reaction times, varying according to Timbe et al. (2014) between 2.1 and 3.9 years. Accordingly, the current findings confirm that the base flow – originating from deeper mineral soil and bedrock layers– is dominating the overall hydrological system in the study area (Crespo et al., 2012; Goller et al., 2005). Apart from this dominating source of base flow, Bücker et al. (2010) identified near-surface lateral flow as a second component to be relevant for the generation of base flow for pasture sites.

2.4 Measurements

2.4.1 Passive capillary fiberglass wick samplers (PCaps)

We installed *passive capillary fiberglass wick samplers* (*wick samplers* for short; designed according to Mertens et al., 2007) as soil water collectors at three locations along an altitudinal transects under pasture vegetation at three soil depths. PCaps maintain a fixed tension based on the type and length of wick (Mertens et al., 2007), require low maintenance and are most suitable to sample mobile soil water without altering its isotopic signature (Frisbee et al., 2010; Landon et al., 1999). We used woven and braided 3/8 in. fiberglass wicks (Amatex Co. Norristown, PA, US). Half (0.75 m) of the 1.5 m wick was unraveled and placed over a $0.30 \times 0.30 \times 0.01\,\text{m}$ square plastic plate, covered with fine-grained parent soil material and then put in contact with the undisturbed soil.

Every collector was designed to sample water from three different soil depths (0.10, 0.25 and 0.40 m) with the same suction, all having the same sampling area of $0.09\,\text{m}^2$, wick type, hydraulic head of 0.3 m (vertical distance) and total wick length of 0.75 m. To simplify the collection of soil water, the wick samplers drained into bottles placed inside a centralized tube with an inner diameter of 0.4 m and a depth of 1.0 m. To avoid any unnecessary alterations of the natural flow above the extraction area of the wick sampler, the centralized tube was placed downhill and the plates were evenly spread uphill around the tube. A flexible silicon tube with a wall thickness of 5 mm was used to house the wick and

Table 1. Soil physical parameters.

Soil code	Clay	Texture	Silt	Porosity	K_{sat}^*	Van Genuchten–Mualem parameters	
	[%]	sand [%]	[%]	[%]	[m d^{-1}]	α	n
A1 & A1 top	34	17	49	81	0.324	0.641	1.16
A2 & A2 top	19	33	49	63	0.324	0.352	1.13
A3 & A3 top	15	34	51	74	0.324	0.221	1.24
B1	8	16	76	66	0.228	1.046	1.19
B2	15	34	51	59	0.228	0.145	1.13
B3	11	18	70	58	0.228	0.152	1.16
C1	15	45	40	55	0.026	0.023	1.12
C2	45	20	35	47	0.026	0.004	1.17

* K_{sat} values are based on values taken within the proximity of the hillslope under similar land use by Crespo et al. (2012) and Zimmermann and Elsenbeer (2008).

to connect it to the 2 L sampling bottles storing the collected soil water. The silicon tube prevents evaporation and contamination of water flowing through the wick. Weekly bulk samples were collected over the period from October 2010 until December 2012 when the sample volume exceeded 2 mL. Soil water and the previously mentioned precipitation samples are analyzed using a CRDS with a precision of 0.1 ‰ for ^{18}O and 0.5 for ^{2}H (Picarro L1102-i, CA, US).

2.4.2 Soil survey

The basic soil and soil hydraulic properties for each distinct soil layer along the hillslope were investigated up to a depth of 2 m. Pürckhauer sampling for soil texture and succession of soil horizons was done every 25 m, while every 100 m soil pits were dug for sampling soil texture, soil water retention curves (pF curves), porosity and succession of soil horizons. The results were grouped into eight classes (Table 1) and assigned to the modeling mesh as shown in Fig. 2. Retention curves (pF curves) were represented by the *van Genuchten–Mualem* function using the parameters α and n.

All soils developed from the same parent material (clay schist) and are classified as Haplic Cambisol with varying soil thickness. Soil thickness generally increased downhill, varying between 0.8 and 1.8 m in depressions. Clay illuviation was more pronounced in the upper part of the hillslope (higher gradient in clay content), indicating lower conductivities in deeper soil layers.

2.5 Modeling

2.5.1 The Catchment Modeling Framework (CMF)

The Catchment Modeling Framework developed by Kraft et al. (2011) is a modular hydrological model based on the concept of finite volume method introduced by Qu and Duffy (2007). Within CMF those finite volumes (e.g., soil water storages, streams) are linked by a series of

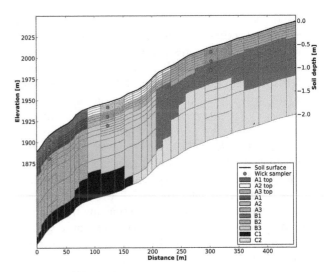

Figure 2. Elevation profile (top black line, left ordinate), succession of soil layer types (color plate) and soil depths assigned to the modeling grid (right ordinate).

flow-accounting equations (e.g., Richards or Darcy equation) to a one- to three-dimensional representation of the real-world hydrological system. The flexible setup of CMF and the variety of available flow-accounting equations allows customizing the setup as required in the presented study. In addition to the water fluxes, the advective movement of tracers within a given system can be accounted for by CMF, making this modeling framework especially suitable to be used in our tracer study (Kraft et al., 2010). Starting with Beven and Germann (1982) scientists over the last decades have frequently argued that the Richards equation along with similar flow-accounting equation assuming a time invariant and well-mixed homogenous flow of water through the soil pore space, similar to those currently implemented in CMF, are not suitable to account for preferential flow relevant for modeling tracer transport (Brooks et al., 2010; Germann et al., 2007; Hrachowitz et al., 2013; Stumpp and Maloszewski, 2010). Being developed for the quantitative representation of soil water flow, these equations cannot distinguish between water stored in different soil compartments (namely the soil matrix and macro-pores) and only artificially try to represent macropore flow, e.g., by favoring high saturated conductivity values or misshaped conductivity curves controlling the flow of water between soil compartments. Even though the capabilities of CMF to account for preferential flow are still in the development phase (e.g., by following the dual-permeability approach in the future) and are not accounted for in the presented setup, our setup will once more highlight potential drawbacks of the modeling approaches relying on the Richards equation while modeling tracer transport at the hillslope scale.

2.5.2   Setup of CMF

To govern the water fluxes within our system, we used the following flow-accounting equations: Manning equation for surface water flow; Richards equation for a full 2-D representation of the subsurface flow; Shuttleworth–Wallace modification (Shuttleworth and Wallace, 1985) of the Penman–Monteith method to control evaporation and transpiration; and constant Dirichlet boundary conditions representing the groundwater table and the outlet of the system as a rectangular ditch with a depth of 1.5 m. The lower boundary condition is only applicable if groundwater table is > 2 m below ground. Preliminary testing revealed that a discretization based on a constant vertical shift (5 m) and alternating cell width increasing width depth (ranging from 1.25 to 83.75 cm) yielded the optimum model performance with regard to computing time and model quality. Based on 5 m contour lines (derived by local lidar measurements with a raster resolution of 1 m; using the Spatial Analyst package of ArcGis 10.1 from ESRI) this hillslope was further separated into 32 cells ranging in size from 16.6 to 2921.6 m^2 (Fig. 1a). To account for small-scale dynamics in the mixing process of stable water isotopes and to be able to run the model with a satisfactory speed, two different horizontal resolutions were used to discretize each layer with depth. Layers encompassing wick samplers and their upslope neighbor were run with a finer resolution of at least 26 virtual soil layers increasing in thickness width depth (1 × 1.25, 13 × 2.5, 7 × 5 and 5 × 10–50 cm). All other cells were calculated with coarser resolution of at least 14 virtual soil layers (1 × 1.25, 1 × 2.5, 6 × 5, 3 × 10 cm and 3 × 15–83.75 cm). When the delineated soil type changed within a soil layer, it was further subdivided according to Fig. 2.

2.5.3   Evapotranspiration

Soil evaporation, evaporation of intercepted water and plant transpiration are calculated separately using the sparse canopy evapotranspiration method by Shuttleworth and Wallace (1985), in its modification by Federer et al. (2003) and Kraft et al. (2011). This approach requires the following parameterizations: soil-surface-wetness-dependent resistance to extract water from the soil (r_{ss}); the plant-type-dependent bulk stomatal resistance to extract water from the leaves (r_{sc}); and the aerodynamic resistance parameters (r_{aa}, r_{as}, and r_{ac}) for sparse crops as described by Shuttleworth and Gurney (1990) and Federer et al. (2003), whereby r_{ac} (resistance canopy atmosphere) restricts the vapor movement between the leaves and the zero plane displacement height and r_{as} (resistance soil atmosphere) restricts the vapor movement between the soil surface and the zero plane displacement height, which is the height of the mean canopy flow (Shuttleworth and Wallace, 1985; Thom, 1972). The aerodynamic resistance parameter r_{aa} refers to the resistance to move vapor between the zero plane displacement height and the reference height at which the available measurements were made. The necessary assumptions to parameterize the plant (*Setaria sphacelata*) and soil-dependent parameters of the Shuttleworth–Wallace equation using the assumptions made by Federer et al. (2003) and Kraft et al. (2011) are listed in Table 2.

Furthermore, soil water extraction by evaporation only affects the top soil layer, and soil water extraction by transpiration is directly controlled by root distribution at a certain soil depth. In accordance with field observations, we assumed an exponential decay of root mass with depth, whereby 90 % of the total root mass is concentrated in the top 0.20 m.

2.5.4   Calibration and validation

For calibration and validation purposes, we compared measured and modeled stable water isotope signatures of ^{2}H and ^{18}O of the soil water at each depths of the each wick sampler along the modeled hillslope. Hourly values of the modeled isotopic soil water signature were aggregated to represent the mean isotopic composition in between measurements (≈ 7 days) and are reported in per mill relative to the Vienna Standard Mean Ocean Water (VSMOW) (Craig, 1961).

Literature and measured values for soil and plant parameters (Tables 1 and 2) were used to derive the initial values for the calibration process. The initial states for calibration were retrieved by artificially running the model with those initial values for the first 2 years of the available data set (Table 3). The results of this pre-calibration run were used as a starting point for all following calibration runs. A warm-up period of 4 months (1 July–31 October 2010) preceded the calibration period (1 November 2010–31 October 2011) to adjust the model to the new parameter set. To simulate a wide range of possible flow conditions and limit the degrees of freedom for the possible model realizations, we selected K_{sat} and porosity for calibration, while the van Genuchten–Mualem parameters remained constant since measured pF curves were available. Even though not all

Table 2. Plant (*Setaria sphacelata*)- and soil-dependent parameters used for the Shuttleworth–Wallace equation.

Parameter	Symbol	Value	Unit	Used to calculate	Source
Potential soil surface resistance	$r_{ss\,pot}$	500	$s\,m^{-1}$	r_{ss}	Federer et al. (2003)
Max stomatal conductivity or max leaf conductance	g_{max}	270	$s\,m^{-1}$	r_{sc}	Körner et al. (1979)
Leaf area index	LAI	3.7	$m^2\,m^{-2}$	r_{sc}	Bendix et al. (2010)
Canopy height	h	0.2	m	r_{aa}, r_{ac} and r_{as}	Estimate based on hand measurements
Representative leaf width	w	0.015	m	r_{ac}	
Extinction coefficient for photosynthetically active radiation in the canopy	CR	70	%	r_{sc}	Federer et al. (2003)
Canopy storage capacity	–	0.15	$mm\,LAI^{-1}$	Interception	Federer et al. (2003)
Canopy closure	–	90	%	Throughfall	Estimate based on image evaluation
Albedo	alb	11,7	%	Net radiation	Bendix et al. (2010)

Table 3. Modeling periods.

Description	Period		Duration
	Start	End	[days]
Initial states	1 July 2010	30 June 2012	730
Warm-up period	1 July 2010	31 October 2010	122
Calibration period	1 November 2010	31 October 2011	364
Validation period	1 November 2011	31 October 2012	365

Table 4. Soil parameter ranges for the Monte Carlo simulations (assuming uniform distribution for each parameter).

Soil code	K_{sat} [$m\,d^{-1}$]		Porosity [$m^3\,m^{-3}$]	
	Min	Max	Min	Max
A1-3 top	0.001	35	0.3	0.9
A1-3	0.001	30	0.3	0.9
B1-3	0.001	12	0.1	0.8
C1-2	0.001	8	0.1	0.8

sensitive parameters of the Richards equation controlling the flow regime were accounted for during the calibration process, we assume that the measured van Genuchten–Mualem parameters alpha and n are in the good agreement with the actual flow characteristics of the soils. As is typical for the application of the van Genuchten–Mualem approach, the tortuosity/connectivity coefficient remained constant throughout all model runs with a value of 0.5. Beside the four soil parameters shown in Table 1 and the upper and lower boundary conditions, only the nine parameters of the Shuttleworth–Wallace equation (Table 2) had to be set prior to each model run. To further control the unknown lower boundary condition and complement the calibration process, the suction induced by groundwater depth was changed for each calibration run.

To increase the efficiency of the calibration runs and evenly explore the given parameter space, we used the Latin Hypercube method presented by McKay et al. (1979). The parameter range of each variable was therefore subdivided into 10 strata and sampled once using uniform distribution. All strata are then randomly matched to get the final parameter sets. A total of 10^5 parameter sets were generated for calibration with varying values for K_{sat} and porosity for all eight soil types as well as different groundwater depths. An initial trial using 10^4 parameter sets was used to narrow down the parameter range as specified in Table 4 for K_{sat}

and porosity for all eight soil types and to 0 to 100 m for the applicable groundwater depths. The performance of each parameter set was evaluated based on the goodness-of-fit criteria Nash–Sutcliffe efficiency (NSE) and the coefficient of determination (R^2). In addition, the bias was calculated as an indicator for any systematic or structural deviation of the model.

After the calibration the best-performing ("behavioral") models according to a NSE > 0.15, an overall bias < ± 20.0‰ δ^2H and a coefficient of determination $R^2 > 0.65$ were used for the validation period (Table 3) using the final states of the calibration period as initial values.

3 Results and discussion

3.1 Model performance

In order to quantify the flow processes, we first validated the overall suitability of the chosen model approach and the performance of the parameter sets. The parameter sets best representing the isotope dynamics of δ^2H (as previously defined as best-performing (behavioral) parameter sets; same accounts for δ^{18}O; results are not shown) during the calibration period explained the observed variation to an even higher

Table 5. Model performance during calibration and validation for all behavioral model runs (based on all calibration runs with NSE > 0.15, bias < ±20.0‰ δ^2H and R^2 > 0.65). Best modeled fit based on NSE.

	Calibration 2010–2011		Validation 2011–2012		Best modeled
	Mean	SD	Mean	SD	fit
NSE	0.19	0.008	0.35	0.029	0.42
R^2	0.67	0.008	0.66	0.020	0.84
Bias	−15.90	0.113	−16.93	0.344	−16.16

degree during the validation period (average NSE 0.19 for calibration versus 0.35 for validation).

The linear correlation between modeled and observed isotope dynamics of δ^2H, for the best-performing parameter sets, were equally good during the calibration and validation period ($R^2 \approx 0.66$) (Table 5). The goodness-of-fit criteria for the single best performing parameter set ("best model fit") show an R^2 of 0.84 and a NSE of 0.42.

Figure 3 depicts the measured and modeled temporal development of the soil water isotope profile along the studied hillslope as well as the δ^2H signature and amount of the incoming rainfall used to drive the model. The measured temporal delay of the incoming signal with depth and the general seasonal pattern of the δ^2H signal are captured by the model (Fig. 3).

The bias was negative throughout all model realizations during calibration and validation (−15.90 (±0.11 SD)‰ δ^2H and −16.93 (±0.34 SD)‰ δ^2H, respectively; see Table 5). Even though the high bias indicates a structural insufficiency of the model, we are confident that this can be mostly attributed to the discrimination of evaporation processes at the soil–atmosphere interface and on the canopy.

Our first hypothesis – that evaporation in general plays only a minor role for the soil water isotope cycle under full vegetation – therefore needs to be reconsidered. Even though hypothesis I has previously been frequently used as an untested assumption for various models (e.g., Vogel et al., 2010; Dohnal et al., 2012), it is rarely scrutinized under natural conditions. A complete rejection of this hypothesis could therefore affect the interpretations in those studies and limit their applicability. However, further studies are needed to support these findings and before finally rejecting this hypothesis. The lateral mixing processes may be obscuring the observed near-surface enrichment, and the effect of preferential flow currently not fully accounted for could further hinder the full interpretation of these findings. It still holds true that

– the quantitative loss due to surface evaporation in areas with a high leaf area index is more or less insignificant

(accounting for 38 mm a^{-1} out of 1896 mm a^{-1}; $\approx 2\,\%$; Fig. 5);

– the isotopic enrichment due to evaporation for vegetated areas is considerably lower than for non-vegetated areas, as previously shown by Dubbert et al. (2013);

– high rainfall intensity constrains any near-surface isotopic enrichment related to evaporation (Hsieh et al., 1998).

However, our results indicate that the contribution of potential canopy evaporation (accounting for 344 mm a^{-1} out of 1896 mm a^{-1}; $\approx 18\,\%$; Fig. 5) to enrich the canopy storage and thereby potential throughfall (discriminating ^{18}O and ^{2}H resulting in more positive isotope signatures) still could partially explain the observed bias.

Nevertheless we presume that fog drip, created by sieving passing clouds or radiation fog frequently occurring in the study area (Bendix et al., 2008), explains the majority of the observed bias. Depending on the climatic processes, generating the fog drip is typically isotopically enriched compared to rainfall, due to different condensation temperatures (Scholl et al., 2009). To get an impression of the magnitude of the possible bias due to throughfall and fog drip compared to direct rainfall, we compare the observed bias with a study presented by Liu et al. (2007) conducted in a tropical seasonal rain forest in China. They observed an average enrichment of +5.5‰ δ^2H for throughfall and +45.3‰ δ^2H for fog drip compared to rainfall. Even though the observed enrichment of fog drip and throughfall by Liu et al. (2007) may not be as pronounced within our study area (Goller et al., 2005), the general tendency could explain the modeled bias. According to Bendix et al. (2008) fog and cloud water deposition within our study area contributes 121 to 210 mm a^{-1} at the respective elevation. Assessing the actual amount fog drip for grass species like *Setaria sphacelata* under natural conditions is challenging and has so far not been accounted for.

If further discrimination below the surface were to substantially alter the isotope signature, the bias would change continuously with depth. Any subsurface flow reaching wick samplers at lower elevations would then further increase the bias. However, the negative bias of −16.19 (±2.80 SD)‰ δ^2H in all monitored top wick samplers during validation accounts for most of the observed bias in the two deeper wick samplers amounting to −17.32 (±2.47 SD)‰ δ^2H. Thus we conclude that the bias is mainly a result of constrains related to modeling surface processes, rather than subsurface ones.

Figure 4 shows the behavior of the chosen parameter sets for saturated hydraulic conductivity and groundwater depth during calibration and validation. The parameter space allows us to assess the range of suitable parameters and their sensitivity over a given parameter range. During calibration the given parameter space could not be constrained to more precise values for all parameters, which in this case should

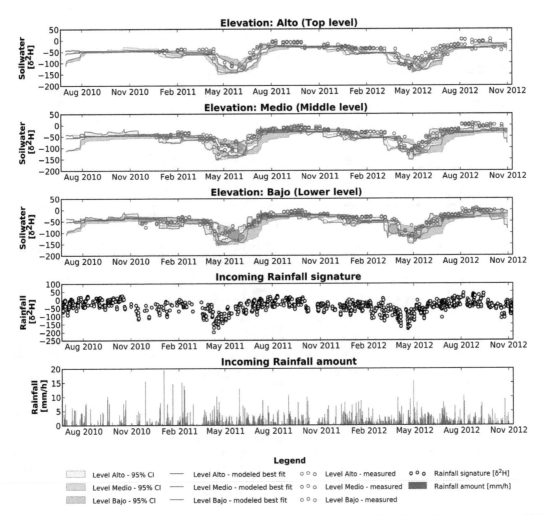

Figure 3. Time series of soil water isotope signatures (top panels 1–3 for each elevation) for all behavioral model runs with NSE > 0.15, bias < ±20.0 ‰ δ^2H and R^2 > 0.65, showing the 95 % confidence interval (CI; transparent areas) and best modeled fit (solid line) vs. measured values (circles) at all three elevations (2010, 1949 and 1904 m a.s.l.) and soil depths below ground (0.10, 0.25 and 0.40 m). Bottom panels 4 and 5: isotopic signature and rainfall amount, respectively.

show a lower SD (Table 6) and narrower box plots (Fig. 4). Especially the K_{sat} values of the soil layers A1, A3 and B1–B3; the porosity for all soil layers (not included in Fig. 4); and the groundwater depth depict a low sensitive over the entire calibration range (indicated by a high SD, wide box plot and evenly scattered points; Table 6 and Fig. 4). In particular the low sensitivity of the model towards groundwater depth seems surprising, but it can be explained by the potentially low saturated hydraulic conductivities of the lower soil layers C1 and C2 limiting the percolation into the lower soil layers outside of the modeling domain. Even an extreme hydraulic potential, induced by a deep groundwater body, can be limited by a low hydraulic conductivity. Nonetheless it is noteworthy that no model run without an active groundwater body as a lower boundary condition (groundwater depth < 2 m) results in a model performance with NSE > 0 (Fig. 4). With a groundwater depth above 2 m the boundary condition would serve as a source of water with

an undefined isotopic signal and prevent any percolation of water into deeper soil layers outside of the modeling domain. The results are therefore in alignment with the topography of the system, indicating an active groundwater body deeper than 2 m, and support our second hypothesis, which we will further discuss in Sect. 3.2. We identified several parameter combinations showing the same model performance, known as equifinality according to Beven and Freer (2001). The observed equifinality can partially be explained by counteracting effects of a decreasing K_{sat} and an increasing pore space, or by the water flow being restrained due to lower hydraulic conductivities at adjoining soil layers. Especially for deeper soil layers the interaction between surrounding layers makes it especially difficult to further constrain the given parameter range. Even though the parameter ranges for all behavioral model realizations are not so well confined, the small confidence intervals indicate a certain degree of robustness towards the predicted flows (Fig. 3). Additional soil moisture

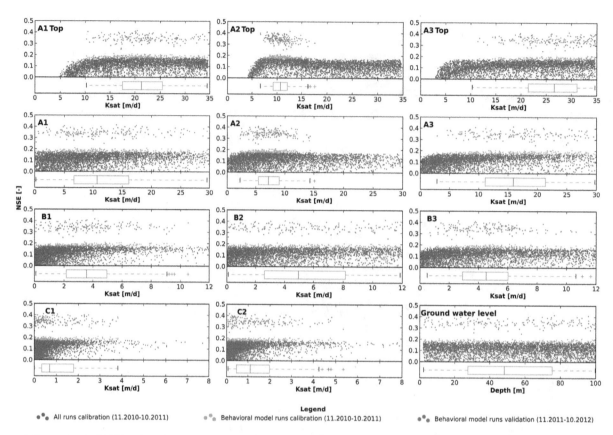

Figure 4. Dotty plots of NSE values (> 0.0) during calibration (blue) and for behavioral model runs (NSE > 0.15, bias $< \pm 20.0\permil$ δ^2H and $R^2 > 0.65$) during calibration (orange) and validation (red) for saturated hydraulic conductivity (K_{sat}) for all soil types and groundwater depth. Box plots show the unweighted parameter distribution of all behavioral model runs (NSE > 0.15, bias $< \pm 20.0\permil$ δ^2H and $R^2 > 0.65$). Results for soil porosity look similar to those of the groundwater and are therefore not shown.

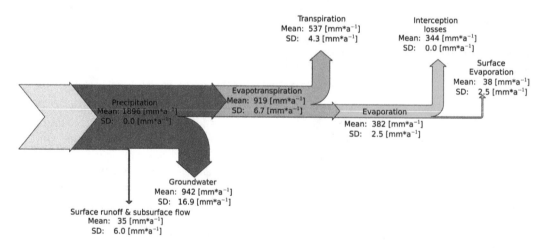

Figure 5. Mean annual flows and standard derivation (SD) of the main flow components at a hillslope scale of all behavioral model runs from 2010 to 2012.

measurements complementing the current setup in the future will allow us to put further confidence in this new approach and the drawn conclusions and allow us to directly compare different calibration targets (i.e., soil moisture vs. soil water isotopic signature).

Initial K_{sat} values based on literature values (see Table 1) deviate to a large extent from those derived through the calibration process. This is attributable to the occurrence of preferential flow within the macro-pores (Bronstert and Plate, 1997) and the sampling method (PCaps) used to extract the

Table 6. Parameter ranges used for validation (all calibration runs with NSE > 0.15, bias < ±20.0‰ δ^2H and R^2 > 0.65) and parameter set for the best modeled fit based on NSE.

	Mean	SD	Best modeled fit
K_{sat} [m d^{-1}]			
A1 top	21.8	5.8	20.4
A2 top	11.0	2.3	12.6
A3 top	25.6	6.3	29.6
A1	11.7	6.6	13.5
A2	7.4	2.8	8.9
A3	15.7	6.4	15.3
B1	4.0	2.4	4.0
B2	5.2	3.2	10.5
B3	4.6	2.2	2.5
C1	1.3	1.2	0.6
C2	1.7	1.4	0.1
Porosity [m^3 m^{-3}]			
A1 top	0.54	0.08	0.44
A2 top	0.56	0.09	0.44
A3 top	0.66	0.09	0.53
A1	0.55	0.08	0.42
A2	0.55	0.09	0.46
A3	0.65	0.09	0.74
B1	0.34	0.09	0.31
B2	0.64	0.09	0.54
B3	0.75	0.09	0.70
C1	0.54	0.09	0.41
C2	0.55	0.09	0.67
Groundwater depth [m]			
	50.5	28.6	76.5

soil water stored in the soil with a matrix potential up to 30 hPa (Landon et al., 1999). It becomes apparent that the mixing processes (based on dispersion and molecular diffusion) are not sufficient to equilibrate the isotope signature over the entire pore space (Landon et al., 1999; Šimůnek et al., 2003) and that the flow through the pore space is not homogenous. Thus the isotopic signature between the sampled pore media and the total modeled pore space differs (Brooks et al., 2010; Hrachowitz et al., 2013; McDonnell and Beven, 2014; McGlynn et al., 2002). The model tries to account for these effects by favoring high K_{sat} values during calibration (McDonnell and Beven, 2014; McGlynn et al., 2002).

Modeling soil water movement under such conditions should therefore be used with caution for models based on the Darcy–Richards equation, which assumes instantaneously homogeneous mixed solutions and uniform flow. In line with the argumentation started by Beven and Germann (1982) and refreshed in their recent paper (Beven and Germann, 2013), we therefore stress the importance

of accounting for preferential flow processes and overcoming the limitation of the Darcy–Richards equation limiting the explanatory power of hydrological models predicting water flow and solute/isotope transport in particular. Like Gerke (2006) and Šimůnek and van Genuchten (2008), among others, we therefore seek to implement a dual-permeability approach accounting for different flow patterns within the soil pore space (Gerke, 2006; Jarvis, 2007; Šimůnek and van Genuchten, 2008; Vogel et al., 2000, 2006, 2010). In the style of existing 1-D models for soil water isotope transport presented by Braud et al. (2005) and Haverd and Cuntz (2010), the inter-soil mixing processes by dispersion and molecular diffusion between different soil pore space compartments shall be accounted for in the future. Based on the presented findings this can now be extended towards the development and application of soil water isotope models under natural conditions. To conclude, the results highlight the general suitability of high-resolution soil water isotope profiles to improve our understanding of subsurface water flux separation implemented in current hillslope model applications and to predict subsurface soil water movement.

3.2 Modeled water fluxes

Acknowledging the general suitability of the model to delineate the prevailing flow patterns, we will now compare those to the current hydrological process understanding presented in the Introduction. Figure 5 depicts the water balance of the modeled hillslope based on all behavioral model realizations, separating the amount of incoming precipitation into the main flow components: surface runoff and subsurface flow directly entering the stream, percolation to groundwater and evapotranspiration.

Evapotranspiration is further subdivided into transpiration and evaporation from the soil surface and the canopy, whereby evaporation from the canopy is designated as interception losses. Due to the small confidence intervals of the behavioral model runs (see Fig. 3) the standard deviations of the model's flow components are relatively small (see Fig. 5; standard deviation and mean value were computed without weighting the likelihood value).

The observed order of magnitude for evapotranspiration is in good agreement with previous values of 945 and 876 mm a^{-1} reported for tropical grasslands by Windhorst et al. (2013a) and Oke (1987), respectively. As previously mentioned, the evaporation of 382 mm a^{-1} is dominated by interception losses accounting for 344 mm a^{-1}. Overall, these results support hypothesis II, which stated that a large share of the incoming precipitation is routed through the deeper soil layer and/or the groundwater body (here 49.7 % or 942 mm a^{-1}) before it enters the stream. This also explains the long mean transit time of water of around 1.0 to 3.9 years (Crespo et al., 2012; Timbe et al., 2014) in comparison to the fast runoff reaction time. Well in agreement with our current process understanding and hypothesis III, we can further show that the occurrence of surface runoff (33 mm a^{-1}) due to Hortonian overland flow is less important. For the graphical representation the surface runoff has therefore been combined with subsurface flow (2 mm a^{-1}) to "surface runoff and subsurface flow", accounting in total for 35 mm a^{-1} (see Fig. 5). A more heterogeneous picture can be depicted if we take a closer look at the flow processes along the studied hillslope and its soil profiles (Fig. 6).

Vertical fluxes still dominate the flow of water (Fig. 6b), but the near-surface lateral flow components predicted by Bücker et al. (2010) become more evident (Fig. 6a). Explained by the high saturated hydraulic conductivities in the top soil layers (Table 6 and Fig. 4) up to 7.3×10^3 m^3 a^{-1} are transported laterally between cells in the top soil layer, referring to 15.6 % of the total flow leaving the system per year. According to the model results, deep lateral flow is minimal, accounting only for < 0.1 % of the total flow. It only occurs on top of the deeper soil horizons with low K_{sat} values. For all behavioral model realizations the groundwater level was > 2 m, thereby limiting the direct contribution of subsurface flow (2 mm a^{-1}) to the tributary, which had a hydraulic potential of only 1.5 m. Over the entire hillslope the importance of overland flow remains below 3 % (≈ 50 mm a^{-1}), of which a part is re-infiltrating, summing up to total overland flow losses of around 2 % at the hillslope scale (35 mm a^{-1}, Fig. 5). These results demonstrate the importance of near-surface lateral flow and hence support hypothesis IV.

4 Conclusions

These data and findings support and complement the existing process understanding mainly gained by Goller et al. (2005), Fleischbein et al. (2006), Boy et al. (2008), Bücker et al. (2010), Crespo et al. (2012) and Timbe et al. (2014) to a large extent. Moreover, it was possible to quantify for the first time the relevance of near-surface lateral flow generation. The observed dominance of vertical percolation into the groundwater body, and thereby the importance of preferential flow seems to be quite common for humid tropical montane regions and has recently been reported by Muñoz-Villers and McDonnell (2012) in a similar environment.

Being aware of the rapid rainfall–runoff response of streams within the catchment of the Rio San Francisco, it has been questioned whether and how the system can store water for several years and still release it within minutes. Throughout the last decades several studies have observed similar hydrological behavior especially for steep humid montane regions (e.g., McDonnell, 1990; Muñoz-Villers and McDonnell, 2012), and concepts have been developed to explain this behavior: e.g., piston flow (McDonnell, 1990), kinematic waves (Lighthill and Whitham, 1955) and transmissivity feedback (Kendall et al., 1999). Due to the limited depth of observations (max depth 0.4 m) and the low overall

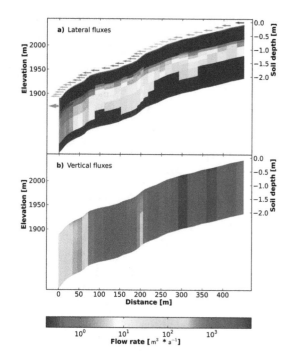

Figure 6. (a) Lateral and **(b)** vertical fluxes for the best modeled fit. Arrows indicate the amount of surface runoff and direct contribution to the outlet through subsurface flow. The maximum flow between storage compartments is 7.3×10^3 m^3 a^{-1}, and the total observed flow leaving as well as entering the system accumulates to 37×10^3 m^3 a^{-1}.

influence of the lateral flows, a more exact evaluation of the fate of the percolated water is still not possible. However, we are confident that in combination with a suitable concept to account for the rapid mobilization of the percolated water into a tributary and experimental findings, further confining possible model realizations, an improved version of the current approach could further close the gap in our current process understanding.

Over decades hydrological models which are based on the Richards or Darcy equation (like the one we used) have been tuned to predict quantitative flow processes and mostly been validated using soil moisture data suitable to account for overall storage changes. Our results imply that doing this considerably well does not necessarily mean that the models actually transport the *right* water at the *right* time. Using tracer data to validate models as we did entails that those models now not only have to transport the correct amount but additionally the right water. Consequently, the relevance of the correct representation of uneven preferential flow through pipes or macropores, which is misleadingly compensated by high conductivities over the entire pore space within models based on the Richards or Darcy equation, becomes immense. Distinguishing between water flowing in different compartments (e.g., pipes, cracks and macro-pores) of the soil is a key task to get a closer and more precise representation of the natural flow processes. Even though the chosen modeling structure currently lacks a sufficient robustness to be widely applicable, it highlights the potential and future research directions for soil water isotope modeling.

Acknowledgements. The current study was conducted within the DFG Research Group FOR 816 "Biodiversity and sustainable management of a megadiverse mountain rain forest in south Ecuador" and the follow-up project PAK 825/3. The authors are very grateful for the funding supplied by the German Research Foundation (DFG) (BR2238/4-2 and BR2238/14-1) and thank Thorsten Peters of the University of Erlangen for providing meteorological data.

Edited by: H. Cloke

References

Barthold, F. K., Tyralla, C., Schneider, K., Vaché, K. B., Frede, H.-G., and Breuer, L.: How many tracers do we need for end member mixing analysis (EMMA)? A sensitivity analysis, Water Resour. Res., 47, W08519, doi:10.1029/2011WR010604, 2011.

Beck, E., Makeschin, F., Haubrich, F., Richter, M., Bendix, J., and Valerezo, C.: The Ecosystem (Reserva Biológica San Francisco), in: Gradients in a Tropical Mountain Ecosystem of Ecuador, Ecological Studies, edited by: Beck, E., Bendix, J., Kottke, I., Makeschin, F., and Mosandl, R., Springer, Berlin, Heidelberg, 1–13, 2008.

Bendix, J., Rollenbeck, R., Richter, M., Fabian, P., and Emck, P.: Climate, in: Gradients in a Tropical Mountain Ecosystem of Ecuador, Ecological Studies, edited by: Beck, E., Bendix, J., Kottke, I., Makeschin, F., and Mosandl, R., Springer, Berlin, Heidelberg, 63–73, 2008.

Bendix, J., Silva, B., Roos, K., Göttlicher, D. O., Rollenbeck, R., Nauß, T., and Beck, E.: Model parameterization to simulate and compare the PAR absorption potential of two competing plant species, Int. J. Biometeorol., 54, 283–295, doi:10.1007/s00484-009-0279-3, 2010.

Beven, K. and Freer, J.: Equifinality, data assimilation, and uncertainty estimation in mechanistic modelling of complex environmental systems using the GLUE methodology, J. Hydrol., 249, 11–29, doi:10.1016/S0022-1694(01)00421-8, 2001.

Beven, K. and Germann, P.: Macropores and water flow in soils, Water Resour. Res., 18, 1311–1325, doi:10.1029/WR018i005p01311, 1982.

Beven, K. and Germann, P.: Macropores and water flow in soils revisited, Water Resour. Res., 49, 3071–3092, doi:10.1002/wrcr.20156, 2013.

Bogner, C., Bauer, F., Trancón y Widemann, B., Viñan, P., Balcazar, L., and Huwe, B.: Quantifying the morphology of flow patterns in landslide-affected and unaffected soils, J. Hydrol., 511, 460–473, doi:10.1016/j.jhydrol.2014.01.063, 2014.

Boy, J., Valarezo, C., and Wilcke, W.: Water flow paths in soil control element exports in an Andean tropical montane forest, Eur. J. Soil Sci., 59, 1209–1227, doi:10.1111/j.1365-2389.2008.01063.x, 2008.

Braud, I., Bariac, T., Gaudet, J. P., and Vauclin, M.: SiSPAT-Isotope, a coupled heat, water and stable isotope (HDO and H218O) transport model for bare soil. Part I. Model description and first verifications, J. Hydrol., 309, 277–300, doi:10.1016/j.jhydrol.2004.12.013, 2005.

Bronstert, A.: Capabilities and limitations of detailed hillslope hydrological modelling, Hydrol. Process., 13, 21–48, doi:10.1002/(SICI)1099-1085(199901)13:1<21::AID-HYP702>3.0.CO;2-4, 1999.

Bronstert, A. and Plate, E. J.: Modelling of runoff generation and soil moisture dynamics for hillslopes and micro-catchments, J. Hydrol., 198, 177–195, doi:10.1016/S0022-1694(96)03306-9, 1997.

Brooks, J. R., Barnard, H. R., Coulombe, R., and McDonnell, J. J.: Ecohydrologic separation of water between trees and streams in a Mediterranean climate, Nat. Geosci., 3, 100–104, doi:10.1038/ngeo722, 2010.

Bücker, A., Crespo, P., Frede, H.-G., Vaché, K., Cisneros, F., and Breuer, L.: Identifying Controls on Water Chemistry of Tropical Cloud Forest Catchments: Combining Descriptive Approaches and Multivariate Analysis, Aquat. Geochem., 16, 127–149, doi:10.1007/s10498-009-9073-4, 2010.

Bücker, A., Crespo, P., Frede, H.-G., and Breuer, L.: Solute behaviour and export rates in neotropical montane catchments under different land-uses, J. Trop. Ecol., 27, 305–317, doi:10.1017/S0266467410000787, 2011.

Colman, R. L. and Wilson, G. P. M.: The Effect of Floods on Pasture Plants, Agric. Gaz. NSW, 71, 337–347, 1960.

Craig, H.: Standard for Reporting Concentrations of Deuterium and Oxygen-18 in Natural Waters, Science, 133, 1833–1834, doi:10.1126/science.133.3467.1833, 1961.

Craig, H. and Gordon, L. I.: Deuterium and oxygen 18 variations in the ocean and the marine atmosphere, in: Proceedings of the Conference on the Stable Isotopes in Oceanographic Studies and Paleotemperatures, edited by: Tongiogi, E. and Lishi, E. F., Pisa, Spoleto, Italy, 9–130, 1965.

Crespo, P., Bücker, A., Feyen, J., Vaché, K. B., Frede, H.-G., and Breuer, L.: Preliminary evaluation of the runoff processes in a remote montane cloud forest basin using Mixing Model Analysis and Mean Transit Time, Hydrol. Process., 26, 3896–3910, doi:10.1002/hyp.8382, 2012.

Davies, J., Beven, K., Rodhe, A., Nyberg, L., and Bishop, K.: Integrated modeling of flow and residence times at the catchment scale with multiple interacting pathways, Water Resour. Res., 49, 4738–4750, doi:10.1002/wrcr.20377, 2013.

Dohnal, M., Vogel, T., Šanda, M., and Jelínková, V.: Uncertainty Analysis of a Dual-Continuum Model Used to Simulate Subsurface Hillslope Runoff Involving Oxygen-18 as Natural Tracer, J. Hydrol. Hydromech., 60, 194–205, doi:10.2478/v10098-012-0017-0, 2012.

Dubbert, M., Cuntz, M., Piayda, A., Maguás, C., and Werner, C.: Partitioning evapotranspiration – Testing the Craig and Gordon model with field measurements of oxygen isotope ratios of evaporative fluxes, J. Hydrol., 496, 142–153, doi:10.1016/j.jhydrol.2013.05.033, 2013.

Federer, C. A., Vörösmarty, C., and Fekete, B.: Sensitivity of Annual Evaporation to Soil and Root Properties in Two Models of Contrasting Complexity, J. Hydrometeorol., 4, 1276–1290, doi:10.1175/1525-7541(2003)004<1276:SOAETS>2.0.CO;2, 2003.

Fleischbein, K., Wilcke, W., Valarezo, C., Zech, W., and Knoblich, K.: Water budgets of three small catchments under montane forest in Ecuador: experimental and modelling approach, Hydrol. Process., 20, 2491–2507, doi:10.1002/hyp.6212, 2006.

Frisbee, M. D., Phillips, F. M., Campbell, A. R., and Hendrickx, J. M. H.: Modified passive capillary samplers for collecting samples of snowmelt infiltration for stable isotope analysis in remote, seasonally inaccessible watersheds 1: laboratory evaluation, Hydrol. Process., 24, 825–833, doi:10.1002/hyp.7523, 2010.

Garvelmann, J., Külls, C., and Weiler, M.: A porewater-based stable isotope approach for the investigation of subsurface hydrological processes, Hydrol. Earth Syst. Sci., 16, 631–640, doi:10.5194/hess-16-631-2012, 2012.

Genereux, D. P. and Hooper, R. P.: Oxygen and hydrogen isotopes in rainfall-runoff studies, in: Isotope Tracers in Catchment Hydrology, edited by: Kendall, C. and McDonnell, J. J., Elsevier, Amsterdam, 319–346, 1999.

Gerke, H. H.: Preferential flow descriptions for structured soils, J. Plant Nutr. Soil Sci., 169, 382–400, doi:10.1002/jpln.200521955, 2006.

Germann, P., Helbling, A., and Vadilonga, T.: Rivulet Approach to Rates of Preferential Infiltration, Vadose Zone J., 6, 207–220, doi:10.2136/vzj2006.0115, 2007.

Goller, R., Wilcke, W., Leng, M. J., Tobschall, H. J., Wagner, K., Valarezo, C., and Zech, W.: Tracing water paths through small catchments under a tropical montane rain forest in south Ecuador by an oxygen isotope approach, J. Hydrol., 308, 67–80, doi:10.1016/j.jhydrol.2004.10.022, 2005.

Hacker, J. B. and Jones, R. J.: The Setaria sphacelata complex – a review, Trop. Grassl., 3, 13–34, 1969.

Haverd, V. and Cuntz, M.: Soil–Litter–Iso: A one-dimensional model for coupled transport of heat, water and stable isotopes in soil with a litter layer and root extraction, J. Hydrol., 388, 438–455, doi:10.1016/j.jhydrol.2010.05.029, 2010.

Hrachowitz, M., Savenije, H., Bogaard, T. A., Tetzlaff, D., and Soulsby, C.: What can flux tracking teach us about water age distribution patterns and their temporal dynamics?, Hydrol. Earth Syst. Sci., 17, 533–564, doi:10.5194/hess-17-533-2013, 2013.

Hsieh, J. C., Chadwick, O. A., Kelly, E. F., and Savin, S. M.: Oxygen isotopic composition of soil water: Quantifying evaporation and transpiration, Geoderma, 82, 269–293, doi:10.1016/S0016-7061(97)00105-5, 1998.

Huwe, B., Zimmermann, B., Zeilinger, J., Quizhpe, M., and Elsenbeer, H.: Gradients and Patterns of Soil Physical Parameters at Local, Field and Catchment Scales, in: Gradients in a Tropical Mountain Ecosystem of Ecuador, Ecological Studies, edited by: Beck, E., Bendix, J., Kottke, I., Makeschin, F., and Mosandl, R., Springer, Berlin, Heidelberg, 375–386, 2008.

Jarvis, N. J.: A review of non-equilibrium water flow and solute transport in soil macropores: principles, controlling factors and consequences for water quality, Eur. J. Soil Sci., 58, 523–546, doi:10.1111/j.1365-2389.2007.00915.x, 2007.

Kendall, K. A., Shanley, J. B., and McDonnell, J. J.: A hydrometric and geochemical approach to test the transmissivity feedback hypothesis during snowmelt, J. Hydrol., 219, 188–205, doi:10.1016/S0022-1694(99)00059-1, 1999.

Kirkby, M.: Hillslope runoff processes and models, J. Hydrol., 100, 315–339, doi:10.1016/0022-1694(88)90190-4, 1988.

Körner, C., Scheel, J., and Bauer, H.: Maximum leaf diffusive conductance in vascular plants, Photosynthetica, 13, 45–82, 1979.

Kraft, P., Multsch, S., Vaché, K. B., Frede, H.-G., and Breuer, L.: Using Python as a coupling platform for integrated catchment models, Adv. Geosci., 27, 51–56, doi:10.5194/adgeo-27-51-2010, 2010.

Kraft, P., Vaché, K. B., Frede, H.-G., and Breuer, L.: CMF: A Hydrological Programming Language Extension For Integrated Catchment Models, Environ. Model. Softw., 26, 828–830, doi:10.1016/j.envsoft.2010.12.009, 2011.

Landon, M. K., Delin, G. N., Komor, S. C., and Regan, C. P.: Comparison of the stable-isotopic composition of soil water collected from suction lysimeters, wick samplers, and cores in a sandy unsaturated zone, J. Hydrol., 224, 45–54, doi:10.1016/S0022-1694(99)00120-1, 1999.

Leibundgut, C., Maloszewski, P., and Külls, C.: Tracers in Hydrology, John Wiley & Sons, Chichester, UK, 2011.

Lighthill, M. J. and Whitham, G. B.: On Kinematic Waves. II. A Theory of Traffic Flow on Long Crowded Roads, Proc. Roy. Soc. Lond. Ser. Math. Phys. Sci., 229, 317–345, doi:10.1098/rspa.1955.0089, 1955.

Liu, W. J., Liu, W. Y., Li, P. J., Gao, L., Shen, Y. X., Wang, P. Y., Zhang, Y. P., and Li, H. M.: Using stable isotopes to determine sources of fog drip in a tropical seasonal rain forest of Xishuangbanna, SW China, Agr. Forest Meteorol., 143, 80–91, doi:10.1016/j.agrformet.2006.11.009, 2007.

McDonnell, J. J.: A Rationale for Old Water Discharge Through Macropores in a Steep, Humid Catchment, Water Resour. Res., 26, 2821–2832, doi:10.1029/WR026i011p02821, 1990.

McDonnell, J. J. and Beven, K.: Debates – The future of hydrological sciences: A (common) path forward? A call to action aimed at understanding velocities, celerities and residence time distributions of the headwater hydrograph, Water Resour. Res., 50, 5342–5350, doi:10.1002/2013WR015141, 2014.

McDonnell, J. J., Sivapalan, M., Vaché, K., Dunn, S., Grant, G., Haggerty, R., Hinz, C., Hooper, R., Kirchner, J., Roderick, M. L., Selker, J., and Weiler, M.: Moving beyond heterogeneity and process complexity: A new vision for watershed hydrology, Water Resour. Res., 43, W07301, doi:10.1029/2006WR005467, 2007.

McGlynn, B. L., McDonnel, J. J., and Brammer, D. D.: A review of the evolving perceptual model of hillslope flowpaths at the Maimai catchments, New Zealand, J. Hydrol., 257, 1–26, doi:10.1016/S0022-1694(01)00559-5, 2002.

McKay, M. D., Beckman, R. J., and Conover, W. J.: A Comparison of Three Methods for Selecting Values of Input Variables in the Analysis of Output from a Computer Code, Technometrics, 21, 239–245, doi:10.2307/1268522, 1979.

McMillan, H., Tetzlaff, D., Clark, M., and Soulsby, C.: Do time-variable tracers aid the evaluation of hydrological model structure? A multimodel approach, Water Resour. Res., 48, W05501, doi:10.1029/2011WR011688, 2012.

Mertens, J., Diels, J., Feyen, J., and Vanderborght, J.: Numerical analysis of Passive Capillary Wick Samplers prior to field installation, Soil Sci. Soc. Am. J., 71, 35–42, doi:10.2136/sssaj2006.0106, 2007.

Muñoz-Villers, L. E. and McDonnell, J. J.: Runoff generation in a steep, tropical montane cloud forest catchment on permeable volcanic substrate, Water Resour. Res., 48, W09528, doi:10.1029/2011WR011316, 2012.

Oke, T. R.: Boundary Layer Climates, 2nd Edn., Methuen, London, UK, 1987.

Qu, Y. and Duffy, C. J.: A semidiscrete finite volume formulation for multiprocess watershed simulation, Water Resour. Res., 43, W08419, doi:10.1029/2006WR005752, 2007.

Rhoades, C. C., Eckert, G. E., and Coleman, D. C.: Soil carbon differences among forest, agriculture, and secondary vegetation in lower montane Ecuador, Ecol. Appl., 10, 497–505, doi:10.1890/1051-0761(2000)010[0497:SCDAFA]2.0.CO;2, 2000.

Scholl, M. A., Shanley, J. B., Zegarra, J. P., and Coplen, T. B.: The stable isotope amount effect: New insights from NEXRAD echo tops, Luquillo Mountains, Puerto Rico, Water Resour. Res., 45, W12407, doi:10.1029/2008WR007515, 2009.

Shuttleworth, W. J. and Gurney, R. J.: The theoretical relationship between foliage temperature and canopy resistance in sparse crops, Q. J. Roy. Meteorol. Soc., 116, 497–519, doi:10.1002/qj.49711649213, 1990.

Shuttleworth, W. J. and Wallace, J. S.: Evaporation from sparse crops-an energy combination theory, Q. J. Roy. Meteorol. Soc., 111, 839–855, doi:10.1002/qj.49711146910, 1985.

Šimůnek, J. and van Genuchten, M. T.: Modeling Nonequilibrium Flow and Transport Processes Using HYDRUS, Vadose Zone J., 7, 782–797, doi:10.2136/vzj2007.0074, 2008.

Šimůnek, J., Jarvis, N. J., and van Genuchten, M. T., and Gärdenäs, A.: Review and comparison of models for describing non-equilibrium and preferential flow and transport in the vadose zone, J. Hydrol., 272, 14–35, doi:10.1016/S0022-1694(02)00252-4, 2003.

Sklash, M. G., Farvolden, R. N., and Fritz, P.: A conceptual model of watershed response to rainfall, developed through the use of oxygen-18 as a natural tracer, Can. J. Earth Sci. 13, 271–283, doi:10.1139/e76-029, 1976.

Soulsby, C., Rodgers, P., Smart, R., Dawson, J., and Dunn, S.: A tracer-based assessment of hydrological pathways at different spatial scales in a mesoscale Scottish catchment, Hydrol. Process., 17, 759–777, doi:10.1002/hyp.1163, 2003.

Stumpp, C. and Maloszewski, P.: Quantification of preferential flow and flow heterogeneities in an unsaturated soil planted with different crops using the environmental isotope δ^{18}O, J. Hydrol., 394, 407–415, doi:10.1016/j.jhydrol.2010.09.014, 2010.

Tetzlaff, D., McDonnell, J. J., Uhlenbrook, S., McGuire, K. J., Bogaart, P. W., Naef, F., Baird, A. J., Dunn, S. M., and Soulsby, C.: Conceptualizing catchment processes: simply too complex?, Hydrol. Process., 22, 1727–1730, doi:10.1002/hyp.7069, 2008.

Thom, A. S.: Momentum, mass and heat exchange of vegetation, Q. J. Roy. Meteorol. Soc., 98, 124–134, doi:10.1002/qj.49709841510, 1972.

Timbe, E., Windhorst, D., Crespo, P., Frede, H.-G., Feyen, J., and Breuer, L.: Understanding uncertainties when inferring mean transit times of water trough tracer-based lumped-parameter models in Andean tropical montane cloud forest catchments, Hydrol. Earth Syst. Sci., 18, 1503–1523, doi:10.5194/hess-18-1503-2014, 2014.

Uhlenbrook, S., Roser, S., and Tilch, N.: Hydrological process representation at the meso-scale: the potential of a distributed, conceptual catchment model, J. Hydrol., 291, 278–296, doi:10.1016/j.jhydrol.2003.12.038, 2004.

Vogel, H.-J., Cousin, I., Ippisch, O., and Bastian, P.: The dominant role of structure for solute transport in soil: experimental evidence and modelling of structure and transport in a field experiment, Hydrol. Earth Syst. Sci., 10, 495–506, doi:10.5194/hess-10-495-2006, 2006.

Vogel, T., Gerke, H. H., Zhang, R., and Van Genuchten, M. T.: Modeling flow and transport in a two-dimensional dual-permeability system with spatially variable hydraulic properties, J. Hydrol., 238, 78–89, doi:10.1016/S0022-1694(00)00327-9, 2000.

Vogel, T., Sanda, M., Dusek, J., Dohnal, M., and Votrubova, J.: Using Oxygen-18 to Study the Role of Preferential Flow in the Formation of Hillslope Runoff, Vadose Zone J., 9, 252–259, doi:10.2136/vzj2009.0066, 2010.

Weiler, M. and McDonnell, J.: Virtual experiments: a new approach for improving process conceptualization in hillslope hydrology, J. Hydrol., 285, 3–18, doi:10.1016/S0022-1694(03)00271-3, 2004.

Wheeler, M. D., Newman, S. M., Orr-Ewing, A. J., and Ashfold, M. N. R.: Cavity ring-down spectroscopy, J. Chem. Soc. Faraday Trans., 94, 337–351, doi:10.1039/A707686J, 1998.

Windhorst, D., Brenner, S., Peters, T., Meyer, H., Thies, B., Bendix, J., Frede, H.-G., and Breuer, L.: Impacts of Local Land-Use Change on Climate and Hydrology, in: Ecosystem Services, Biodiversity and Environmental Change in a Tropical Mountain Ecosystem of South Ecuador, Ecological Studies Vol. 221, edited by: Bendix, J., Beck, E., Bräuning, A., Makeschin, F., Mosandl, R., Scheu, S., and Wilcke, W., Springer, Berlin, Heidelberg, New York, 275–286, 2013a.

Windhorst, D., Waltz, T., Timbe, E., Frede, H.-G., and Breuer, L.: Impact of elevation and weather patterns on the isotopic composition of precipitation in a tropical montane rainforest, Hydrol. Earth Syst. Sci., 17, 409–419, doi:10.5194/hess-17-409-2013, 2013b.

Zimmermann, B. and Elsenbeer, H.: Spatial and temporal variability of soil saturated hydraulic conductivity in gradients of disturbance, J. Hydrol., 361, 78–95, doi:10.1016/j.jhydrol.2008.07.027, 2008.

Zimmermann, U., Ehhalt, D., and Muennich, K. O.: Soil-Water Movement and Evapotranspiration: Changes in the Isotopic Composition of the Water, in: Isotopes in Hydrology, International Atomic Energy Agency, Vienna, 567–585, 1968.

12

Estimation of effective porosity in large-scale groundwater models by combining particle tracking, auto-calibration and ^{14}C dating

Rena Meyer[1], **Peter Engesgaard**[1], **Klaus Hinsby**[2], **Jan A. Piotrowski**[3], and **Torben O. Sonnenborg**[2]

[1]Department of Geosciences and Natural Resources Management, University of Copenhagen, Øster Voldgade 10, 1350 Copenhagen, Denmark
[2]Geological Survey of Denmark and Greenland, Øster Voldgade 10, 1350 Copenhagen, Denmark
[3]Department of Geosciences, Aarhus University, Høegh-Guldbergs Gade 2, 8000 Aarhus, Denmark

Correspondence: Rena Meyer (reme@ign.ku.dk)

Abstract. Effective porosity plays an important role in contaminant management. However, the effective porosity is often assumed to be constant in space and hence heterogeneity is either neglected or simplified in transport model calibration. Based on a calibrated highly parametrized flow model, a three-dimensional advective transport model (MODPATH) of a 1300 km² coastal area of southern Denmark and northern Germany is presented. A detailed voxel model represents the highly heterogeneous geological composition of the area. Inverse modelling of advective transport is used to estimate the effective porosity of 7 spatially distributed units based on apparent groundwater ages inferred from 11 ^{14}C measurements in Pleistocene and Miocene aquifers, corrected for the effects of diffusion and geochemical reactions. By calibration of the seven effective porosity units, the match between the observed and simulated ages is improved significantly, resulting in a reduction of ME of 99 % and RMS of 82 % compared to a uniform porosity approach. Groundwater ages range from a few hundred years in the Pleistocene to several thousand years in Miocene aquifers. The advective age distributions derived from particle tracking at each sampling well show unimodal (for younger ages) to multimodal (for older ages) shapes and thus reflect the heterogeneity that particles encounter along their travel path. The estimated effective porosity field, with values ranging between 4.3 % in clay and 45 % in sand formations, is used in a direct simulation of distributed mean groundwater ages. Although the absolute ages are affected by various uncertainties, a unique insight into the complex three-dimensional age distribution pattern and potential advance of young contaminated groundwater in the investigated regional aquifer system is provided, highlighting the importance of estimating effective porosity in groundwater transport modelling and the implications for groundwater quantity and quality assessment and management.

1 Introduction

The age of groundwater, i.e. the time elapsed since the water molecule entered the groundwater (Cook and Herczeg, 2000; Kazemi et al., 2006), is useful (i) to infer recharge rates (e.g. Sanford et al., 2004; Wood et al., 2017) and hence to sustainably exploit groundwater resources; (ii) to evaluate contaminant migration, fate and history (Bohlke and Denver, 1995; Hansen et al., 2012) and predict spread of pollutants and timescales for intrinsic remediation (Kazemi et al., 2006); (iii) to analyse aquifer vulnerability or protection to surface-derived contaminants (e.g. Manning et al., 2005; Bethke and Johnson, 2008; Molson and Frind, 2012; Sonnenborg et al., 2016) and indicate the advance of modern contaminated groundwater (Hinsby et al., 2001b; Gleeson et al., 2015; Jasechko et al., 2017) and groundwater quality in general (Hinsby et al., 2007); and (iv) to contribute to the understanding of the flow system, e.g. in complex geological settings (Troldborg et al., 2008; Eberts et al., 2012).

The groundwater science community (de Dreuzy and Ginn, 2016) has a continued interest in the topic of residence time distributions (RTDs) in the subsurface. Turnadge and Smerdon (2014) reviewed different methods for modelling environmental tracers in groundwater, including lumped parameter models (e.g. Maloszewski and Zuber, 1996), mixing-cell models (e.g. Campana and Simpson, 1984; Partington et al., 2011) and direct age models (e.g. Cornaton, 2012; Goode, 1996; Woolfenden and Ginn, 2009). Here, we focus on three different approaches with specific benefits and disadvantages that are commonly applied to simulate groundwater age in 3-D distributed groundwater flow and transport models (Castro and Goblet, 2005; Sanford et al., 2017). Particle-based advective groundwater age calculation utilizing travel time analysis is computationally easy, but neglects diffusion and dispersion. The full advection–dispersion transport simulation of a solute or an environmental tracer is computationally expensive and limited to the specific tracer characteristics (McCallum et al., 2015; Salmon et al., 2015), but accounts for diffusion, dispersion and mixing. The tracer-independent direct simulation of groundwater mean age (Goode, 1996; Engesgaard and Molson, 1998; Bethke and Johnson, 2002) includes advection, diffusion and dispersion processes and yields a spatial distribution of mean ages. A comparison of ages simulated using any of these methods with ages determined from tracer observations, referred to as apparent ages, is desirable as it can improve the uniqueness in flow model calibration and validation (Castro and Goblet, 2003; Ginn et al., 2009) and it potentially informs about transport parameters such as effective porosity, diffusion and dispersion that are otherwise difficult to estimate. However, the approach is far from straightforward as environmental tracers undergo non-linear changes in their chemical species (McCallum et al., 2015) and groundwater models only represent a simplification and compromise on structural and/or parameter heterogeneity. In a 2-D synthetic model, McCallum et al. (2014) investigated the bias of apparent ages in heterogeneous systems systematically. McCallum et al. (2015) applied correction terms, e.g. diffusion correction for radioactive tracers, on apparent ages to improve the comparability to mean advective ages. They concluded that with increasing heterogeneity the width of the residence time distribution increases and that apparent ages would only represent mean ages if this distribution is narrow and has a small variance. It is important here to distinguish between mean and radiometric ages, as defined by Varni and Carrera (1998) for example. The only way they can be directly compared in reality is if no mixing is taking place, i.e. if the flow field can be regarded as pure piston flow, which will give the kinematic age.

Flow and transport parameters such as hydraulic conductivity, conductance of streambeds and drains, recharge, and dispersivities have gained more and more focus in calibration of groundwater models, recently also on large scales, where of head, flow and tracer observations are widely used as targets (McMahon et al., 2010). However, effective porosity has not received nearly as much attention, and its spatial variability in particular is often neglected, except for Starn et al. (2014). The lack of focus on calibrating distributed effective porosity on a regional scale might be related to the common assumption that recharge in humid climates can be precisely estimated and porosity of porous media is relatively well known from the literature (Sanford, 2011). However, for steady-state flow (Ginn et al., 2009) in a layered aquifer system, Bethke and Johnson (2002) concluded that the mean groundwater age exchange between flow and stagnant zones is only a function of the volume of stored water (Harvey and Gorelick, 1995; Varni and Carrera, 1998). Thus, the groundwater age exchange is directly related to the porosity. Yet, the calibration of a spatially distributed porosity field and its application to simulate groundwater ages and infer capture zones has not gained much attention.

The uniqueness of the presented study lies in the calibration of a three-dimensional, spatially distributed, effective porosity field in a regional-scale complex multi-layered heterogeneous coastal aquifer system. The aims are as follows: (i) to use apparent ages inferred from dissolution- and diffusion-corrected ^{14}C measurements from different aquifer units as targets in auto-calibration with PEST of seven unit-specific effective porosities in an advective (particle tracking) transport model – it would have been optimal to use RTD analysis (de Dreuzy and Ginn, 2016) to compare modelled and inferred groundwater ages in this study, but due to the rather complex nature of our hydrogeological flow model, the inherent uncertainties associated with inferring an apparent age from ^{14}C analysis and the long computer runtimes, we have chosen to use the particle-based kinematic approach of simulating a mixed age at the well screen (or numerical cell with a screen); (ii) to assess the advective age distributions at the sampling locations to obtain information on the age spreading; (iii) to apply the estimated seven porosities in a direct age simulation (Goode, 1996) to gain insight in the three-dimensional age pattern of the investigation area; and (iv) to assess the effect of using the heterogeneous porosity model compared to a homogeneous porosity model for differences in capture zones via particle back-tracking, which is a water management approach to define wellhead protection areas or optimize pump-and-treat locations for remediation of pollution (Anderson et al., 2015).

2 Study area

The $1300\,\mathrm{km}^2$ investigation area is located adjacent to the Wadden Sea in the border region between southern Denmark and northern Germany (Fig. 1). During the Last Glacial Maximum (LGM; 22 to 19 ka ago, Stroeven et al., 2016), the area was the direct foreland of the Scandinavian ice sheet. The low-lying marsh areas (with elevations below mean sea level) in the west were reclaimed from the Wadden Sea over the last few centuries and protected from flooding by a dike for the last ≈ 200 a. A dense network of drainage channels keeps the groundwater level constantly below the ground surface and thus mostly below sea level. The water divide near the

Jutland ridge with elevations of up to 85 m a.s.l. defines the eastern boundary of the study area.

The aquifer systems are geologically complex and highly heterogeneous, spanning Miocene through Holocene deposits. The bottom of the aquifer system is defined by low-permeability Palaeogene marine clay. The overlying Miocene deposits consist of alternating marine clay and deltaic silt and sand (Rasmussen et al., 2010). The Maade formation, an upper Miocene marine clay unit, with a relatively large thickness in the west while thinning out to the east, is located below the Pleistocene and Holocene deposits. Buried valleys filled with glacial deposits, mainly from the Saalian glaciation, cut through the Miocene and reach depths up to 450 m below the surface. They are important hydrogeological features as they may constitute preferential flow paths and locally connect the Pleistocene and Miocene aquifers.

In our previous studies (Jørgensen et al., 2015; Høyer et al., 2017; Meyer et al., 2018a), the available geological and geophysical information including borehole lithology, airborne electromagnetic (AEM) and seismic data were assembled into a heterogeneous geological voxel model comprising 46 geological units with raster sizes of 100 m × 100 m × 5 m. Manual and automatic modelling strategies, such as clay fraction (CF), multi-point simulation (MPS) and a cognitive layer approach, were complementarily applied. Meyer et al. (2018a) investigated the regional flow system and identified the most dominant mechanisms governing the flow system, comprising geological features and land management that are visualized in a conceptual model in Fig. 2. Extensive clay layers separate the Miocene and Pleistocene aquifers; buried valleys locally cut through the Maade formation and connect Miocene and Pleistocene aquifers, allowing groundwater exchange and mixing. The large drainage network, established in the reclaimed terrain and keeping the groundwater table constantly below the sea level, acts as a large sink for the entire area. In the deeper aquifers, significant inflow from the ocean occurs at the coast near the marsh area as a result of a landward head gradient induced by the drainage.

3   Methods

The age simulation and calibration of effective porosities builds upon the calibrated regional-scale groundwater flow model (MODFLOW) of a highly heterogeneous coastal aquifer system by Meyer et al. (2018a). First, an advective transport simulation using MODPATH (Pollock, 2012) was used for the calibration of effective porosities of seven different geological units. ^{14}C observations were corrected for carbon dissolution and diffusion and subsequently used as calibration targets during inverse modelling with PEST. It would have been optimal to use RTD analysis (de Dreuzy and Ginn, 2016) to compare modelled and inferred groundwater ages in this study. But, due to the rather complex nature of our hydrogeological flow model, the inherent uncer-

tainties associated with inferring an apparent age from ^{14}C analysis, and the long computer runtimes, we have chosen to use the particle-based kinematic approach of simulating a mixed age at the well screen (or numerical cell with a screen). Secondly, the analysis of advective age distributions at ^{14}C sampling locations provided an insight in the ranges of travel times and distances and hereby the complexity of groundwater age mixing. Thirdly, the estimated effective porosities were used in a direct age simulation (Goode, 1996) in order to investigate the spatial groundwater age distribution in the regional aquifer system. Finally, the impact of using a seven-porosity model compared to a constant porosity model on capture zone delineation at two well locations was assessed.

3.1 ^{14}C measurements

During a field campaign in February 2015, 18 groundwater samples were collected from wells at seven sites with screens at different depths and in different aquifers (Fig. 1, Table 1). The wells were pumped clean with 3 times their volume to prevent the influence of mixing with stagnant water. In situ parameters (pH, EC, O_2) were measured and, after they stabilized, samples for radiocarbon analyses were collected in 1 L opaque glass bottles. The 18 groundwater samples were analysed for $\delta^{13}C_{\%oVPBD}$ with an isotope ratio mass spectrometer (IRMS) and for ^{14}C with an accelerator mass spectrometer (AMS) at the AGH University of Science and Technology, Kraków, Poland, and in the Poznań Radiocarbon Laboratory, Poznań, Poland, respectively, in September 2015.

3.1.1 ^{14}C correction for dissolution and diffusion

The ^{14}C activity (A_m) was measured in the dissolved inorganic carbon (DIC) content of the groundwater. Uncertainties arise from geochemical and hydrodynamic processes that change the ^{14}C content in the aquifer (e.g. Bethke and Johnson, 2008; Sudicky and Frind, 1981). The dissolution of "dead" fossil (^{14}C-free) carbon dilutes the ^{14}C content in groundwater and results in lower ^{14}C concentrations (Appelo and Postma, 2005). Diffusion into aquitards also reduces the ^{14}C concentration in the aquifer (Sanford, 1997). Both processes reduce the ^{14}C concentration and result in an apparent groundwater age that is older than the true age. Consequently, the measured ^{14}C activities were corrected for carbonate dissolution as well as aquitard diffusion prior to use in the calibration.

A modified chemical correction was applied that takes into account the effect of dissolution as described by Boaretto et al. (1998). This method was successfully used in Danish geological settings similar to those investigated in the present study (Eq. 2; Boaretto et al., 1998; Hinsby et al., 2001a). The initial ^{14}C activities were corrected for fossil carbon dissolution (Pearson and Hanshaw, 1970) assuming an atmospheric ^{14}C activity (A_0) of 100 pMC (percent modern carbon) and a $\delta^{13}C$ concentration of $-25\%o$ in the soil CO_2,

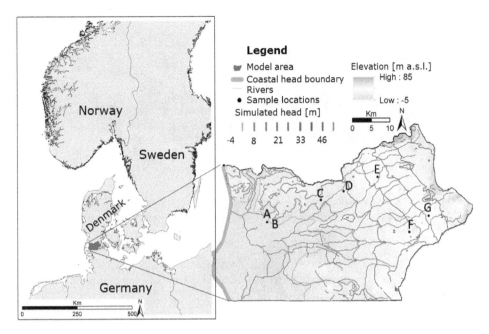

Figure 1. Investigation area at the border between Denmark and Germany. Simulated hydraulic heads are from the shallow aquifer (Meyer et al., 2018a). Topography, ^{14}C sample locations (A–G), river network and coastal head boundary are indicated.

and a ^{14}C activity of 0 pMC and δ^{13}C concentration of 0‰ in the dissolved carbonate. With a decay rate constant (λ) of 1.21×10^{-4} a^{-1} for the ^{14}C decay, the dissolution-corrected age τ_C was calculated as (e.g. Bethke and Johnson, 2008)

$$\tau_C = \frac{-1}{\lambda} \ln\left(\frac{A_m}{A_0}\right), \tag{1}$$

with

$$A_0 = \frac{\delta^{13}C}{-25} \times 100. \tag{2}$$

Subsequently, a diffusion correction was made to take into account diffusion loss into low-permeability layers (Sanford, 1997). Aquitard diffusion is sensitive to porosity, diffusion coefficient and the thicknesses of the active flow (aquifer) and stagnant (aquitard) zones (Sudicky and Frind, 1981). Because of the geological complexity, the sand-to-clay ratio based on voxel lithology was used to calculate the relative aquifer / aquitard ($a/b = 0.72$) thicknesses. Diffusion-corrected groundwater ages were calculated for three different diffusion coefficients (D): 1.26×10^{-9} $m^2 s^{-1}$ (Jaehne et al., 1987) representing the CO_2 diffusion in water, 1×10^{-10} $m^2 s^{-1}$ as an average for clay deposits (Freeze and Cherry 1979; Sanford, 1997) and 2.11×10^{-10} $m^2 s^{-1}$ as calculated by Scharling (2011), using aquifer effective porosities (n_e) ranging from 0.16 to 0.35 and aquitard (b) thicknesses between 10 and 50 m. Based on the ranges of variables, an average corrected age and the corresponding standard deviation were obtained for each sample (Table 1). Corrected groundwater sample age (τ_D), also referred to as the

apparent age, was calculated as

$$\tau_D = \tau_C \times \left(\frac{\lambda}{\lambda + \lambda'}\right), \tag{3}$$

with

$$\lambda' = 2 \times \tanh\left[\left(\frac{b}{2}\right) \times \left(\frac{\lambda}{D}\right)^{\frac{1}{2}}\right] \times \frac{(\lambda D)^{\frac{1}{2}}}{n_e a}. \tag{4}$$

3.2 Groundwater flow model

Meyer et al. (2018a) simulated the 3-D steady-state regional groundwater flow using MODFLOW 2000 (Harbaugh et al., 2000). A brief description of the model set up and calibration results are presented here; further details can be found in Meyer et al. (2018a). The model was discretized horizontally by 200 m × 200 m in the west and 400 m × 200 m in the east and vertically by 5 m above 150 m b.s.l. and 10 m below 150 m b.s.l., resulting in 1.2 million active cells. The voxel geology was interpolated to the MODFLOW grid and 46 hydrogeological units were defined. No-flow boundaries were used along the flow lines in the north and south, along a water divide in the east and at the bottom, where the Palaeogene clay constitutes the base of the aquifer system. At the western coast a density-corrected constant head boundary was applied (Fig. 1; Guo and Langevin, 2002; Post et al., 2007; Morgan et al., 2012). Distributed net recharge, averaged over the years 1991–2010, was extracted from the national water resources model (Henriksen et al., 2003) and included as

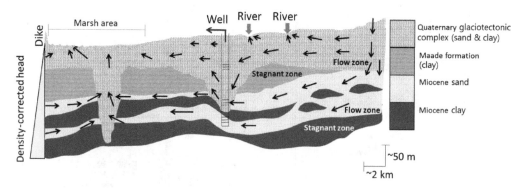

Figure 2. Conceptual regional model showing a simplified geology featuring buried valleys, groundwater flow and stagnant zones (as used for the diffusion correction). Arrows indicate general flow field of groundwater. Also shown are the boundary conditions, i.e. density corrected coastal boundary, drained marsh area, rivers.

Table 1. Sampling wells, uncorrected and corrected groundwater ages. Bold indicates samples used for calibration. Note that lower numbers of the wells indicate deeper locations (m b.s. = metres below ground surface, SD = standard deviation, pMC = percent modern carbon).

Well	DGU no.	Screen depth (m b.s.)	Aquifer geology	Measured ^{14}C (pMC)	Uncorrected ^{14}C-derived groundwater ages (years)	$\delta^{13}C_m$ (‰ VDPD)	Groundwater ages corrected for dissolution and diffusion (SD) (years)
A1	**166.761-1**	**246–252**	**Buried valley**	**46.44**	**6161**	**−13.2**	**344 (59)**
A2	**166.761-2**	**204–210**	**Buried valley**	**49.95**	**5576**	**−13**	**108 (19)**
B1	**166.762-1**	**160–166**	**Buried valley**	**49.84**	**5593**	**−13.9**	**293 (50)**
B2	**166.762-2**	**102–108**	**Buried valley**	**51.9**	**5268**	**−13.2**	**46 (8)**
C1	167.1545-1	306–312	Buried valley	0.48	42 889	−5.9	10 429 (1789)
C2	167.1545-2	273–276	Buried valley	1.03	36 755	−7.7	9097 (1569)
C3	167.1545-3	215–218	Buried valley	0.16	51 714	−11	15 038 (2593)
C4	**167.1545-4**	**142–149**	**Buried valley**	**33.84**	**8703**	**−13.2**	**1191 (205)**
C5	**167.1545-5**	**116–123**	**Buried valley**	**43.18**	**6746**	**−13.1**	**518 (89)**
D1	159.1335-1	290–295	Miocene	1.8	32 271	−7.9	7671 (1323)
D2	159.1335-2	277–282	Miocene	1.35	34 582	−10.6	9229 (1591)
E1	**159.1444-1**	**194–200**	**Buried valley**	**31.34**	**9320**	**−12**	**1141 (197)**
E3	**159.1444-3**	**81–87**	**Buried valley**	**40.29**	**7302**	**−12.8**	**642 (111)**
F1	168.1378-1	372–378	Miocene	46.12	6216	−12.3	173 (30)
F2	168.1378-2	341–345	Miocene	2.85	28 580	−13.3	7836 (1351)
F3	**168.1378-3**	**208–214**	**Miocene**	**25.73**	**10 904**	**−12.6**	**1800 (310)**
G1	**168.1546-1**	**110–120**	**Miocene**	**42.57**	**6860**	**−12.3**	**388 (67)**
G2	**168.1546-2**	**74–84**	**Pleistocene/ Miocene**	**45.33**	**6355**	**−12**	**153 (26)**

a specified flux condition. Internal specified boundaries include abstraction wells with a total flux of $26 \times 10^6 \, \text{m}^3 \, \text{year}^{-1}$ (averaged over the years 2000–2010, corresponding to 4 % of the total recharge) as well as rivers and drains.

Horizontal hydraulic conductivities (one for each hydrogeological unit), two anisotropy factors (K_h/K_v, one for sand and one for clay units), and river and drain conductances were calibrated using a multi-objective regularized inversion scheme (PEST; Doherty, 2016) with head and mean stream flow observations as targets. The resulting head distribution is shown in Fig. 1. Horizontal hydraulic conductivities were estimated in a range of $K_h \in [1 \, \text{m} \, \text{d}^{-1}; 83 \, \text{m} \, \text{d}^{-1}]$ for Pleis-

tocene sand units, $K_h \in [0.028 \, \text{m} \, \text{d}^{-1}; 0.19 \, \text{m} \, \text{d}^{-1}]$ for Pleistocene clay units, $K_h \in [0.008 \, \text{m} \, \text{d}^{-1}; 0.016 \, \text{m} \, \text{d}^{-1}]$ for the Maade formation, $K_h \in [16 \, \text{m} \, \text{d}^{-1}; 46 \, \text{m} \, \text{d}^{-1}]$ for Miocene sand and $K_h \in [0.14 \, \text{m} \, \text{d}^{-1}; 0.23 \, \text{m} \, \text{d}^{-1}]$ for Lower Miocene clay.

The steady-state MODFLOW flow solution (calibration results summarized in Fig. 3; Meyer et al., 2018a, also contains an identifiability and uncertainty analysis of the estimated parameters as well as an evaluation and discussion of the non-uniqueness of the flow model) forms the basis for the advective transport simulation using MODPATH.

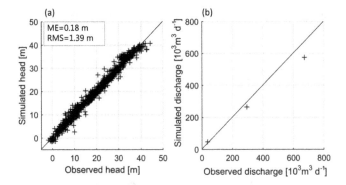

Figure 3. Calibration results of steady-state groundwater flow model that forms the basis for the advective transport model (modified after Meyer et al. (2018a). Panel **(a)**: simulated versus observed hydraulic head; panel **(b)**: simulated versus observed stream discharge. ME = mean error, RMS = root mean square.

3.3   Advective transport model

Advective transport simulation was performed using MODPATH (Pollock, 2012) in particle back-tracking mode. Hereby, the travel time of a particle, released in a cell, is calculated based on the MODFLOW cell-by-cell flow rates (q). The advective travel time (t) along the travel paths in 3-D (x) is calculated as

$$t(x) = \int_{x_0}^{x} \frac{n_e(x)}{q(x)} dx. \tag{5}$$

In addition to the input data required by MODFLOW to generate the flow solution, MODPATH requires a value for effective porosity (n_e) to calculate the seepage velocity.

The groundwater age can be seen as the backward integration of travel times along the travel path back to its recharge location. Hence, the simulated groundwater age is a function of the ratio of flux to effective porosity and the travel distance. In this study, the total flux is controlled by prescribed recharge and heterogeneous distribution of hydrogeological parameters (e.g. hydraulic conductivity, porosity).

In order to ensure stability (Konikow et al., 2008), 1000 particles were distributed evenly in the cell of the well screen and their average simulated particle age was compared with apparent groundwater ages (derived from Eq. 3).

The corrected ^{14}C ages were used as targets in the objective function (see below) of the simulated average travel time during calibration. The particle-based approach used in this study computes the kinematic age at a point. With 1000 particles released in each cell with a screen, we essentially get an age distribution of kinematic ages by perturbing the measurement location within the cell, reflecting the mixing of waters from different origins. The ^{14}C ages have also been diffusion-corrected (Sect. 3.1.1) so that dilution or mixing due to loss of ^{14}C into the stagnant zones have been accounted for.

3.3.1   Calibrating porosity

The flow solution of the calibrated flow model (Meyer et al., 2018a) constitutes the base for the 3-D advective transport model. Depending on the depositional environment and clay or sand content, effective porosities of seven units corresponding to two Pleistocene sand, two Pleistocene clay, one Miocene sand and two Miocene clay units were estimated using a regularized (Tikhonov) inversion with PEST (Tikhonov and Arsenin, 1977; Doherty, 2016). As the calibration approach is similar to the one of Meyer et al. (2018a), only additional characteristics are described in the following. Average corrected ^{14}C groundwater ages from 11 samples with a ^{14}C activity higher than 5 pMC (Table 1) were used as calibration targets. ^{14}C activity lower than 5 pMC was not used as it was assumed that the boundary conditions of the flow model (e.g. sea level, recharge, head gradients) were not representative for pre-Holocene conditions. Moreover, the data from well F1 were excluded from calibration as an age inversion with F2 was observed here (Table 1), probably due to local heterogeneity or contamination of water with a higher ^{14}C concentration, which cannot be reproduced by the model. The average uncertainty of apparent ages was estimated to be about 102 years. This value was based on the average of the standard deviation of the diffusion correction for the selected 11 samples and was used for weighting of the individual ages.

When Tikhonov regularization is applied, a regularized objective function (Φ_r) is added to the measurement objective function (Φ_m) in the form of the weighted least-squares of the residuals of preferred parameter values and parameter estimates. Within the limits of the user-defined objective function (PHIMLIM) and the acceptable objective function (PHIMACCEPT), the weight of the regularized objective function (μ) increases and the parameter estimates are directed towards the preferred values.

Calibration settings such as initial and preferred values and final parameter estimates are shown in Table 2. Values for PHIMLIM and PHIMACCEPT were set to 60 and 100, respectively. The total objective function (Φ_{tot}), minimized by PEST, is then the sum of the measurement objective function (Φ_m) and the regularized objective function (Φ_r):

$$\Phi_{tot} = \Phi_m + \mu^2 \Phi_r, \tag{6}$$

with

$$\Phi_m = \sum (\omega_a (a_{obs} - a_{sim}))^2, \tag{7}$$

where a_{obs} and a_{sim} are observed and simulated groundwater ages, respectively, and the weight ω_a is the inverse of the standard deviation of the observed age. The calibration is evaluated based on the mean error (ME) and the root mean square (RMS) between apparent (corrected ^{14}C ages) and advective groundwater ages. Parameter identifiability (Doherty and Hunt, 2009) is used to investigate to what extent the effective porosities were constrained through model calibration. Identifiability close to 1 means that the information content of the observations used during calibration can constrain the parameter. Parameters with an identifiability close to zero cannot be constrained.

3.4 Direct age

To visualize the mean groundwater age pattern in the regional 3-D aquifer system, direct simulation of mean groundwater age was performed with MT3DMS (standard finite difference solver with upstream weighting) and the chemical reaction package using a zeroth-order production term (Goode, 1996; Bethke and Johnson, 2008). Hereby, mean groundwater age is simulated in analogy to solute transport as an "age mass" (Bethke and Johnson, 2008). For each elapsed time unit (day) the water "age mass" increases by 1 day in each cell. Increases or decreases in ages are a result of diffusion, dispersion and advection (Bethke and Johnson, 2008). The transient advection–dispersion equation of solute transport of "age mass" in three dimensions and with varying density and porosity is given by Goode (1996):

$$\frac{\partial a n_e \rho}{\partial t} = n_e \rho - \nabla a \rho \mathbf{q} + \nabla n_e \rho \mathbf{D} \times \nabla a + F, \tag{8}$$

where F is an internal net source of mass age, $\mathbf{q}$ the Darcy flux ($\mathrm{m\,d^{-1}}$), a a mean age (d), n_e the effective porosity, ρ the density of water ($\mathrm{kg\,m^{-3}}$) and $\mathbf{D}$ the dispersion tensor ($\mathrm{m^2\,d^{-1}}$), including molecular diffusion and hydrodynamic dispersion. The initial concentration of the "age mass" was set to zero, while a constant age of zero was assigned to the recharge boundary and the constant head boundary at the coast. Steady-state conditions were evaluated based on the change in mass storage in a 40 000-year simulation. The age mass storage (m) in the whole model was calculated for each time step as the sum of mass in each cell (m_i). The latter was calculated by multiplying the cell dimensions ($\Delta z, \Delta x, \Delta y$) with porosity ($n_e$) and age ($a_s$):

$$m = \sum m_i, \tag{9}$$

with

$$m_i = \Delta z \times \Delta x \times \Delta y \times n_e \times a_s. \tag{10}$$

The percentage change in mass storage (Δm_t) per time step (Δt) was calculated as

$$\frac{\Delta m_t}{\Delta t} = \left(\frac{m_t - m_{t-1}}{m_{t-1}} \right) \times 100. \tag{11}$$

The integral of the change in mass storage over time was used to define quasi-steady-state conditions. This was reached when

$$\int_{t1}^{t} \Delta m(t) \mathrm{d}t \geq 0.95 \times \int_{t1}^{\infty} \Delta m(t) \mathrm{d}t. \tag{12}$$

Dispersion experiments were carried out for longitudinal dispersivity α_L values of 0, 5, 20, 50 and 500 m, while the horizontal transversal α_{TH} and vertical transversal α_{TV} dispersivities were specified to 10 % and 1 % of α_L, respectively. A diffusion coefficient of $1 \times 10^{-9}\,\mathrm{m^2\,s^{-1}}$ was used to account for self-diffusion of the water molecule at about 10 °C (Harris and Woolf, 1980).

3.5 Capture zones

Well capture zones are used in water management to define areas of groundwater protection, where human actions, such as agricultural use, are restricted. Simulated by the uniform effective porosity and distributed effective porosities model, the capture zones of one existing well (Abild, abstraction rate $27\,\mathrm{m^3\,d^{-1}}$) located in a buried valley and one virtual well (AW, abstraction rate $280\,\mathrm{m^3\,d^{-1}}$) located in a Miocene sand aquifer were evaluated and compared for different back-tracking times using 100 particles per well. Given that the hydraulic conductivity field is unchanged, no differences in the area of the whole capture zone are expected as porosity does not impact the trajectory of the particle path (Hill and Tiedeman, 2007) and only affects the travel time. Hence, the capture zone areas at different times were compared.

4 Results

4.1 ^{14}C corrections

Figure 4 shows the corrected and uncorrected ^{14}C ages over depth. Except for well F, ages increase with depth at each multi-screen location. Otherwise, no clear trend between age and depth can be identified on the regional scale. Uncorrected ages range from 5000 to 50 000 years (Table 1). After correction, all ages decrease and the relative difference between the corrected ages increase, now within a range from 46 to 15 000 years. Hence, it is expected that the oldest water recharged the groundwater at the end of the last glacial period. The majority of the samples represent younger waters, with 12 out of 18 samples being less than 2000 years old.

4.2 Calibration results

The match between the average of simulated groundwater ages (particle tracking with MODPATH) and corrected ^{14}C ages is shown in Fig. 5a.

Results from the calibrated distributed effective porosities model were compared to those from a uniform effective porosity model with an effective porosity of 0.3, which is a typical textbook value for porous media (Hölting and Coldewey, 2013; Anderson et al., 2015) and often used in groundwater modelling studies (e.g. Sonnenborg et al., 2016). The calibrated distributed effective porosity model is able to match all the observations reasonably. This is not the case for the single effective porosity model, where one sample

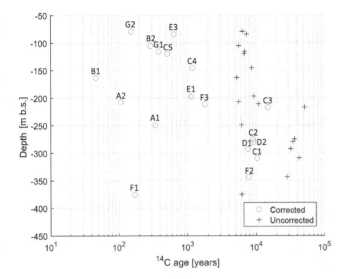

Figure 4. Apparent groundwater ^{14}C ages as a function of groundwater sampling depth: black crosses indicate ages without correction for dissolution and diffusion, blue circles show ages with correction. Labels indicate well location and screen number (see Fig. 1 and Table 1).

in particular is poorly simulated with an estimate of more than 5500 years, whereas the corresponding observation only reach about 1200 years. The ME and RMS of the calibrated distributed effective porosity model were −2.3 years and 267 years, respectively, which correspond to a reduction in ME of 99 % and RMS of 82 % compared to the single effective porosity model. Considering the uncertainties involved in estimation of apparent age (see uncertainty estimates in Table 1, column to the right), the match is found to be acceptable. Comparison of the average uncertainty in apparent ages used for calibration of 102 years with the achieved RMS of 267 years indicates that no overfitting occurred and that mismatches can be a result of small-scale heterogeneity below grid resolution, errors in the model structure or uncertainties of parameters, for example.

The estimated effective porosities of the seven hydrogeological units are listed in Table 2. Realistic values are found for all parameters and the values of the sand units are generally higher than those of the clay units. However, the effective porosity estimate of 0.13 for Pleistocene sand 1 is relatively low. This may be explained by the fact that this unit does not represent sand exclusively everywhere. The Pleistocene deposits in the area are highly heterogeneous (Jørgensen et al., 2015) and it is therefore difficult to identify units exclusively composed of sand, partly due to the difficulties in using AEM data to guide the distinction between sand and clay at a relatively small scale. Hence, Pleistocene sand 1 may to some extent represent a mixture of sand and clay. The relatively small effective porosities for clay units might be due to compaction as a result of glacial loading in the course of several glacial periods during the Pleistocene. Additionally,

uncertainties in the estimates of hydraulic conductivity from Meyer et al. (2018a) will translate into errors in seepage flux and hence ages. Uncertainties and errors in hydraulic conductivity may therefore be partly compensated by estimates of effective porosity that are somewhat different from the expected value.

The parameter identifiability (Fig. 5b) shows that the corrected ^{14}C ages may constrain four out of seven estimated effective porosities, i.e. of Pleistocene sand 1, Pleistocene clay 2, Miocene sand and Miocene clay. The warmer colours (red–yellow) indicate that the parameter is less influenced by measurement noise (Doherty, 2015, Fig. 5b). Where the parameter identifiability is relatively low (< 0.8), i.e. for effective porosities of Pleistocene sand 2, Pleistocene clay 1 and Miocene clay (Maade), the estimated parameter value is more constrained by the regularization and hence stays close to the preferred value (Table 2). The low identifiability is a result of the distribution (or density) of observations compared to the particle travel paths. Figure 6 shows the pathlines of particle back-tracking (for better visualization only one path line is shown per well screen). As mentioned above, only ^{14}C observations with an activity higher than 5 pMC (Table 1) were used, which excludes results from well screens C1, C2, C3, D1, D2, F1 and F2. The recharge area is mostly located to the east (Fig. 6). The Maade formation is more dominant towards the west while it is patchy in the east. Consequently, it does not affect the particle tracking as much in the east. Only effective porosities of geological units through which particles actually travel are well informed by the observations. The low-permeability Maade unit acts as an obstacle to the travel paths and since the particles circumvent the Maade formation the actual value of porosity has no impact on the age. The Maade unit significantly affects the age distribution due to its influence on travel paths, but no sensitivity to the porosity of the unit is found.

Pleistocene sand 2 represents less than 5 % of the total amount of cells (Table 2) and it occurs mostly in the west. Pleistocene clay 1 is mostly shallow, patchy and located far away from the well locations. Hence, the impact of these two geological units on the particle tracks is also relatively small and results in low identifiability (Fig. 5b).

4.3 Advective age distribution at observation wells

Figure 7 shows the simulated advective age distribution at the sampling locations (A–G, Fig. 1).

The results show a wide variety of mean particle ages (Table 3) and the shape of the age distributions is very different (Fig. 7). The well screens with mean particle ages less than 1000 years (except for C4 and F3 which have slightly higher mean particle ages) show particle age distributions that are mostly narrow and unimodal (except E3 and G1), which is also reflected in a small standard deviation (smaller than 20 % of the mean age, except E3 and G1), see column B in Table 3. The particle age distributions of older waters

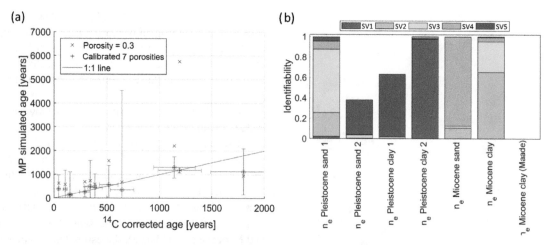

Figure 5. Calibration results: **(a)** "x" shows apparent ages simulated with MODPATH (MP) and a porosity of 0.3 as often used in porous media models and "+" shows MP ages simulated based on the seven calibrated porosities (Table 2); standard deviations based on MP and correction terms (see Sect. 3.1.1) are shown. **(b)** Parameter identifiability of effective porosities (warmer colours correspond to singular values (SV) of a lower index, cooler colour to SV of higher index) of the different geological formations; the identifiability of the Maade porosity is close to zero.

Table 2. Calibration settings and results: parameters with initial, preferred and estimated values for effective porosity.

Parameter (n_e)	Initial/preferred value	Estimated value	% of cells	Objective function	
Pleistocene sand 1	0.3	0.130	24.4	PHIMLIM	60
Pleistocene sand 2	0.3	0.263	2.5	PHIMACCEPT	100
Pleistocene clay 1	0.1	0.085	11.6	ϕ_m achieved	74
Pleistocene clay 2	0.05	0.043	4.8		
Miocene sand	0.3	0.450	15.1		
Miocene clay	0.1	0.102	22.8		
Miocene clay (Maade formation)	0.05	0.049	18.8		

with a mean particle age higher than 1000 years (Table 3) tend to have broader and/or multi-modal shapes (Fig. 7c, d) and large standard deviations (Table 3, column B).

The mean distance that particles travel from their recharge points to the sampling well (Table 3, column D) ranges between 3 and 28 km. The younger waters (mean particle ages < 1000 years) show path lengths less than 10 km (except at well location E3 and F3; Fig. 8), while most of the older waters travel more than 20 km. However, the relation between path length and travel time is far from linear. At some well locations (e.g. well locations A2, B2, C4, C5) the relation between path length and travel time forms a few distinct small clouds without much spread, indicating that the particles follow alternative large-scale preferential flow paths. At other locations a larger and more diffusive spread is found, either in travel times (e.g. well locations C1, C2, C3, D1, E1) or path lengths (e.g. well locations F3, G2, E3). The large spread in travel times indicates that some particles travel slowly through clay units of various thicknesses. The large spread in path lengths originates from long and quick

or short and slow travel paths through or around clay units and reflects the geological heterogeneity.

4.4 Regional age distribution based on direct age simulation

Figure 9 shows the ME and RMS of the direct mean age and the apparent age (corrected ^{14}C) for different α_L values (α_{TH} and α_{TV} are tied to α_L, see Sect. 3.4).

Minimum ME and RMS values are achieved for longitudinal dispersivities $\alpha_L < 5$ m. For lower α_L the effect on ME and RMS is insignificant as numerical dispersion is expected to dominate at this scale. With higher α_L values ME and RMS increase significantly. Other regional-scale studies (e.g. Sonnenborg et al., 2016) have used longitudinal dispersivity in the magnitude of tens of metres or more to account for geological heterogeneities at formation scale. The very detailed voxel geological model resolves heterogeneities at a scale of 200 m × 200 m. Hence, it is assumed that mixing at scales larger than 200 m is accounted for by the geological

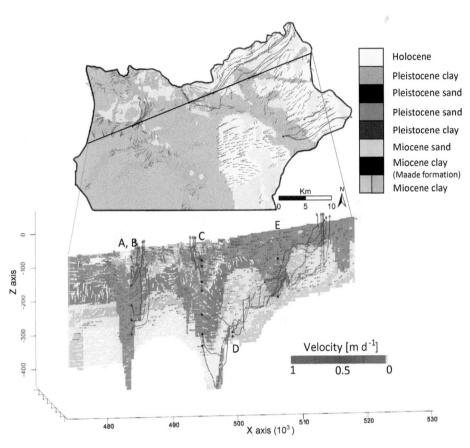

Figure 6. Horizontal geological cross section at an elevation of −100 m a.s.l. and a SW–NE cross section through sampling well locations (A–E). Bluish (cold) colours represent pre-Pleistocene sediments (dark blue = clay, light blue = sand), while warm colours represent Pleistocene deposits (red = sand, brown = clay). Also shown are MODPATH back-tracking lines (1 per cell) and groundwater flow velocity vectors. The units for the x and z axes are meters (m).

model. Therefore, the dispersivity should only describe the heterogeneity at a flow scale of several hundreds of metres, which justifies the use of a relatively small α_L. In accordance with Gelhar et al. (1992) flow scales of hundreds of metres result in α_L of magnitudes in the range of a few metres, which is also in line with studies in the Dutch polder system where dispersivity values of 2 m were applied in similar-sized models (e.g. Oude Essink et al., 2010; Pauw et al., 2012). Similar to Weissmann et al. (2002) and LaBolle and Fogg (2001) the simulations showed little sensitivity to local-scale dispersivity because at the modelling scale of tens of kilometres, dispersion is dominated by facies-scale heterogeneity, which is captured by the detailed, highly resolved geological model. On the grid scale of 200 to 400 m and with the standard difference solver for the advection–dispersion equation, a substantial numerical dispersion is expected. Choosing the TVD or MOC solver scheme for the advection–dispersion equation would have been more accurate in terms of less numerical dispersion, but would have required excessive running times, which made it impractical to use in this study. Since there is no sensitivity for lower α_L (numerical dispersion dominates

at this scale), physical dispersivity was set to zero in the following simulations of direct age.

The directly simulated mean age distribution on a regional scale (Fig. 10) shows a general age evolution from young water in the recharge area in the east towards older water in the west (Fig. 10b, e, f). Young water also enters the system through the coastal boundary in the west (Fig. 10b, e, f). The age distribution is strongly affected by geology and is therefore in good agreement with the interpretation of the flow system by Meyer et al. (2018a). Two main aquifers are present on a regional scale: a shallow Pleistocene sand aquifer and a deep Miocene sand aquifer, separated by the Maade formation and locally connected through buried valleys (conceptual model in Figs. 2, 10g, h). The regional mean age distribution also reflects this system. Younger waters dominate the shallow Pleistocene aquifers (Fig. 10a, e, f), where the flow regime can be described as mostly local and intermediate (Tóth, 1963). The separating Maade formation with its increasing thickness towards the west (Fig. 10d) acts as a stagnant zone where groundwater age increases (Fig. 10c). The underlying Miocene sand shows the mean age evolution from young water in the recharge areas in the

Table 3. Results of the analysis of particle age distributions and path lengths. Bold indicates well screens used for calibration.

Well	A Mean particle age (years)	B SD particle age (years)	C Median particle age (years)	D Mean path length (km)	E SD path length (km)
A1	**536**	**72**	**503**	**7.50**	**0.22**
A2	**392**	**16**	**387**	**6.17**	**0.36**
B1	**400**	**71**	**367**	**6.28**	**0.41**
B2	272	**31**	277	**3.69**	**0.61**
C1	7232	2814	6503	26.68	1.64
C2	3654	2816	2818	27.52	0.94
C3	2640	608	2542	27.83	0.86
C4	**1038**	**45**	**1036**	**3.24**	**0.07**
C5	**542**	**116**	**512**	**3.19**	**0.17**
D1	14 122	7563	13 479	22.12	1.05
D2	5028	4498	3064	21.94	0.97
E1	**1306**	**1508**	**908**	**13.56**	**0.82**
E3	**404**	**448**	**300**	**11.50**	**1.30**
F1	6649	1405	6394	17.01	0.47
F2	2950	1584	2768	16.34	0.78
F3	**1129**	**60**	**1120**	**14.83**	**0.57**
G1	470	258	514	6.82	0.62
G2	135	17	130	6.02	0.46

east to older water towards the discharge zones in the west (Fig. 10b, e, f). Here the flow regime is dominated by regional flow (Tóth, 1963). Special features are the buried valleys where downward flow of young waters, upwelling of old waters and mixing occurs (Fig. 10e, f, g, h). At the coastal boundary in the west young water enters the system and due to the density-corrected head boundary a wedge is formed with young waters in the wedge and old water accumulating in the transition zone (Fig 10e, f). The two cross sections (Fig. 10e and f) differ in their geological connection to the sea boundary (compare geological sections in Fig. 10g and h). In (e) a buried valley connects the inland aquifer with the sea and here younger waters reach further inland due the relatively higher hydraulic conductivity and the inland head gradient as a result of the drainage system. Moreover, buried valleys constitute locations where the deep aquifer system, bearing old waters, connects with the shallow one, and here upwelling of older waters occurs due to the higher heads in the deep semi-confined (by the Maade aquitard) Miocene aquifer. In cross section (f) where the buried valley occurs further inland, the young ocean water penetrates the higher permeable Miocene aquifer but is impeded in the low permeable sections and hence does not reach as far inland. Another feature is the human land use change including an extensive drainage network with drain elevations below the sea level in the marsh area. There, old groundwater is forced upward, partly through buried valleys, before it could discharge to the sea.

4.4.1 Direct simulated mean age distribution in geological units

The steady-state distribution of direct simulated mean groundwater age was reached after 26 000 years. Over this time span the system has been exposed to transient stresses from human activity and climatic changes (glacial cover, sea level, etc.). Therefore, the steady-state assumption is a notable simplification, which is further discussed in Sect. 5.1.

In Fig. 11 the normalized direct age distributions are shown for (a) the whole model, (b) the Pleistocene aquifer, (c) the Maade clay formation that acts as an aquitard and (d) the Miocene sand aquifer (compare the geological setting with conceptual model in Fig. 2). The directly simulated mean groundwater ages for the whole model, the Pleistocene sand, the Maade formation and the Miocene sand were determined by a moment analysis (Levenspiel and Sater, 1966) as 2574, 1009, 3883 and 2087 years, respectively. The shape of the age distribution in these units varies significantly. The Pleistocene sand shows a unimodal distribution with one peak at ≈ 100 years and a tail (Fig. 11b). The age distribution is governed by recharge of young water and discharge through rivers and drains, which are fed by the upwelling older groundwater (Fig. 10a). The age distribution in the Maade formation is multi-modal, with five peaks at about 600, 1400, 3900, 6500 and 7600 years (Fig. 11c). Comparison of Fig. 10c and d reveals a positive relation between age and thickness of the Maade formation. The age distribution in the underlying Miocene sand has one peak at 200 years followed by a plateau between 1600 and 3100 years and a

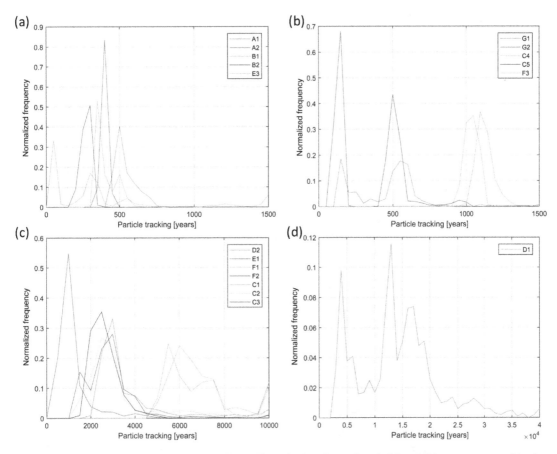

Figure 7. Particle age distributions at sampling wells A–G (see Fig. 1 for locations). Panels **(a)** and **(b)**: young waters (bin size = 50 years) show a narrow, unimodal distribution; panel **(c)** old waters (bin size = 500 years) have broader and often multimodal distributions; panel **(d)**: multi-modal age distribution at sample location D1 (bin size = 1000 years), which shows the longest travel times.

small peak at 7800 years (Fig. 11d). This distribution is controlled by the overlying and separating Maade formation in the west and the inter-layering with Miocene clay.

4.4.2 Advective and directly simulated ages

The comparison of the advective ages with the direct simulated ages at the sampling well locations shows a good match for advective ages with a small variance and worsens when the variance increases (Fig. 12). Older ages are generally associated with larger variances. Where the mismatch between advective and direct ages is large, the direct simulated mean ages are consistently lower than mean ages derived from particle back-tracking (see discussion below) because of diffusion into clay units. However, most of them lie within 1 standard deviation; but please observe that the standard deviation spans several thousands of years at some locations, where particle travel time distributions show a multi-modal shape.

4.4.3 Capture zones: effect of porosity

Figure 11 shows the capture zones at the Abild well for 1500 and 2000 years and for the virtual well (AW) for 1000,

2000 and 3000 years for a constant effective porosity of 0.3 (solid line) and the calibrated distributed effective porosities model (dashed line), respectively. The capture zones of the two models vary both in extent and shape. The areas of the capture zone differ by up to 50 %. Interestingly, it is not always the same effective porosity model that has the smaller capture zone, but it changes due to the heterogeneity in the geological model and the assigned effective porosities. However, the results illustrate the importance of reliable estimates of effective porosity when delineating the capture zone of an abstractions well.

5 Discussion

[14]C observations were used to constrain the estimation of effective porosities of a large-scale coastal aquifer system using an approach similar to Konikow et al. (2008), Weissmann et al. (2002) and Starn et al. (2014). Advective transport modelling and direct age simulations were applied to gain insight into the regional age structure of this highly heterogeneous geological system. In the following, limitations, uncertainties and simplifications of the model structure as well as esti-

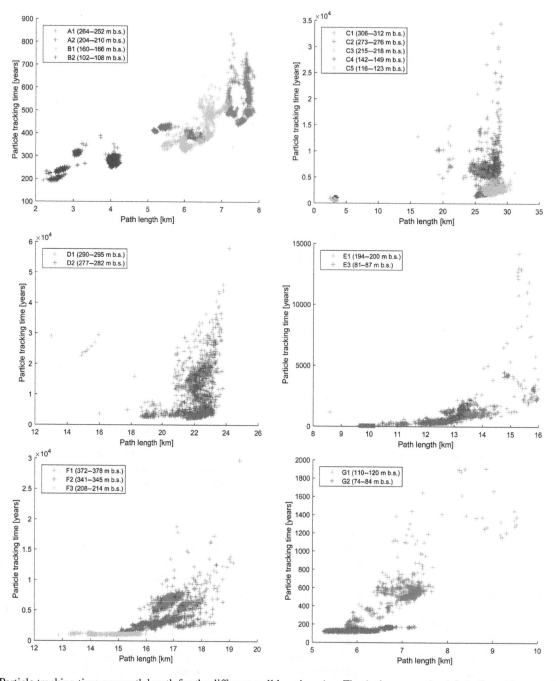

Figure 8. Particle tracking time over path length for the different well locations (see Fig. 1; the screen depth is indicated in parentheses).

mated parameters and resulting interpretations are discussed. A detailed description of the age distribution is provided to highlight the relevant physical processes and their interactions.

5.1 Uncertainties

5.1.1 Boundary conditions

Uncertainties in model results originate partly from simplifications in boundary conditions and geological heterogeneities that are not resolved at the grid scale. Groundwater recharge, drain levels, well abstractions and sea levels were assumed to be constant over time for practical reasons and to reduce computational time. However, Karlsson et al. (2014)

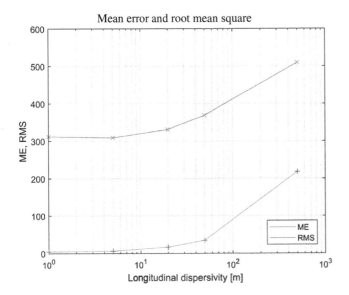

Figure 9. Mean error (ME) and root mean square (RMS) between corrected ^{14}C ages (shaded in gray in Table 1) and directly simulated ages as a function of longitudinal dispersivity.

showed that recharge has changed significantly in Denmark during the last centuries. Changes in recharge could result in different age patterns (Goderniaux et al., 2013). Similarly, sea level changes that were disregarded in this study would have an effect on the groundwater age distribution in the coastal areas (Delsman et al., 2014). Prescribing a vertical coastal age boundary of zero years is another simplification that neglects the vertical mixing and dispersion, which would result in an increase of age with depth (Post et al., 2013). However, since these physical processes were difficult to quantify, estimating age at this boundary would be highly uncertain. Thus, a constant age of zero years was applied.

The area close to the coast has not only been affected by changing sea levels during the past thousands of years, but also by saltwater intrusion. In this study, the density effects on flow were accounted for in a simplified way by using a density-corrected constant head boundary at the coast. Both sea level changes and density effects would also have affected the age distribution. The impact on age calculations due to density effects would be largest close to the coast. However, most of the groundwater samples used for age estimations were collected several tens of kilometres inland and are therefore expected to be affected to a minor extent. To quantify the impact of boundary conditions and saltwater intrusion on the particle tracking, the differences of particle travel path lengths for a 200-year period, investigated based on the present model and a preliminary density-driven model (SEAWAT) accounting for non-stationary and density effects (similar to the one presented in Meyer, 2018), are computed. The relative differences are below 10 % (except at location A and B). Also, the uncertainties introduced by simplifying the density boundary effects are likely to be less impor-

tant compared to other uncertainties associated, for example, with estimating the groundwater age by the procedures for correcting ^{14}C activities. A solution would, of course, be to use a fully density-driven model such as SEAWAT as in Meyer (2018) or Delsman et al. (2014). But the very long computer runtimes for these kinds of models and the need for several thousand model runs during calibration made it infeasible to use a variable-density flow model.

5.1.2 Apparent age as calibration target

Uncertainties in the use of ^{14}C as a groundwater dating tool and as calibration target arise at different levels. First, sampling of well screens with a length of 6–10 m would encompass a range of groundwater ages as a result of mixing of groundwater of different ages. Thereby younger waters, corresponding to DIC with a higher ^{14}C content, would dominate older water (Park et al., 2002). The ^{14}C content is measured in the DIC of the groundwater. In order to obtain a reliable age estimate, the origin of DIC in groundwater is important. For the different processes that can affect the DIC and change its ^{14}C content (e.g. dissolution, precipitation, isotopic exchange) a variety of correction models exists (see overview of correction models in IAEA, 2013). For the investigated system, corrections for carbonate dissolution and diffusion were applied, but it cannot be ruled out that other chemical processes might also have changed the ^{14}C content over the past few thousands of years. The ^{14}C correction for diffusion into stagnant zones is sensitive to aquifer porosity, aquitard thickness and diffusion constant. The geology is highly complex and aquitard thickness and porosity distribution change spatially over the entire region, whereas the correction terms were based on the properties averaged over hydrogeological units. Hence, average values of diffusion corrections were applied with parameters varying in ranges realistic for an aquifer system at this scale. However, in reality a groundwater particle would have been exposed to a variety of aquifer and aquitard thicknesses and porosities along its flow path, implying smaller or larger diffusion. The correction results show that both carbonate dissolution and diffusion into stagnant zones reduce the apparent groundwater age considerably, both at a similar magnitude as observed by Scharling (2011) and Hinsby et al. (2001a).

As mentioned in the introduction, the apparent age (or radiometric age) is not equal to the mean particle-based kinematic age. This introduces additional, but unknown uncertainties. Ideally, one could develop an advection–dispersion equation for the second moment and solve for the variance of ages (Varni and Carrera, 1998) and use that together with the directly simulated mean age (or first moment) to establish a relation between radiometric and mean ages. This has not been pursued as we believe the benefits from this would be masked by uncertainty in age dating ^{14}C (i.e. uncertainty on analyses, and corrections for effects of geochemical and physical processes).

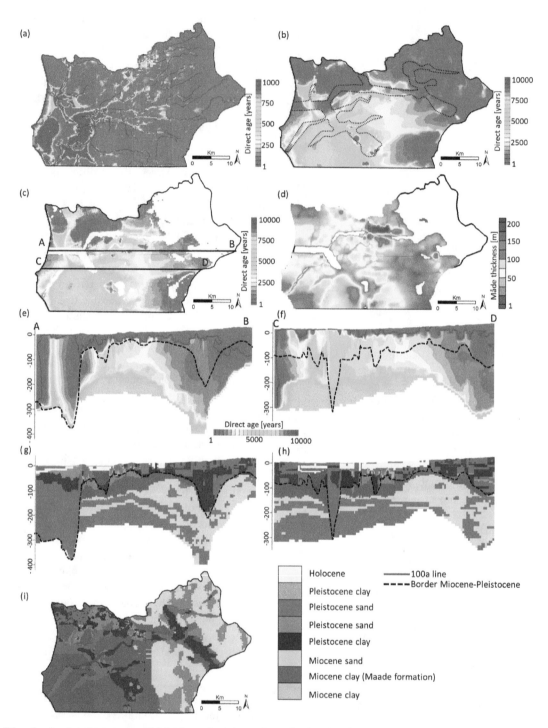

Figure 10. Directly simulated mean ages: **(a)** horizontal section at layer 2, also showing river network; **(b)** horizontal section at a depth of 100 m a.s.l. (buried valleys indicated with dotted lines); **(c)** horizontal section at the top of the Maade formation; **(d)** extent and thickness of the Maade formation; **(e)** cross sections A–B and **(f)** C–D; Pleistocene-Miocene boundary indicated with dashed lines (buried valleys), 100 year lines; **(g)** and **(h)** geological cross section and **(i)** horizontal geological section, main geological units indicated (a detailed geological description is given in Meyer et al., 2018a). Notice that the colour scheme in **(a)** is different in order to better resolve younger ages close to the surface.

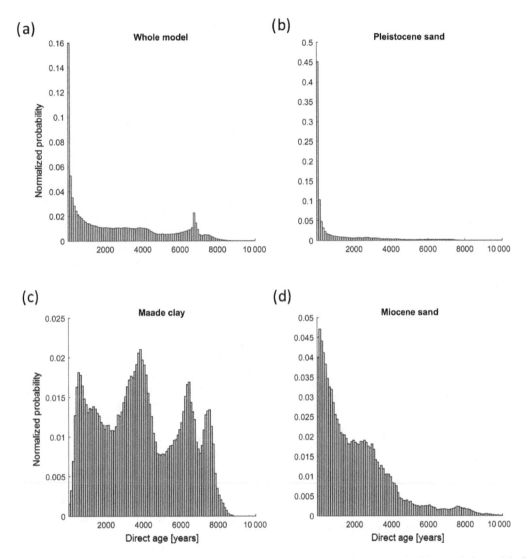

Figure 11. Frequency distributions (bin size = 100 years) of directly simulated groundwater ages in (a) the whole model, (b) the shallow Pleistocene aquifer, (c) the separating Miocene clay (Maade formation) and (d) the deep Miocene aquifer.

Finally, the calibration of effective porosity using an advective transport model relies on a calibrated 3-D flow solution that already bears uncertainties with respect to structure and parameters, as addressed by Meyer et al. (2018a). The number and position of the released particles contribute to the uncertainty, especially in heterogeneous systems, as pointed out by Konikow et al. (2008) and Varni and Carrera (1998). The use of a high number of particles – here 1000 particles were distributed in one cell – generally reduces the uncertainty and enhances stability of the solution. The arithmetic mean of the 1000 released particles evenly distributed in the sampling cells resulted in estimates of effective porosities in the range of 0.13 to 0.45 for sand and 0.043 to 0.1 for clay units, which is significantly different to porosities of 0.25 or 0.30 that are often used in porous media (e.g. Sonnenborg et al., 2016). The reliability of the estimated effective porosities was assessed through the identifiability that

depends on the observation density (see Sect. 4.2) and is high for four out of the seven estimated porosities.

5.1.3 Commensurability

The comparison of groundwater ages, estimated from tracer concentration in a water sample, and simulated groundwater ages, either derived by particle tracking or direct age modelling, bears the problem of commensurability, the comparison of a point measurement relative to the modelling scale. The water sample represents the age distribution in the direct surrounding of the well screen, which only makes up a few percent of the water in one model cell.

The differences between mean advective ages and directly simulated mean ages as described in Sect. 4.4 can be related to the simulation methods. While particle tracking neglects dispersion, but allows an age distribution in a cell to be sim-

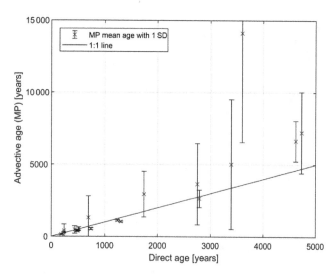

Figure 12. Mean advective age (MODPATH (MP) particle back-tracking) compared to directly simulated mean groundwater age at sampling well locations; error bars on advective age represent 1 standard deviation.

ulated (by perturbing the measurement location so to speak), direct age modelling allows for dispersion/diffusion to be accounted for, resulting in only the mean age at a cell. The mismatches between advective and direct age can be related to the diffusion and dispersion processes (here represented by numerical dispersion as dispersivity was set to zero), which are included in the direct age approach, but neglected in simulating advective ages.

5.2 Flow system and age distribution interpretation

5.2.1 Advective age distribution

The analysis of the advective age and travel distance distributions (Figs. 7 and 8, Table 3) revealed a larger variance of ages for waters with a higher mean age. Following the pathlines of wells with younger waters (e.g. Fig. 6, well locations A, B, C4, C5 and G), recharge areas are more proximal (path length < 10 km, Fig. 8, Table 3). Consequently, the particles pass through fewer hydrogeological units and hence the flow path is less influenced by heterogeneous geology, which results in a smaller variance in ages and path lengths (Fig. 8, Table 3). Particles travelling to well locations C1, C2, C3, D, E and F (e.g. Fig. 6) have to travel through a variety of hydrogeological units, characterized by different hydraulic conductivities and effective porosities and hence showing a broader age distribution and larger variance as well as longer travel distances. Their broad and multi-modal age distributions reflect the up-gradient heterogeneity in fluxes, related to hydraulic conductivity and effective porosity. This behaviour is in accordance with conclusions by Weissmann et al. (2002), who investigated groundwater ages in a hetero-

geneous 3-D alluvial aquifer based on particle tracking and CFC-derived ages.

5.2.2 Regional age pattern

The regional age pattern derived from direct age simulation is consistent with the findings of Meyer et al. (2018a) about the flow system. The two-aquifer system is separated by a confining aquitard in the west. The shallow aquifer system consisting of glaciotectonically disturbed Pleistocene sands mixed with clays is dominated by local and intermediate flow regimes and contains water of younger ages. The confining aquitard (Maade formation) shows older waters and a positive relation between ages and aquitard thickness what agrees with Bethke and Johnson (2008). In the deep Miocene sand aquifer that is interbedded with Miocene clay, regional flow regimes dominate and groundwater ages vary from young waters in the recharge areas in the east, where the overlying confining aquitard does not exist, to very old waters (up to 10 000 years) in the west. The confining Miocene aquitard (Maade formation) influences the age distribution pattern in the underlying Miocene sand in two ways. First, it limits the ability of deeper groundwater to seep upward and mix with the younger waters in the shallow aquifer. Secondly, the age flux from the aquitard to the aquifer shows a positive correlation with the ratio between aquitard thickness and aquifer thickness (Bethke and Johnson, 2008).

At the buried valleys, groundwater exchange and hence age mixing occurs. Upwelling of the older groundwater from the deeper aquifer happens preferentially through these buried valleys. The dense drainage network in the west close to the coast acts as a regional sink, with younger groundwater flowing horizontally and older water flowing vertically and discharging to the drains. At the coastal boundary in the west, where a constant concentration of an "age mass" zero was assigned to the density-corrected constant head boundary, an age wedge characterized by waters of contrasting ages is established as a result of intruding young ocean water that meets old waters in the transition zone. This agrees with the findings by Post et al. (2013) based on simulation of synthetic groundwater age patterns in coastal aquifers using density-driven flow.

The results of our study differ significantly from findings by Sonnenborg et al. (2016), who investigated a regional aquifer system with a similar geological setting located a few hundred kilometres north of the present study area. Their direct simulation of groundwater ages shows a pattern of much younger water than here, rarely exceeding 700 years even in the deepest aquifers, while in our study ages exceeding 10 000 years occur. The discrepancies may arise from differences in the geological models. In the area of Sonnenborg et al. (2016) the thickness of the Miocene sand units decreases towards the west and disappears before reaching the west coast. Sonnenborg et al. (2016) conclude that rivers control the age distribution even in deep aquifers. Based

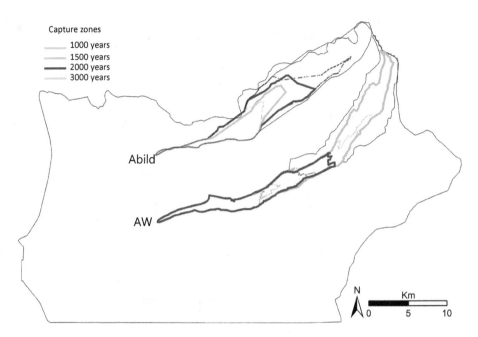

Figure 13. Capture zones at a well in Abild and a virtual well (AW) with a comparison of capture zones for a model with homogeneous porosity of 0.3 in all geological units (solid lines) and one with seven different porosities (dashed lines).

on particle tracking they found that the flow regimes were dominated by local and intermediate flow (see Tóth, 1963), with flow lengths not exceeding 15 km. In contrast, in the study presented here, the Miocene sand extends to the coast and probably beyond. While the age pattern in the shallow aquifers is controlled by rivers and drains (similarly to Sonnenborg et al., 2016), the age pattern in the deep aquifers is dominated by the extent and thickness of the Maade formation, the marsh area as a location of preferred discharge and the occurrence of buried valleys as locations of groundwater exchange, especially upwelling of old groundwater. Particle path lengths reach up to 30 km and regional flow dominates in the Miocene aquifer.

5.3 Perspectives of using spatial and temporal groundwater age distributions in groundwater quantity and quality assessment and management

The groundwater age distribution in aquifers is closely related to the distribution of physical (e.g. hydraulic conductivity and porosity) and chemical parameters (e.g. concentrations of contaminants and natural geogenic elements) of the aquifers and aquitards. Hence, tracer- and model-estimated groundwater age distributions provide important information for the assessment of the hydraulic properties of the subsurface as demonstrated in this study, and as an indicator of groundwater quality and vulnerability (Hinsby et al., 2001a; Sonnenborg et al., 2016), including contaminant migration (Hinsby et al., 2001a), contents of harmful geogenic elements such as arsenic and molybdenum (Edmunds and Smedley, 2000; Smedley and Kinniburgh, 2002, 2017),

and the risk of saltwater intrusion (MacDonald et al., 2016; Larsen et al., 2017; Meyer et al., 2018b). Groundwater age distributions in time and space are therefore important pieces of information for groundwater status assessment and the development of proper water management strategies that consider and protect both water resources quality and quantity (MacDonald et al., 2016). Water quality issues are often related to human activities such as contamination or over-abstraction (MacDonald et al., 2016) and are typically found in waters younger than 100 years to depth of about 100 m (Seiler and Lindner, 1995; Hinsby et al., 2001a), although deep subsurface activities may threaten deeper and older resources (Harkness et al., 2017). Deeper and older water is generally not contaminated or affected by human activities, but the impact of natural processes and contents of dissolved trace elements increases with depth and transport times (Edmunds and Smedley, 2000). Similarly, the risk of saltwater intrusion from fossil seawater in old marine sediments increases with depth in inland aquifers and reduces the amount of available high-quality groundwater resources (MacDonald et al., 2016; Larsen et al., 2017; Meyer et al., 2018b). Furthermore, old groundwater resources that are only slowly replenished are more vulnerable to over-exploitation, which leads to declining water tables, increasing hydraulic gradients and long-term non-steady-state conditions that change the regional flow pattern (Seiler and Lindner, 1995) and potentially result in contamination of deeper groundwater resources by shallow groundwater leaking downward. The presented modelling results show that the Miocene sand aquifer is protected by the overlying Maade formation over a wide area. The Miocene aquifer bears old waters (> 100 a, see Fig. 10e, f) of high quality (Hinsby and Rasmussen, 2008), especially in the east and the central part of the area, as the risk of seawater intrusion increases towards the west. However, caution should be shown as the shallow and the deep aquifers are naturally connected through buried valleys, where groundwater exchange occurs in both direction (Meyer et al., 2018a). In these geological features, young and possibly contaminated water can be found to greater depth (Seifert et al., 2008; Fig. 10e). Moreover, deep, old waters are vulnerable to contamination by modern pollutants as a result of the construction of wells with long screens, connecting different aquifers separated by aquitards (Seiler and Lindner, 1995; Jasechko et al., 2017; Fig. 2).

6 Conclusions

The originality of this study comes from a 3-D multi-layer coastal regional advective transport model, where heterogeneities are resolved on a grid scale. The distributed effective porosity field was found by parameter estimation based on apparent ages determined from ^{14}C activities, corrected for dissolution and diffusion. Based on regularized inversion seven effective porosities were estimated. Four of these were found to have high identifiability, indicating that they are well constrained by the age data. The remaining three have moderate to low identifiability, implying that they are less or poorly constrained by the data. In the latter case, parameter estimates close to the preferred values were obtained because of the use of Tikhonov regularization. By using a distributed effective porosity field, it was possible to match the observed age data significantly better than if effective porosity was assumed to be homogeneous and represented by a single value.

The advective age distributions at the well locations show a wide range of ages from a few hundred to several thousand years. Younger waters show narrower unimodal age distribution with small variances while older waters have wide age distributions and are often multi-modal with large variances. The variances in age distribution reflect the spatial heterogeneity encountered by the groundwater when travelling from the recharge location to the sampling point.

The estimated effective porosity field was subsequently applied in a direct age simulation that provided insight into the 3-D groundwater age pattern in a regional multi-layered aquifer system and the probable advance of modern potentially contaminated groundwater. Large areas in the shallow Pleistocene aquifer is dominated by young recharging groundwater (< 200 a) while older water is upwelling into rivers and drains in the marsh area. Hence, the upper aquifer is prone to contamination. In large areas the deeper Miocene aquifer is separated and protected by the Maade formation bearing old water, whereas young and possibly contaminated water is located in the recharge area in the east and in the buried valleys where the shallow and deep aquifer systems are connected.

The study clearly demonstrates the governing effect of the highly complex geological architecture of the aquifer system on the age pattern. Even though there are multiple uncertainties and assumptions related to groundwater age and its use in calibration, the results demonstrate that it is possible to estimate transport parameters that contain valuable information for assessment of groundwater quantity and quality issues. This can be used in groundwater management problems in general, as demonstrated in an example of capture zone delineation where a heterogeneous distributed effective porosity field resulted in a 50 % change in the capture zone area compared to the case of homogeneous effective porosity. The adopted approach is easy to implement even in large-scale models where auto-calibration of transport parameters using models based on the advection–dispersion equation might be restricted by computer runtime.

Author contributions. RM, PE, JAP, KH and TOS contributed to the conception of the work. RM, PE and TOS designed the modelling approach. RM and KH planned and conducted the groundwater sampling campaign, and the use and correction of ^{14}C. RM carried out the simulations, analysis of results and discussion, and prepared the paper. Drafts of the paper have been substantially revised and discussed with PE, TOS, JAP and KH.

Competing interests. The authors declare that they have no conflict of interest.

Acknowledgements. This study stems from the SaltCoast project generously funded by GeoCenter Denmark. The authors extend sincere thanks to all individuals and institutions whose collaboration and support at various stages facilitated completion of this study. The authors thank the editor Graham Fogg as well as Timothy Ginn and one anonymous reviewer for their comments that helped to improve the paper substantially.

Edited by: Graham Fogg

References

Anderson, M., Woessner, W. W., and Hunt, R.: Applied Groundwater Modeling: Simulation of Flow and Advective Transport, 2nd Edn., Elsevier, 2015.

Appelo, C. A. J. and Postma, D.: Geochemistry, Groundwater and Pollution, Balkema Publishers, Amsterdam, 2005.

Bethke, C. M. and Johnson, T. M.: Paradox of groundwater age, Geology, 30, 107–110, https://doi.org/10.1130/0091-7613(2002)030<0107:POGA>2.0.CO;2, 2002.

Bethke, C. M. and Johnson, T. M.: Groundwater Age and Groundwater Age Dating, Annu. Rev. Earth Pl. Sc., 36, 121–152, https://doi.org/10.1146/annurev.earth.36.031207.124210, 2008.

Boaretto, E., Thorling, L., Sveinbjörnsdóttir, Á. E., Yechieli, Y., and Heinemeier, J.: Study of the effect of fossil organic carbon on 14C in groundwater from Hvinningdal, Denmark, in: Proceedings of the 16th International 14C Conference, vol. 40, edited by: Mook, W. G. and van der Plicht, J., 915–920, 1998.

Bohlke, J. K. and Denver, J. M.: Combined use of ground- water dating, chemical, and isotopic analyses to resolve the history and fate of nitrate contamination in two agricultural watersheds, atlantic coastal Plain, Maryland, Water Resour. Res., 31, 2319–2339, https://doi.org/10.1029/95WR01584, 1995.

Campana, M. E. and Simpson, E. S.: Groundwater residence times and recharge rates using a discrete-state compartment model and 14C data, J. Hydrol., 72, 171–185, https://doi.org/10.1016/0022-1694(84)90190-2, 1984.

Castro, M. C. and Goblet, P.: Calibration of regional groundwater flow models: Working toward a better understanding of site-specific systems, Water Resour. Res., 39, 1172, https://doi.org/10.1029/2002WR001653, 2003.

Castro, M. C. and Goblet, P.: Calculation of ground water ages-a comparative analysis, Groundwater, 43, 368–380, https://doi.org/10.1111/j.1745-6584.2005.0046.x, 2005.

Cook, P. G. and Herczeg, A. L.: Environmental tracers in subsurface hydrology, Springer Science and Business, 2000.

Cornaton, F. J.: Transient water age distributions in environmental flow systems: The time-marching Laplace transform solution technique, Water Resour. Res., 48, 1–17, https://doi.org/10.1029/2011WR010606, 2012.

de Dreuzy, J.-R. and Ginn, T. R.: Residence times in subsurface hydrological systems, introduction to the Special Issue, J. Hydrol., 543, 1–6, https://doi.org/10.1016/j.jhydrol.2016.11.046, 2016.

Delsman, J. R., Hu-a-ng, K. R. M., Vos, P. C., de Louw, P. G. B., Oude Essink, G. H. P., Stuyfzand, P. J., and Bierkens, M. F. P.: Paleo-modeling of coastal saltwater intrusion during the Holocene: an application to the Netherlands, Hydrol. Earth Syst. Sci., 18, 3891–3905, https://doi.org/10.5194/hess-18-3891-2014, 2014.

Doherty, J.: Calibration and uncertainty analysis for complex environmental models, Wartermark Numerical Computing, Brisbane, Australia, ISBN: 978-0-9943786-0-6, 2015.

Doherty, J.: Model-Independent Parameter Estimation I, Watermark Numerical Computing, 366, available at: http://www.pesthomepage.org/Downloads.php (last access: 1 August 2017), 2016.

Doherty, J. and Hunt, R. J.: Two statistics for evaluating parameter identifiability and error reduction, J. Hydrol., 366, 119–127, https://doi.org/10.1016/j.jhydrol.2008.12.018, 2009.

Eberts, S. M., Böhlke, J. K., Kauffman, L. J., and Jurgens, B. C.: Comparison of particle-tracking and lumped-parameter age-distribution models for evaluating vulnerability of production wells to contamination, Hydrogeol. J., 20, 263–282, https://doi.org/10.1007/s10040-011-0810-6, 2012.

Edmunds, W. M. and Smedley, P. L.: Residence time indicators in groundwater: The East Midlands Triassic sandstone aquifer, Appl. Geochem., 15, 737–752, https://doi.org/10.1016/S0883-2927(99)00079-7, 2000.

Engesgaard, P. and Molson, J.: Direct simulation of ground water age in the Rabis Creek Aquifer, Denmark, Ground Water, 36, 577–582, https://doi.org/10.1111/j.1745-6584.1998.tb02831.x, 1998.

Freeze, R. A. and Cherry, J. A.: Groundwater, Prentice-Hall, ISBN 9780133653120, 1979.

Gelhar, L. W., Welty, C., and Rehfeldt, K. R.: A Critical Review of Data on Field-Scale Dispersion in Aquifers, Water Resour. Res., 28, 1955–1974, https://doi.org/10.1029/92WR00607, 1992.

Ginn, T. R., Haeri, H., Massoudieh, A., and Foglia, L.: Notes on groundwater age in forward and inverse modeling, Transport Porous Med., 79, 117–134, https://doi.org/10.1007/s11242-009-9406-1, 2009.

Gleeson, T., Befus, K. M., Jasechko, S., Luijendijk, E., and Cardenas, M. B.: The global volume and distribution of modern groundwater, Nat. Geosci., 9, 161–167, https://doi.org/10.1038/ngeo2590, 2015.

Goderniaux, P., Davy, P., Bresciani, E., De Dreuzy, J. R., and Le Borgne, T.: Partitioning a regional groundwater flow system into shallow local and deep regional flow compartments, Water Resour. Res., 49, 2274–2286, https://doi.org/10.1002/wrcr.20186, 2013.

Goode, J. D.: Direct Simulation of Groundwater age, Water Resour. Res., 32, 289–296, 1996.

Guo, W. and Langevin, C. D.: User's Guide to SEAWAT: A Computer Program For Simulation of Three-Dimensional Variable-Density Ground-Water Flow, Tallahassee, 2002.

Hansen, B., Dalgaard, T., Thorling, L., Sørensen, B., and Erlandsen, M.: Regional analysis of groundwater nitrate concentrations and trends in Denmark in regard to agricultural influence, Biogeosciences, 9, 3277–3286, https://doi.org/10.5194/bg-9-3277-2012, 2012.

Harbaugh, A. W., Banta, E. R., Hill, M. C., and McDonald, M. G.: MODFLOW-2000, The U. S. Geological Survey modular ground-water model-user guide to modularization concepts and the ground-water flow process, USGS Open-File Rep. 00-92, 2000.

Harkness, J. S., Darrah, T. H., Warner, N. R., Whyte, C. J., Moore, M. T., Millot, R., Kloppmann, W., Jackson, R. B., and Vengosh, A.: The geochemistry of naturally occurring methane and saline groundwater in an area of unconventional shale gas development, Geochim. Cosmochim. Ac., 208, 302–334, https://doi.org/10.1016/j.gca.2017.03.039, 2017.

Harris, K. R. and Woolf, L. A.: Pressure and temperature dependence of the self diffusion coefficient of water and oxygen-18 water, J. Chem. Soc. Faraday Trans., 76, 377–385, https://doi.org/10.1039/f19807600377, 1980.

Harvey, C. F. and Gorelick, S. M.: Temporal moment-generating equations: Modeling transport and mass transfer in heterogenous aquifers, Water Resour. Manag., 31, 1895–1911, 1995.

Henriksen, H. J., Troldborg, L., Nyegaard, P., Sonnenborg, T. O., Refsgaard, J. C., and Madsen, B.: Methodology for construction, calibration and validation of a national hydrological model for Denmark, J. Hydrol., 280, 52–71, https://doi.org/10.1016/S0022-1694(03)00186-0, 2003.

Hill, M. C. and Tiedeman, C. R.: Effective Groundwater Model Calibration: with analysis of data, sensitivities, predictions and uncertainties, Wiley, 2007.

Hinsby, K. and Rasmussen, E. S.: The Miocene sand aquifers, Jutland, Denmark, in: Natural groundwater quality, edited by: Edmunds, W. M. and Shand, P., Blackwell, 2008.

Hinsby, K., Harrar, W. G., Nyegaard, P., Konradi, P. B., Rasmussen, E. S., Bidstrup, T., Gregersen, U., and Boaretto, E.: The Ribe Formation in western Denmark – Holocene and Pleistocene groundwaters in a coastal Miocene sand aquifer, Geol. Soc. London, Spec. Publ. 189, 29–48, https://doi.org/10.1144/GSL.SP.2001.189.01.04, 2001a.

Hinsby, K., Edmunds, W. M., Loosli, H. H., Manzano, M., Condesso De Melo, M. T., and Barbecot, F.: The modern water interface: recognition, protection and development – advance of modern waters in European aquifer systems, Geol. Soc. London, Spec. Publ., 189, 271–288, https://doi.org/10.1144/gsl.sp.2001.189.01.16, 2001b.

Hinsby, K., Purtschert, R. and Edmunds, W. M.: Groundwater age and quality, in: Groundwater Science and Policy, edited by: Quevauviller, P., Royal Society of Chemistry, London, 2007.

Hölting, B. and Coldewey, W. G.: Hydrogeologie, 8th Edn., Springer-Verlag, Berin, Heidelberg, 2013.

Høyer, A.-S., Vignoli, G., Hansen, T. M., Vu, L. T., Keefer, D. A., and Jørgensen, F.: Multiple-point statistical simulation for hydrogeological models: 3-D training image development and conditioning strategies, Hydrol. Earth Syst. Sci., 21, 6069–6089, https://doi.org/10.5194/hess-21-6069-2017, 2017.

IAEA: Isotope Methods for Dating Old Groundwater, Internation Atomic Agency, Vienna, 2013.

Jaehne, B., Heinz, G., and Dietrich, W.: Measurement of the diffusion coefficients of sparingly soluble gases in water, J. Geophys. Res., 92, 10767–10776, 1987.

Jasechko, S., Perrone, D., Befus, K. M., Cardenas, M. B., Ferguson, G., Gleeson, T., Luijendijk, E., McDonnell, J. J., Taylor, R. G., Wada, Y., and Kirchner, J. W.: Global aquifers dominated by fossil groundwaters but wells vulnerable to modern contamination, Nat. Geosci., 10, 425–429, https://doi.org/10.1038/ngeo2943, 2017.

Jørgensen, F., Høyer, A.-S., Sandersen, P. B. E., He, X., and Foged, N.: Combining 3D geological modelling techniques to address variations in geology, data type and density – An example from Southern Denmark, Comput. Geosci., 81, 53–63, https://doi.org/10.1016/j.cageo.2015.04.010, 2015.

Karlsson, I. B., Sonnenborg, T. O., Jensen, K. H., and Refsgaard, J. C.: Historical trends in precipitation and stream discharge at the Skjern River catchment, Denmark, Hydrol. Earth Syst. Sci., 18, 595–610, https://doi.org/10.5194/hess-18-595-2014, 2014.

Kazemi, G. A., Lehr, J. H., and Perrochet, P.: Groundwater Age, Wiley – Interscience, https://doi.org/0.1002/0471929514, 2006.

Konikow, L. F., Hornberger, G. Z., Putnam, L. D., Shapiro, A. M., and Zinn, B. A.: The use of groundwater age as a calibration target, in: Proceedings of ModelCare: Calibration and Reliability in Groundwater Modelling: Credibility of Modelling, IAHS, 250–256, 2008.

LaBolle, E. M. and Fogg, G. E.: Role of Molecular Diffusion in Contaminant Migration and Recovery in an Alluvial Aquifer System, Transport Porous Med., 42, 155–179, https://doi.org/10.1023/A:1006772716244, 2001.

Larsen, F., Tran, L. V., Van Hoang, H., Tran, L. T., Christiansen, A. V., and Pham, N. Q.: Groundwater salinity influenced by Holocene seawater trapped in incised valleys in the Red River delta plain, Nat. Geosci., 10, 376–381, https://doi.org/10.1038/ngeo2938, 2017.

Levenspiel, O. and Sater, V. E.: Two-phase flow in packed beds, I EC Fundam., 5, 86–92, 1966.

MacDonald, A. M., Bonsor, H. C., Ahmed, K. M., Burgess, W. G., Basharat, M., Calow, R. C., Dixit, A., Foster, S. S. D., Gopal, K., Lapworth, D. J., Lark, R. M., Moench, M., Mukherjee, A., Rao, M. S., Shamsudduha, M., Smith, L., Taylor, R. G., Tucker, J., van Steenbergen, F., and Yadav, S. K.: Groundwater quality and depletion in the Indo-Gangetic Basin mapped from in situ observations, Nat. Geosci., 9, 762–766, https://doi.org/10.1038/ngeo2791, 2016.

Maloszewski, P. and Zuber, A.: Lumped parameter modles for the interpretation of environmental tracer data, in: Manual on Mathematical Models in Isotope Hydrology, IAEA-TECDOC 910, 9–50, 1996.

Manning, A. H., Solomon, D. K., and Thiros, S. A.: 3 H/ 3 He Age Data in Assessing the Susceptibility of Wells to Contamination, Ground Water, 43, 353–367, https://doi.org/10.1111/j.1745-6584.2005.0028.x, 2005.

McCallum, J. L., Cook, P. G., Simmons, C. T., and Werner, A. D.: Bias of Apparent Tracer Ages in Heterogeneous Environments, Groundwater, 52, 239–250, https://doi.org/10.1111/gwat.12052, 2014.

McCallum, J. L., Cook, P. G., and Simmons, C. T.: Limitations of the Use of Environmental Tracers to Infer Groundwater Age, Groundwater, 53, 56–70, https://doi.org/10.1111/gwat.12237, 2015.

McMahon, P. B., Carney, C. P., Poeter, E. P., and Peterson, S. M.: Use of geochemical, isotopic, and age tracer data to develop models of groundwater flow for the purpose of water management, northern High Plains aquifer, USA, Appl. Geochem., 25, 910–922, https://doi.org/10.1016/j.apgeochem.2010.04.001, 2010.

Meyer, R.: Large scale hydrogeological modelling of a low-lying complex coastal aquifer system, University of Copenhagen, 184 pp., 2018.

Meyer, R., Engesgaard, P., Høyer, A.-S., Jørgensen, F., Vignoli, G., and Sonnenborg, T. O.: Regional flow in a complex coastal aquifer system: combining voxel geological modelling with regularized calibration, J. Hydrol., 562, 544–563, https://doi.org/10.1016/j.jhydrol.2018.05.020, 2018a.

Meyer, R., Engesgaard, P., and Sonnenborg, T. O.: Origin and dynamics of saltwater intrusions in regional aquifers; combining 3D saltwater modelling with geophysical and geochemical data, Water Resour. Res., in review, 2018b.

Molson, J. W. and Frind, E. O.: On the use of mean groundwater age, life expectancy and capture probability for defining aquifer vulnerability and time-of-travel zones for source water protection, J. Contam. Hydrol., 127, 76–87, https://doi.org/10.1016/j.jconhyd.2011.06.001, 2012.

Morgan, L. K., Werner, A. D., and Simmons, C. T.: On the interpretation of coastal aquifer water level trends and water balances: A precautionary note, J. Hydrol., 470–471, 280–288, https://doi.org/10.1016/j.jhydrol.2012.09.001, 2012.

Oude Essink, G. H. P., Van Baaren, E. S., and De Louw, P. G. B.: Effects of climate change on coastal groundwater systems: A modeling study in the Netherlands, Water Resour. Res., 46, 1–16, https://doi.org/10.1029/2009WR008719, 2010.

Park, J., Bethke, C. M., Torgersen, T., and Johnson, T. M.: Transport modeling applied to the interpretation of groundwater 36 Cl age, Water Resour. Res., 38, 1–15, https://doi.org/10.1029/2001WR000399, 2002.

Partington, D., Brunner, P., Simmons, C. T., Therrien, R., Werner, A. D., Dandy, G. C., and Maier, H. R.: A hydraulic mixing-cell method to quantify the groundwater component of streamflow within spatially distributed fully integrated surface water-groundwater flow models, Environ. Model. Softw., 26, 886–898, https://doi.org/10.1016/j.envsoft.2011.02.007, 2011.

Pauw, P., De Louw, P. G. B., and Oude Essink, G. H. P.: Groundwater salinisation in the Wadden Sea area of the Netherlands: Quantifying the effects of climate change, sea-level rise and anthropogenic interferences, Neth. J. Geosci., 91, 373–383, https://doi.org/10.1017/S0016774600000500, 2012.

Pearson, F. J. and Hanshaw, B. B.: Sources of Dissolved Carbonate Species in Groundwater and their Effects on Carbon-14 Dating, Symposium on Iostopic Hydrology, International Atomic Energy Agency (IAEA), 1970.

Pollock, D. W.: User Guide for MODPATH Version 6 – A Particle-Tracking Model for MODFLOW, U.S. Geol. Surv. Tech. Methods 6-A41, 2012.

Post, V. E. A., Vandenbohede, A., Werner, A. D., and Teubner, M. D.: Groundwater ages in coastal aquifers, Adv. Water Resour., 57, 1–11, https://doi.org/10.1016/j.advwatres.2013.03.011, 2013.

Post, V. E. A., Kooi, H., and Simmons, C.: Using hydraulic head measurements in variable-density ground water flow analyses, Ground Water, 45, 664–71, https://doi.org/10.1111/j.1745-6584.2007.00339.x, 2007.

Rasmussen, E. S., Dybkjær, K., and Piasecki, S.: Lithostratigraphy of the Upper Oligocene-Miocene succession of Denmark, Geol. Surv. Denmark Greenl. Bull., 22, ISBN 9788778712912, 2010.

Salmon, S. U., Prommer, H., Park, J., Meredith, K. T., Turner, J. V., and McCallum, J. L.: A general reactive transport modeling framework for simulating and interpreting groundwater 14C age and delta 13C, Water Resour. Res., 51, 359–376, https://doi.org/10.1002/2014WR015779, 2015.

Sanford, W.: Calibration of models using groundwater age, Hydrogeol. J., 19, 13–16, https://doi.org/10.1007/s10040-010-0637-6, 2011.

Sanford, W. E.: Correcting for Diffusion in Carbon-14 Dating of Ground Water, Ground Water, 35, 357–361, 1997.

Sanford, W. E., Plummer, L. N., McAda, D. P., Bexfield, L. M., and Anderholm, S. K.: Hydrochemical tracers in the middle Rio Grande Basin, USA: 2. Calibration of a groundwater-flow model, Hydrogeol. J., 12, 389–407, https://doi.org/10.1007/s10040-004-0326-4, 2004.

Sanford, W. E., Plummer, L. N., Busenber, G. C., Nelms, D. L., and Schlosser, P.: Using dual-domain advective-transport simulation to reconcile multiple-tracer ages and estimate dual-porosity transport parameters, Water Resour. Res., 53, 5002–5016, https://doi.org/10.1002/2016WR019469, 2017.

Scharling, P. B.: Hydrogeological modeling and multiple environmental tracer analysis of the deep Miocene aquifers within the Skjern and Varde river catchment, University of Copenhagen, 2011.

Seifert, D., Sonnenborg, T. O., Scharling, P., and Hinsby, K.: Use of alternative conceptual models to assess the impact of a buried valley on groundwater vulnerability, Hydrogeol. J., 16, 659–674, https://doi.org/10.1007/s10040-007-0252-3, 2008.

Seiler, K. and Lindner, W.: Near-surface and deep groundwaters, J. Hydrol., 165, 33–44, https://doi.org/10.1016/0022-1694(95)92765-6, 1995.

Smedley, P. L. and Kinniburgh, D. G.: A review of the source, behaviour and distribution of arsenic in natural waters, Appl. Geochem., 17, 517–568, https://doi.org/10.1016/S0883-2927(02)00018-5, 2002.

Smedley, P. L. and Kinniburgh, D. G.: Molybdenum in natural waters: A review of occurrence, distributions and controls, Appl. Geochem., 84, 387–432, https://doi.org/10.1016/j.apgeochem.2017.05.008, 2017.

Sonnenborg, T. O., Scharling, P. B., Hinsby, K., Rasmussen, E. S., and Engesgaard, P.: Aquifer Vulnerability Assessment Based on Sequence Stratigraphic and 39 Ar Transport Modeling, Groundwater, 54, 214–230, https://doi.org/10.1111/gwat.12345, 2016.

Starn, J. J., Green, C. T., Hinkle, S. R., Bagtzoglou, A. C., and Stolp, B. J.: Simulating water-quality trends in public-supply wells in transient flow systems, Ground Water, 52, 53–62, https://doi.org/10.1111/gwat.12230, 2014.

Stroeven, A. P., Hättestrand, C., Kleman, J., Heyman, J., Fabel, D., Fredin, O., Goodfellow, B. W., Harbor, J. M., Jansen, J. D., Olsen, L., Caffee, M. W., Fink, D., Lundqvist, J., Rosqvist, G. C., Strömberg, B., Jansson, K. N.: Deglaciation of Fennoscandia, Quaternary Sci. Rev., 147, 91–121, https://doi.org/10.1016/j.quascirev.2015.09.016, 2016.

Sudicky, E. A. and Frind, E. O.: Carbon 14 dating of groundwater in confined aquifers: Implications of aquitard diffusion, Water Resour. Res., 17, 1060–1064, https://doi.org/10.1029/WR017i004p01060, 1981.

Tikhonov, A. N. and Arsenin, V. Y.: Solutions of ill-posed problems, Winston, Washington, DC, 1977.

Tóth, J.: A theoretical analysis of groundwater flow in small drainage basins, J. Geophys. Res., 68, 4795–4812, https://doi.org/10.1029/JZ068i016p04795, 1963.

Troldborg, L., Jensen, K. H., Engesgaard, P., Refsgaard, J. C., and Hinsby, K.: Using Environmental Tracers in Modeling Flow in a Complex Shallow Aquifer System, J. Hydrol. Eng., 13, 199–204, https://doi.org/10.1061/(ASCE)1084-0699(2008)13:11(1037), 2008.

Turnadge, C. and Smerdon, B. D.: A review of methods for modelling environmental tracers in groundwater: Advantages of tracer concentration simulation, J. Hydrol., 519, 3674–3689, https://doi.org/10.1016/j.jhydrol.2014.10.056, 2014.

Varni, M. and Carrera, J.: Simulation of groundwater age distributions, Water Resour. Res., 34, 3271–3281, https://doi.org/10.1029/98WR02536, 1998.

Weissmann, G. S., Zhang, Y., LaBolle, E. M., and Fogg, G. E.: Dispersion of groundwater age in an alluvial aquifer system, Water Resour. Res., 38, 16.1–16.13, https://doi.org/10.1029/2001WR000907, 2002.

Wood, C., Cook, P. G., Harrington, G. A., and Knapton, A.: Constraining spatial variability in recharge and discharge in an arid environment through modeling carbon-14 with improved boundary conditions, Water Resour. Res., 53, 142–157, https://doi.org/10.1002/2015WR018424, 2017.

Woolfenden, L. R. and Ginn, T. R.: Modeled ground water age distributions, Ground Water, 47, 547–557, https://doi.org/10.1111/j.1745-6584.2008.00550.x, 2009.

13

Stable oxygen isotope variability in two contrasting glacier river catchments in Greenland

Jacob C. Yde[1], **Niels T. Knudsen**[2], **Jørgen P. Steffensen**[3], **Jonathan L. Carrivick**[4], **Bent Hasholt**[5], **Thomas Ingeman-Nielsen**[6], **Christian Kronborg**[2], **Nicolaj K. Larsen**[2], **Sebastian H. Mernild**[1,7], **Hans Oerter**[8], **David H. Roberts**[9], **and Andrew J. Russell**[10]

[1]Faculty of Engineering and Science, Sogn og Fjordane University College, Sogndal, Norway
[2]Department of Geoscience, University of Aarhus, Aarhus, Denmark
[3]Centre for Ice and Climate, University of Copenhagen, Copenhagen, Denmark
[4]School of Geography and water@leeds, University of Leeds, Leeds, UK
[5]Department of Geosciences and Natural Resource Management, University of Copenhagen, Copenhagen, Denmark
[6]Arctic Technology Centre, Technical University of Denmark, Kgs. Lyngby, Denmark
[7]Antarctic and Sub-Antarctic Program, Universidad de Magallanes, Punta Arenas, Chile
[8]Alfred Wegener Institute, Helmholtz Centre for Polar and Marine Research, Bremerhaven, Germany
[9]Department of Geography, University of Durham, Durham, UK
[10]School of Geography, Politics & Sociology, Newcastle University, Newcastle upon Tyne, UK

Correspondence to: Jacob C. Yde (jacob.yde@hisf.no)

Abstract. Analysis of stable oxygen isotope ($\delta^{18}O$) characteristics is a useful tool to investigate water provenance in glacier river systems. In order to attain knowledge on the diversity of $\delta^{18}O$ variations in Greenlandic rivers, we examined two contrasting glacierised catchments disconnected from the Greenland Ice Sheet (GrIS). At the Mittivakkat Gletscher river, a small river draining a local temperate glacier in southeast Greenland, diurnal oscillations in $\delta^{18}O$ occurred with a 3 h time lag to the diurnal oscillations in run-off. The mean annual $\delta^{18}O$ was -14.68 ± 0.18‰ during the peak flow period. A hydrograph separation analysis revealed that the ice melt component constituted 82 ± 5 % of the total run-off and dominated the observed variations during peak flow in August 2004. The snowmelt component peaked between 10:00 and 13:00 local time, reflecting the long travel time and an inefficient distributed subglacial drainage network in the upper part of the glacier. At the Kuannersuit Glacier river on the island Qeqertarsuaq in west Greenland, the $\delta^{18}O$ characteristics were examined after the major 1995–1998 glacier surge event. The mean annual $\delta^{18}O$ was -19.47 ± 0.55‰. Despite large spatial variations in the $\delta^{18}O$ values of glacier ice on the newly formed glacier tongue, there were no diurnal os-

cillations in the bulk meltwater emanating from the glacier in the post-surge years. This is likely a consequence of a tortuous subglacial drainage system consisting of linked cavities, which formed during the surge event. Overall, a comparison of the $\delta^{18}O$ compositions from glacial river water in Greenland shows distinct differences between water draining local glaciers and ice caps (between -23.0 and -13.7‰) and the GrIS (between -29.9 and -23.2‰). This study demonstrates that water isotope analyses can be used to obtain important information on water sources and the subglacial drainage system structure that is highly desired for understanding glacier hydrology.

1 Introduction

There is an urgent need for improving our understanding of the controls on water sources and flow paths in Greenland. As in other parts of the Arctic, glacierised catchments in Greenland are highly sensitive to climate change (Milner et al., 2009; Blaen et al., 2014). In recent decades freshwater run-off from the Greenland Ice Sheet (GrIS) to adjacent seas has increased significantly (Hanna et al., 2005, 2008; Bamber et al., 2012; Mernild and Liston, 2012), and the total ice mass loss from the GrIS contributes with 0.33 mm sea level equivalent yr^{-1} to global sea level rise (1993–2010; Vaughan et al., 2013). In addition, ice mass loss from local glaciers (i.e. glaciers and ice caps peripheral to the GrIS; Weidick and Morris, 1998) has resulted in a global sea level rise of 0.09 mm sea level equivalent yr^{-1} (1993–2010; Vaughan et al., 2013). The changes in run-off are coupled to recent warming in Greenland (Hanna et al., 2012, 2013; Mernild et al., 2014), an increasing trend in precipitation and changes in precipitation patterns (Bales et al., 2009; Mernild et al., 2015a), and a decline in albedo (Bøggild et al., 2010; Tedesco et al., 2011; Box et al., 2012; Yallop et al., 2012; Mernild et al., 2015b). Also, extreme surface melt events have occurred in recent years (Tedesco et al., 2008, 2011; van As et al., 2012), and in July 2012 more than 97 % of the GrIS experienced surface melting (Nghiem et al., 2012; Keegan et al., 2014). In this climate change context, detailed catchment-scale studies on water source and water flow dynamics are urgently needed to advance our knowledge of the potential consequences of future hydrological changes in Greenlandic river catchments.

Analysis of stable oxygen isotopes is a very useful technique to investigate water provenance in glacial river systems. Stable oxygen isotopes are natural conservative tracers in low-temperature hydrological systems (e.g. Moser and Stichler, 1980; Gat and Gonfiantini, 1981; Haldorsen et al., 1997; Kendall et al., 2014). Consequently, oxygen isotopes can be applied to determine the timing and origin of changes in water sources and flow paths because different water sources often have isotopically different compositions due to their exposure to different isotopic fractionation processes. Since the 1970s, this technique has been widely used for hydrograph separation (Dinçer et al., 1970). Most often a conceptual two-component mixing model is applied, where an old-water component (e.g. groundwater) is mixed with a new-water component (e.g. rain or snowmelt), assuming that both components have spatial and temporal homogeneous compositions. The general mixing model is given by the equation

$$QC = Q_1C_1 + Q_2C_2 + ..., \tag{1}$$

where the discharge Q and the isotopic value C are equal to the sum of their components. This simplified model has limitations when a specific precipitation event is analysed because the water isotope composition in precipitation (new water) may vary considerably during a single event (e.g. McDonnell et al., 1990) and changes in contributions from secondary old-water reservoirs may occur (e.g. Hooper and Shoemaker, 1986). Nevertheless, water isotope mixing models still provide valuable information on spatial differences in hydrological processes on diurnal to annual timescales (Kendall et al., 2014).

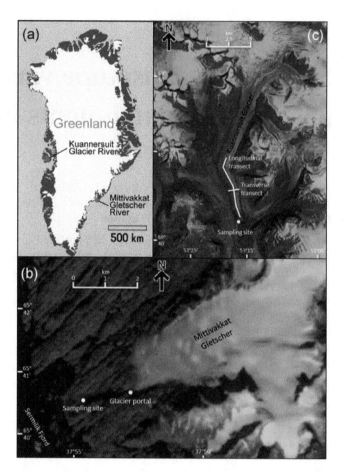

Figure 1. Location map **(a)** of the study areas at **(b)** the Mittivakkat Gletscher river, southeast Greenland (image from Landsat 8 OLI on 3 September 2013), and at **(c)** the Kuannersuit Glacier river, west Greenland (image from Landsat 8 OLI on 8 July 2014).

In glacier-fed river systems, the principal water sources to bulk run-off derive from ice melt, snowmelt, rainfall and groundwater components. Depending on the objectives of the study and on the environmental setting, hydrograph separation of glacial rivers has been based on assumed endmember isotope mixing between two or three prevailing components (Behrens et al., 1971, 1978; Fairchild et al., 1999; Mark and Seltzer, 2003; Theakstone, 2003; Yde and Knudsen, 2004; Mark and McKenzie, 2007; Yde et al., 2008; Bhatia et al., 2011; Kong and Pang, 2012; Ohlanders et al., 2013; Blaen et al., 2014; Dahlke et al., 2014; Hindshaw et al., 2014; Meng et al., 2014; Penna et al., 2014; Rodriguez et al., 2014; Zhou et al., 2014). As glacierised catchments vary in size, altitudinal range, hypsometry, degree of glaciation, and thermal and morphological glacier types, isotope hydrograph separation often requires that the primary local controls on run-off generation are identified in order to analyse the variability in isotope time series. In detailed studies it may even be necessary to divide a main component, such as ice melt, into several ice facies sub-components (Yde and Knudsen, 2004).

However, in highly glacierised catchments the variability in oxygen isotope composition is generally controlled by seasonal snowmelt and ice melt with episodic inputs of rainwater, whereas contributions from shallow groundwater flow may become important in catchments, where glaciers comprise a small proportion of the total area (e.g. Blaen et al., 2014).

In this study, we examine the stable oxygen isotope composition in two Greenlandic glacier river systems, namely the Mittivakkat Gletscher river ($13.6\,\mathrm{km^2}$), which drains a local non-surging glacier in southeast Greenland, and the Kuannersuit Glacier river ($258\,\mathrm{km^2}$), which drains a local glacier on the island Qeqertarsuaq, west Greenland. The latter experienced a major glacier surge event in 1995–1998. Our aim is to gain insights into the variability and controls of the oxygen isotope composition in contrasting glacierised river catchments located peripheral to the GrIS (i.e. the river systems do not drain meltwater from the GrIS). Besides a study by Andreasen (1984) at the glacier Killersuaq in west Greenland, this is the first study of oxygen isotope dynamics in rivers draining glacierised catchments peripheral to the GrIS.

2 Study sites

2.1 Mittivakkat Gletscher river, Ammassalik Island, southeast Greenland

Mittivakkat Gletscher ($65°41'$ N, $37°50'$ W) is the largest glacier complex on Ammassalik Island, southeast Greenland (Fig. 1). The entire glacier covered an area of $26.2\,\mathrm{km^2}$ in 2011 (Mernild et al., 2012) and has an altitudinal range between 160 and 880 m a.s.l. (Mernild et al., 2013a). Bulk meltwater from the glacier drains primarily westwards to the proglacial Mittivakkat Valley and flows into the Sermilik Fjord. The sampling site is located at a hydrometric station 1.3 km down-valley from the main subglacial meltwater portal. The hydrological catchment has an area of $13.6\,\mathrm{km^2}$, of which $9.0\,\mathrm{km^2}$ is glacierised (66 %). The maritime climate is Low Arctic with annual precipitation ranging from 1400 to 1800 mm water equivalent (w.e.) $\mathrm{yr^{-1}}$ (1998–2006) and a mean annual air temperature (MAAT) at 515 m a.s.l. of $-2.2\,°\mathrm{C}$ (1993–2011; updated from Mernild et al., 2008a). There are no observations of contemporary permafrost in the area, and the proglacial vegetation cover is sparse.

The glacier has undergone continuous recession since the end of the Little Ice Age (Knudsen et al., 2008; Mernild et al., 2011). In recent decades the recession has accelerated and the glacier has lost approximately 29 % of its volume between 1994 and 2012 (Yde et al., 2014), and surface mass balance measurements indicate a mean thinning rate of 1.01 m w.e. $\mathrm{yr^{-1}}$ between 1995–1996 and 2011–2012 (Mernild et al., 2013a). Similar to other local glaciers in the Ammassalik region, Mittivakkat Gletscher is severely out of contemporary climatic equilibrium (Mernild et al., 2012,

Table 1. Summary of $\delta^{18}O$ mean and range in bulk water samples at the Mittivakkat Gletscher river.

Year	Campaign period	n	$\delta^{18}O_{mean}$	$\delta^{18}O_{max}$	$\delta^{18}O_{min}$
2003	11–13 Aug	4	−14.42	−14.30	−14.65
2004	8–22 Aug	103	−14.55	−14.19	−14.91
2005	30 May–12 Jun	29	−14.71	−14.35	−15.16
	23–26 Jul	19	−14.10	−13.74	−14.41
	11–19 Aug	44	−14.73	−14.13	−16.43
2006	11–16 Aug	11	−14.85	−14.26	−15.42
2007	2–10 Aug	17	−14.69	−14.07	−15.11
2008	29 May–11 Jun*	28	−16.92	−15.92	−17.35
	10–16 Aug	15	−14.84	−14.47	−15.20
2009	8–16 Aug	17	−14.88	−14.56	−15.13

* Collected at a sampling site ca. 500 m closer to the glacier front.

2013b) and serves as a representative location for studying the impact of climate change on glacierised river catchments in southeast Greenland (e.g. Mernild et al., 2008b, 2015b; Bárcena et al., 2010, 2011; Kristiansen et al., 2013; Lutz et al., 2014).

2.2 Kuannersuit Glacier river, Qeqertarsuaq, west Greenland

Kuannersuit Glacier ($69°46'$ N, $53°15'$ W) is located in central Qeqertarsuaq (formerly Disko Island), west Greenland (Fig. 1). It is an outlet glacier descending from the Sermersuaq ice cap and belongs to the Qeqertarsuaq–Nuussuaq surge cluster (Yde and Knudsen, 2007). In 1995, the glacier started to surge down the Kuannersuit Valley with a frontal velocity up to 70 m per day (Larsen et al., 2010). By the end of 1998 or beginning of 1999, the surging phase terminated and the glacier went into its quiescent phase, which is presumed to last more than 100 years (Yde and Knudsen, 2005a). The 1995–1998 surge of Kuannersuit Glacier is one of the largest land-terminating surge events ever recorded; the glacier advanced 10.5 km down-valley, and approximately $3\,\mathrm{km^3}$ of ice was moved to form a new glacier tongue (Larsen et al., 2010).

The Kuannersuit Glacier river originates from a portal at the western side of the glacier terminus, and the sampling site is located 200 m down-stream (Yde et al., 2005a). The catchment area has an altitude range of 100–1650 m a.s.l. and covers $258\,\mathrm{km^2}$, of which Kuannersuit Glacier constitutes $103\,\mathrm{km^2}$ of the total glacierised area of $168\,\mathrm{km^2}$ (Yde and Knudsen, 2005a). The valley floor consists of unvegetated outwash sediment; dead-ice deposits; and ice-cored, vegetated terraces. The proglacial area of the catchment is situated in the continuous permafrost zone (Yde and Knudsen, 2005b), and the climate is polar continental (Humlum, 1999). There are no meteorological observations from the area, but at the coastal town of Qeqertarsuaq (formerly Godhavn), located 50 km to the southwest, the MAAT was −2.7 and $-1.7\,°\mathrm{C}$ in 2011 and 2012, respectively (Cappelen, 2013).

Table 2. Summary of $\delta^{18}O$ mean and range in bulk water samples at the Kuannersuit Glacier river.

Year	Campaign period	n	$\delta^{18}O_{mean}$	$\delta^{18}O_{max}$	$\delta^{18}O_{min}$
2000	24–27 Jul	21	−19.80	−19.47	−19.97
2001	14–31 Jul	109	−19.25	−17.82	−19.55
2002	14–15 Jul	21	−19.01	−18.75	−19.39
2003	18–26 Jul	27	−20.43	−19.03	−21.88
2005	19–24 Jul	2	−19.42	−19.32	−19.51

3 Methods

3.1 Sampling protocol and isotope analyses

In total, 287 oxygen isotope samples were collected from the Mittivakkat Gletscher river during the years 2003–2009 (Table 1). Most of the sampling campaigns were conducted in August at the end of the peak flow period (i.e. the summer period with relatively high run-off). The most intensively sampled period was from 8 to 22 August 2004, where sampling was conducted with a 4 h frequency supplemented by short periods of higher frequency sampling. In the years 2005 and 2008, meltwater was also collected during the early melt season (i.e. the period before the subglacial drainage system is well established) to evaluate the seasonal variability in the $\delta^{18}O$ signal. An additional 40 river samples were collected for multi-sampling tests.

During five field seasons in July 2000, 2001, 2002, 2003 and 2005, a total of 180 oxygen isotope samples were collected from the Kuannersuit Glacier river (Table 2), and another 44 river samples were collected for multi-sampling tests. In addition, 13 ice samples were obtained along a longitudinal transect at the centreline of the newly formed glacier tongue with 500 m sampling increments in July 2001, and 23 ice samples were collected along a transverse transect with 50 m sampling increments in July 2003. The transverse transect crossed the longitudinal transect at a distance of 3250 m from the glacier front. Seven samples of rainwater were collected in a Hellmann rain gauge located in the vicinity of the glacier terminus in July 2002.

All water samples were collected manually in 20 mL vials. Ice samples were collected in 250 mL polypropylene bottles or plastic bags before being slowly melted and decanted to 20 mL vials. The vials were stored in cold ($\sim 5\,°C$) and dark conditions to avoid fractionation related to biological activity.

The relative deviations (δ) of water isotope compositions ($^{18}O/^{16}O$) were expressed in per mil (‰) relative to Vienna Standard Mean Ocean Water (0 ‰; Coplen, 1996). The stable oxygen isotope analyses were performed at the Niels Bohr Institute, University of Copenhagen, Denmark, using mass spectrometry with an instrumental precision of ±0.1 ‰ in the oxygen isotope ratio ($\delta^{18}O$) value.

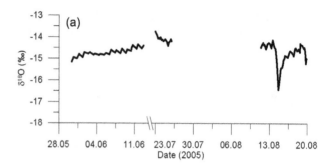

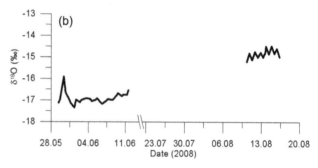

Figure 2. $\delta^{18}O$ time series of meltwater draining Mittivakkat Gletscher in **(a)** 2005 and **(b)** 2008.

The oxygen isotope data from this study are available in the Supplement (Tables S1–S6).

3.2 Multi-sample tests

In the Mittivakkat Gletscher river, we conducted three multi-sample tests at 14:00 local time on 9, 15 and 21 August 2004 to determine the combined uncertainty related to sampling and analytical error. During the multi-sample tests samples were collected simultaneously (within 3 min). The tests show standard deviations of 0.08 ($n = 25$), 0.06 ($n = 5$) and 0.04 ‰ ($n = 10$), respectively, which are lower than the instrumental precision (±0.1 ‰).

In the Kuannersuit Glacier river, multi-sample tests were conducted in 2001, 2002 and 2003, showing a standard deviation of ±0.16 ($n = 5$), ±0.13 ($n = 17$) and ±0.44 ‰ ($n = 22$), respectively. The multi-sample test in 2003 showed a standard deviation significantly larger than the instrumental precision (±0.1 ‰). This deviation cannot be explained by the presence of a few high $\delta^{18}O$ values. The most plausible explanation is that the glacier run-off was not well mixed in 2003, possibly because different parts of the drainage system merged close to the glacier portal.

3.3 Run-off measurements

Stage–discharge relationships were used to determine run-off at each study site. The accuracy of individual run-off measurements is within ±7 % (e.g. Herschy, 1999). For details on run-off measurements we refer to Hasholt and

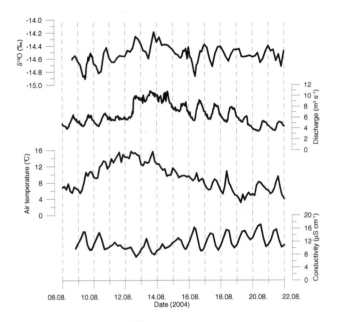

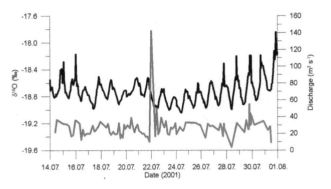

Figure 4. Time series of $\delta^{18}O$ (red curve) and discharge (black curve) in the Kuannersuit Glacier river during the period 14–31 July 2001.

Figure 3. Time series of $\delta^{18}O$, discharge, air temperature and electric conductivity in meltwater draining Mittivakkat Gletscher during the period 8–21 August 2004.

Mernild (2006) for the Mittivakkat Gletscher river and Yde et al. (2005a) for the Kuannersuit Glacier river. In short, at the Mittivakkat Gletscher river the run-off measurements were conducted at a hydrometric monitoring station located after the braided river system had changed into a single river channel about 500 m from the river outlet. The station was installed in August 2004 and recorded water stage every 10 min during the peak flow period. At the Kuannersuit Glacier river the run-off measurements were obtained at a hydrometric monitoring station installed in July 2001 at a location where the river merges to a single channel. Water stage was recorded every hour during the peak flow period. The station was destroyed during the spring river break-up in 2002.

4 Results

4.1 $\delta^{18}O$ characteristics

At the Mittivakkat Gletscher river, the early melt season is characterised by an increasing trend in $\delta^{18}O$. In 2005 the $\delta^{18}O$ values in the early melt season were coincident with the $\delta^{18}O$ values during the peak flow period (Fig. 2a; Table 1). This indicates that the onset of ice melt commenced before the early melt season sampling campaign. In contrast, the 2008 onset of ice melt was delayed, and snowmelt totally dominated the bulk composition of the river water except on 30 May 2008, when a rainfall event (19 mm in the nearby town of Tasiilaq, located 10 km to the southeast of the Mittivakkat Gletscher river catchment; Cappelen, 2013) caused a positive peak in $\delta^{18}O$ of $\sim 1\,‰$ (Fig. 2b). This difference

between the early ablation seasons in 2005 and 2008 is consistent with the meteorological record from Tasiilaq, which shows that the region received a large amount of precipitation in May 2008 (140 mm) compared to a dry May 2005 (17 mm; Cappelen, 2013). Episodic effects on $\delta^{18}O$ by precipitation seem common throughout the ablation season. For instance, another short-term change occurred on 14–15 August 2005 (Fig. 2a), when a negative peak in $\delta^{18}O$ of $\sim 2\,‰$ coincided with a snowfall event (14 mm in Tasiilaq; Cappelen, 2013) and subsequent elevated contribution from snowmelt.

During the peak flow periods, the mean annual $\delta^{18}O$ was $-14.68 \pm 0.18\,‰$ (Table 1). We use the 2004 time series to assess oxygen isotope dynamics in the Mittivakkat Gletscher river during the peak flow period when the subglacial drainage system is assumed to be well established, transporting the majority of meltwater in a channelised network (Mernild, 2006). In Fig. 3, the 2004 $\delta^{18}O$ time series is shown together with run-off (at the hydrometric station), air temperature (at a nunatak at 515 m a.s.l.) and electrical conductivity (at the hydrometric station; corrected to 25 °C). There was no precipitation during the entire sampling period, except for some drizzle on 8 August prior to the collection of the first sample. The time series shows characteristic diurnal variations in $\delta^{18}O$ composition, e.g. on 9–10 and 16–18 August 2004. However, the diurnal pattern was severely disturbed at around 03:00 on 11 August 2004. The hydrograph shows that during the falling limb the diurnal trend in run-off was interrupted, coinciding with an air temperature increase and a change in $\delta^{18}O$ from decreasing to slightly increasing values. The run-off stayed almost constant until a rapid 39 % increase in run-off occurred at 13:00 on 12 August 2004, accompanied by an increase in $\delta^{18}O$ and decrease in electrical conductivity. Thereafter, run-off remained at an elevated level for more than 2 days before returning to a diurnal oscillation of run-off. Hydrograph separation of water sources is a helpful tool to elucidate the details of this event (see Sect. 4.3).

In the Kuannersuit Glacier river, the sample-weighted mean annual $\delta^{18}O$ was $-19.47 \pm 0.55\,‰$ during the peak

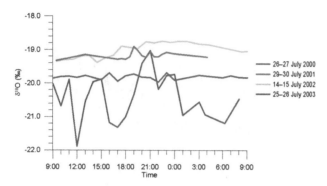

Figure 5. Diurnal $\delta^{18}O$ variations in the Kuannersuit Glacier river on studied days in July in the post-surge years 2000–2003. Multi-sample tests conducted in 2001, 2002 and 2003 showed standard deviations of ± 0.16, ± 0.13 and $\pm 0.44\,‰$, respectively.

flow period (a sample-weighted value is applied because the number of samples per year deviated between 2 and 109). In Fig. 4, the variations in $\delta^{18}O$ are presented together with run-off for the period 14–31 July 2001. The 2001 run-off measurements showed diurnal oscillations with minimums around 10:00–12:00 and maximums at 19:00–20:00, correlating well with reversed oscillations in solutes (Yde et al., 2005a) and poorly with suspended sediment concentrations (Knudsen et al., 2007). However, the variability of $\delta^{18}O$ did not correlate with run-off or any of these variables. While some of the episodic damming and meltwater release events appear as peaks on the run-off time series, the peaks in the $\delta^{18}O$ time series coincided with rainfall events (e.g. on the nights of 21 and 29 July 2001). Besides these episodic peaks, a lack of diurnal fluctuations in $\delta^{18}O$ characterised the $\delta^{18}O$ time series.

Figure 5 shows the diurnal $\delta^{18}O$ variations during 4 days in July without rainfall in the years 2000–2003. There were no diurnal oscillations in 2000, 2001 and 2002. In 2003, the fluctuations were much larger than in the preceding years, but the highest $\delta^{18}O$ ($-19.03\,‰$) was measured at 21:00, and low $\delta^{18}O$ prevailed during the night ($\sim -21.0\,‰$). This diurnal variability was also reflected in the standard deviations of the measurements taken over the 24 h periods, which increased from ± 0.07 in 2000 to ± 0.11, ± 0.23 and $\pm 0.70\,‰$ in 2001, 2002 and 2003, respectively. The corresponding diurnal amplitudes for 2000–2003 were 0.28, 0.42, 0.64 and 2.85‰, respectively. Although these measurements from a single day each year are insufficient to represent the conditions for the entire peak flow period, they may indicate post-surge changes in the structure of subglacial hydrological system which are worth addressing in detail in future studies of the hydrological system of surging glaciers.

4.2 $\delta^{18}O$ endmember components

On Mittivakkat Gletscher, three snow pits (0.1 m sampling increments) were excavated at different altitudes in May 1999, showing a mean $\delta^{18}O$ composition of $-16.5 \pm 0.6\,‰$ (hereafter the uncertainty of $\delta^{18}O$ is given by the standard deviation) in winter snow (Dissing, 2000). The range of individual samples in each snow pit varied between -14.5 and $-19.5\,‰$ (269 m a.s.l.; mean $\delta^{18}O$ $=-16.24 \pm 1.35$; $n = 36$), -13.8 and $-21.2\,‰$ (502 m a.s.l.; mean $\delta^{18}O = -17.11 \pm 2.13$; $n = 21$), and -11.9 and $-21.6\,‰$ (675 m a.s.l.; mean $\delta^{18}O = -16.18 \pm 2.70$; $n = 26$; Dissing, 2000). Also, two ice-surface $\delta^{18}O$ records of 2.84 and 1.05 km in length (10 m sampling increments) were obtained from the glacier terminus towards the equilibrium line (Boye, 1999). The glacier ice $\delta^{18}O$ ranged between -15.0 and $-13.3\,‰$ with a mean $\delta^{18}O$ of $-14.1\,‰$ (Boye, 1999), and the theoretical altitudinal effect (Dansgaard, 1964) of higher $\delta^{18}O$ towards the equilibrium line altitude (ELA) was not observed. The reasons for an absence of a $\delta^{18}O$ lapse rate are most likely the limited size and altitudinal range (160–880 m a.s.l.) of Mittivakkat Gletscher, but ice dynamics, ice age and meteorological conditions such as frequent inversion (Mernild and Liston, 2010) may also have an impact. The $\delta^{18}O$ of summer rain has not been determined in this region, but at the coastal village of Ittoqqortoormiit, located ~ 840 km to the north of Mittivakkat Gletscher, observations show monthly mean $\delta^{18}O$ in rainwater of -12.8, -9.1 and $-8.8\,‰$ in June, July and August, respectively (data available from the International Atomic Energy Agency database WISER). Based on these observations it is evident that endmember snowmelt has a relatively low $\delta^{18}O$ compared to endmember ice melt and that these two water source components can be separated. Contributions from rainwater will likely result in episodic increase in the $\delta^{18}O$ of bulk meltwater.

In the Kuannersuit Glacier river system, the glaciological setting differed from the Mittivakkat Gletscher river system. During the surge event of Kuannersuit Glacier, the glacier front advanced from ~ 500 down to 100 m a.s.l., while a significant part of the glacier surface in the accumulation area was lowered by more than 100 m to altitudes below the ELA (~ 1100–1300 m a.s.l.). A helicopter survey in July 2002 revealed that the post-surge accumulation area ratio was less than 20 % (Yde et al., 2005a). Hence, we assume that the primary post-surge water source during the peak flow period is ice melt, particularly from ablation of the new glacier tongue. The mean $\delta^{18}O$ value of glacier ice collected along the longitudinal and transverse transects was $-20.5 \pm 1.0\,‰$ ($n = 36$). This is consistent with $\delta^{18}O$ values of glacier ice located near the glacier front, showing mean $\delta^{18}O$ of $-19.4 \pm 0.9\,‰$ ($n = 20$) in a section with debris layers formed by thrusting and $-19.8 \pm 1.1\,‰$ ($n = 37$) in a section without debris layers (Larsen et al., 2010). In contrast to the setting at the Mittivakkat Gletscher river, it was likely that another ice melt component in bulk run-off from Kuannersuit Glacier comprised water from several ice facies sub-component sources with various $\delta^{18}O$ values and spatial variability. During the surge event, a thick debris-rich basal

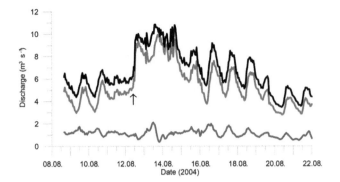

Figure 6. Hydrograph showing the separation of the discharge in the Mittivakkat Gletscher river (black curve) into an ice melt component (red curve) and a snowmelt component (blue curve) during the period 8–21 August 2004. The error of the ice melt and snowmelt components depends on the constant endmember estimates and the cubic spline interpolation. The arrow indicates the onset of the abrupt change in discharge.

ice sequence was formed beneath the glacier and exposed along the glacier margins and at the glacier terminus (Yde et al., 2005b; Roberts et al., 2009; Larsen et al., 2010). The basal ice consisted of various genetic ice facies, where different isotopic fractionation processes during the basal ice formation resulted in variations in the $\delta^{18}O$ composition. The $\delta^{18}O$ in massive stratified ice was $-16.6 \pm 1.9\%$ ($n = 10$); in laminated stratified ice it was $-19.6 \pm 0.7\%$ ($n = 9$); and in dispersed ice it was $-18.8 \pm 0.6\%$ ($n = 41$; Larsen et al., 2010). Also, during the termination of the surge event in winter 1998–1999, proglacial naled was stacked into ~ 3 m thick sections of thrust-block naled at the glacier front, as the glacier advanced into the naled (Yde and Knudsen, 2005b; Yde et al., 2005b; Roberts et al., 2009). Naled is an extrusive ice assemblage formed in front of the glacier by rapid freezing of winter run-off and/or proglacial upwelling water mixed with snow. A profile in a thrust-block naled section showed a $\delta^{18}O$ of $-20.1 \pm 0.5\%$ ($n = 60$; excluding an outlier polluted by rainwater; Yde and Knudsen, 2005b). With regard to the endmember compositions of snowmelt and rainwater at the Kuannersuit Glacier river, it was not possible to access snow on the upper part of the glacier, so no $\delta^{18}O$ values on snowmelt were measured. Rainwater was collected during rainfall events in July 2002, showing a wide range in $\delta^{18}O$ between -18.78 and -6.57% and a median $\delta^{18}O$ of $-10.32 \pm 4.49\%$ ($n = 7$; Table S6).

4.3 Hydrograph separation

The conditions for conducting hydrograph separation during the peak flow period were different for the two study catchments. At the Mittivakkat Gletscher river it was possible to distinguish between the $\delta^{18}O$ values of endmember ice melt and snowmelt components, and there were diurnal oscillations in $\delta^{18}O$. In contrast, the available data from the Kuan-

nersuit Glacier river did not allow hydrograph separation in the years following the surge event. Here, there were no diurnal oscillations in $\delta^{18}O$, and the composition and importance of the snowmelt component were unknown. Hence, we will continue by using the 2004 time series to construct a two-component hydrograph separation (Eq. 1) during a period without precipitation for the Mittivakkat Gletscher river.

First, we apply time-series cubic spline interpolation to estimate $\delta^{18}O$ at 1 h time-step increments, matching the temporal resolution of the run-off observations. This approach allows a better assessment of the diurnal $\delta^{18}O$ signal. For instance, a best-fit analysis shows that overall the $\delta^{18}O$ signal lags 3 h behind run-off ($r^2 = 0.66$; linear correlation without lag shows $r^2 = 0.58$), indicating the combined effect of the two primary components, snowmelt and ice melt, on the $\delta^{18}O$ variations. The diurnal amplitude in $\delta^{18}O$ ranged between 0.11 (11 August 2004) and 0.49‰ (16 August 2004). However, there was no statistical relation between diurnal $\delta^{18}O$ amplitude and daily air temperature amplitude ($r^2 = 0.28$), indicating that other forcings than variability in surface melting may have a more dominant effect on the responding variability in $\delta^{18}O$.

Based on the assumption that snowmelt and ice melt reflect their endmember $\delta^{18}O$ compositions (-16.5 and -14.1%, respectively), a hydrograph showing contributions from snowmelt and ice melt is constructed for the 2004 sampling period (Fig. 6). The ice melt component constituted $82 \pm 5\%$ (where $\pm$ indicates the standard deviation of the hourly estimates) of the total run-off and dominated the observed variations in total run-off ($r^2 = 0.99$). This is expected late in the peak flow period, when the subglacial drainage mainly occurs in a channelised network in the lower part of the glacier (Mernild, 2006). The slightly decreasing trend in the daily snowmelt component was likely a consequence of the diminishing snow cover on the upper part of the glacier. The snowmelt component peaked around 10:00–13:00 each day, reflecting the long distance from the melting snowpack to the proglacial sampling site and the possible existence of an inefficient distributed subglacial drainage network in the upper part of the glacier.

The most likely reason for an abrupt change in glacial run-off, such as the one observed during the early morning of 11 August 2004 followed by the sudden release of water 34 h later, is a roof collapse causing ice-block damming of a major subglacial channel. The hydrograph separation (Fig. 6) shows that the proportion between ice melt and snowmelt remained almost constant after the event commenced, indicating that the bulk water derived from a well-mixed part of the drainage system, which was unaffected by the large diurnal variation in ice melt generation. This suggests that the functioning drainage network transported meltwater from the upper part of the glacier with limited connection to the drainage network in the lower part. Meanwhile, ice melt was stored in a dammed section of the subglacial network located in the lower part of the glacier and suddenly released when the dam broke at 13:00 on 12 August (Fig. 6). In the following hours ice melt comprised up to 94 % of the total run-off. On 13 August the snowmelt component peaked at noon but then dropped markedly, and in the evening it only constituted 4 % of the total run-off. On 14 August there were still some minor disturbances in the lower drainage network, but from 15 August the drainage system had stabilised and the characteristic diurnal glacionival oscillations had taken over (Figs. 3 and 6).

4.4 Uncertainties in $\delta^{18}O$ hydrograph separation models

The accuracy of endmember hydrograph separation models is limited by the uncertainties of the estimated values of each endmember component, the uncertainty of the cubic spline interpolation at each data point and the uncertainty of $\delta^{18}O$ in the river. While the uncertainty of $\delta^{18}O$ in the river is likely to be relatively small, the uncertainties of each endmember component must be kept in mind (e.g. Cable et al., 2011; Arendt et al., 2015). The assumption of discrete values of each endmember component is unlikely to reflect the spatial and temporal changes in bulk $\delta^{18}O$ of snowmelt, ice melt and rainwater. For instance, Raben and Theakstone (1998) found a seasonal increase in mean $\delta^{18}O$ in snow pits on Austre Okstindbreen, Norway, and episodic events such as passages of storms (e.g. McDonnell et al., 1990; Theakstone, 2008) or melting of fresh snow in the late ablation season may cause temporal changes in one component. Also, snowpacks have a non-uniform layered structure with heterogeneous $\delta^{18}O$ composition, and isotopic fractionation is likely to occur as melting progresses and the snowpack is mixed with rainwater (e.g. Raben and Theakstone, 1998; Lee et al., 2010). It is also difficult to assess how representative snow pits and ice transects are for the bulk $\delta^{18}O$ value of each component. Spatial differences in $\delta^{18}O$ may exist within and between snow pits, but the overall effect on the isotopic composition of the water leaving the melting snowpack at a given time is unknown.

4.5 Longitudinal and transverse $\delta^{18}O$ transects

Glacier ice samples were collected on the surface of Kuannersuit Glacier to gain insights into the spatial variability of $\delta^{18}O$ on the newly formed glacier tongue. Both the longitudinal and transverse transects showed large spatial fluctuations in $\delta^{18}O$ (Fig. 7). The longitudinal transect was sampled along the centreline but showed unsystematic fluctuations on a 500 m sampling increment scale. In contrast, the transverse transect, which was sampled 3250 m up-glacier with 50 m increments, showed a more systematic trend where relatively high $\delta^{18}O$ values were observed along both lateral margins. From the centre towards the western margin an increasing trend of 0.46‰ per 100 m prevailed, whereas the eastern central part showed large fluctuations in $\delta^{18}O$ be-

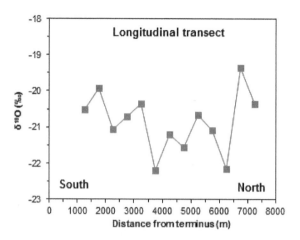

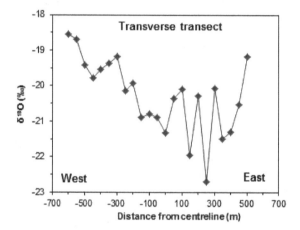

Figure 7. Variations in $\delta^{18}O$ of glacier ice along a longitudinal transect and a transverse transect on Kuannersuit Glacier. The transverse transect crosses the longitudinal transect at a distance of 3250 m from the glacier terminus.

tween −22.69 and −20.08‰. The total range of measured $\delta^{18}O$ in glacier ice along the transverse transect was 4.14‰. A possible explanation of this marked spatial variability may be that the ice forming the new tongue derived from different pre-surge reservoirs on the upper part of the glacier. If so, it is very likely that the marginal glacier ice was formed at relatively low elevations (high $\delta^{18}O$ signal), whereas the glacier ice in the western central part mainly derived from high-elevation areas of Sermersuaq ice cap (low $\delta^{18}O$ signal). At present, there are only a few comparable studies on transverse variations in $\delta^{18}O$ across glacier tongues. Epstein and Sharp (1959) found a decrease in $\delta^{18}O$ towards the margins of Saskatchewan Glacier, Canada. Hambrey (1974) measured a similar decrease in $\delta^{18}O$ towards the margins of Charles Rabots Bre, Norway, in an upper transect, whereas a lower transect showed wide unsystematic variations in $\delta^{18}O$. Hambrey (1974) concluded that in the upper transect the marginal ice derived from higher altitudes than ice in the centre, whereas in the lower transect the wide variations were related to structural complexity of the glacier. However, both

Table 3. Maximum and minimum $\delta^{18}O$ in glacier rivers.

Site	Sampling period	Latitude	Longitude	Maximum (‰)	Minimum (‰)	Reference
Greenland						
Mittivakkat Gletscher (local glacier)	2003–2009	65°41′ N	37°50′ W	−13.7	−17.4	This paper
Kuannersuit Glacier (ice cap outlet)	2000–2005	69°46′ N	53°15′ W	−17.8	−21.9	This paper
Hobbs Gletscher (local glacier)	2004	65°46′ N	38°11′ W	−14.7	−15.1	Yde (unpublished data)
Imersuaq (GrIS outlet)	2000	66°07′ N	49°54′ W	−24.3	−29.9	Yde and Knudsen (2004)
Killersuaq (ice cap outlet)	1982–1983	66°07′ N	50°10′ W	−19.5	−23.0	Andreasen (1984)
Leverett Glacier (GrIS outlet)	2009	67°04′ N	50°10′ W	−23.2	−24.2	Hindshaw et al. (2014)
Isunnguata Sermia (GrIS outlet)	2008	67°11′ N	50°20′ W	−26.2[a]		Yde (unpublished data)
"N" Glacier (GrIS outlet)	2008	68°03′ N	50°16′ W	∼ −23.3	∼ −28.3	Bhatia et al. (2011)
Scandinavia and Svalbard						
Austre Okstindbreen, Norway	1980–1995	66°00′ N	14°10′ E	−11.8	−14.4	Theakstone (2003)
Storglaciären, Sweden	2004 & 2011	67°54′ N	18°38′ E	−10.9	−15.9	Dahlke et al. (2014)
Austre Grønfjordbreen, Svalbard	2009	77°56′ N	14°19′ E	−11.2[a]		Yde et al. (2012)
Dryadbreen, Svalbard	2012	78°09′ N	15°27′ E	−13.0	−15.5	Hindshaw et al. (2016)
Longyearbreen, Svalbard	2004	78°11′ N	15°30′ E	−12.3	−16.7	Yde et al. (2008)
European Alps						
Glacier de Tsanfleuron, Switzerland	1994	46°20′ N	07°15′ E	∼ −7.8	−12.2	Fairchild et al. (1999)
Dammagletscher, Switzerland	2008	46°38′ N	08°27′ E	−13.3	−17.3	Hindshaw et al. (2011)
Hintereisferner, Austria	1969–1970	46°49′ N	10°48′ E	∼ −13.8	∼ −19.4	Behrens et al. (1971)
Kesselwandferner, Austria	1969–1970	46°50′ N	10°48′ E	∼ −14.8	∼ −18.1	Behrens et al. (1971)
Andes						
Cordillera Blanca catchments, Peru	2004–2006	9–10° S	77–78° W	−13.3	−15.3	Mark and McKenzie (2007)
Juncal River, Chile	2011–2012	32°52′ S	70°10′ W	∼ −16.4	∼ −18.0	Ohlanders et al. (2013)
Asia						
Hailuogou Glacier river, China	2008–2009	29°34′ N	101°59′ E	−13.7	−17.6	Meng et al. (2014)
Kumalak Glacier no. 72, China	2009	41°49′ N	79°51′ E	−9.8[a]		Kong and Pang (2012)
Urumqi Glacier no. 1, China	2009	43°07′ N	86°48′ E	−8.7[a]		Kong and Pang (2012)

[a] Single sample.

of these studies are based on few samples. Hence, it therefore remains unknown whether a high spatial variability in $\delta^{18}O$ is a common phenomenon or related to specific circumstances such as surge activity or presence of tributary glaciers.

5 Discussion

5.1 Differences in $\delta^{18}O$ between the Mittivakkat Gletscher river and Kuannersuit Glacier river

A significant difference between the $\delta^{18}O$ dynamics in the Mittivakkat Gletscher river and Kuannersuit Glacier river is the marked diurnal oscillations in the former and the lack of a diurnal signal in the latter during the peak flow period. At the Mittivakkat Gletscher river, the 2004 hydrograph separation analysis showed a 3 h lag of $\delta^{18}O$ to run-off caused by the difference in travel time for ice melt and snowmelt. Meltwater in the early melt season was dominated by snowmelt with relatively high $\delta^{18}O$ and weak diurnal oscillations; whereas diurnal oscillations with amplitudes between 0.11 and 0.49‰ existed during the peak flow period due to mixing of a dominant ice melt component and a secondary snowmelt component. Diurnal oscillations in $\delta^{18}O$ are common in meltwater from small, glacierised catchments; for instance, at Aus-

tre Okstindbreen, Norway, the average diurnal amplitude is approximately 0.2‰ (Theakstone, 1988, 2003; Theakstone and Knudsen, 1989, 1996a, b). The largest diurnal amplitudes in $\delta^{18}O$ (up to 4.3‰) have been observed in small-scale GrIS catchments, such as at Imersuaq and "N Glacier", where large differences in $\delta^{18}O$ exist between various ice facies and snowmelt (Yde and Knudsen, 2004; Bhatia et al., 2011).

The lack of strong diurnal oscillations as observed in the post-surge years at the Kuannersuit Glacier river indicates a mono-source system, a well-mixed drainage network or a multi-source system, where the primary components have similar $\delta^{18}O$ compositions. The expected primary component, glacier ice melt, has lower $\delta^{18}O$ than bulk run-off, and there must be additional contributions from basal ice melt (similar $\delta^{18}O$ composition to run-off), snowmelt (unknown $\delta^{18}O$ composition) or rainwater (higher $\delta^{18}O$ composition than run-off). We therefore hypothesise that the presence of a well-mixed drainage network is the most likely reason for the observed $\delta^{18}O$ signal in the bulk run-off from Kuannersuit Glacier. During the surge event the glacier surface became heavily crevassed and the pre-existing drainage system collapsed (Yde and Knudsen, 2005a). It is a generally accepted theory that the drainage system of surging glaciers transforms into a distributed network where meltwater is routed via a system of linked cavities (Kamb et al., 1985; Kamb, 1987), but little is known about how subglacial drainage systems evolve into discrete flow systems in the years following a surge event. In the initial quiescent phase at Kuannersuit Glacier, frequent loud noises interpreted as drainage system roof collapses were observed, in addition to episodic export of ice blocks from the portal, suggesting ongoing changes to the englacial and subglacial drainage system. A consequence of these processes is also visible on the glacier surface, where circular collapse chasms formed above marginal parts of the subglacial drainage system (Yde and Knudsen, 2005a).

Lack of diurnal oscillations in $\delta^{18}O$ has previously been related to other causes at non-surging glaciers. At Glacier de Tsanfleuron, Switzerland, sampling in the late melt season (23–27 August 1994) showed no diurnal variations in $\delta^{18}O$, which was interpreted by Fairchild et al. (1999) as a consequence of limited altitudinal range (less than 500 m) of the glacier. An alternative explanation may be that snowmelt only constituted so small a proportion of the total run-off in the late melt season that discrimination between snowmelt and ice melt was impossible. At the glacier Killersuaq, an outlet glacier from the ice cap Amitsulooq in west Greenland, Andreasen (1984) found that diurnal oscillations in $\delta^{18}O$ were prominent during the relatively warm summer of 1982, whereas no diurnal $\delta^{18}O$ oscillations were observed in 1983 because the glacier was entirely snow-covered throughout the ablation season, due to low summer surface mass balance caused by the 1982 El Chichón eruption (Ahlstrøm et al., 2007).

5.2 $\delta^{18}O$ compositions in glacier rivers

It is clear from the studies at Mittivakkat Gletscher and Kuannersuit Glacier that glacier rivers have different $\delta^{18}O$ compositions. The bulk meltwater from Mittivakkat Gletscher has a $\delta^{18}O$ composition similar to the water draining the nearby local glacier Hobbs Gletscher and to waters from studied valley and outlet glaciers in Scandinavia, Svalbard, the European Alps, the Andes and Asia (Table 3). The $\delta^{18}O$ composition of Kuannersuit Glacier is lower and similar to the $\delta^{18}O$ composition of the glacier Killersuaq (Table 3). Currently, the lowest $\delta^{18}O$ compositions are found in bulk meltwater draining the GrIS in west Greenland (Table 3), but there is a lack of $\delta^{18}O$ data from Antarctic rivers. Estimations of $\delta^{18}O$ based on δD measurements suggest $\delta^{18}O$ values of −32.1, −34.4 and −41.9‰ in waters draining Wilson Piedmont Glacier, Rhone Glacier and Taylor Glacier, respectively (Henry et al., 1977).

The differences in $\delta^{18}O$ in glacial rivers are due to a combination of geographical effects related to altitude, continentality and latitude (Dansgaard et al., 1973) and temporal effects that work on various timescales and in specific environments. These temporal effects include a seasonal effect (Dansgaard, 1964), a monsoonal effect (Tian et al., 2001;

Kang et al., 2002), a precipitation amount effect (Holdsworth et al., 1991) and a palaeoclimatic effect (Reeh et al., 2002). For instance, the altitude and continentality effects cause low $\delta^{18}O$ in rivers draining the GrIS compared to rivers draining valley glaciers at similar latitudes (Table 3). More data on the $\delta^{18}O$ composition and dynamics in glacial rivers are needed to improve the understanding of how the relative influence of geographical and temporal effects varies on local and regional scales.

6 Conclusions

In this study, we have examined the oxygen isotope hydrology in two of the most studied glacierised river catchments in Greenland to improve our understanding of the prevailing differences between contrasting glacial environments. This study has provided insights into the variability and composition of $\delta^{18}O$ in river water draining glaciers and ice caps adjacent to the GrIS.

The following results were found:

– The Mittivakkat Gletscher river on Ammassalik Island, southeast Greenland, has a mean annual $\delta^{18}O$ of −14.68 ± 0.18‰ during the peak flow period, which is similar to the $\delta^{18}O$ composition in glacier rivers in Scandinavia, Svalbard, the European Alps, the Andes and Asia. The Kuannersuit Glacier river on the island Qeqertarsuaq, west Greenland, has a lower mean annual $\delta^{18}O$ of −19.47 ± 0.55‰, which is similar to the $\delta^{18}O$ composition in bulk meltwater draining an outlet glacier from the ice cap Amitsulooq but higher than the $\delta^{18}O$ composition in bulk meltwater draining the GrIS.

– In the Mittivakkat Gletscher river the diurnal oscillations in $\delta^{18}O$ were conspicuous. This was due to the presence of an efficient subglacial drainage system and diurnal variations in the ablation rates of snow and ice that had distinguishable oxygen isotope compositions. The diurnal oscillations in $\delta^{18}O$ lagged behind the diurnal oscillations in run-off by approximately 3 h. A hydrograph separation analysis revealed that the ice melt component constituted 82 ± 5 % of the total run-off and dominated the observed variations in total run-off during the peak flow period in 2004. The snowmelt component peaked between 10:00 and 13:00, reflecting the long travel time and a possibly inefficiently distributed subglacial drainage network in the upper part of the glacier.

– In contrast to the Mittivakkat Gletscher river, the Kuannersuit Glacier river showed no diurnal oscillations in $\delta^{18}O$. This is likely a consequence of glacier surging. In the years following a major surge event, where Kuannersuit Glacier advanced 10.5 km, meltwater was routed through a tortuous subglacial conduit network of linked cavities, mixing the contributions from glacier ice, basal ice, snow and rainwater.

– This study has shown that environmental and physical contrasts in glacier river catchments influence the spatio-temporal variability of the $\delta^{18}O$ compositions. In Greenlandic glacier rivers, the variability in $\delta^{18}O$ composition is much higher than previously known ranging from relatively high $\delta^{18}O$ values in small-scale coastal glacierised catchments to relatively low $\delta^{18}O$ values in GrIS catchments. This study demonstrates that water isotope analyses can be used to obtain important information on water sources and subglacial drainage system structure that is highly desired for understanding glacier hydrology.

Acknowledgements. We thank all the students who have participated in the fieldwork over the years. We are also grateful to the University of Copenhagen for allowing us to use the facilities at the Arctic Station and Sermilik Station and to the Niels Bohr Institute, University of Copenhagen, for processing the isotope samples. We thank Andreas Peter Bech Mikkelsen and four reviewers for valuable comments on the manuscript.

Edited by: M. Hrachowitz

References

Ahlstrøm, A. P., Bøggild, C. E., Olesen, O. B., Petersen, D., and Mohr, J. J.: Mass balance of the Amitsulôq ice cap, West Greenland, IAHS-AISH P., 318, 107–115, 2007.

Andreasen, J.-O.: Ilt-isotop undersøgelse ved Kidtlessuaq, Vestgrønland, PhD Dissertation, Aarhus University, Aarhus, 33 pp., 1984.

Arendt, C. A., Aciego, A. M., and Hetland, E. A.: An open source Bayesian Monte Carlo isotope mixing model with applications in Earth surface processes, Geochem. Geophy. Geosy., 16, 1274–1292, doi:10.1002/2014GC005683, 2015.

Bales, R. C., Guo, Q., Shen, D., McConnell, J. R., Du, G., Burkhart, J. F., Spikes, V. B., Hanna, E., and Cappelen, J.: Annual accumulation for Greenland updated using ice core data developed during 2000–2006 and analysis of daily coastal meteorological data, J. Geophys. Res., 114, D06116, doi:10.1029/2008JD011208, 2009.

Bamber, J., van den Broeke, M., Ettema, J., Lenaerts, J., and Rignot, E.: Recent large increase in freshwater fluxes from Greenland into the North Atlantic, Geophys. Res. Lett., 39, L19501, doi:10.1029/2012GL052552, 2012.

Bárcena, T. G., Yde, J. C., and Finster, K. W.: Methane flux and high-affinity methanotrophic diversity along the chronosequence of a receding glacier in Greenland, Ann. Glaciol., 51, 23–31, 2010.

Bárcena, T. G., Finster, K. W., and Yde, J. C.: Spatial patterns of soil development, methane oxidation, and methanotrophic diversity along a receding glacier forefield, Southeast Greenland, Arct. Antarct. Alp. Res., 43, 178–188, 2011.

Behrens, H., Bergmann, H., Moser, H., Rauert, W., Stichler, W., Ambach, W., Eisner, H., and Pessl, K.: Study of the discharge of Alpine glaciers by means of environmental isotpes and dye tracers, Z. Gletscherkunde Glazialgeologie, 7, 79–102, 1971.

Behrens, H., Moser, H., Oerter, H., Rauert, W., and Stichler, W.: Models for the runoff from a glaciated catchment area using measurements of environmental isotope contents, in: Proceedings of the International Symposium on Isotope Hydrology, International Atomic Energy Agency, Vienna, 829–846, 1978.

Bhatia, M. P., Das, S. B., Kujawinski, E. B., Henderson, P., Burke, A., and Charette, M. A.: Seasonal evolution of water contributions to discharge from a Greenland outlet glacier: insight from a new isotope-mixing model, J. Glaciol., 57, 929–941, 2011.

Blaen, P. J., Hannah, D. M., Brown, L. E., and Milner, A. M.: Water source dynamics of high Arctic river basins, Hydrol. Process., 28, 3521–3538, 2014.

Bøggild, C. E., Brandt, R. E., Brown, K. J., and Warren, S. G.: The ablation zone in Northeast Greenland: ice types, albedos and impurities, J. Glaciol., 56, 101–113, 2010.

Box, J. E., Fettweis, X., Stroeve, J. C., Tedesco, M., Hall, D. K., and Steffen, K.: Greenland ice sheet albedo feedback: thermodynamics and atmospheric drivers, The Cryosphere, 6, 821–839, doi:10.5194/tc-6-821-2012, 2012.

Boye, B.: En undersøgelse af variationer i $\delta^{18}O$-indholdet i prøver indsamlet på Mittivakkat gletscheren i Østgrønland, Ms Thesis, Aarhus University, Aarhus, 101 pp., 1999.

Cable, J., Ogle, K., and Williams, D.: Contribution of glacier meltwater to streamflow in the Wind River Range, Wyoming, inferred via a Bayesian mixing model applied to isotopic measurements, Hydrol. Process., 25, 2228–2236, doi:10.1002/hyp.7982, 2011.

Cappelen, J.: Weather observations from Greenland 1958–2012, Danish Meteorological Institute Technical Report 13–11, Danish Meteorological Institute, Copenhagen, 23 pp., 2013.

Coplen, T. B.: New guidelines for reporting stable hydrogen, carbon, and oxygen isotope-ratio data, Geochim. Cosmochim. Ac., 60, 3359–3360, 1996.

Dahlke, H. E., Lyom, S. W., Jansson, P., Karlin, T., and Rosquist, G.: Isotopic investigation of runoff generation in a glacierized catchment in northern Sweden, Hydrol. Process., 28, 1383–1398, 2014.

Dansgaard, W.: Stable isotopes in precipitation, Tellus, 16, 436–468, 1964.

Dansgaard, W., Johnsen, S. J., Clausen, H. B., and Gundestrup, N.: Stable isotope glaciology, Medd. Grønland, 197, 53 pp., 1973.

Dinçer, T., Payne, B. R., Florkowski, T., Martinec, J., and Tongiorgi, E.: Snowmelt runoff from measurements of tritium and oxygen-18, Water Resour. Res., 6, 110–124, 1970.

Dissing, L.: Studier af kemiske forhold i sne aflejret på og ved Mittivakkat-gletscheren i Østgrønland, Ms Thesis, Aarhus University, 102 pp., 2000.

Epstein, S. and Sharp, R. P.: Oxygen-isotope variations in the Malaspina and Saskatchewan Glaciers, J. Geol., 67, 88–102, 1959.

Fairchild, I. J., Killawee, J. A., Sharp, M. J., Spiro, B., Hubbard, B., Lorrain, R., and Tison, J.-L.: Solute generation and transfer from a chemically reactive Alpine glacial-proglacial system, Earth Surf. Proc. Land., 24, 1189–1211, 1999.

Gat, J. R. and Gonfiantini, R.: Stable isotope hydrology. Deuterium and oxygen-18 in the water cycle, Technical Report Series 210, International Atomic Energy Agency, Vienna, 334 pp., 1981.

Haldorsen, S., Riise, G., Swensen, B., and Sletten, R. S.: Environmental isotopes as tracers in catchments, in: Geochemical processes, weathering and groundwater recharge in catchments, edited by: Saether, O. M. and de Caritat, P., Balkema, Rotterdam, 185–210, 1997.

Hambrey, M. J.: Oxygen isotope studies at Charles Rabots Bre, Okstindan, northern Norway, Geogr. Ann. A, 56, 147–158, 1974.

Hanna, E., Huybrechts, P., Janssens, I., Cappelen, J., Steffen, K., and Stephens, A.: Runoff and mass balance of the Greenland ice sheet: 1958–2003, J. Geophys. Res., 110, D13108, doi:10.1029/2004JD005641, 2005.

Hanna, E., Huybrechts, P., Steffen, K., Cappelen, J., Huff, R., Shuman, C., Irvine-Fynn, T., Wise, S., and Griffiths, M.: Increased runoff from melt from the Greenland ice sheet: A response to global warming, J. Climate, 21, 331–341, 2008.

Hanna, E., Mernild, S. H., Cappelen, J., and Steffen, K.: Recent warming in Greenland in a long-term instrumental (1881–2012) climatic context, Part 1: evaluation of surface air temperature records, Environ. Res. Lett., 7, 045404, doi:10.1088/1748-9326/7/4/045404, 2012.

Hanna, E., Jones, J. M., Cappelen, J., Mernild, S. H., Wood, L., Steffen, K., and Huybrechts, P.: Discerning the influence of North Atlantic atmospheric and oceanic forcing effects on 1900–2012 Greenland summer climate and melt, Int. J. Climatol., 33, 862–888, 2013.

Hasholt, B. and Mernild, S. H.: Glacial erosion and sediment transport in the Mittivakkat Glacier catchment, Ammassalik Island, southeast Greenland, 2005, IAHS-AISH P., 306, 45–55, 2006.

Henry, C. H., Wilson, A. T., Popplewell, K. B., and House, D. A.: Dating of geochemical events in Lake Bonney, Antarctica, and their relation to glacial and climate changes, New Zeal. J. Geol. Geop., 20, 1103–1122, 1977.

Herschy, R. W.: Hydrometry: Principles and practice, 2nd edition, Wiley, 384 pp., 1999.

Hindshaw, R. S., Tipper, E. T., Reynolds, B. C., Lemarchard, E., Wiederhold, J. G., Magnusson, J., Bernasconi, S. M., Kretzschmar, R., and Bourdon, B.: Hydrological control of stream water chemistry in a glacial catchment (Damma Glacier, Switzerland), Chem. Geol., 285, 215–230, doi:10.1016/j.chemgeo.2011.04.012, 2011.

Hindshaw, R. S., Rickli, J., Leuthold, J., Wadham, J., and Bourdon, B.: Identifying weathering sources and processes in an outlet glacier of the Greenland Ice Sheet using Ca and Sr isotope ratios, Geochim. Cosmochim. Ac., 145, 50–71, 2014.

Hindshaw, R. S., Heaton, T. H. E., Boyd, E. S., Lindsay, M. R., and Tipper, E. T.: Influence of glaciation on mechanisms of mineral weathering in two high Arctic catchments, Chem. Geol., 420, 37–50, doi:10.1016/j.chemgeo.2015.11.004, 2016.

Holdsworth, G., Fogarasi, S., and Krouse, H. R.: Variation of the stable isotopes of water with altitude in the Saint Elias Mountains of Canada, J. Geophys. Res. 96, 7483–7494, 1991.

Hooper, R. P. and Shoemaker, C. A.: A comparison of chemical and isotopic hydrograph separation, Water Resour. Res., 106, 233–244, 1986.

Humlum, O.: Late-Holocene climate in central West Greenland: meteorological data and rock-glacier isotope evidence, Holocene, 9, 581–594, 1999.

Kamb, B.: Glacier surge mechanism based on linked cavity configuration of the basal water conduit system, J. Geophys. Res., 92, 9083–9100, 1987.

Kamb, B., Raymond, C. F., Harrison, W. D., Engelhardt, H., Echelmeyer, K. A., Humphrey, N., Brugman, M. M., and Pfeffer, T.: Glacier surge mechanism: 1982–1983 surge of Variegated Glacier, Alaska, Science, 227, 469–479, 1985.

Kang, S., Kreutz, K. J., Mayewski, P. A., Qin, D., and Yao, T.: Stable-isotopic composition of precipitation over the northern slope of the central Himalaya, J. Glaciol., 48, 519–526, 2002.

Keegan, K. M., Albert, M. R., McConnell, J. R., and Baker, I.: Climate change and forest fires synergistically drive widespread melt events of the Greenland Ice Sheet, P. Natl. Acad. Sci. USA, 111, 7964–7967, 2014.

Kendall, C., Doctor, D. H., and Young, M. B.: Environmental isotope applications in hydrological studies, in: Treatise on Geochemistry, second edition, edited by: Holland, H. D. and Turekian, K. K., Elsevier, Oxford, 7, chapter 9, 273–327, 2014.

Knudsen, N. T., Yde, J. C., and Gasser, G.: Suspended sediment transport in glacial meltwater during the initial quiescent phase after a major surge event at Kuannersuit Glacier, Greenland, Geogr. Tidsskr., 107, 1–7, 2007.

Knudsen, N. T., Nørnberg, P., Yde, J. C., Hasholt, B., and Heinemeier, J.: Recent marginal changes of the Mittivakkat Glacier, Southeast Greenland and the discovery of remains of reindeer (Rangifer tarandus), polar bear (Ursus maritimus) and peaty material, Geogr. Tidsskr., 108, 137–142, 2008.

Kong, Y. and Pang, Z.: Evaluating the sensitivity of glacier rivers to climate change based on hydrograph separation of discharge, J. Hydrol., 434–435, 121–129, 2012.

Kristiansen, S. M., Yde, J. C., Bárcena, T. G., Jakobsen, B. H., Olsen, J., and Knudsen, N. T.: Geochemistry of groundwater in front of a warm-based glacier in Southeast Greenland, Geogr. Ann. A, 95, 97–108, 2013.

Larsen, N. K., Kronborg, C., Yde, J. C., and Knudsen, N. T.: Debris entrainment by basal freeze-on and thrusting during the 1995–1998 surge of Kuannersuit Glacier on Disko Island, west Greenland, Earth Surf. Proc. Land., 35, 561–574, 2010.

Lee, J., Feng, X., Faiia, A., Posmentier, E., Osterhuber, R., and Kirchner, J.: Isotopic evolution of snowmelt: A new model incorporating mobile and immobile water, Water Resour. Res., 46, W11512, doi:10.1029/2009WR008306, 2010.

Lutz, S., Anesio, A. M., Villar, S. E. J., and Benning, L. G.: Variations of algal communities cause darkening of a Greenland glacier, FEMS Microbiol. Ecol., 89, 402–414, 2014.

Mark, B. G. and Seltzer, G. O.: Tropical glacier meltwater contribution to stream discharge: a case study in the Cordillera Blanca, Peru, J. Glaciol., 49, 271–281, 2003.

Mark, B. G. and McKenzie, J. M.: Tracing increasing tropical Andean glacier melt with stable isotopes in water, Environ. Sci. Technol., 41, 6955–6960, 2007.

McDonnell, J. J., Bonell, M., Stewart, M. K., and Pearce, A. J.: Deuterium variations in storm rainfall – implications for stream hydrograph separation, Water Resour. Res., 26, 455–458, 1990.

Meng, Y., Liu, G., and Zhang, L.: A comparative study on stable isotopic composition in waters of the glacial and nonglacial rivers in Mount Gongga, China, Water Environ. J., 28, 212–221, 2014.

Mernild, S. H.: The internal drainage system of the lower Mittivakkat Glacier, Ammassalik Island, SE Greenland, Geogr. Tidsskr., 106, 13–24, 2006.

Mernild, S. H. and Liston, G. E.: The influence of air temperature inversions on snowmelt and glacier mass balance simulations, Ammassalik Island, Southeast Greenland, J. Appl. Meteorol. Clim., 49, 47–67, 2010.

Mernild, S. H. and Liston, G. E.: Greenland freshwater runoff, Part II: Distribution and trends, 1960–2010, J. Clim., 25, 6015–6035, 2012.

Mernild, S. H., Hasholt, B., Jakobsen, B. H., and Hansen, B. U.: Meteorological observations 2006 and ground temperature variations over 12-year at the Sermilik Station, Ammassalik Island, Southeast Greenland, Geogr. Tidsskr., 108, 153–161, 2008a.

Mernild, S. H., Liston, G. E., and Hasholt, B.: East Greenland freshwater runoff to the Greenland-Iceland-Norwegian Seas 1999–2004 and 2071–2100, Hydrol. Process., 22, 4571–4586, 2008b.

Mernild, S. H., Knudsen, N. T., Lipscomb, W. H., Yde, J. C., Malmros, J. K., Hasholt, B., and Jakobsen, B. H.: Increasing mass loss from Greenland's Mittivakkat Gletscher, The Cryosphere, 5, 341–348, doi:10.5194/tc-5-341-2011, 2011.

Mernild, S. H., Malmros, J. K., Yde, J. C., and Knudsen, N. T.: Multi-decadal marine- and land-terminating glacier recession in the Ammassalik region, southeast Greenland, The Cryosphere, 6, 625–639, doi:10.5194/tc-6-625-2012, 2012.

Mernild, S. H., Pelto, M., Malmros, J. K., Yde, J. C., Knudsen, N. T., and Hanna, E.: Identification of snow ablation rate, ELA, AAR and net mass balance using transient snowline variations on two Arctic glaciers, J. Glaciol., 59, 649–659, 2013a.

Mernild, S. H., Knudsen, N. T., Hoffman, M. J., Yde, J. C., Hanna, E., Lipscomb, W. H., Malmros, J. K., and Fausto, R. S.: Volume and velocity changes at Mittivakkat Gletscher, southeast Greenland, J. Glaciol., 59, 660–670, 2013b.

Mernild, S. H., Hanna, E., Yde, J. C., Cappelen, J., and Malmros, J. K.: Coastal Greenland air temperature extremes and trends 1890–2010: annual and monthly analysis, Int. J. Climatol., 34, 1472–1487, 2014.

Mernild, S. H., Hanna, E., McConnell, J. R., Sigl, M., Beckerman, A. P., Yde, J. C., Cappelen, J., Malmros, J. K., and Steffen, K.: Greenland precipitation trends in a long-term instrumental climate context (1890–2012): evaluation of coastal and ice core records, Int. J. Climatol., 35, 303–320, 2015a.

Mernild, S. H., Malmros, J. K., Yde, J. C., Wilson, R., Knudsen, N. T., Hanna, E., Fausto, R. S., and van As, D.: Albedo decline on Greenland's Mittivakkat Gletscher in a warming climate, Int. J. Climatol., 35, 2294–2307, 2015b.

Milner, A. M., Brown, L. E., and Hannah, D. M.: Hydroecological response of river systems to shrinking glaciers, Hydrol. Process., 23, 62–77, 2009.

Moser, H. and Stichler, W.: Environmental isotopes in ice and snow, in: Handbook of environmental isotope geochemistry, The terrestrial environment, A, edited by: Fritz, P. and Fontes, J. C., Elsevier, Amsterdam, Volume 1, 141–178, 1980.

Nghiem, S. V., Hall, D. K., Mote, T. L., Tedesco, M., Albert, M. R., Keegan, K., Shuman, C. A., DiGirolamo, N. E., and Neumann, G.: The extreme melt across the Greenland ice sheet in 2012, Geophys. Res. Lett., 39, L20502, doi:10.1029/2012GL053611, 2012.

Ohlanders, N., Rodriguez, M., and McPhee, J.: Stable water isotope variation in a Central Andean watershed dominated by glacier and snowmelt, Hydrol. Earth Syst. Sci., 17, 1035–1050, doi:10.5194/hess-17-1035-2013, 2013.

Penna, D., Engel, M., Mao, L., Dell'Agnese, A., Bertoldi, G., and Comiti, F.: Tracer-based analysis of spatial and temporal variations of water sources in a glacierized catchment, Hydrol. Earth Syst. Sci., 18, 5271–5288, doi:10.5194/hess-18-5271-2014, 2014.

Raben, P. and Theakstone, W. H.: Changes of ionic and oxygen isotopic composition of the snowpack at the glacier Austre Okstindbreen, Norway, 1995, Nord. Hydrol., 29, 1–20, 1998.

Reeh, N., Oerter, H., and Thomsen, H. H.: Comparison between Greenland ice-margin and ice-core oxygen-18 records, Ann. Glaciol., 35, 136–144, 2002.

Roberts, D. H., Yde, J. C., Knudsen, N. T., Long, A. J., and Lloyd, J. M.: Ice marginal dynamics and sediment delivery mechanisms during surge activity, Kuannersuit Glacier, Disko Island, West Greenland, Quaternary Sci. Rev., 28, 209–222, 2009.

Rodriguez, M., Ohlanders, N., and McPhee, J.: Estimating glacier and snowmelt contributions to stream flow in a Central Andes catchment in Chile using natural tracers, Hydrol. Earth Syst. Sci. Discuss., 11, 8949–8994, doi:10.5194/hessd-11-8949-2014, 2014.

Tedesco, M., Serreze, M., and Fettweis, X.: Diagnosing the extreme surface melt event over southwestern Greenland in 2007, The Cryosphere, 2, 159–166, doi:10.5194/tc-2-159-2008, 2008.

Tedesco, M., Fettweis, X., van den Broeke, M. R., van de Wal, R. S. W., Smeets, C. J. P. P., van de Berg, W. J., Serreze, M. C., and Box, J. E.: The role of albedo and accumulation in the 2010 melting record in Greenland, Environ. Res. Lett., 6, 014005, doi:10.1088/1748-9326/6/1/014005, 2011.

Theakstone, W. H.: Temporal variations of isotopic composition of glacier-river water during summer: observations at Austre Okstindbreen, Okstindan, Norway, J. Glaciol., 34, 309–317, 1988.

Theakstone, W. H.: Oxygen isotopes in glacier-river water, Austre Okstindbreen, Okstidan, Norway, J. Glaciol., 49, 282–298, 2003.

Theakstone, W. H.: Dating stratigraphic variations of ions and oxygen isotopes in a high-altitude snowpack by comparison with daily variations of precipitation chemistry at a low-altitude site, Hydrol. Res., 39, 101–112, 2008.

Theakstone, W. H. and Knudsen, N. T.: Temporal changes of glacier hydrological systems indicated by isotopic and related observations at Austre Okstindbreen, Okstindan, Norway, 1976–87, Ann. Glaciol., 13, 252–256, 1989.

Theakstone, W. H. and Knudsen, N. T.: Isotopic and ionic variations in glacier river water during three contrasting ablation seasons, Hydrol. Process., 10, 523–539, 1996a.

Theakstone, W. H. and Knudsen, N. T.: Oxygen isotope and ionic concentrations in glacier river water: multi-year observations in the Austre Okstindbreen basin, Norway, Nord. Hydrol., 27, 101–116, 1996b.

Tian, L., Masson-Delmotte, V., Stiévenard, M., Yao, T., and Jouzel, J.: Tibetan Plateau summer monsoon northward extent revealed by measurements of water stable isotopes, J. Geophys. Res., 106, 28081–28088, 2001.

van As, D., Hubbard, A. L., Hasholt, B., Mikkelsen, A. B., van den Broeke, M. R., and Fausto, R. S.: Large surface meltwater discharge from the Kangerlussuaq sector of the Greenland ice sheet during the record-warm year 2010 explained by detailed energy balance observations, The Cryosphere, 6, 199–209, doi:10.5194/tc-6-199-2012, 2012.

Vaughan, D. G., Comiso, J. C., Allison, J., Carrasco, J., Kaser, G., Kwok, R., Mote, P., Murray, T., Paul, F., Ren, J., Rignot, E., Solomina, O., Steffen, K., and Zhang, T.: Observations: Cryosphere, in: Climate Change 2013: The physical science basis, Contribution of working group I to the Fifth Assessment Report of the Intergovernmental Panel on Climate Change, edited by: Stocker, T. F., Qin, D., Plattner, G.-K., Tignor, M., Allen, S. K., Boschung, J., Nauels, A., Xia, Y., Bex, V., and Midgley, P. M., Cambridge University Press, Cambridge, United Kingdom and New York, NY, USA, 2013.

Weidick, A. and Morris, E.: Local glaciers surrounding the continental ice sheets, in: Into the second century of world glacier monitoring – prospects and strategies, edited by: Haeberli, W., Hoelzle, M., and Suter, S., UNESCO Studies and Reports in Hydrology, 56, 167–176, 1998.

Yallop, M. L., Anesio, A. M., Perkins, R. G., Cook, J., Telling, J., Fagan, D., MacFarlene, J., Stibal, M., Barker, G., Bellas, C., Hodson, A., Tranter, M., Wadham, J., and Roberts, N. W.: Photophysiology and albedo-changing potential of the ice algal community on the surface of the Greenland ice sheet, ISME J., 6, 2302–2313, 2012.

Yde, J. C. and Knudsen, N. T.: The importance of oxygen isotope provenance in relation to solute content of bulk meltwaters at Imersuaq Glacier, West Greenland, Hydrol. Process., 18, 125–139, 2004.

Yde, J. C. and Knudsen, N. T.: Glaciological features in the initial quiescent phase of Kuannersuit Glacier, Greenland, Geogr. Ann. A, 87, 473–485, 2005a.

Yde, J. C. and Knudsen, N. T.: Observations of debris-rich naled associated with a major glacier surge event, Disko Island, West Greenland, Permafrost Periglac., 16, 319–325, 2005b.

Yde, J. C. and Knudsen, N. T.: 20th-century glacier fluctuations on Disko Island (Qeqertarsuaq), Greenland, Ann. Glaciol., 46, 209–214, 2007.

Yde, J. C., Knudsen, N. T., and Nielsen, O. B.: Glacier hydrochemistry, solute provenance, and chemical denudation at a surge-type glacier in Kuannersuit Kuussuat, Disko Island, West Greenland, J. Hydrol., 300, 172–187, 2005a.

Yde, J. C., Knudsen, N. T., Larsen, N. K., Kronborg, C., Nielsen, O. B., Heinemeier, J., and Olsen, J.: The presence of thrust-block naled after a major surge event: Kuannersuit Glacier, West Greenland, Ann. Glaciol., 42, 145–150, 2005b.

Yde, J. C., Riger-Kusk, M., Christiansen, H. H., Knudsen, N. T., and Humlum, O.: Hydrochemical characteristics of bulk meltwater from an entire ablation season, Longyearbreen, Svalbard, J. Glaciol., 54, 259–272, 2008.

Yde, J. C., Hodson, A. J., Solovjanova, I., Steffensen, J. P., Nørnberg, P., Heinemeier, J., and Olsen, J.: Chemical and isotopic characteristics of a glacier-derived naled in front of Austre Grønfjordbreen, Svalbard, Polar Res., 31, 17628, doi:10.3402/polar.v31i0.17628, 2012.

Yde, J. C., Kusk Gillespie, M., Løland, R., Ruud, H., Mernild, S. H., de Villiers, S., Knudsen, N. T., and Malmros, J. K.: Volume measurements of Mittivakkat Gletscher, Southeast Greenland, J. Glaciol., 60, 1199–1207, 2014.

Zhou, S., Wang, Z., and Joswiak, D. R.: From precipitation to runoff: stable isotopic fractionation effect of glacier melting on a catchment scale, Hydrol. Process., 28, 3341–3349, 2014.

Permissions

The contributors of this book come from diverse backgrounds, making this book a truly international effort. This book will bring forth new frontiers with its revolutionizing research information and detailed analysis of the nascent developments around the world.

We would like to thank all the contributing authors for lending their expertise to make the book truly unique. They have played a crucial role in the development of this book. Without their invaluable contributions this book wouldn't have been possible. They have made vital efforts to compile up to date information on the varied aspects of this subject to make this book a valuable addition to the collection of many professionals and students.

This book was conceptualized with the vision of imparting up-to-date information and advanced data in this field. To ensure the same, a matchless editorial board was set up. Every individual on the board went through rigorous rounds of assessment to prove their worth. After which they invested a large part of their time researching and compiling the most relevant data for our readers.

The editorial board has been involved in producing this book since its inception. They have spent rigorous hours researching and exploring the diverse topics which have resulted in the successful publishing of this book. They have passed on their knowledge of decades through this book. To expedite this challenging task, the publisher supported the team at every step. A small team of assistant editors was also appointed to further simplify the editing procedure and attain best results for the readers.

Apart from the editorial board, the designing team has also invested a significant amount of their time in understanding the subject and creating the most relevant covers. They scrutinized every image to scout for the most suitable representation of the subject and create an appropriate cover for the book.

The publishing team has been an ardent support to the editorial, designing and production team. Their endless efforts to recruit the best for this project, has resulted in the accomplishment of this book. They are a veteran in the field of academics and their pool of knowledge is as vast as their experience in printing. Their expertise and guidance has proved useful at every step. Their uncompromising quality standards have made this book an exceptional effort. Their encouragement from time to time has been an inspiration for everyone.

The publisher and the editorial board hope that this book will prove to be a valuable piece of knowledge for researchers, students, practitioners and scholars across the globe.

List of Contributors

J. Wenninger and S. Uhlenbrook
UNESCO-IHE Institute for Water Education, Department of Water Science and Engineering, 2601 DA Delft, the Netherlands
Delft University of Technology, Faculty of Civil Engineering and Applied Geosciences, Water Resources Section, 2600 GA Delft, the Netherlands

S. Tekleab
UNESCO-IHE Institute for Water Education, Department of Water Science and Engineering, 2601 DA Delft, the Netherlands
Addis Ababa University, Institute for Environment, Water and Development, Addis Ababa, Ethiopia
Hawassa University, Institute of Technology, Department of Irrigation and Water Resources Engineering, Hawassa, Ethiopia
Delft University of Technology, Faculty of Civil Engineering and Applied Geosciences, Water Resources Section, 2600 GA Delft, the Netherlands

Rabia Slimani and Belhadj Hamdi-Aïssa
Ouargla University, Fac. des Sciences de la Nature et de la Vie, Lab. Biochimie des Milieux Désertiques, 30000 Ouargla, Algeria

Abdelhamid Guendouz
Blida University, Science and Engineering Faculty, Soumaâ, Blida, Algeria

Fabienne Trolard and Guilhem Bourrié
INRA – UMR1114 EMMAH, Avignon, France

Adnane Souffi Moulla
Algiers Nuclear Research Centre, Alger-RP, 16000 Algiers, Algeria

A. J. Zurek, S. Witczak, J. Kania, A. Postawa, J. Karczewski and W. J. Moscicki
AGH University of Science and Technology, Faculty of Geology, Geophysics and Environmental Protection, Krakow, Poland

M. Dulinski, P. Wachniew and K. Rozanski
AGH University of Science and Technology, Faculty of Physics and Applied Computer Science, Krakow, Poland

Natalie Orlowski
Institute for Landscape Ecology and Resources Management (ILR), Research Centre for BioSystems, Land Use and Nutrition (iFZ), Justus Liebig University Gießen, Gießen, Germany

Global Institute for Water Security, University of Saskatchewan, Saskatoon, Canada

Philipp Kraft and Jakob Pferdmenges
Institute for Landscape Ecology and Resources Management (ILR), Research Centre for BioSystems, Land Use and Nutrition (iFZ), Justus Liebig University Gießen, Gießen, Germany

Lutz Breuer
Institute for Landscape Ecology and Resources Management (ILR), Research Centre for BioSystems, Land Use and Nutrition (iFZ), Justus Liebig University Gießen, Gießen, Germany
Centre for International Development and Environmental Research, Justus Liebig University Gießen, Gießen, Germany

Stephen D. Parkes
Water Desalination and Reuse Centre, King Abdullah University of Science and Technology (KAUST), Jeddah, Saudi Arabia

Josiah Strauss
Department of Civil and Environmental Engineering, University of New South Wales, Sydney, Australia

Matthew F. McCabe and Ali Ershadi
Water Desalination and Reuse Centre, King Abdullah University of Science and Technology (KAUST), Jeddah, Saudi Arabia
Department of Civil and Environmental Engineering, University of New South Wales, Sydney, Australia

Alan D. Griffiths, Scott Chambers, Alastair G. Williams and Adrian Element
Australian Nuclear Science and Technology Organization, Sydney, New South Wales, Australia

Lixin Wang
Department of Earth Sciences, Indiana University – Purdue University Indianapolis (IUPUI), Indianapolis, IN, USA

X. Song, L. Yang, Y. Zhang, D. Han, Y. Ma and H. Bu
Key Laboratory of Water Cycle and Related Land Surface Processes, Institute of Geographic Sciences and Natural Resources Research, Chinese Academy of Sciences, 11 A, Datun Road, Chaoyang District, Beijing, 100101, China

F. Liu
Key Laboratory of Water Cycle and Related Land Surface Processes, Institute of Geographic Sciences and Natural Resources Research, Chinese Academy of Sciences, 11 A, Datun Road, Chaoyang District, Beijing, 100101, China University of Chinese Academy of Sciences, Beijing, 100049, China

Jana von Freyberg and James W. Kirchner
Department of Environmental Systems Science, ETH Zurich, Zurich, Switzerland
Swiss Federal Institute for Forest, Snow and Landscape Research (WSL), Birmensdorf, Switzerland

Bjørn Studer
Department of Environmental Systems Science, ETH Zurich, Zurich, Switzerland

Jan Schmieder, Florian Hanzer, Thomas Marke and Ulrich Strasser
Institute of Geography, University of Innsbruck, 6020 Innsbruck, Austria

Jakob Garvelmann, Michael Warscher and Harald Kunstmann
Institute of Meteorology and Climate Research – Atmospheric Environmental Research, Karlsruhe Institute of Technology, 82467 Garmisch-Partenkirchen, Germany

Bojie Fu
Research Center for Eco-Environmental Sciences, Chinese Academy of Sciences, Beijing, 100085, China

Yonggang Yang
Institute of Loess Plateau, Shanxi University, Taiyuan, Shanxi, 030006, China
Research Center for Eco-Environmental Sciences, Chinese Academy of Sciences, Beijing, 100085, China

V. V. Camacho Suarez
Department of Water Science and Engineering, UNESCO-IHE Institute for Water Education, 2601 DA Delft, the Netherlands

A. M. L. Saraiva Okello, J. W. Wenninger and S. Uhlenbrook
Department of Water Science and Engineering, UNESCO-IHE Institute for Water Education, 2601 DA Delft, the Netherlands
Section of Water Resources, Delft University of Technology, 2600 GA Delft, the Netherlands

D. Windhorst, H.-G. Frede and L. Breuer
Institute for Landscape Ecology and Resources Management (ILR), Research Centre for BioSystems, Land Use and Nutrition (IFZ), Justus-Liebig-Universität Gießen, Gießen, Germany

P. Kraft and E. Timbe
Institute for Landscape Ecology and Resources Management (ILR), Research Centre for BioSystems, Land Use and Nutrition (IFZ), Justus-Liebig-Universität Gießen, Gießen, Germany
Departamento de Recursos Hídricos y Ciencias Ambientales, Universidad de Cuenca, Cuenca, Ecuador

Rena Meyer and Peter Engesgaard
Department of Geosciences and Natural Resources Management, University of Copenhagen, Øster Voldgade 10, 1350 Copenhagen, Denmark

Klaus Hinsby and Torben O. Sonnenborg
Geological Survey of Denmark and Greenland, Øster Voldgade 10, 1350 Copenhagen, Denmark

Jan A. Piotrowski
Department of Geosciences, Aarhus University, Høegh-Guldbergs Gade 2, 8000 Aarhus, Denmark

Jacob C. Yde
Faculty of Engineering and Science, Sogn og Fjordane University College, Sogndal, Norway

Niels T. Knudsen, Christian Kronborg and Nicolaj K. Larsen
Department of Geoscience, University of Aarhus, Aarhus, Denmark

Jørgen P. Steffensen
Centre for Ice and Climate, University of Copenhagen, Copenhagen, Denmark

Jonathan L. Carrivick
School of Geography and water@leeds, University of Leeds, Leeds, UK

Bent Hasholt
Department of Geosciences and Natural Resource Management, University of Copenhagen, Copenhagen, Denmark

Thomas Ingeman-Nielsen
Arctic Technology Centre, Technical University of Denmark, Kgs. Lyngby, Denmark

Sebastian H. Mernild
Faculty of Engineering and Science, Sogn og Fjordane University College, Sogndal, Norway
Antarctic and Sub-Antarctic Program, Universidad de Magallanes, Punta Arenas, Chile

Hans Oerter
Alfred Wegener Institute, Helmholtz Centre for Polar and Marine Research, Bremerhaven, Germany

David H. Roberts
Department of Geography, University of Durham, Durham, UK

Andrew J. Russell
School of Geography, Politics & Sociology, Newcastle University, Newcastle upon Tyne, UK

Index

Printed in the USA
CPSIA information can be obtained
at www.ICGtesting.com
LVHW051519091023
760584LV00008B/75

9 781641 165785